Mathematical Recreations from the Tournament of the Towns

Mathematical Recreations from the Tournament of the Towns contains the complete list of problems and solutions to the International Mathematics Tournament of the Towns from Fall 2007 to Spring 2021. The primary audience for this book is the army of recreational mathematicians united under the banner of Martin Gardner. It should also have great value to students preparing for mathematics competitions and trainers of such students. This book also provides an entry point for students in upper elementary schools.

Features

- Huge recreational value to mathematics enthusiasts
- Accessible to upper-level high school students
- Problems classified by topics such as two-player games, weighing problems, mathematical tasks etc.

AK Peters/CRC Recreational Mathematics Series

Series Editors

Robert Fathauer
Snezana Lawrence
Jun Mitani
Colm Mulcahy
Peter Winkler
Carolyn Yackel

Six Simple Twists
The Pleat Pattern Approach to Origami Tessellation Design, Second Edition
Benjamin DiLeonardo-Parker

Tessellations
Mathematics, Art, and Recreation
Robert Fathauer

Mathematics of Casino Carnival Games
Mark Bollman

Mathematical Puzzles
Peter Winkler

X Marks the Spot
The Lost Inheritance of Mathematics
Richard Garfinkle, David Garfinkle

Luck, Logic, and White Lies
The Mathematics of Games, Second Edition
Jörg Bewersdorff

Mathematics of The Big Four Casino Table Games
Blackjack, Baccarat, Craps, & Roulette
Mark Bollman

Star Origami
The Starrygami™ Galaxy of Modular Origami Stars, Rings and Wreaths
Tung Ken Lam

Mathematical Recreations from the Tournament of the Towns
Andy Liu, Peter Taylor

For more information about this series please visit: https://www.routledge.com/AK-PetersCRC-Recreational-Mathematics-Series/book-series/RECMATH?pd=published,forthcoming&pg=2&pp=12&so=pub&view=list

Mathematical Recreations from the Tournament of the Towns

Andy Liu
Peter Taylor

CRC Press
Taylor & Francis Group
Boca Raton London New York

CRC Press is an imprint of the
Taylor & Francis Group, an **informa** business
AN A K PETERS BOOK

First edition published 2023
by CRC Press
6000 Broken Sound Parkway NW, Suite 300, Boca Raton, FL 33487-2742

and by CRC Press
4 Park Square, Milton Park, Abingdon, Oxon, OX14 4RN

CRC Press is an imprint of Taylor & Francis Group, LLC

Library of Congress Cataloging-in-Publication Data

Names: Liu, A. C. F. (Andrew Chiang-Fung), author. | Taylor, P. J. (Peter James), 1947- author. | Konstantinov, N. N. (Nikolaĭ Nikolaevich), 1932-2021, honoree.
Title: Mathematical recreations from the Tournament of the Towns / Andy Liu, Peter Taylor.
Description: First edition. | Boca Raton : AK Peters/CRC Press, 2023. | Series: AK Peters/CRC recreational mathematics series | "Dedicated to the memory of Nikolay N. Konstantinov"-- title page verso. | Includes bibliographical references and index.
Identifiers: LCCN 2022023876 (print) | LCCN 2022023877 (ebook) | ISBN 9781032352923 (hardback) | ISBN 9781032352909 (paperback) | ISBN 9781003326212 (ebook)
Subjects: LCSH: International Mathematics Tournament of the Towns (Competition) | Mathematical recreations.
Classification: LCC QA95 .L765 2023 (print) | LCC QA95 (ebook) | DDC 793.74--dc23/eng20221007
LC record available at https://lccn.loc.gov/2022023876
LC ebook record available at https://lccn.loc.gov/2022023877

ISBN: 978-1-032-35292-3 (hbk)
ISBN: 978-1-032-35290-9 (pbk)
ISBN: 978-1-003-32621-2 (ebk)

DOI: 10.1201/ 9781003326212

Typeset in Latin Roman
by KnowledgeWorks Global Ltd.

Publisher's note: This book has been prepared from camera-ready copy provided by the authors.

Dedicated to the Memory of

NIKOLAY N. KONSTANTINOV

Table of Contents

A Brief History of the Tournament of Towns

The International Mathematics Tournament of the Towns is a mathematical problem-solving competition which started in the former Soviet Union (USSR) in 1980.

In the years immediately preceding 1980, there was considerable concern about the opportunities available to students from the larger cities, such as Moscow, Leningrad and Kiev, to participate in the USSR All-Union Olympiad, under the format of this event. The All-Union Olympiad had, for many years been organised by A. N. Kolmogorov and his jury. Due to the difficulty encountered by Kolmogorov and his colleagues in changing the format to allow more students from the larger cities to participate, a new competition was organised.

In the first year (1980) this competition was known as the "Olympiad of Three Towns" (it was held just in Moscow, Leningrad and Riga). Other towns were allowed to participate after the first year and the competition's name was changed to "Tournament of the Towns". In the early years the competition had no official status, but in 1984 the competition became officially recognised and the organising committee became a subcommittee of the USSR Academy of Sciences.

At about this time towns from other countries were invited to participate. Towns from Bulgaria first took part in the 6th Tournament (1984 to 1985). The infrastructure for supporting mathematics competitions was so strong in Bulgaria that most major towns participated and a national committee was established to coordinate Bulgarian entries. Bulgarian towns were frequently among the most successful participants in the Tournament. During this time towns from other countries in the (then) Eastern Bloc also participated. These countries included Romania and Poland.

In 1988 I attended ICME-6 in Budapest and received an invitation via two Bulgarian mathematicians, Jordan Tabov and Petar Kenderov, to enter Australian teams in the Tournament. At the time Australia had become under increasing notice for its successful Australian Mathematics Competition and hosting the 1988 International Mathematical Olympiad, both under the direction of my colleague at the University of Canberra, Peter O'Halloran. Under Peter's influence the World Federation of National Mathematics Competitions (WFNMC) had also been founded in 1984. So I think Jordan and Petar were acting on behalf of the organisers wanting to spread the Tournament to the West, and they felt that Australia might have the infrastructure to start.

I looked at the problems from the previous Tournament. They looked very creative, not purely technical as in most Olympiads, but requiring good structural thinking, as well as technical knowledge. The problems were often presented in real-life settings which could be explained to people who were not mathematicians and having unexpected results. It looked very interesting, enough to form a mathematics circle in Canberra which could continue that which had been run for some time by colleagues Mike Newman and Laci Kovacs. So we reassembled their circle with the Tournament problems as material.

So then at short notice students from Canberra did participate in the 10th Tournament (1988 to 1989), becoming the first Western students and first English-speaking students to participate in the Tournament. The 11th Tournament (1989 to 1990) saw further growth in the Tournament. More Australian towns participated, and a national committee was established to coordinate Australian participation. Entries were also received from Hamburg (then West Germany) and colourado Springs (USA).

In May 1989, for the first time I met Andy Liu, whom we had invited to Canberra to help set the 1990 papers of the Australian Mathematics Competition. I acquainted him with the Tournament, and he was obviously just as impressed by the quality of the problems as I had been. He and I became close working colleagues on the Tournament in the years to come.

In 1990 I visited Moscow for discussions, mainly with Konstantinov, problems committee chairman Nikolay Vasiliev and others. I explained to Konstantinov that we were enthusiastic, and had started organising the Tournament, but in order to have complete integrity in English we needed all the back papers in English. He went to a filing cabinet and gave me a complete set, but of course all in Russian. I undertook to translate all the questions into English and to publish them with solutions via the Australian Mathematics Trust.

Many of my colleagues contributed the solutions, and we published the best, but Andy Liu was now working closely with me, and he produced the greatest number of solutions, and some of really outstanding quality.

The 12th Tournament saw even further growth. Towns from Canada, Spain, Czechoslovakia, Iran, Yugoslavia, New Zealand, the United Kingdom, Zimbabwe, French Polynesia, India, Indonesia, Singapore, Israel, Colombia, Hong Kong and the Philippines either entered or were gearing up for entering in the near future.

In August 1990, the First Conference of the WFNMC was held at the University of Waterloo, Canada, though Andy did not attend. At a meeting, the International Mathematics Tournament of the Towns was formally set up, and the following committee was elected.

Patron:	Ludwig Fadeev, Moscow.
President:	Nikolay Konstantinov, Moscow.
Vice-Presidents:	Agnis Andjans, Riga, Helmut Müller, Hamburg, Jordan Tabov, Sofia, Peter Taylor, Canberra, and Alexey Tolpygo, Kiev.
Members:	Kiril Bankov, Sofia, Lubomir Lubenov, Stara Zagora and Alexey Sosinski, Moscow.
Problems Committee:	Nikolay Vasiliev (Chairman), Moscow, Dimitry Fomin and Sergey Fomin, Leningrad.

It should be noted that Professor Konstantinov, the President of this Committee and the Moscow organising committee, had a long history in the Tournament, dating back to being a member of Kolmogorov's All-Union Olympiad jury from 1969 to 1979.

The next meeting of the international committee was held in Quebec City, Canada, in conjunction with ICME-7 in 1992. Andy Liu was this time in attendance, and his absence from the founding committee was rectified by his appointment there as a Vice President.

The Tournament complements the national and international mathematics Olympiads, and it has the advantage of not requiring travel, thus enabling all talented students to participate.

In Russia the Tournament has additional activity, including a summer research conference, to which Russian and international students are invited, and where students are introduced to problems with known and unknown solutions to work on. They usually take on the problems in groups, and just as in mathematical circles, where students from different schools meet each other and become friends, this also happens, and there are often discussions on matters other than mathematics. This has helped create valuable friendships and collaborations. For an account of the 1993 conference in Beloretsk, Russia, see "A Mathematical Journey," by Andy Liu, available at http://cms.math.ca/publications/crux/issue/?volume=20&issue=1.

A more detailed history of the Tournament is to be found in the article: "Birth of the Tournament of the Towns," by N. N. Konstantinov, J. B. Tabov and P. J. Taylor, Journal of the World Federation of National Mathematics Competitions, Volume 4, Number 2, 1991. It is available at http://www.wfnmc.org/mc19912.html.

Peter Taylor, University of Canberra, 2022.

Preface

Klaus Peters and I first met at the annual joint meeting of the American Mathematical Society and the Mathematical Association of America. He had a booth featuring fantastic publications from the company, *A. K. Peters Publishers*, which he founded with his wife **Alice**. We also met at the biannual Gathering for Gardner, and had become good friends. The company became a sponsor of the *Alberta High School Mathematics Competition*. Later, I bought a few shares in Klaus's company.

Klaus passed away shortly after. I was delighted when **Colm Mulcahy**, the current president of the Gathering for Gardner, informed me recently that **CRC Press**, under the banner of **Taylor & Francis Publishers**, was planning an *A. K. Peters/CRC Recreational Mathematics Series.* He suggested that I should contribute a volume.

I am no general in the army of recreational mathematicians, but I am a reasonably competent chief of staff. Over my long involvement with mathematics outreach and mathematics competitions, I have come across many outstanding problems. Invariably, they all have a recreational element in them. So I turn to what I consider the best collection, the *International Mathematics Tournament of the Towns.*

I first became aware of the Tournament when I visited Australia in 1989. It marked the beginning of my long friendship with **Peter Taylor**. We collaborated closely on many fronts, but principally on the Tournament. I met the great **Nikolay Konstantinov** in person in 1993 at the Tournament's Summer Seminar in Beloretsk, Russia, and again in 2005 at the Seminar in Mir Town, Belarus. In between, I invited Nikolay for a visit to Edmonton.

The problems and solutions to the first 28 Tournaments have been published by the Australian Mathematics Trust, with Peter at the helm. After he stepped down as the Executive Director of the Trust, the project was abandoned. Part III of this book continues from Tournament 29 to Tournament 42.

To justify the inclusion of this book in the series, Peter and I have identified 162 Tournament problems with distinctly recreational flavour. They are loosely organized into 18 sets of 9 problems with similar themes and presented in Part I. We also find that many Tournament problems are very suitable for introducing mathematical ideas to young minds, and put them together in Part II.

Andy Liu, University of Alberta, 2022.

Acknowledgement

I am grateful to **Kate Jones** and **George Sicherman** who have read the manuscript and made many valuable suggestions. I am grateful to an anonymous reviewer who had made numerous astute corrections and valuable suggestions.

I thank **Colm Mulcahy** and **Callum Fraser** for encouraging Peter and me to submit this book to the *A. K. Peters/CRC Recreational Mathematics Series*. I thank **Callum Fraser**, **Mansi Kabra**, **Robin Lloyd Starkes**, **Shashi Kumar** and other staff members of **CRC Press/Taylor & Francis** for their administrative and technical support.

Authors Bio

Peter Taylor obtained his doctorate in mathematics from the University of Adelaide in 1972. He was an academic at the University of Canberra (UC) for over forty years. In 1976, he was a founder of what was to become the Australian Mathematics Trust (AMT) and served as its Executive Director from 1994 to 2012. He is currently a Professor Emeritus at UC.

Peter had been a member of the Australian Mathematical Olympiad Committee and of the Problem Committee for the Australian Mathematics Competitions. He brought the International Mathematics Tournament of the Towns (IMTT) to western countries in 1988, translating the early problems into English by learning Russian. Under his helm, the AMT published six books covering the first 28 IMTT. He was an author or a co-author for five of them.

Andy Liu obtained his doctorate in mathematics as well as a professional diploma in elementary education in 1976, qualifying him officially to teach from kindergarten to graduate school. He was an academic at his alma mater for over thirty years, where he is currently a Professor Emeritus. During that period, he ran a mathematics circle for students in upper elementary or junior high schools. The members had published over fifty papers in scientific journals.

Andy had served as the deputy leader of the USA team and the leader of the Canadian team in the International Mathematical Olympiad. He had chaired its Problem Committee once and was a member three other times. He had authored or co-authored eighteen books in mathematics, and had been credited as an editor or a co-editor for several others.

Part I

Mathematical Recreations

There are two major world events in mathematical recreations, the **International Puzzle Party** (IPP) and the **Gathering for Gardner** (G4G).

The IPP was founded by **Jerry Slocum** in 1978. It is held every year, with the location rotating among North America, Asia and Europe. The emphasis is on mechanical puzzles, the best-known example being Rubik's Cube, designed by the Hungarian engineer **Ernő Rubik**. Jerry is a puzzle collector and historian. He is both the authority and an expert in this field. He had donated his collection of mechanical puzzles, one of the largest in the world, to the Lilly Library of Indiana University.

Noted designers, mainly of packing and assembly puzzles in three dimensions, are **Oskar van Deventer**, **Nobuyuki Yoshigahara**, **Stewart Coffin**, **Akio Kamei** and **William Cutler**. Among the younger generation are **Tom van der Zanden**, **Hajime Katsumoto** and **Koichi Miura**.

The G4G was founded by **Tom Rodgers** in 1993. It is held every two years, always in Atlanta. It focuses on three areas in which **Martin Gardner** was interested and had significant contributions. They are mathematical recreations, magic and debunking of pseudo-science. More than half of the attendees come for mathematical recreations, primarily in the form of pencil and paper puzzles.

In the nineteenth century, the dominant puzzlists were **Henry Dudeney** and **Sam Loyd**. Their work was popularized by Martin Gardner in the twentieth century. It should be mentioned that earlier in that century, the Russian author **Yakov Perelman** was writing recreational books in mathematics and physics. They were translated into other languages.

One of the books edited by Martin is the English translation of "The Moscow Puzzles" by **Boris Kordemsky**. More recent books in mathematical recreations are "Mathematical Puzzles" by **Peter Winkler**, published by the CRC in 2021, and "Arithmetical, Geometrical and Combinatorial Puzzles from Japan" by **Tadao Kitazawa**, published by the AMS/MAA in 2021. See also many puzzle books by **Richard I. Hess**.

Bridging the gap between the two types of mathematical recreations are the mechanical puzzles in two dimensions. The most popular family of planar puzzle pieces are the *polyominoes*, consisting of unit squares joined side to side. The name is derived from the domino which consists of two unit squares. Thus the monomino is just a unit square. There are two different trominoes. The 1×3 rectangle is called the I-tromino, while the V-tromino is a 2×2 square with a unit square missing from one corner.

The diagram below shows the five tetrominoes, which are shaded, along with the twelve pentominoes. Each has a letter code.

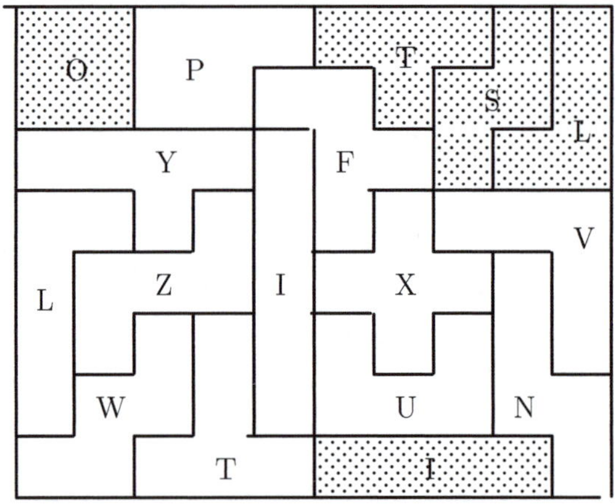

The standard reference in this field is *Polyominoes: Puzzles, Patterns, Problems & Packings* by **Solomon Golomb**, published by Princeton University Press in 1994. He was a world-class scientist whose pioneering research spearheaded the conversion from analog computing to digital computing. He won numerous awards, including a Gold Medal of Science from President **Barack Obama** in 2012.

Many colourful laser-cut acrylic and wooden sets, of polyominoes and other geometric figures, are produced by **Kate Jones**, a Hungarian-American artist. They may be ordered from the website of **Kadon Enterprises**:

http://gamepuzzles.com

A remarkable example is the set *Shardinaires-9*, produced by Kadon Enterprises and designed by computer wizard **George Sicherman**, of the Sicherman Dice fame. The nine pieces may be packaged in a 4×5 rectangle as shown in the diagram below.

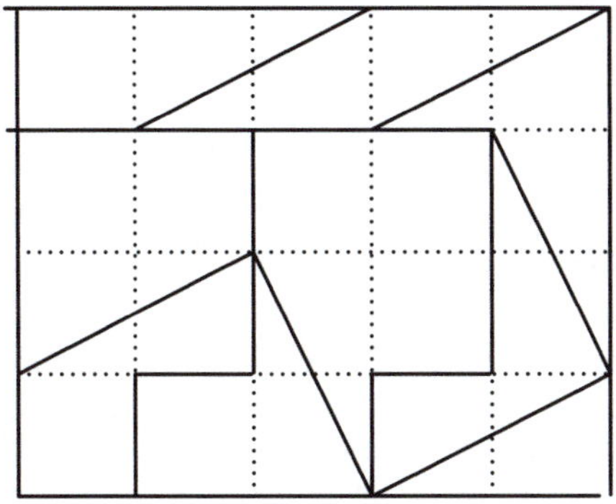

The principal task is to use all nine pieces to construct, one at a time, each of the tetrominoes and the pentominoes. In other words, the challenge is to build both four squares and five squares with the same nine pieces.

In this book, we focus on mathematical recreations of the type promoted by the G4G. **David Singmaster** has published many scholarly papers in this field. He is also very knowledgeable in mechanical puzzles. In fact, he was among the first who studied Rubik's cube and wrote about it.

Many problems from the International Mathematics Tournament of the Towns have very strong recreational flavour, making it a fertile source for mathematical recreations. We have selected 162 of them and grouped them into 18 sets of 9 loosely related problems. Each problem is identified by the round and the year in which it occurred, followed by the level (O for ordinary and A for advanced) and its number in that paper. The reader can track down its solution in Part III.

Problems

Set 1.

1. (Fall 2010 Senior O-5)
   In a tournament with 55 participants, one match is played at a time, with the loser dropping out. In each match, the numbers of wins so far of the two participants differ by not more than 1. What is the maximal number of matches for the winner of the tournament?

2. (Fall 2020 Junior O-2)
   Eight players participated in several tournaments. In each, they paired themselves in four quarterfinals. The winners move onto to two semifinals, and eventually, two of them reach the final. Each player had played every other players during these tournaments.
   (a) Must each player reach the semifinals more than once?
   (b) Must each player reach the finals at least once?

3. (Fall 2012 Junior O-1)
   The family names of Clark, Donald, Jack, Robin and Steven are Clarkson, Donaldson, Jackson, Robinson and Stevenson, but not in that order. Clark is 1 year older than Clarkson, Donald is 2 years older than Donaldson, Jack is 3 years older than Jackson and Robin is 4 years older than Robinson. Who is older, Steven or Stevenson, and what is the difference in their ages?

4. (Spring 2014 Junior A-1)
   During the Christmas Party, Santa hands out 47 chocolates and 74 marshmallows. Each girl gets 1 more chocolate than each boy, and each boy gets 1 more marshmallow than each girl. How many children are at the party?

5. (Spring 2008 Junior A-5)
   Standing in a circle are 99 girls, each with a candy. In each move, each girl gives her candy to either neighbour. If a girl receives two candies in the same move, she eats one of them. What is the minimum number of moves after which only one candy remains?

6. (Spring 2013 Senior A-2)
   A boy and a girl are sitting at opposite ends of a long bench. One at a time, twenty other children take seats on the bench. If a boy takes a seat between two girls or if a girl takes a seat between two boys, he or she is said to be brave. At the end, the boys and girls are sitting alternately. What is the number of brave children?

7. (Fall 2014 Senior O-5)
   At the beginning, there are some silver coins on a table. In each move, we can either add a gold coin and record the number of silver coins on a blackboard, or remove a silver coin and record the number of gold coins on a whiteboard. At the end, only gold coins remain on the table. Must the sum of the numbers on the blackboard be equal to the sum of the numbers on the whiteboard?

8. (Fall 2018 Junior A-2)
   On an island with 2018 inhabitants each person is either a knight, a knave or a kneejerk. Everyone knows who everyone else is. Each inhabitant is asked in turn whether there are more knights than knaves on the island. A knight always tells the truth, a knave always lies, and a kneejerk agrees with the majority of those who speak before him. In case of a tie, such as when a kneejerk speaks first, he answers "Yes" and "No" at random. If exactly 1009 answers are "Yes", at most how many kneejerks are on the island?

9. (Spring 2021 Junior A-4)
   Fifty natives stand in a circle. Each announces the age of his left neighbour. Then each announces the age of his right neighbour. Each native is either a knight who tell both numbers correctly, or a knucklehead who increases one of the numbers by 1 and decreases the other by 1. Each knucklehead chooses which number to increase and which to decrease independently. Is it always possible to determine which of the natives are knights and which are knuckleheads?

Set 2.

1. (Fall 2008 Junior O-2)
   Twenty-five of the numbers 1, 2, ..., 50 are chosen. Twenty-five of the numbers 51, 52, ..., 100 are also chosen. No two chosen numbers differ by 0 or 50. Find the sum of all 50 chosen numbers.

2. (Spring 2014 Junior O-1)
   When each of 100 numbers was increased by 1, the sum of their squares remained unchanged. Each of the new numbers is increased by 1 once more. How will the sum of their squares change this time?

3. (Fall 2014 Junior O-2)
   Do there exist ten pairwise distinct positive integers such that their average divided by their greatest common divisor is equal to
   (a) 6;
   (b) 5?

4. (Spring 2015 Junior O-3)
Basil computes the sum of several consecutive positive integers starting from 1. Patty computes the sum of 10 consecutive positive powers of 2, not necessarily starting from 1. Can the two sums be equal?

5. (Fall 2014 Senior O-3)
Peter writes down the sum of every subset of size 7 of a set of 15 distinct integers, and Betty writes down the sum of every subset of size 8 of the same set. If they arrange their numbers in non-decreasing order, can the two lists turn out to be identical?

6. (Spring 2013 Junior A-1)
The sum of any two of n distinct numbers is a positive integer power of 2. What is the maximum value of n?

7. (Fall 2015 Junior A-2)
From $\{1,2,3,\ldots,100\}$, k integers are removed. Among the numbers remaining, do there always exist k distinct integers with sum 100 if

   (a) $k = 9$;

   (b) $k = 8$?

8. (Spring 2019 Junior A-1)
Do there exist seven distinct positive integers with sum 100 such that they are determined uniquely by the fourth largest among them?

9. (Spring 2014 Junior A-4)
On each of 100 cards, Anna writes down a positive integer. These numbers are not necessarily distinct. On $\binom{100}{2}$ cards, Anna writes down the sums of these numbers taken 2 at a time. On $\binom{100}{3}$ cards, she writes down the sums taken 3 at a time. She continues until she finally writes down the sum of all 100 numbers on 1 card. She is allowed to send some of these $2^{100} - 1$ cards to Boris, no two of which may contain the same number. Boris knows the rules by which the cards are prepared. What is the minimum number of cards Anna must send to Boris in order for him to determine the original 100 numbers?

Set 3.

1. (Spring 2016 Junior O-3)
Do there exist 2016 integers whose sum and product are both 2016?

2. (Fall 2017 Junior O-1)
Five nonzero numbers are added in pairs. Five of the sums are positive and the other five are negative. If they are mutiplied in pairs, find the numbers of positive and negative products.

3. (Fall 2018 Junior O-2)
Determine all positive integers n for which the numbers $1, 2, \ldots, 2n$ can be arranged in pairs so that if the sum of each pair is computed, the product of the sums is the square of an integer.

4. (Spring 2021 Senior O-3)
For which positive integers n do there exist n consecutive positive integers whose product is equal to the sum of n other consecutive positive integers?

5. (Fall 2014 Junior A-3)
Is it possible to divide all positive divisors of 100!, including 1 and 100!, into two groups of equal size such that the product of the numbers in each group is the same?

6. (Spring 2021 Senior A-2)
Does there exist a positive integer n such that for any real numbers x and y, there exist n real numbers such that x is equal to their sum and y is equal to the sum of their reciprocals?

7. (Spring 2014 Senior A-1)
Anna writes down several 1s, puts either a $+$ sign or a $\times$ sign between every two adjacent 1s, adds several pairs of brackets and gets an expression equal to 2014. Boris takes Anna's expression and interchanges all the $+$ signs with all the $\times$ signs. Is it possible that his expression is also equal to 2014?

8. (Fall 2014 Senior A-3)
Gregory writes down 100 numbers on a blackboard and calculates their product. In each move, he increases each number by 1 and calculates their product. What is the maximum number of moves Gregory can make if the product after each move does not change?

9. (Fall 2016 Senior A-4)
The 2016 pairwise sums of 64 numbers are recorded on one piece of paper. They are distinct and positive. The 2016 pairwise products of the same 64 numbers are recorded on another piece of paper. They are also distinct and positive. Later, it is forgotten which piece is which. Is it still possible to determine which piece is which?

Set 4.

1. (Fall 2020 Junior O-5)
Eight elephants are lined up in increasing order of weight. Starting from the third, the weight of each elephant is equal to the total weight of the two elephants immediately in front.

One of them tests positive for Covid, and may have lost some weight. The others test negative and their weights remain the same. Is it possible, in two weighings using a balance, to test if there has been a weight loss, and if so, to determine the elephant which has lost weight?

2. (Fall 2017 Junior O-3)
Among 100 coins in a row are 26 fake ones which form a consecutive block. The other 74 coins are real, and they have the same weight. All fake coins are lighter than real ones, but their weights are not necessarily equal. What is the minimum number of weighings on a balance to guarantee finding at least one fake coin?

3. (Fall 2019 Junior A-3)
The weight of each of 100 coins is 1 gram, 2 grams or 3 grams, and there is at least one of each kind. Is it possible to determine the weight of each coin using at most 101 weighings on a balance?

4. (Fall 2012 Senior O-5)
Among 239 coins, there are two fake coins of the same weight, and 237 real coins of the same weight but different from that of the fake coins. Is it possible, in three weighings on a balance, to determine whether the fake coins are heavier or lighter than the real coins? It is not necessary to identify the fake coins.

5. (Fall 2018 Junior O-4)
Kate has three real coins of the same weight and two fake coins whose total weight is the same as that of two real coins. However, one of them is heavier than a real coin while the other is lighter. Can Kate identify the heavier coin as well as the lighter coin in three weighings on a balance? She must decide in advance which coins are to be weighed without waiting for the result of any weighing.

6. (Spring 2021 Junior O-3)
There are four coins of weights 1001, 1002, 1004 and 1005 grams, respectively. Is it possible to determine the weight of each coin using a balance at most four times?

7. (Spring 2021 Senior O-2)
There are four coins of weights 1000, 1002, 1004 and 1005 grams, respectively. Is it possible to determine the weight of each coin using a balance at most four times, if

   (a) the balance is normal;

   (b) the balance is faulty, in that its left pan is 1 gram lighter than its right pan?

8. (Fall 2008 Junior A-2)
Each of four coins weighs an integral number of grams. Available for use is a balance which shows the difference of the weights between the objects in the left pan and those in the right pan. Is it possible to determine the weight of each coin by using this balance four times, if it may make a mistake of 1 gram either way in at most one weighing?

9. (Fall 2017 Junior A-1)
We have a faulty balance with which equilibrium may be obtained only if the ratio of the total weights in the left pan and in the right pan is 3:4. We have a token of weight 6 kg, a sufficient supply of sugar and bags of negligible weight to hold the sugar. In each weighing, you may put the token or any bags of sugar of known weight on the balance, and add sugar to a bag so that equilibrium is obtained. Is it possible to obtain a bag of sugar of weight 1 kg?

Set 5.

1. (Spring 2009 Junior O-4)
When a positive integer is increased by 10%, the result is another positive integer whose digit-sum has decreased by 10%. Is this possible?

2. (Spring 2013 Junior O-2)
Start with an non-negative integer n. In each move, we may add 9 to the current number or, if the number contains a digit 1, we may delete it. If there are leading 0s as a result, they are also deleted. Is it always possible to obtain the number $n + 1$ in a finite number of steps?

3. (Fall 2013 Junior O-2)
Can the ten digits 0, 1, 2, 3, 4, 5, 6, 7, 8 and 9 be arranged in a row so that no matter which six digits are removed, the remaining four digits, without changing their order, form a composite number?

4. (Fall 2016 Junior O-1)
Do there exist five positive integers such that their ten pairwise sums end in different digits?

5. (Fall 2020 Senior O-3)
A positive multiple of 2020 has distinct digits, and if any two of them switch positions, the resulting number is not a multiple of 2020. How many digits can such a number have?

6. (Spring 2010 Junior A-4)
Is it possible that the sum of the digits of a positive integer n is 100 while the sum of the digits of the number n^3 is 100^3?

7. (Spring 2017 Junior O-1)
Find the smallest positive multiple of 2017 such that its first four digits are 2016.

8. (Spring 2020 Junior A-1)
Does there exist a positive multiple of 2020 which contains each of the ten digits the same number of times?

9. (Spring 2017 Senior A-6)
Find all positive integers n which have a multiple with digit sum k for any integer $k \geq n$.

Set 6.

1. (Fall 2007 Junior O-1)
Black and white counters are placed on an 8×8 board, with at most one counter on each square. What is the maximum number of counters that can be placed such that each row and each column contains twice as many white counters as black ones?

2. (Spring 2009 Senior A-3)
There is a counter in each square of a 10×10 board. We may choose a diagonal containing an even number of counters and remove any counter from it. What is the maximum number of counters which can be removed from the board by these operations?

3. (Fall 2012 Junior O-3)
The game Minesweeper is played on a 10×10 board. Each square either contains a bomb or is vacant. On each vacant square is recorded the number of bombs among the eight adjacent squares. Then all the bombs are removed, and new bombs are placed in all squares which were previously vacant. Then numbers are recorded on vacant squares as before. Can the sum of all numbers on the board now be greater than the sum of all numbers on the board before?

4. (Spring 2018 Junior O-1)
Six rooks are placed on the squares of a 6×6 board so that no two of them attack each other. Is it possible for every empty square to be attacked by two rooks
   (a) at the same distance away;
   (b) at different distances away?

5. (Fall 2018 Senior O-5)
Pete is placing 500 kings on a 100×50 board so that no two attack each other. Basil is placing 500 kings on the white squares of a 100×100 board so that no two attack each other. Who has more ways to place the kings?

6. (Spring 2020 Senior A-7)
For which k is it possible to place a finite number of queens on the squares of an infinite board, so that the number of queens in each row, each column and each diagonal is either 0 or exactly k?

7. (Spring 2010 Junior A-7)
A number of ants are on a 10×10 board, each in a different square. Every minute, each ant crawls to the adjacent square either to the east, to the south, to the west or to the north. It continues to crawl in the same direction as long as this is possible, but reverses direction if it has reached the edge of the board. In one hour, no two ants ever occupy the same square. What is the maximum number of ants on the board?

8. (Spring 2011 Junior A-6)
Two ants crawl along the sides of the 49 squares of a 7×7 board. Each ant passes through all 64 vertices exactly once and returns to its starting point. What is the smallest possible number of sides covered by both ants?

9. (Fall 2017 Senior O-5)
On a 100×100 board, an ant starts from the bottom left corner, visits the top left corner and ends on the top right corner. It goes between squares sharing a common side. The moves are alternately vertical and horizontal, with the first move horizontal. Must there exist two adjacent squares such that the ant has gone from one to the other at least twice?

Set 7.

1. (Fall 2020 Junior O-3)
Anna and Boris play a game with n counters. Anna goes first, and turns alternate thereafter. In each move, a player takes either 1 counter or a number of counters equal to a prime divisor of the current number of counters. The player who takes the last counter wins. For which n does Anna have a winning strategy?

2. (Fall 2008 Senior A-3)
Anna and Boris play a game with $n > 2$ piles each initially consisting of a single counter. The players take turns, Anna going first. In each move, a player chooses two piles containing numbers of counters relatively prime to each other, and merge the two piles into one. The player who cannot make a move loses the game. For each n, determine the player with a winning strategy.

3. (Fall 2013 Junior A-7)
Peter and Betty are playing a game with 10 stones in each of 11 piles. Peter moves first, and turns alternate thereafter. In his turn, Peter must take 1, 2 or 3 stones from any one pile. In her turn, Betty must take one stone from 1, 2 or 3 piles. Whoever takes the last stone overall is the winner. Which player has a winning strategy?

4. (Fall 2007 Junior A-4)
Two players take turns entering a symbol in an empty square of a $1 \times n$ board, where n is an integer greater than 1. Aaron always enters the symbol X and Betty always enters the symbol O. Two identical symbols may not occupy adjacent squares. A player without a move loses the game. If Aaron goes first, which player has a winning strategy?

5. (Spring 2008 Junior A-3)
Alice and Brian are playing a game on a $1 \times (n+2)$ board. To start the game, Alice places a counter on any of the n interior squares. In each move, Brian chooses a positive integer k. Alice must move the counter to the k-th square on the left or the right of its current position. If the counter moves off the board, Alice wins. If it lands on either of the end squares, Brian wins. If it lands on another interior square, the game proceeds to the next move. For which values of n does Brian have a strategy which allows him to win the game in a finite number of moves?

6. (Spring 2008 Senior A-2)
Alice and Brian are playing a game on the real line. To start the game, Alice places a counter on a number x where $0 < x < 1$. In each move, Brian chooses a positive number d. Alice must move the counter to either $x + d$ or $x - d$. If it lands on 0 or 1, Brian wins. Otherwise the game proceeds to the next move. For which values of x does Brian have a strategy which allows him to win the game in a finite number of moves?

7. (Spring 2012 Senior O-5)
In an 8×8 board, the rows are numbered from 1 to 8 and the columns are labeled from a to h. In a two-player game on this board, the first player has a white rook which starts on the square b2, and the second player has a black rook which starts on the square c4. The two players take turns moving their rooks. In each move, a rook lands on another square in the same row or the same column as its starting square. However, that square cannot be under attack by the other rook, and cannot have been landed on before by either rook. The player without a move loses the game. Which player has a winning strategy?

8. (Spring 2020 Junior O-5)
On an 8×8 board, there is a rook in each of the squares $(a, 1)$ and $(c, 3)$. Alice moves first, followed by Bob, and turns alternate thereafter. In each turn, the player moves one of the rooks any number of squares upwards or to the right. The rook may not move through or stop at the square of the other rook. The player who moves either rook into the square $(h, 8)$ wins. Which of Alice and Bob has a winning strategy?

9. (Fall 2020 Junior A-3)
Anna and Boris play a game in which Anna goes first and turns alternate thereafter. In her turn, Anna suggests an integer. In his turn, Boris writes down on a whiteboard either that number or the sum of that number with all previously written numbers. Is it always possible for Anna to ensure that at some moment among the written numbers there are one hundred copies of the number

   (a) 5;

   (b) 10?

Set 8.

1. (Fall 2009 Junior O-4)
On a lottery ticket, a number consisting of seven different digits is to be written. On the draw date, an official number with seven different digits is revealed. A ticket wins a prize if it matches the official number in at least one digit. Is it possible to guarantee winning a prize by buying at most six tickets?

2. (Fall 2011 Senior O-2)
Peter buys a lottery ticket on which he enters an n-digit number, none of the digits being 0. On the draw date, the lottery administrators will reveal an $n \times n$ table, each square containing one of the digits from 1 to 9. A ticket wins a prize if it does *not* match any row or column of this table, read in either direction. Peter wants to bribe the administrators to reveal the digits on some squares chosen by Peter, so that Peter can guarantee to have a winning ticket. What is the minimum number of digits Peter has to know?

3. (Spring 2012 Junior A-6)
A bank has one million clients, one of whom is Inspector Gadget. Each client has a unique PIN number consisting of six digits. Dr. Claw has a list of all the clients. He is able to break into the account of any client, choose any n digits of the PIN number and copy them. The n digits he copies from different clients need not be in the same n positions. He can break into the account of each client, but only once.

What is the smallest value of n which allows Dr. Claw to determine the complete PIN number of Inspector Gadget?

4. (Spring 2020 Junior A-2)
A dragon has 41! heads and the knight has three swords. The gold sword cuts off half of the current number of heads of the dragon plus one more. The silver sword cuts off one third of the current number of heads plus two more. The bronze sword he cuts off one fourth of the current number of heads plus three more. The knight can can use any of the three swords in any order. However, if the current number of heads of the dragon is not a multiple of 2 or 3, the swords do not work and the dragon eats the knight. Will the knight be able to kill the dragon by cutting off all its heads?

5. (Spring 2011 Junior O-5)
A dragon gives a captured knight 100 coins. Half of them are magical, but only the dragon knows which they are. Each day, the knight divides the coins into two piles which are not necessarily equal in size. If each pile contains the same number of magic coins, or the same number of non-magic coins, the knight will be set free. Can the knight guarantee himself freedom in at most

   (a) 50 days;

   (b) 25 days?

6. (Fall 2009 Senior A-1)
After a gambling session, each of one hundred pirates calculates the amount he has won or lost. Money can only change hands in the following way. Either one pirate pays an equal amount to every other pirate, or one pirate receives the same amount from every other pirate. Each pirate has enough money to make any payment. Is it always possible, after several such steps, for all the winners to receive exactly what they have won and for all losers to pay exactly what they have lost?

7. (Spring 2020 Senior O-3)
There are 41 letters on a circle; each letter is A or B. We may replace ABA by B and vice versa, as well as replace BAB by A and vice versa. Is it always possible, using these replacements, to obtain a circle containing a single letter?

8. (Fall 2019 Junior O-3)
Counters numbered 1 to 100 are arranged in order in a row. It costs 1 dollar to interchange two adjacent counters, but nothing to interchange two counters with exactly 3 other counters between them. What is the minimum cost for rearranging the 100 counters in reverse order?

9. (Fall 2019 Senior O-2)
Counters numbered 1 to 100 are arranged in order in a row. It costs 1 dollar to interchange two adjacent counters, but nothing to interchange two counters with exactly 4 other counters between them. What is the minimum cost for rearranging the 100 counters in reverse order?

Set 9.

1. (Fall 2007 Junior O-4)
Each square of a 29×29 table contains one of the integers 1, 2, 3, ..., 29, and each of these integers appears 29 times. The sum of all the numbers above the main diagonal is equal to three times the sum of all the numbers below this diagonal. Determine the number in the central square of the table.

2. (Spring 2014 Junior O-3)
Each of the squares in a 5×7 table contains a number. Peter knows only that the sum of the numbers in the 6 squares of any 2×3 or 3×2 rectangle is 0. He is allowed to ask for the number in any position in the table. What is the minimum number of questions he needs to ask in order to be able to determine the sum of all 35 numbers in the table?

3. (Spring 2018 Junior O-3)
In a 4×4 square, the sum of the numbers in each row and in each column is the same. Nine of the numbers are erased, leaving behind the seven shown below. Is it possible to recover uniquely

   (a) one;

   (b) two

   of the erased numbers?

1			2
	4	5	
	6	7	
3			

4. (Spring 2020 Senior O-1)
Is it possible to fill a 40×41 table with integers so that for each n, there are n copies of n among the squares which share common sides with its square?

5. (Spring 2017 Senior O-3)
In a 1000×1000 table of numbers, the sum of the numbers in each rectangle consisting of n of the squares is the same. Find all values of n for which all the numbers in the table must be equal.

6. (Spring 2020 Junior A-4)
For which integers $n \geq 2$ is it possible to write real numbers into the squares of an $n \times n$ table, so that every integer from 1 to $2n(n-1)$ appears exactly once as the sum of the numbers in two adjacent squares?

7. (Spring 2015 Junior A-5)

   (a) A $2 \times n$ table of numbers with $n > 2$ is such that the sums of all the numbers in the columns are different. Is it always possible to permute the numbers in the table so that the column sums are still different and now the row sums are also different?

   (b) Is this always possible for a 10×10 table?

8. (Fall 2016 Junior A-2)
There are 64 positive integers in the squares of an 8×8 board. Whenever the board is covered by 32 dominoes, the sum of the two integers covered by each domino is unique. Is it possible that the largest integer on the board does not exceed 32?

9. (Spring 2018 Senior A-3)
One hundred different numbers are written in the squares of a 10×10 table. In each move, you can take out a rectangle formed by the squares, perform a half-turn and then put it back. Is it always possible to arrange the numbers in the square so that they increase in every row from left to right, and in every column from top to bottom, in no more than 99 moves?

Set 10.

1. (Spring 2018 Junior A-1)
Thirty-nine nonzero numbers are written in a row. The sum of any two adjacent numbers is positive, while the sum of all the numbers is negative. Is the product of all these numbers negative or positive?

2. (Spring 2020 Junior O-2)
What is the maximum number of distinct integers in a row such that the sum of any 11 adjacent integers is either 100 or 101?

3. (Spring 2019 Senior O-5)
Several positive integers are arranged in a row. Their sum is 2019. None of them is equal to 40, and the sum of any block of adjacent numbers is not equal to 40. What is the maximum length of this row?

4. (Spring 2008 Senior A-5)
The positive integers are arranged in a row in some order, each occurring exactly once. Does there always exist an adjacent block of at least two numbers somewhere in this row such that the sum of the numbers in the block is a prime number?

5. (Fall 2011 Junior O-4)
Along a circle are the integers from 1 to 33 in some order. The sum of every pair of adjacent numbers is computed. Is it possible for these sums to consist of 33 consecutive numbers?

6. (Spring 2015 Senior O-5)
Along a circle are 2015 positive integers such that the difference between any two adjacent numbers is equal to their greatest common divisor. Determine the maximal value of a positive number which divides the product of these 2015 numbers.

7. (Spring 2010 Junior A-3)
Along a circle are placed 999 numbers, each 1 or -1, and there is at least one of each. The product of each block of 10 adjacent numbers along the circle is computed. Let S denote the sum of these 999 products.
   (a) What is the minimum value of S?
   (b) What is the maximum value of S?

8. (Spring 2013 Junior A-4)
On a circle are 1000 non-zero numbers painted alternately black and white. Each black number is the sum of its two neighbours while each white number is the product of its two neighbours. What are the possible values of the sum of these 1000 numbers?

9. (Spring 2013 Senior A-4)
Is it possible to arrange the numbers $1, 2, \ldots, 100$ on a circle in some order so that the absolute value of the difference between any two adjacent numbers is at least 30 and at most 50?

Set 11.

1. (Fall 2009 Junior O-1)
Is it possible to dissect a square into nine squares, with five of them of one size, three of them of another size and one of them a third size?

2. (Fall 2011 Junior O-3)
From the 9×9 board, all 16 unit squares whose row numbers and column numbers are both even have been removed. The punctured board is dissected into rectangular pieces. What is the minimum number of square pieces?

3. (Spring 2015 Junior O-4)
The figure in the diagram below is to be dissected along the dotted lines into a number of squares which are not necessarily all of the same size or all of different sizes. What is the minimum number of such squares?

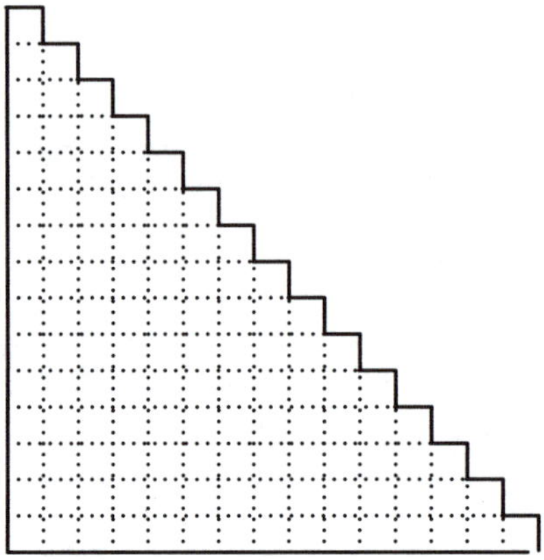

4. (Spring 2016 Junior O-4)
In a 10×10 board, the 25 squares in the upper left 5×5 subboard are black while all remaining squares are white. The board is divided into a number of connected pieces of various shapes and sizes such that the number of white squares in each piece is three times the number of black squares in that piece. What is the maximum number of pieces?

5. (Spring 2009 Junior A-2)

   (a) Does there exist a non-convex polygon which can be dissected into two congruent parts by a line segment which cuts one side of the original polygon in half and another side in the ratio 1:2?

   (b) Can such a polygon be convex?

6. (Spring 2011 Junior A-1)
Does there exist a hexagon that can be dissected into four congruent triangles by a straight cut?

7. (Spring 2021 Junior O-4)
Is it possible to dissect a polygon into four isosceles triangles no two of which are congruent, if the polygon is

   (a) a square;

   (b) an equilateral triangle?

8. (Spring 2010 Junior O-2)
Karlsson and Lillebror are dividing a square cake. Karlsson chooses a point P of the cake which is not on the boundary. Lillebror makes a straight cut from P to the boundary of the cake, in any direction he chooses. Then Karlsson makes a straight cut from P to the boundary, at a right angle to the first cut. Lillebror will get the smaller of the two pieces. Can Karlsson prevent Lillebror from getting at least one quarter of the cake?

9. (Spring 2017 Senior A-3)
Is it possible to dissect a cube into two pieces which can be reassembled into a convex polyhedron with only triangular and hexagonal faces?

Set 12.

1. (Fall 2015 Senior A-2)
A 10×10 board is partitioned into 20 pentominoes by 80 unit segments lying on common sides of adjacent unit squares. What is the maximum number of different pentominoes among these 20?

2. (Fall 2018 Junior O-3)
A 7×14 board is constructed from copies of the O-tetromino and the V-tromino. Is it possible that
   (a) the same number of copies of each piece is used;
   (b) more copies of the O-tetromino are used than copies of the V-tromino?

3. (Fall 2016 Senior O-4)
A 100-omino can be dissected into two congruent 50-ominoes as well as 25 congruent tetrominoes. Is it always possible to dissect it into 50 dominoes?

4. (Spring 2014 Junior A-2)
Anna paints several squares of a 5×5 board. Boris' task is to cover up all of them by placing non-overlapping copies of the V-tromino so that each copy covers exactly three squares of the board. What is the minimum number of squares Anna must paint in order to prevent Boris from succeeding in his task?

5. (Spring 2018 Senior O-4)
A corner square of an 8×8 board is painted, and a counter is put on it. Peter and Basil take turns moving the counter, Peter going first. In his turn, Peter moves the counter once as a queen to an unpainted square, and Basil moves the counter twice as a king to unpainted squares. The square visited by Peter and both squares visited by Basil are then painted. The player without a move loses the game. Which player has a winning strategy?

6. (Fall 2015 Junior A-1)
A polyomino is said to be *amazing* if it is not a rectangle and several copies of it can be assembled into a larger copy of it. The diagram below shows that the V-tromino is amazing.

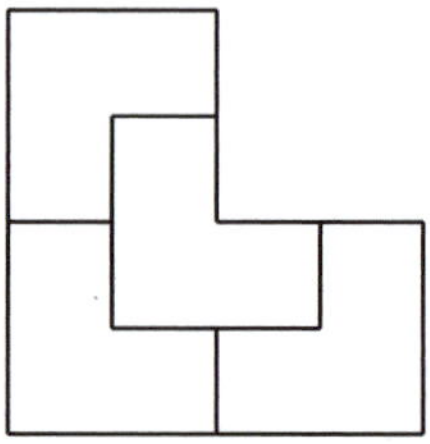

(a) Does there exist an amazing tetromino?

(b) Determine all $n > 4$ such that there exists an amazing n-omino.

7. (Fall 2016 Junior A-4)
In a 7×7 box, each of the 49 pieces is either dark chocolate or white chocolate. In each move, Alex may eat two adjacent pieces along a row, a column or a diagonal, provided that they are of the same kind.

What is the maximum number of pieces Alex can guarantee to be able to eat, regardless of the initial arrangement the pieces?

8. (Fall 2018 Junior A-4)
An invisible alien spaceship in the shape of an O-tetromino may land on our 7×7 airfield, occupying four of the 49 squares. Sensors are to be placed in certain squares of the airfield. A sensor will send a signal if it is in a square on which the spaceship lands. From the locations of all the sensors which send signals, we must be able to determine exactly on which four squares the spaceship has landed. What is the smallest number of sensors we have to place?

9. (Fall 2019 Junior O-5)
Basil has sufficiently many copies of the I-tricube, which is a $1 \times 1 \times 3$ block, and of the V-tricube, which is a $1 \times 2 \times 2$ block with a unit cube missing at a corner. Basil builds with these pieces a solid rectangular box each dimension of which is at least 2. What is the minimum number of copies of the I-tricube which Basil must use?

Set 13.

1. (Fall 2016 Junior A-1)
Each of ten boys has 100 cards. In each move, one of the boys gives one card to each of the other boys. What is the minimal number of moves before each boy has a different number of cards?

2. (Fall 2010 Senior A-1)
   There are 100 points on the plane. All 4950 pairwise distances between two points have been recorded.

   (a) A single record has been erased. Is it always possible to restore it using the remaining records?

   (b) Suppose no three points are on a line, and k records were erased. What is the maximum value of k such that restoration of all the erased records is always possible?

3. (Spring 2017 Junior O-4)
   One hundred children of distinct heights stand in a line. At each step, a group of 50 consecutive children is chosen and they are rearranged in an arbitrary way. Is it always possible, in 6 such steps, to arrange the children so that their heights decrease from left to right?

4. (Spring 2017 Senior O-4)
   Ten children of distinct heights stand in a circle. In each step, one of them moves to a new place in the circle between two children. What is the minimum number of steps in which the children can always be arranged so that their heights increase in clockwise order?

5. (Spring 2012 Junior O-1)
   Buried under each square of an 8×8 board is either a treasure or a message. There is only one square with a treasure. A message indicates the minimum number of steps needed to go from its square to the treasure square. Each step takes one from a square to another square sharing a common side. What is the minimum number of squares one must dig up in order to bring up the treasure for sure?

6. (Spring 2011 Senior O-3)
   An integer k is given, where $2 \leq k \leq 50$. Along a circle are 100 white points. In each move, we choose a block of k adjacent points such that the first and the last are white, and we paint both of them black. For which values of k is it possible for us to paint all 100 points black after 50 moves?

7. (Fall 2016 Junior O-5)
   One hundred bear-cubs have 1, 2, ..., 2^{99} berries, respectively. A fox chooses two bear-cubs and divides their berries equally between them. If a berry is left over, the fox eats it. What is the least number of berries the fox can leave for the bear-cubs?

8. (Fall 2010 Junior A-7)
Merlin summons the n knights of Camelot for a conference. Each day, he assigns them to the n seats at the Round Table. From the second day on, any two neighbours may interchange their seats if they were not neighbours on the first day. The knights try to sit in some cyclic order which has already occurred before on an earlier day. If they succeed, then the conference comes to an end when the day is over. What is the maximum number of days for which Merlin can guarantee that the conference will last?

9. (Spring 2014 Senior A-6)
A computer directory lists all pairs of cities connected by direct flights. Anna hacks into the computer and permutes the names of the cities. It turns out that no matter which other city is renamed Moscow, she can rename the remaining cities so that the directory is perfectly correct. Later, Boris does the same thing. However, he insists on exchanging the names of Moscow with another city. Is it always possible for him to rename the remaining cities so that the directory is perfectly correct?

Set 14.

1. (Spring 2012 Junior O-2)
The number 4 has an odd number of odd positive divisors, namely 1, and an even number of even positive divisors, namely 2 and 4. Is there a number with an odd number of even positive divisors and an even number of odd positive divisors?

2. (Spring 2018 Junior O-4)
Each of three positive integers is a multiple of the greatest common divisor of the other two, and a divisor of the least common multiple of the other two. Must these three numbers be the same?

3. (Spring 2012 Junior A-5)
Let p be a prime number. A set of $p+2$ positive integers, not necessarily distinct, is called *interesting* if the sum of any p of them is divisible by each of the other two. Determine all interesting sets.

4. (Fall 2007 Junior A-2)
   (a) Each of Peter and Basil thinks of three positive integers. For each pair of his numbers, Peter writes down the greatest common divisor of the two numbers. For each pair of his numbers, Basil writes down the least common multiple of the two numbers. If both Peter and Basil write down the same three numbers, must these three numbers be equal to each other?
   (b) Is the analogous result true if each of Peter and Basil thinks of four positive integers instead?

5. (Spring 2009 Junior O-2)
Let $a \wedge b$ denote the number a^b. The order of operations in the expresson $7 \wedge 7 \wedge 7 \wedge 7 \wedge 7 \wedge 7 \wedge 7$ must be determined by inserting five pairs of brackets. Is it possible to put brackets in two distinct ways so that the expressions have the same value?

6. (Spring 2012 Junior O-4)
Brackets are to be inserted into the expression $10\div9\div8\div7\div6\div5\div4\div3\div2$ so that the resulting number is an integer.

   (a) Determine the maximum value of this integer.

   (b) Determine the minimum value of this integer.

7. (Spring 2015 Junior O-5)
We have one copy of 0 and two copies of each of $1, 2, \dots, n$. For which n can we arrange these $2n+1$ numbers in a row, such that for $1 \leq k \leq n$, there are exactly k other numbers between the two copies of the number m?

8. (Spring 2011 Junior A-3)
Baron Munchausen has a set of 50 coins. The mass of each is a distinct positive integer not exceeding 100, and the total mass is even. The Baron claims that it is not possible to divide the coins into two piles with equal total mass. Can the Baron be right?

9. (Fall 2011 Junior A-3)
A set of at least two objects with pairwise different weights has the property that for any pair of objects from this set, we can choose a subset of the remaining objects so that their total weight is equal to the total weight of the given pair. What is the minimum number of objects in this set?

Set 15.

1. (Spring 2011 Junior O-4)
Each diagonal of a convex quadrilateral divides it into two isosceles triangles. The two diagonals of the same quadrilateral divide it into four isosceles triangles. Must this quadrilateral be a square?

2. (Fall 2016 Senior O-2)
Of the triangles determined by 100 points on a line plus an extra point not on the line, at most how many of them can be isosceles?

3. (Spring 2010 Senior A-1)
Is it possible to divide all the lines in the plane into pairs of perpendicular lines so that every line belongs to exactly one pair?

4. (Spring 2020 Junior O-1)
In a 6×6 county, 27 of the squares are cities and the other 9 villages. Each village is claimed by a city if and only if it shares at least one vertex with the city. Is it possible that the number of cities claiming a village is different for each village?

5. (Spring 2013 Junior A-3)
Is it possible to mark some of the squares of a 19×19 board so that each 10×10 board contains a different number of marked squares?

6. (Fall 2019 Junior A-7)
Peter has an $n \times n$ stamp, $n > 10$, such that 102 of the unit squares are coated with black ink. He presses this stamp 100 timies on a 101×101 grid, each time leaving a black imprint on 102 unit squares of the grid. Is it possible that the grid is black except for one unit square at a corner?

7. (Spring 2014 Senior O-5)
A park is in the shape of a convex quadrilateral $ABCD$. Alex, Ben and Chris are jogging there, each at his own constant speed. Alex and Ben start from A at the same time, Alex jogging along AB and Ben along AC. When Alex arrives at B, he immediately continues on along BC. At the time Alex arrives at B, Chris starts from B, jogging along BD. Alex and Ben arrive at C at the same time, and lex immediately continues on along CD. He and Chris arrive at D at the same time. Can it happen that Ben and Chris meet each other at the point of intersection of AC and BD?

8. (Fall 2011 Junior O-5)
On a highway, a pedestrian and a cyclist are going in the same direction, while a cart and a car are coming from the opposite direction. All are travelling at constant speeds, not necessarily equal to one another. The cyclist catches up with the pedestrian at 10 o'clock. After a time interval, the cyclist meets the cart, and after another time interval equal to the first, she meets the car. After a third time interval, the car meets the pedestrian, and after another time interval equal to the third, the car catches up with the cart. If the pedestrian meets the car at 11 o'clock, when does he meet the cart?

9. (Fall 2013 Senior O-5)
A spaceship lands on an asteroid, which is known to be either a sphere or a cube. The spaceship sends out an explorer which crawls on the surface of the asteroid.

The explorer continuously transmits its current position in space to the spaceship, until it reaches the point which is symmetric to the landing site relative to the centre of the asteroid. Thus the spaceship can trace the path along which the explorer is moving. Can it happen that these data are not sufficient for the spaceship to determine whether the asteroid is a sphere or a cube?

Set 16.

1. (Fall 2019 Junior O-1)
   A magician lays the 52 cards of a standard deck in a row, and announces in advance that the Three of Clubs will be the only card remaining after 51 steps. In each step, the audience points to any card. The magician can either remove that card or remove the card in the corresponding position counting from the opposite end. What are the possible positions for the Three of Clubs at the start in order to guarantee the success of this trick?

2. (Fall 2007 Senior O-4)
   The audience chooses two of twenty-nine cards, numbered from 1 to 29. The assistant of a magician chooses two of the remaining twenty-seven cards, and asks a member of the audience to take them to the magician, who is in another room. The two cards are presented to the magician in an arbitrary order. Does there exist an arrangement with the assistant beforehand such that the magician can deduce which two cards the audience has chosen?

3. (Spring 2019 Junior O-5)
   As the assistant watches, the audience puts a coin in each of two of 12 boxes in a row. The assistant opens one box that does not contain a coin and exits. The magician enters and opens four boxes simultaneously. Does there exist a method that will guarantee that both coins are in the four boxes opened by the magician?

4. (Spring 2019 Senior O-4)
   As the assistant watches, the audience puts a coin in each of two of 13 boxes in a row. The assistant opens one box that does not contain a coin and exits. The magician enters and opens four boxes simultaneously. Does there exist a method that will guarantee that both coins are in the four boxes opened by the magician?

5. (Fall 2007 Junior A-6)
   The audience arranges n coins in a row. The sequence of heads and tails is chosen arbitrarily. The audience also chooses a number between 1 and n inclusive.

Then the assistant turns one of the coins over, and the magician is brought in to examine the resulting sequence. By an agreement with the assistant beforehand, the magician tries to determine the number chosen by the audience.

(a) If this is possible for some n, is it also possible for $2n$?

(b) Determine all n for which this is possible.

6. (Spring 2021 Junior A-6)
A hotel has n unoccupied rooms upstairs, k of which are under renovation. All doors are closed, and it is impossible to tell if a room is occupied or under renovation without opening the door. There are 100 tourists in the lobby downstairs. Each in turn goes upstairs to open the door of some room. If it is under renovation, she closes its door and opens the door of another room, continuing until she reaches a room not under renovation. She moves in that room and then closes the door. Each tourist chooses the doors she opens. For each k, determine the smallest n for which the tourists can agree on a strategy while in the lobby, so that no two of them move into the same room.

7. (Spring 2008 Senior A-6)
Seated in a circle are 11 wizards. A different positive integer not exceeding 1000 is pasted onto the forehead of each. A wizard can see the numbers of the other 10, but not his own. Simultaneously, each wizard puts up either his left hand or his right hand. Then each declares the number on his forehead at the same time. Is there a strategy on which the wizards can agree beforehand, which allows each of them to make the correct declaration?

8. (Spring 2013 Senior A-7)
One thousand wizards are standing in a column. Each is wearing one of the hats numbered from 1 to 1001 in some order, one hat not being used. Each wizard can see the number of the hat of any wizard in front of him, but not that of any wizards behind. Starting from the back, each wizard in turn calls out a number from 1 to 1001 so that every other wizard can hear it. Each number can be called out at most once. At the end, a wizard who fails to call out the number on his hat is removed from the Council of Wizards. This procedure is known to the wizards in advance, and they have a chance to discuss strategy. Is there a strategy which can keep in the Council of Wizards

(a) more than 500 of these wizards;

(b) at least 999 of these wizards?

9. (Spring 2018 Senior A-7)
Each of n wizards in a column wears a white hat or a black hat chosen with equal probability. Each can see the hats of the wizards in front of him, but not his own.

Starting from the last wizard, each in turn guesses the colour of his own hat. Also, each wizard except the first one announces a positive integer which is heard by everyone. The wizrds can agree in advance on a collective strategy on what number each should announce, in order to maximize the number of correct guesses. Unfortunately, some of the wizards do not care for this, and may do as they like. It is not known who they are, but that there are exactly k of them. What is the maximal number of correct guesses which can be guaranteed by some collective strategy, despite the possible actions of the uncaring wizards?

Set 17.

1. (Spring 2009 Senior A-6)
   A positive integer n is given. Anna and Boris take turns marking points on a circle. Anna goes first and uses the red colour while Boris uses the blue colour. When n points of each colour have been marked, the game is over, and the circle has been divided into $2n$ arcs. The winner is the player who has the longest arc both endpoints of which are of this player's colour. Which player can always win, regardless of any action of the opponent?

2. (Spring 2014 Senior A-5)
   A scalene triangle is given. In each move, Anna chooses a point on the plane, and Boris decides whether to paint it red or blue. Anna wins if she can get three points of the same colour forming a triangle similar to the given one. What is the minimum number of moves Anna needs to force a win, regardless of the shape of the given triangle?

3. (Spring 2020 Junior A-6)
   Alice has a deck of 36 cards, 4 suits of 9 cards each. She picks any 18 cards and gives the rest to Bob. In each of 18 turns, Alice plays one of her cards first and then Bob plays one of his cards. If the two cards are of the same suit or of the same value, Bob gains a point. What is the maximum number of points he can guarantee regardless of Alice's actions?

4. (Fall 2020 Junior A-6)
   Anna and Boris play a game involving two round tables at each of which n children are seated. Each child is a sworn friend of both neighbours but no others. Anna can make two children agree to be sworn friends, whether they sit at the same table or at different tables. She can do so with $2n$ pairs. Boris can then make n of those pairs change their minds. Anna wins if the $2n$ children can be seated at one large round table so that each is a sworn friend of both neighbours. For which n does Anna have a winning strategy?

5. (Fall 2018 Senior A-7)
There are $n \geq 2$ cities, each with the same number of citizens. Initially, every citizen has exactly 1 dollar. In a game between Anna and Boris, turns alternate. Anna chooses one citizen from every city, and Boris redistributes their wealth so that the at least one citizen has a different number of dollars from before.

   Anna wins if in every city, there is at least one citizen with no money. Which player has a winning strategy if the number of citizens in each city is

   (a) $2n$;

   (b) $2n - 1$?

6. (Fall 2009 Senior A-6)
Olga and Max visit a certain archipelago with 2009 islands. Some pairs of islands are connected by boats which run both ways. Olga choose the first island on which they land. Then Max choose the next island which they can visit. Thereafter, the two take turns choosing an accessible island which they have not yet visited. When they arrive at an island which is connected only to islands they have already visited, whoever's turn to choose next will be the loser. Can Olga always win, regardless of how Max plays and regardless of the way the islands are connected?

7. (Spring 2021 Senior A-4)
There is a row of $100n$ tuna sandwiches. A boy and his cat take alternate turns, with the cat going first. In her turn, the cat eats the tuna from one sandwich anywhere in the row if there is any tuna left. In his turn, the boy eats the first sandwich from either end, and continues until he has eaten 100 of them, switching ends at any time. Can the boy guarantee that, for every positive integer n, the last sandwich he eats contains tuna?

8. (Spring 2011 Senior A-7)
Among a group of programmers, every two either know each other or do not know each other. Eleven of them are geniuses. Two companies hire them one at a time, alternately, and may not hire someone already hired by the other company. There are no conditions on which programmer a company may hire in the first round. Thereafter, a company may only hire a programmer who knows another programmer already hired by that company. Is it possible for the company which hires second to hire ten of the geniuses, no matter what the hiring strategy of the other company may be?

9. (Fall 2010 Senior A-4)
Two dueling wizards are at an altitude of 100 metres above the sea. They cast spells in turn, and each spell is of the form "decrease the altitude by a metres for me and by b metres for my rival," where a and b are real numbers such that $0 < a < b$. Different spells have different values for a and b. The set of spells is the same for both wizards, the spells may be cast in any order, and the same spell may be cast many times. A wizard wins if after some spell, he is still above water but his rival is not. Does there exist a set of spells such that the second wizard has a guaranteed win, if the number of spells is
   (a) finite;
   (b) infinite?

Set 18.

1. (Spring 2011 Junior O-3)
Worms grow at the rate of 1 meter per hour. When they reach their maximum length of 1 meter, they stop growing. A full-grown worm may be dissected into two new worms of arbitrary lengths totalling 1 meter. Starting with 1 full-grown worm, can one obtain 10 full-grown worms in less than 1 hour?

2. (Fall 2010 Senior O-3)
From a police station situated on a straight road infinite in both directions, a thief has stolen a police car. Its maximal speed equals 90% of the maximal speed of a police cruiser. When the theft is discovered some time later, a policeman starts to pursue the thief on a cruiser. However, he does not know in which direction along the road the thief has gone, nor does he know how long ago the car has been stolen. Can the policeman be sure of catching the thief?

3. (Spring 2014 Junior O-2)
Olga's mother bakes 7 apple pies, 7 banana pies and 1 cherry pie. They are arranged in that exact order on a the rim of a round plate when they are put into the microwave oven. All the pies look alike, but Olga knows only their relative positions on the plate because it has rotated. She wants to eat the cherry pie. She is allowed to taste three of them, one at a time, before making up her mind which one she will take. Can she guarantee that she can take the cherry pie?

4. (Spring 2014 Junior O-5)
Forty Thieves are ranked from 1 to 40, and Ali Baba is also given the rank 1. They want to cross a river using a boat. Nobody may be in the boat alone, and no two people whose ranks differ by more than 1 may be in the boat at the same time. Is this task possible?

5. (Fall 2015 Junior O-5)
In a country there are 100 cities. Every two cities are connected by direct flight in both directions, both costing the same amount. We wish to visit all the other 99 cities and then return to our home city, so that the average cost per flight on our trip is not greater than the average cost of all flights.
   (a) Is it always possible to do so?
   (b) Is it always possible to do so if we leave out one of the other 99 cities?

6. (Fall 2013 Junior A-4)
Penny chooses an interior point of one of the squares of an 8×8 board. Basil draws a subboard, consisting of one or more squares, whose boundary is a single closed polygonal line which does not intersect itself. Penny will then tell Basil whether the chosen point is inside or outside this subboard. What is the minimum number of times Basil has to do this in order to determine whether the chosen point is black or white?

7. (Fall 2017 Junior A-4)
One hundred doors and one hundred keys are numbered 1 to 100 respectively. Each door is opened by a unique key whose number differs from the number of the door by at most one. Is it possible to match the keys with the doors in n attempts, where
   (a) $n = 99$;
   (b) $n = 75$;
   (c) $n = 74$?

8. (Fall 2008 Senior A-7)
A contest consists of 30 true or false questions. Victor knows nothing about the subject matter. He may write the contest several times, with exactly the same questions, and is told how many questions he has answered correctly each time. Can he be sure that he will answer all 30 questions correctly
   (a) on his 30th attempt;
   (b) on his 25th attempt?

9. (Fall 2009 Senior A-7)
At the entrance to a cave is a rotating round table. On top of the table are n identical barrels, evenly spaced along its circumference. Inside each barrel is a herring either with its head up or its head down. In a move, Ali Baba chooses from 1 to n of the barrels and turns them upside down. Then the table spins around. When it stops, it is impossible to tell which barrels have been turned over. The cave will open if the heads of the herrings in all n barrels are all up or are all down. Determine all values of n for which Ali Baba can open the cave in a finite number of moves.

Answers

Problem 1.

Set 1. 8. **Set 2.** 2525. **Set 3.** Yes. **Set 4.** Yes. **Set 5.** Yes. **Set 6.** 48
Set 7. $n \not\equiv 0 \pmod 4$. **Set 8.** Yes. **Set 9.** 15. **Set 10.** Positive.
Set 11. Yes. **Set 12.** 1. **Set 13.** 45. **Set 14.** No. **Set 15.** No.
Set 16. First and last positions. **Set 17.** Boris. **Set 18.** Yes.

Problem 2.

Set 1. (a) Yes; (b) Yes. **Set 2.** Increases by 200.
Set 3. 4 positive, 6 negative. **Set 4.** 1. **Set 5.** Yes. **Set 6.** 90.
Set 7. Always Boris. **Set 8.** n. **Set 9.** 1. **Set 10.** 22. **Set 11.** 0.
Set 12. (a) Yes; (b) No. **Set 13.** (a) No; (b) 96. **Set 14.** Yes.
Set 15. 150. **Set 16.** Yes. **Set 17.** 5. **Set 18.** Yes.

Problem 3.

Set 1. Stevenson older by 10 years. **Set 2.** (a) Yes; (b) No.
Set 3. $n \geq 2$. **Set 4.** Yes. **Set 5.** Yes. **Set 6.** Yes. **Set 7.** Betty.
Set 8. 3. **Set 9.** (a) Yes; (b) No. **Set 10.** 1019. **Set 11.** 15. **Set 12.** Yes.
Set 13. Yes. **Set 14.** All $p+2$ numbers are d for some d, with one of them possibly replaced by pd. **Set 15.** Yes. **Set 16.** Yes. **Set 17.** 15.
Set 18. Yes.

Problem 4.

Set 1. 1, 11 or 121. **Set 2.** Yes. **Set 3.** Odd n. **Set 4.** Yes. **Set 5.** No.
Set 6. (a) Yes; (b) Yes. **Set 7.** Betty. **Set 8.** Yes. **Set 9.** Yes.
Set 10. Yes. **Set 11.** 9. **Set 12.** 9. **Set 13.** 8. **Set 14.** (a) Yes; (b) No.
Set 15. Yes. **Set 16.** Yes. **Set 17.** Odd n. **Set 18.** Yes.

Problem 5.

Set 1. 98. **Set 2.** Yes. **Set 3.** Yes. **Set 4.** Yes. **Set 5.** 6. **Set 6.** Basil.
Set 7. $2k-1$ for some k. **Set 8.** (a) Yes; (b) Yes. **Set 9.** 1. **Set 10.** Yes.
Set 11. (a) Yes; (b) Yes. **Set 12.** Basil. **Set 13.** 3. **Set 14.** Yes.
Set 15. Yes. **Set 16.** (a) Yes; (b) 2^k for any positive k.
Set 17. (a) Anna; (b) Anna. **Set 18.** (a) Yes; (b) No.

Problem 6.

Set 1. 10. **Set 2.** 2. **Set 3.** Yes. **Set 4.** Yes. **Set 5.** Yes. **Set 6.** All k.
Set 7. $\frac{m}{2^n}$, $0 < m < 2^n$. **Set 8.** Yes. **Set 9.** All n. **Set 10.** 3×2^{1009}.
Set 11. Yes. **Set 12.** All $n > 4$. **Set 13.** $2 \leq k \leq 50$ except 5, 9, 13, 17, 21, 25, 29, 33, 37, 41, 45 and 49. **Set 14.** (a) 44800; (b) 7. **Set 15.** Yes.
Set 16. $50k+51$ if k odd, $50k+100$ if k even. **Set 17.** Yes. **Set 18.** 2.

Problem 7.

Set 1. Yes. **Set 2.** (a) No; (b) Yes. **Set 3.** Yes. **Set 4.** (a) No; (b) Yes. **Set 5.** 20161932. **Set 6.** 40. **Set 7.** Bob. **Set 8.** Yes. **Set 9.** (a) Yes; (b) No. **Set 10.** (a) -997; (b) 995. **Set 11.** (a) Yes; (b) Yes. **Set 12.** 32. **Set 13.** 100. **Set 14.** All n. **Set 15.** No. **Set 16.** Yes. **Set 17.** No. **Set 18.** (a) Yes; (b) Yes; (c) No.

Problem 8.

Set 1. 1009. **Set 2.** Yes. **Set 3.** 99. **Set 4.** Yes. **Set 5.** Yes. **Set 6.** 16. **Set 7.** Bob. **Set 8.** $50. **Set 9.** Yes. **Set 10.** 375. **Set 11.** No. **Set 12.** 16. **Set 13.** n **Set 14.** Yes. **Set 15.** 10:40. **Set 16.** (a) Yes; (b) Yes. **Set 17.** Yes. **Set 18.** (a) Yes; (b) Yes.

Problem 9.

Set 1. Yes. **Set 2.** 101. **Set 3.** Yes. **Set 4.** Yes. **Set 5.** $n \not\equiv 0 \pmod 3$. **Set 6.** Yes. **Set 7.** (a) No; (b) Yes. **Set 8.** $61. **Set 9.** Yes. **Set 10.** No. **Set 11.** Yes. **Set 12.** 0. **Set 13.** No. **Set 14.** 6. **Set 15.** Yes. **Set 16.** $n-k-1$. **Set 17.** (a) No; (b) Yes. **Set 18.** 2^k for any k.

Part II

Mathematics Education

The levels of the problems from the International Mathematics Tournament of the Towns range from the relative simple to the most complex, but there is an elegant idea behind each of them. University faculty members supervising undergraduate research will find many problems that can be bases for projects. We leave the choice to the individual instructor.

At the other end of the scale, many of the problems are accessible to upper elementary school students (from Grade 4 to Grade 6). They provide an ideal entry point to recreational mathematics, and introduce important mathematical ideas to these students.

Martin Gardner strongly believed that recreational mathematics has an important role to play in mathematics education. Here we collect 48 such problems and classify them by topic. Some of the problems are presented as they are, while others are downsized. A few are modified. Full solutions with discussions are provided, so that there is no need to refer to the material in Part III.

The reader will discover two recurring themes in the solutions of the problems.

The first one goes by the title of the **Pigeonhole Principle**. It considers the situation where some holes are occupied by pigeons. If there are more pigeons than holes, some hole will be occupied by two or more pigeons. If there are more holes than pigeons, some hole will not be occupied. If the two numbers are equal, it does not necessarily mean that each hole is occupied by a pigeon. However, when that happens, we have the important concept of a *one-to-one correspondence* between the holes and the pigeons.

The second is the concept of *parity*. For whole numbers, this means the state of being odd or being even. Clearly, an odd number cannot be equal to an even number. This simple fact is very useful in establishing that some tasks are impossible. Painting objects in two colours often goes hand in hand with parity arguments.

Arithmetical Recreations

Set A1.

Operations on the whole numbers form the principal part of the elementary school mathematics curriculum. We quote one famous example. The story is told of the German prodigy Karl Frederick Gauss being asked by his teacher to compute the sum of the first 100 whole numbers. He was able to produce the correct answer of 5050 immediately. How did he do it?

Most likely, little Gauss wrote down the first 100 whole numbers in order on one line. On the line below, he wrote them down in the reverse order. Then he added each pair of numbers and found the total of these sums.

$$\begin{array}{rrrcrr} 1 & 2 & 3 & \cdots & 99 & 100 \\ 100 & 99 & 98 & \cdots & 2 & 1 \\ \hline 101 & 101 & 101 & \cdots & 101 & 101 \end{array}$$

Since all the sums are 101, the total is $101 \times 100 = 10100$. Since each whole number from 1 to 100 is added twice, the answer wanted by the teacher is $10100 \div 2 = 5050$.

Problem A1-1.

At the start, the number 7 is on the board. In each move, Alice either adds 9 to the number on the board or, if this number contains a digit 1, she may delete the 1. If there are leading 0s as a result, they are deleted too. Is it possible for Alice to get the number 100 on the board after some moves?

Problem A1-2.

At the start, the number 97 is on the board. In each move, Bob chooses a number which is greater than 1 and divides the number on the board. Bob either adds it to or subtracts it from that number, and replaces it by the result. Is it possible for Bob to get the number 41 on the board after some moves?

Problem A1-3.

Five different numbers are chosen from 1, 2, ..., 10. Five different numbers are also chosen from 11, 12, ..., 20. No two of them differ by exactly 10. Find the sum of all ten chosen numbers.

Problem A1-4.

Ten different numbers are chosen from 1, 2, ..., 20. The sum of these ten numbers is even. Can it happen that it is impossible to divide these numbers into two groups with equal sum? The sizes of the groups need not be the same.

Set A2.

There are many puzzles involving arranging whole numbers along a line or around a circle. When the first fifteen whole numbers are arranged along a line in the order 9, 7, 2, 14, 11, 5, 4, 12, 13, 3, 6, 10, 15, 1 and 8, the sum of every two adjacent numbers is the square of a whole number.

Problem A2-1.

Alice has five cards, each with a different whole number written on it. Betty only knows that their sum is 30. Alice arranges the cards face down in a row, in increasing order. Bob turns over the card in the middle of the row, and can then tell what the other four numbers are. Is this possible?

Problem A2-2.

There are several positive whole numbers in a row. None of them is a 3, and no two or three adjacent numbers add up to 3. The sum of all the numbers is 20. Can there be more than 10 numbers in the row?

Problem A2-3.

We have one copy of 0 and two copies of each of 1, 2, 3 and 4. Can we arrange these nine numbers in a row, such that there is exactly one other numbers between the two copies of 1, two other numbers between the two copies of 2, three other numbers between the two copies of 3, and four other numbers between the two copies of 4?

Problem A2-4.

The numbers from 1 to 17 are written in some order around a circle. The sum of each pair of adjacent numbers is computed. Is it possible for these sums to be 17 consecutive numbers?

Set A3.

There are many puzzles involving whole numbers in a table. The best-known examples are the magic squares, containing consecutive whole numbers from 1 on, such that the sum of all the numbers in each row, in each column and in each of the two long diagonals is constant. The 3×3 magic square is shown in the diagram below.

6	1	8
7	5	3
2	9	4

Problem A3-1.

Each of the squares in a 5×7 table contains a whole number. Peter knows only that the sum of the numbers in the 6 squares of any 2×3 or 3×2 rectangle is 10. He is allowed to ask for the number in any square in the table. What is the minimum number of questions he needs to ask in order to be able to determine the sum of all 35 numbers in the table?

Problem A3-2.

In a 4×4 table, the sum of the numbers in each row and in each column is the same. Nine of the numbers are erased, leaving behind the seven shown in the diagram below. Is it possible to recover uniquely

(a) one;

(b) two

of the erased numbers?

1			2
	4	5	
	6	7	
3			

Problem A3-3.

Each square of a 5×5 table contains one of the numbers 1, 2, 3, 4 and 5, and each of these numbers appears 5 times. The squares on the main diagonal are shaded in the diagram below. The sum of all the numbers on one side of the main diagonal is equal to three times the sum of all the numbers on the other side. Determine the number in the central square of the table, marked by a black dot in the diagram below.

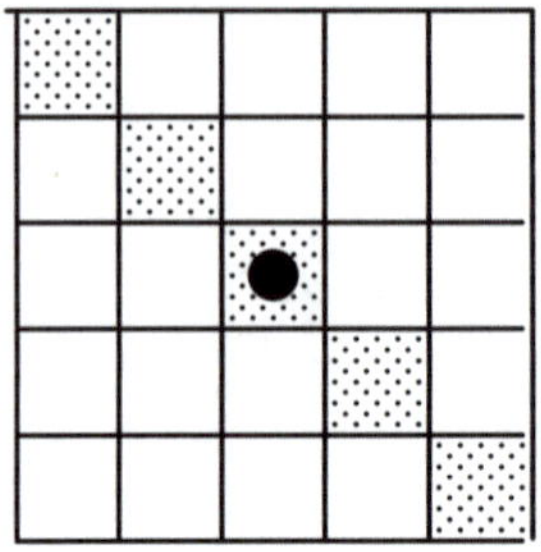

Problem A3-4.

In each square of a 6×6 board is a positive whole number which is at most 18. Whenever the board is covered by 18 dominoes, the sum of the two numbers in the squares covered by each domino is different from the sum of the two numbers in the squares covered by any other domino. Is this possible?

Set A4.

This set contains some miscellaneous arithmetical puzzles.

Problem A4-1.

The family names of Clark, Donald, Jack, Robin and Steven are Clarkson, Donaldson, Jackson, Robinson and Stevenson, but not in that order. Clark is 1 year older than Clarkson, Donald is 2 years older than Donaldson, Jack is 3 years older than Jackson and Robin is 4 years older than Robinson. Who is older, Steven or Stevenson, and what is the difference between their ages?

Problem A4-2.

During a party, 47 chocolates and 74 marshmallows are handed out. Each girl gets 1 more chocolate than each boy, and each boy gets 1 more marshmallow than each girl. There is at least one girl and at least one boy, and each child gets at least one candy of each kind. How many girls and how many boys are at the party?

Problem A4-3.

A tennis club with ten girls is entering teams for the doubles event in a tournament. Each girl has a list of five preferred partners. Two girls can be considered for a pair if each is preferred by the other. What is the

(a) minimum;

(b) maximum

number of pairs that may be considered?

Problem A4-4.

Brackets are to be inserted into the expression $7 \div 6 \div 5 \div 4 \div 3 \div 2$ so that the value of the resulting expression is a whole number.

(a) Determine the maximum value of this number.

(b) Determine the minimum value of this number.

Geometrical Recreations

Set G1.

One of the most popular geometrical recreation is dissection. There are two important types.

In a first, a given figure is to be dissected into two identical pieces. The pieces may be turned or flipped over. The diagram below shows an example on the left, and the solution on the right.

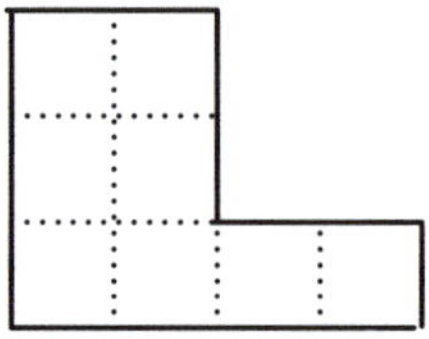

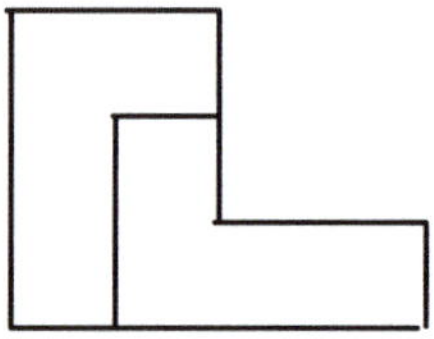

In a second, a given figure is to be dissected into two pieces, which are then assembled into a square. Again, the pieces may be turned or flipped over. The diagram below shows an example on the left, and the solution on the right.

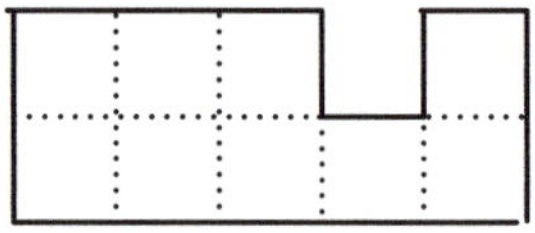

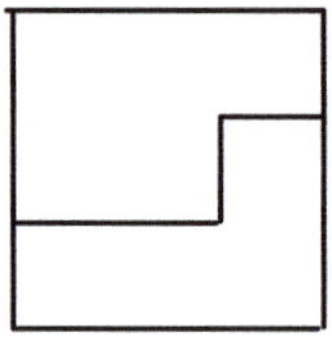

The following problems are different from those above and from one another.

Problem G1-1.

The diagram below shows a 5×5 piece of paper with four unit square holes. Is it possible to dissect this punctured piece of paper into rectangles none of which is a square?

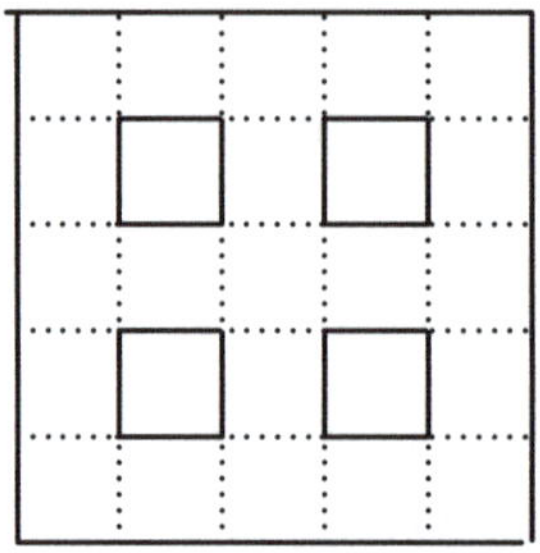

Problem G1-2.

Is it possible to dissect a square into nine squares, with five of them of one size, three of them of another size and one of them of a third size?

Problem G1-3.

In a 6×6 board, the 9 squares in the top left 3×3 subboard are black while all remaining squares are white. The board is dissected into a number of connected pieces of various shapes and sizes such that the number of white squares in each piece is three times the number of black squares in that piece. What is the maximum number of pieces?

Problem G1-4.

The figure in the diagram below is dissected along the dotted lines into squares which may be of any sizes and not necessarily all the same size. What is the minimum number of squares in such a dissection?

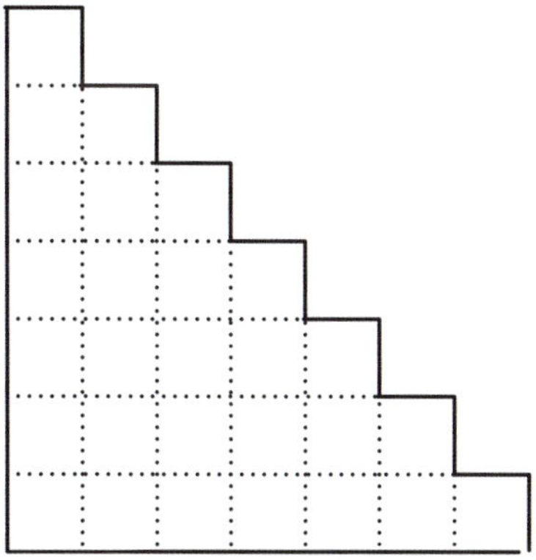

Set G2.

In the standard pattern on a chessboard, the squares are painted black and white such that two squares sharing a common side always have different colours. This is illustrated with the 4×4 chessboard in the diagram below.

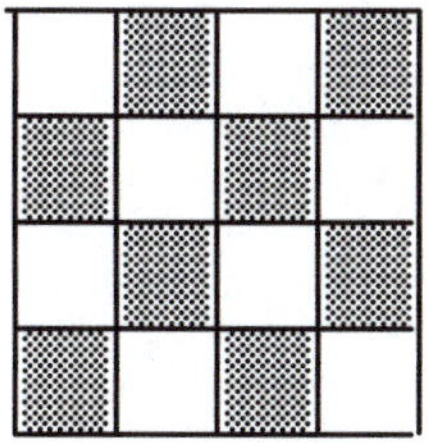

The following problems all involve colouring. In the last two, it is the surfaces of cubes that are painted.

Problem G2-1.

Is it possible to paint some of the squares of a 5×5 board gray so that each 3×3 subboard contains a different number of gray squares?

Problem G2-2.

Peter has a 3×5 stamp such that 6 of the squares are coated with black ink. He presses this stamp 4 times on a 5×5 board, each time leaving a black imprint on 6 squares of the board. Is it possible that every square of the board is black except for one square at a corner?

Problem G2-3.

Is it possible to paint each of the six faces of a cube in one of red, yellow and blue so that each colour is used, and any three faces sharing a common vertex do not all have different colours?

Problem G2-4.

Each of the 54 squares on the faces of a $3 \times 3 \times 3$ cube is painted white, gray or black. Two squares sharing an edge, whether they lie on the same face or not, must have different colours. What is the minimum number of black squares?

Set G3.

One of the most popular families of geometric figures is the *polyominoes*. The name is derived from the domino, which consists of two unit squares joined side to side. Thus a monomino is simply a unit square. Those consisting of three or four unit squares are the V-tromino, the I-tromino, the I-tetromino, the O-tetromino, the L-tetromino, the S-tetromino and the T-tetromino, shown in the diagram below.

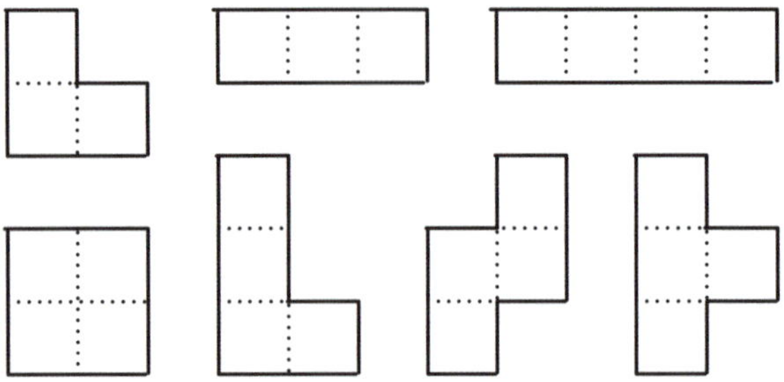

Problem G3-1.

It is easy to put together four copies of any rectangular polyominoes to form a larger copy of itself.

(a) Show that this can also be done with the V-tromino.

(b) Can this be done with any of the non-rectangular tetrominoes?

Problem G3-2.

Anna paints several squares of a 5×5 board. Boris tries to cover up all of them by placing non-overlapping copies of the V-tromino so that each copy covers exactly three squares of the board. What is the minimum number of squares Anna must paint in order to prevent Boris from succeeding?

Problem G3-3.

An invisible alien spaceship in the shape of an O-tetromino may land on our 5×5 airfield, occupying four of the 25 squares. Sensors are to be placed in certain squares of the airfield. A sensor will send a signal if it is in a square on which the spaceship lands. From the locations of all the sensors which send signals, we must be able to determine exactly on which four squares the spaceship has landed. What is the smallest number of sensors required?

Problem G3-4.

In a 5×5 box, each of the 25 pieces is either dark chocolate or white chocolate. In each move, Alex may eat two adjacent pieces along a row, a column or a diagonal, provided that they are of the same kind. What is the maximum number of pieces Alex can guarantee to be able to eat, regardless of the initial arrangement the pieces?

Set G4.

This set contains some miscellaneous geometrical puzzles.

Problem G4-1.

Buried under each square of a 6×6 board is either a treasure or a message. There is only one square with a treasure. A message indicates the minimum number of steps needed to go from its square to the treasure square. Each step takes one from a square to another square sharing a common side. What is the minimum number of squares one must dig up in order to bring up the treasure for sure?

Problem G4-2.

Black and white counters are placed on different squares of a 4×4 board. What is the maximum number of counters if each row and each column contains twice as many white counters as black ones?

Problem G4-3.

Several ants are on different squares of a 6×6 board. Each ant chooses independently whether to crawl along a row or along a column, and in which direction it starts. It crawls one square per move in the chosen direction. When it reaches an edge of the board, it reverses directions in the next move. What is the maximum number of ants on the board if no two of them ever occupy the same square at the same time?

Problem G4-4.

Is it possible to dissect a cube into two pieces which can be reassembled into a convex polyhedron with only triangular and hexagonal faces?

Combinatorial Recreations

Set C1.

In this set, we study moves of chess pieces. A king may move one square in any of eight directions. A queen may move any number of squares in any of eight directions. A rook may move any number of square in only four directions, along a row or along a column. No piece may leave the board.

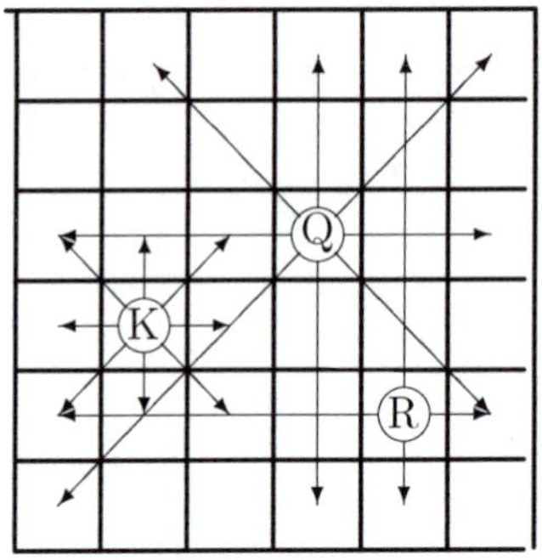

Problem C1-1.

Nine kings are placed on different squares of a 6×6 board. Is it possible that the number of moves to empty squares is different for each king?

Problem C1-2.

Six queens are placed on different squares of a 6×6 board such that they do not attack one another. What is the minimum number of queens in the 3×3 subboard at the top left corner?

Problem C1-3.

Six rooks are placed on different squares of a 6×6 board so that they do not attack one another. Is it possible for every empty square to be attacked by two rooks

(a) at the same distance away;

(b) at different distances away?

Problem C1-4.

A rook visits every square of a 6×6 board exactly once and returns to its starting square. Another rook also makes such a tour. What is the minimum number of moves between the same two adjacent squares that must be made by both rooks?

Set C2.

Magic shows are a favourite entertainment. The story is told of the American puzzlist Sam Loyd who put on a show with his son. The boy was blind-folded. The audience drew a card from a standard deck and showed it to Sam. The boy immediately and correctly named the card. This happened every time.

This was a trick. The boy never said a word. Sam was a ventriloquist! The following puzzles are not tricks, but are based on sound mathematical principles.

Problem C2-1.

To train her assistant, a magician secretly chooses a number consisting of seven different digits. The assistant is allowed to write down several numbers each consisting of seven different digits. He passes the test if at least one of his numbers matches with her number in at least one digit. Can the assistant pass the test by writing down at most six numbers?

Problem C2-2.

A magician lays the 52 cards of a standard deck in a row, and announces in advance that the Three of Clubs will be the only card remaining after 51 steps. In each step, the audience points to any card. The magician can either remove that card or remove the card in the corresponding position counting from the opposite end. What are the possible positions for the Three of Clubs at the start in order to guarantee the success of this trick?

Problem C2-3.

The audience chooses two of five cards numbered from 1 to 5. The assistant of a magician chooses two of the remaining three cards, and asks a member of the audience to take them to the magician, who is in another room. The two cards are presented to the magician in an arbitrary order. By an arrangement with her assistant beforehand, the magician is able to deduce which two cards the audience has chosen. Explain how this may be done.

Problem C2-4.

The mayor was among the one million spectators in a magic show. Each has a ticket with a different six-digit number from 000000 to 999999. The ticket number of each spectator is known to the assistant of the magician. The magician has a list of the spectators, and must determine the ticket number of the mayor. She is allowed to ask her assistant to reveal to her a particular digit of the ticket number of each spectator. This digit is chosen by the magician, and does not have to be the same for each spectator. What is the minimum number of digits she must ask the assistant to reveal in order to succeed?

Set C3.

Many combinatorial puzzles are algorithmic puzzles. An *algorithm* is simply a set of procedures, and is used to attain certain results. Perhaps the best-known example is the Tower of Hanoi.

There are three pegs in the playing board, and three disks of increasing sizes stacked on the first peg. The objective is to transfer this tower to the third peg. In each move, we may only transfer a disk on top of a peg to the top of another peg, and a disk may not be placed on top of a smaller disk.

Suppose the disks are A, B and C in increasing order of size. The task can be accomplished in seven steps.
1. Transfer A to the third peg.
2. Transfer B to the second peg.
3. Transfer A to the second peg.
4. Transfer C to the third peg.
5. Transfer A to the first peg.
6. Transfer B to the third peg.
7. Transfer A to the third peg.

Problem C3-1.

Four doors and four keys are numbered 1 to 4 respectively. Each door is opened by a unique key whose number differs from the number of the door by at most one. Is it possible to match the keys with the doors in three attempts?

Problem C3-2.

Alice's mother bakes 3 apple pies, 3 banana pies and 1 cherry pie. They are arranged in that exact order on a the rim of a round plate when they are put into the microwave oven. All the pies look alike, but Alice knows only their relative positions on the plate because it has rotated. She wants to eat the cherry pie. She is allowed to taste two of them, one at a time, before making up her mind which one she will eat. Can she guarantee that she can eat the cherry pie?

Problem C3-3.

A dragon has 36 heads and the knight has three swords. The gold sword cuts off one half of the current number of heads plus one more. The silver sword cuts off one third of the current number of heads plus two more. The bronze sword cuts off one fourth of the current number of heads plus three more. The knight can use any of the three swords in any order. The dragon is killed if it has no heads left. However, if the current number of its heads is not a multiple of 2 or 3, the swords do not work, and the dragon eats the knight. Will the knight be able to kill the dragon with a final cut using

(a) the gold sword;

(b) the silver sword;

(c) the bronze sword?

Problem C3-4.

Alice and Betty are sixteen, Carla is fifteen, Debra is fourteen and Ellen is thirteen. They want to cross a river in a boat. No girl may be in the boat alone, and no two girls whose ages differ by more than 2 may be in the boat at the same time. Is this task possible?

Set C4.

This set contains some miscellaneous combinatorial puzzles.

Problem C4-1.

In a wrestling tournament, there are 10 participants, all of different strengths. Each wrestler participates in two matches. In each match, the stronger wrestler always wins. A wrestler who wins both matches is given an award. What is the least possible number of wrestlers who win awards?

Problem C4-2.

Each of four boxes contains a different number of pencils. No two pencils in the same box are of the same colour. Can one choose one pencil from each box so that no two chosen pencils are of the same colour?

Problem C4-3.

Ten natives stand in a circle. Each announces the age of his left neighbour. Then each announces the age of his right neighbour. Each native is either a knight who tells both numbers correctly, or a knucklehead who increases one of the numbers by 1 and decreases the other by 1. Each knucklehead chooses which number to increase and which to decrease independently. Is it always possible to determine which of the natives are knights and which are knuckleheads?

Problem C4-4.

On an island with ten inhabitants each person is either a knight, a knave or a kneejerk. Everyone knows who everyone else is. Each inhabitant is asked in turn whether there are more knights than knaves on the island. A knight always tells the truth, a knave always lies, and a kneejerk agrees with the majority of those who speak before him. In case of a tie, such as when a kneejerk speaks first, he answers "Yes" or "No" at random. If exactly five of the ten answers are "Yes", at most how many kneejerks are on the island?

Solutions to Problems in Arithmetical Recreations

Set A1.

Problem A1-1.

If Alice can get 1 on the board, she can keep adding 9s to reach 100. Starting with 7, she can get only 16 by adding 9. If she deletes the 1 now, she can get 6. Adding 9 and deleting 1 alternatingly, she obtains in turn 15, 5, 14, 4, 13, 3, 12, 2, 11 and 1.

Problem A1-2.

The only divisors of 97 are 1 and 97, and Bob may not choose 1. So he must choose 97. Subtracting 97 from 97 leaves him with 0, which is not good. So he adds 97 to get 194. Now if Bob continues to add 97, he will reach 41×97. Subtracting 41 96 times, he will end up with 41.

Problem A1-3.

From each of the pairs (1,11), (2,12), ..., (10,20), we can choose at most one number. Since we are choosing ten numbers from ten pairs, we must choose exactly one number from each pair. Consider first the units digits of the numbers chosen. All ten of them must be different. Hence their sum is $0+1+2+3+4+5+6+7+8+9=45$. Consider now their tens digit. Five of them are 0s and five of them are 1s. Hence their sum is $5 \times 10 = 50$. The overall sum is $45+50=95$.

Problem A1-4.

This cannot happen. The ten numbers may be 2, 4, 6, 8, 10, 12, 14, 16, 18 and 20. Their sum is $10 \times 51 = 110$. If the division is possible, the sum of the numbers in each group must be 55, which is odd. This is impossible since all the numbers are even.

Set A2.

Problem A2-1

This is possible if the numbers on Alice's cards are 1, 2, 8, 9 and 10. When Bob finds out that the number on the middle card is 8, he knows that the sum of the numbers on the two cards on one side is at least $1+2=3$, and the sum of the numbers on the two cards on the other side is at least $9+10=19$. Since $3+8+19=30$, the numbers on the cards must be 1, 2, 8, 9 and 10.

Problem A2-2.

If this is possible, we have to use many 1s. We may start with two of them since we cannot have three 1s in a row. A 1 cannot be next to a 2, and we cannot use any 3s. Hence the next number should be a 4. Continuing this way, we obtain the row of the following eleven numbers 1, 1, 4, 1, 1, 4, 1, 1, 4, 1, 1, and the sum is indeed 20.

Problem A2-3.

Up to symmetry, there are two possible placements for the two 4s, as shown in the diagram below.

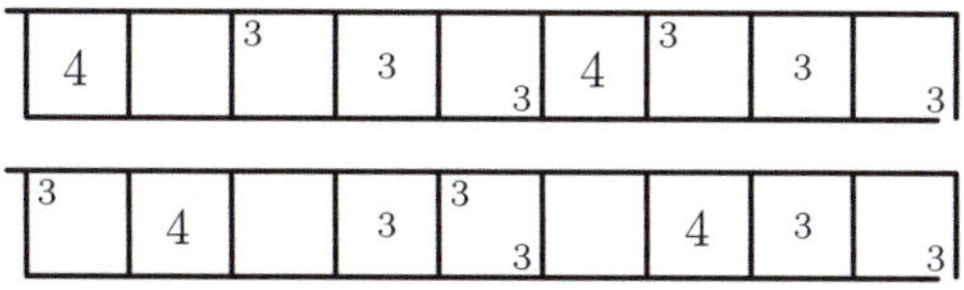

In the first case, there are three possible placements for the two 3s. One of them leads to the solution 413124320, another one leads to 420324131 while the third one does not lead to any solutions. In the second case, there are also three possible placements for the two 3s. One of them leads to two solutions 141302432 and 041312432, another one leads to 240231413 while the third one does not lead to any solution. In summary, there are five different possible placements, not counting the same placements in the reverse order.

Problem A2-4.

We have $1 + 2 + \cdots + 17 = 17 \times (17 + 1) \div 2 = 153$. The total of the 17 sums around the circle is $153 \times 2 = 306$ because each of the numbers 1, 2, ..., 17 is added twice. Now 306 is the sum of 17 consecutive numbers. Hence the middle number is $306 \div 17 = 18$, and the 17 sums are 10, 11, ..., 26. From 26=17+9 and 25=17+8, the neighbours of 17 are 9 and 8. From 24=8+16, the other neighbour of 8 is 16. From 23=16+7, the other neighbour of 16 is 7. Continuing this reasoning, we find that the 17 numbers should be arranged along the circle in the order 9, 1, 10, 2, 11, 3, 12, 4, 13, 5, 14, 6, 15, 7, 16, 8 and 17.

Set A3.

Problem A3-1.

The diagram below shows a partition of the 5×7 table into six 2×3 or 3×2 rectangles, two of which overlap at the central square which is shaded. Hence the sum of 35 numbers is $6 \times 10 = 60$ minus the number in the central square. Peter only needs to ask for the number in that square. This is clearly the minimum value as it cannot be lowered to zero.

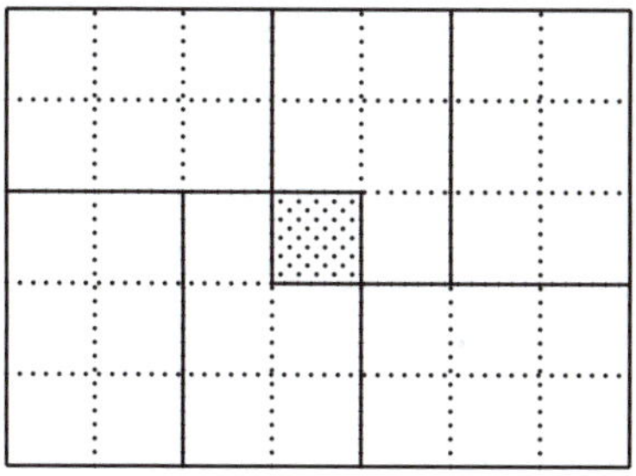

We remark that to solve a problem involving a maximum or a minimum value, we have to do three things. First, we must find the value. Second, we must show that this value can be attained. Third, we must show that this value cannot be improved.

Problem A3-2

(a) By the magic constant, we mean the constant sum of the four numbers in each row and column. We can determine uniquely the number in the square marked with a black dot in the diagram below. Adding the numbers in the second row, the third row, the second column and the third column, we have four times the magic constant. This is the sum of twice 4+5+6+7=22 plus the sum of the numbers in the eight shaded squares. Adding the numbers in the first row, the fourth row, the first column and the fourth column, we also have four times the magic constant. This is the sum of twice 1+2+3=6 plus twice the number we wish to determine plus the sum of the numbers in the eight shaded squares. It follows that the number we wish to determine is equal to $22 - 6 = 16$.

1			2
	4	5	
	6	7	
3			●

(b) It is not possible to uniquely determine any of the numbers in the eight shaded squares. The diagram below shows two possible ways of completing the table. In the second way, each of the numbers in the shaded squares is 1 more than the corresponding number in the first way.

1	10	9	2
10	4	5	3
8	6	7	1
3	2	1	16

1	11	10	2
11	4	5	4
9	6	7	2
3	3	2	16

Problem A3-3.

There are ten numbers on each side of the main diagonal. The maximum sum of ten of the numbers is $5 \times (5 + 4) = 45$ and the minimum sum of ten of the numbers is $5 \times (2 + 1) = 15$. Since 45 is three times 15, we must have all the 4s and 5s on one side of the diagonal and all the 1s and 2s on the other side. This means that every number on the main diagonal, including the one at the central square, must be a 3.

Problem A3-4.

This is possible, as shown in the diagram below. For any two dominoes, the numbers in the white squares which they cover are different while the numbers in the black squares which they cover are the same. Hence the sums are different.

1	1	2	1	3	1
1	4	1	5	1	6
7	1	8	1	9	1
1	10	1	11	1	12
13	1	14	1	15	1
1	16	1	17	1	18

Set A4.

Problem A4-1.

The total age of Clark, Donald, Jack, Robin and Steven must be the same as the total age of Clarkson, Donaldson, Jackson, Robinson and Stevenson. Since Clark, Donald, Jack and Robin are 1+2+3+4=10 years older than Clarkson, Donaldson, Jackson and Robinson, Stevenson must be 10 years older than Steven.

Problem A4-2

The total number of candies is 47+74=121. Since each child gets the same number of candies, the number of children must divide 121. Hence this number is 1, 11 or 121, so that the total number of children must be 11. Dividing 47 by 11, we have a quotient of 4 and a remainder of 3. Hence there are 3 girls picking up the extra chocolates, so that the number of boys is 8. Each girl gets 5 chocolates and each boy 4. From $74 = 6 \times 11 + 8$, each girl gets 6 marshmallows and each boy gets 7.

Problem A4-3

(a) There are $10 \times 9 \div 2 = 45$ pairs of girls. The total number of preferred partners is $10 \times 5 = 50$. It follows that there must be at least $50 - 45 = 5$ teams to be considered. It is possible that there are exactly 5 such teams. Let the ten girls be seated at a round table. Suppose that each prefers the next five in clockwise order. Then each girl forms only one team to be considered, with the girl in the opposite seat as her partner.

(b) Let the ten girls be seated at two tables, with five at each. Suppose each of the five girls at one table prefers a different girl from the other table, along with the other four girls at the same table. Then the number of teams to be considered is $5+2\times(5\times 4\div 2)=25$. If we consider at least 26 teams, then there are at least 52 spots to be filled by the ten girls. This means that some girl has to fill at least six spots, contrary to the condition that each girl prefers only five others.

Problem A4-4.

Bracketing simply separates the factors 7, 6, 5, 4, 3 and 2 into the numerator and the denominator of the overall expression.

(a) We have

$$7\div((((6\div 5)\div 4)\div 3)\div 2)=\frac{7\times 5\times 4\times 3\times 2}{6}=140.$$

Since 6 is the second number in the expression, it must be in the denominator. Hence the maximum value cannot be higher than 140.

(b) Since 7 and 5 are not divisible by any other numbers in the expression, they must be in the numerator. Hence the minimum value cannot be lower than 35. We have

$$7\div(((6\div 5)\div 4)\div(3\div 2))=\frac{7\times 5\times 4\times 3}{6\times 2}=35.$$

Solutions to Problems in Geometrical Recreations

Set G1.

Problem G1-1

A possible dissection is shown in the diagram below.

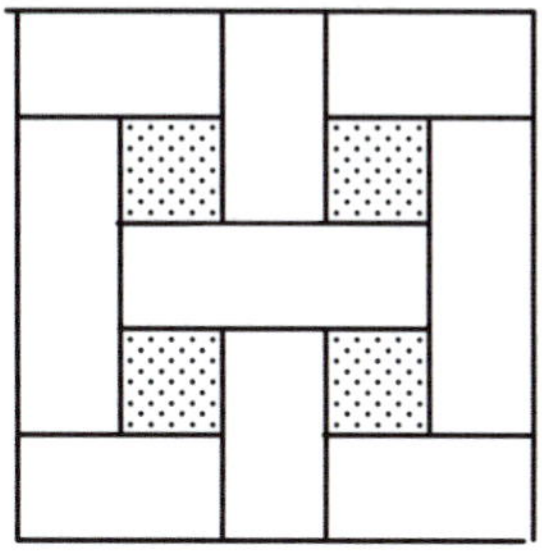

Problem G1-2

The diagram below shows that a 6×6 square can be cut into one 2×2 square, three 3×3 squares and five 1×1 squares.

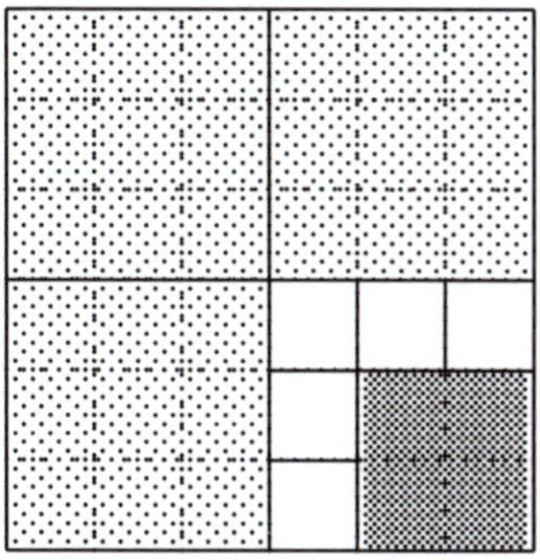

Problem G1-3.

Since there are only 5 black squares which share common sides with white squares, the number of pieces is at most 5. The diagram below shows that we can have 5 pieces.

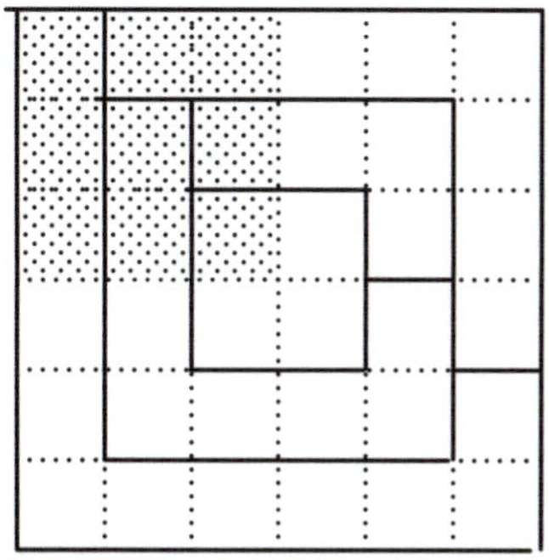

Problem G1-4.

No two of the 7 shaded unit squares can belong to the same square. The diagram below shows a dissection of the figure into exactly 7 squares.

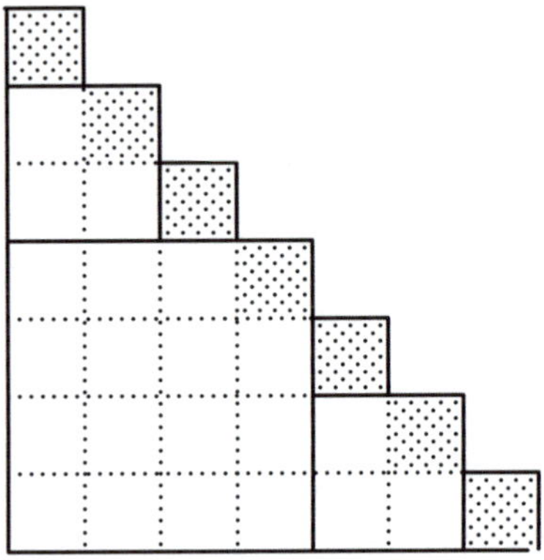

Set G2.

Problem G2-1.

This is possible, as shown in the diagram below. There are nine 3×3 subboard. The numbers of gray squares in them are 8, 7, 6, 5, 4, 3, 2, 1 and 0 respectively.

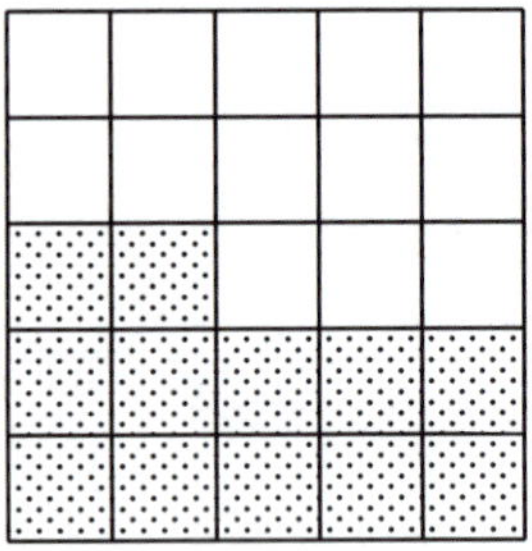

Problem G2-2

The 6 squares coated with black ink are shown in the diagram below on the left. The diagram below on the right shows how every square of the board, except for the one at the top left corner, can turn black after 4 uses of the stamp.

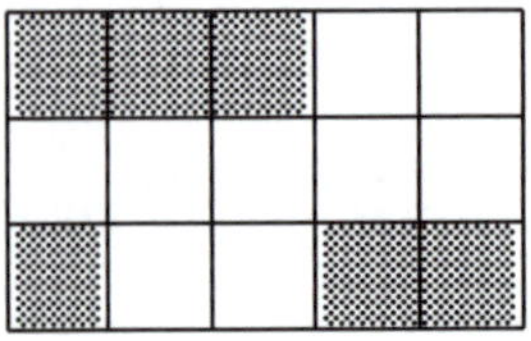

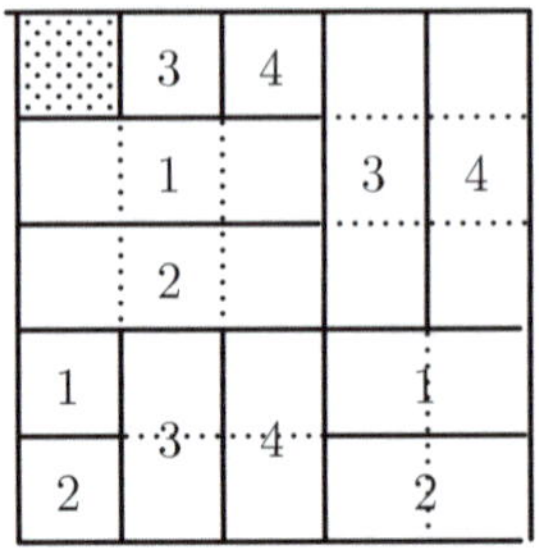

Problem G2-3

Paint one face red, the opposite face blue and the remaining faces yellow.

Problem G2-4

The squares around a vertex of the cube form a three-cycle so that exactly one of them must be black. Since there are 8 three-cycles all independent of one another, we need at least 8 black squares. The remaining squares contain two independent nine-cycles, requiring another 2 black squares. The colouring scheme in the diagram below shows that 10 black squares are sufficient.

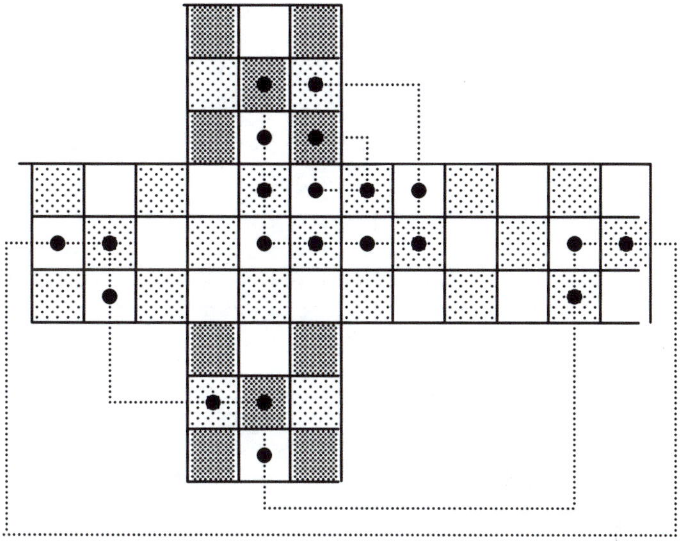

Set G3.

G3-1.

(a) This is shown in the diagram below on the left.

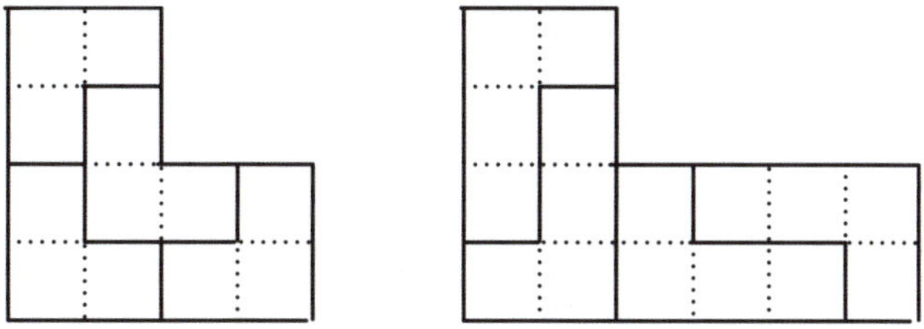

(b) Only the L-tetromino has this property, as shown in the diagram above on the right. A larger copy of a T-tetromino can be formed from 16 copies of a smaller one. No combinations of smaller copies of an S-tetromino can form a larger copy of an S-tetromino.

G3-2.

Anna paints 9 squares as shown in the diagram below at the top left. Boris can put at most 8 copies of the V-tromino on the board because $9 \times 3 > 5 \times 5$. Each copy can cover at most 1 gray square. Hence Boris cannot succeed in his task.

Suppose Anna paints 8 squares or less. Then at least one of the 9 gray squares in the diagram above at the top left is now unpainted. This square can be in the position shown in the diagram above on the bottom left, top right or bottom right. In each case, Boris can succeed in his task.

Problem G3-3

The diagram below on the left shows a placement of 4 sensors which works. Suppose we place only 3 sensors. Divide the airfield into the central square and four 2×3 regions. One of the regions will contain no sensors, and we cannot determine the exact location of an alien spaceship which lands within that region.

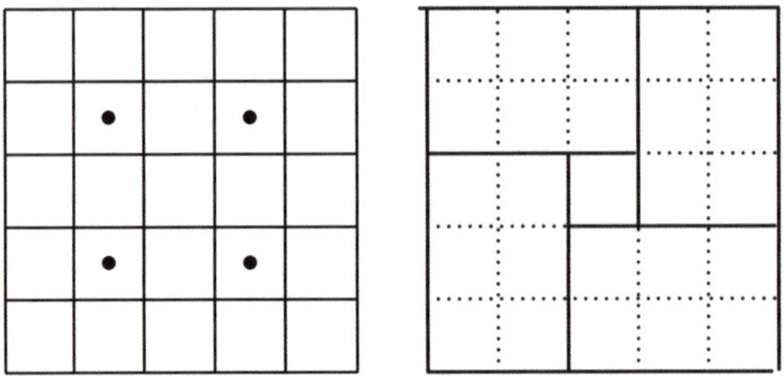

Problem G3-4.

If the initial arrangement is as shown in the diagram below on the left, Alice cannot eat any of the 9 dark chocolates. Since there are 16 milk chocolates, the most she can eat is 16 pieces. She can guarantee to eat 16 pieces because the box contains 8 non-overlapping V-trominoes, as shown in the diagram below on the right. The 3 pieces in each V-tromino are adjacent to one another, and 2 of the pieces must be of the same kind. Hence Alice can eat 2 pieces from each of the 8 V-trominoes.

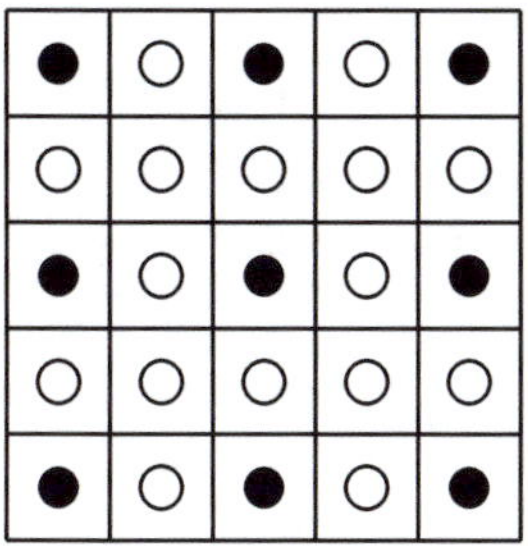

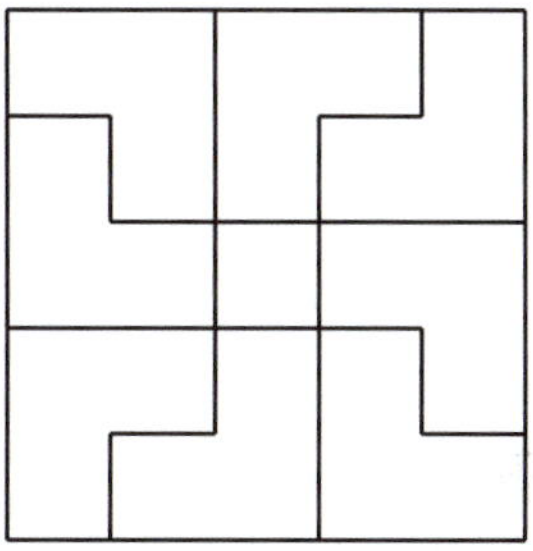

Set G4.

Problem G4-1.

If we dig up only one square, there is no guarantee that we will find the treasure. Suppose we find only a message in the first square we dig up. Whatever the message is, there are at least two possible locations for the treasure. If we dig up only one more square, again there is no guarantee. It follows that we must dig up at least three squares.

Here is a method which requires digging up only three squares. The first two squares we dig up are those at the bottom left and the bottom right corners of the board. In the diagram below, the first number in each square indicates the possible location of the treasure if that number is in the message under the bottom left corner square. The second number applies to the bottom right corner square. Since the pairs of numbers for any two squares are different, the location of the treasure is uniquely determined. The treasure can be brought up by digging at the indicated square.

5,10	6,9	7,8	8,7	9,6	10,5
4,9	5,8	6,7	7,6	8,5	9,4
3,8	4,7	5,6	6,5	7,4	8,3
2,7	3,6	4,5	5,4	6,3	7,2
1,6	2,5	3,4	4,3	5,2	6,1
0,5	1,4	2,3	3,2	4,1	5,0

Problem G4-2.

Since each row contains twice as many white counters as black ones, it is either empty or contains exactly 3 counters. Since there are 4 rows, the maximum number of counters is $4 \times 3 = 12$. The diagram below shows a possible placement of 12 counters with the desired property.

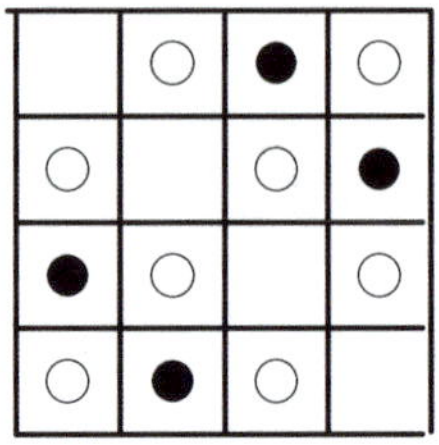

Problem G4-3.

Paint the squares black and white in the standard alternating pattern. If at least three ants choose to move along the same row, two of them will be on squares of the same colour. They will always be on squares of the same colour, and sooner or later, they must occupy the same square at the same time. Hence we can have at most two ants moving along the same row, and similarly along the same column. It follows that the maximum number of ants is 24. The diagram below shows that we can indeed have 24 ants. The two ants moving along the same row or along the same column are on squares of different colours. They will always be on squares of different colours, and can therefore never occupy the same squares at the same time. Finally, consider two ants on squares of the same colour, one moving along a row and the other along a column. Since each dimension of the board is five, each ant repeats the same consecutive ten moves in a cycle. It is routine to check that these two ants do not meet during the first cycle. Hence they cannot meet in any subsequent cycle either.

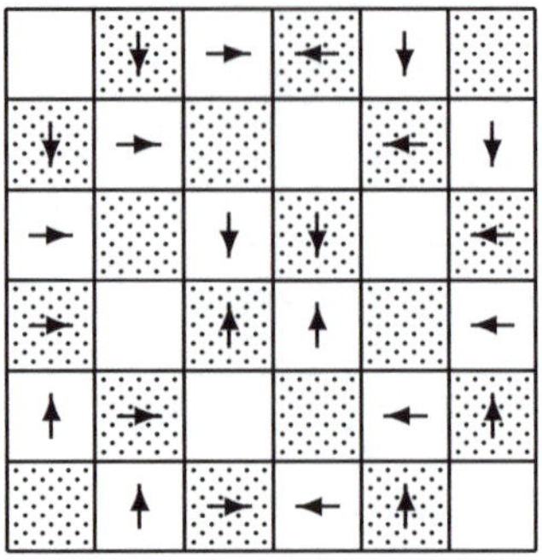

Problem G4-4.

A cube can be dissected into two congruent pieces as shown in the diagram below. Each consists of one hexagonal face, three pentagonal faces (hidden) and three triangular faces.

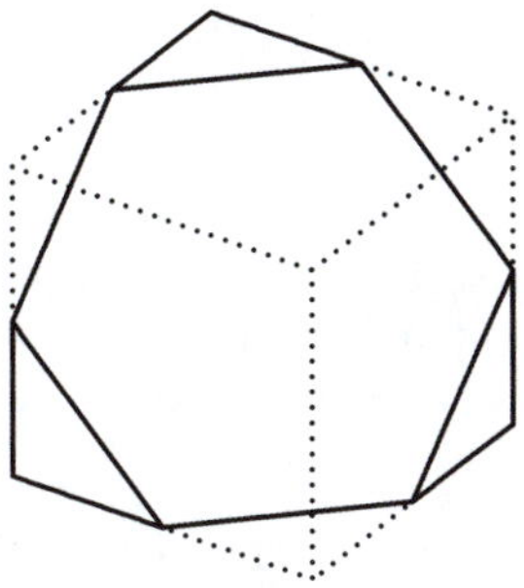

When the two pieces are reassembled by abutting one pentagonal face from each piece, two pairs of pentagonal faces merge into two hexagonal faces (hidden) and two pairs of triangular faces merge into two triangular faces. The resulting convex polyhedron, as shown in the diagram below, has four hexagonal faces and four triangular faces.

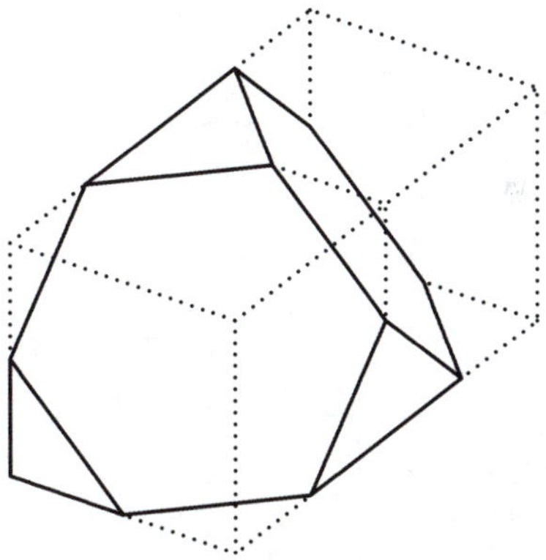

Solutions to Problems in Combinatorial Recreations

Set C1.

Problem C1-1.

We can have as many as nine kings, with 0, 1, ..., 8 possible moves respectively. The king with 0 moves must be at a corner square, surrounded by three other Kings each with a small number of possible moves. Starting from here, we can place the kings one at a time as shown in the diagram below. The kings are represented by white circles, and the numbers inside indicate the numbers of possible moves.

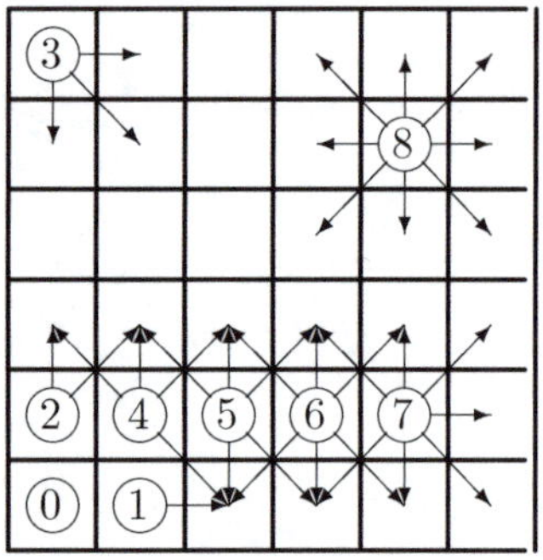

Problem C1-2.

The diagram below on the left shows that this number can be as low as 1.

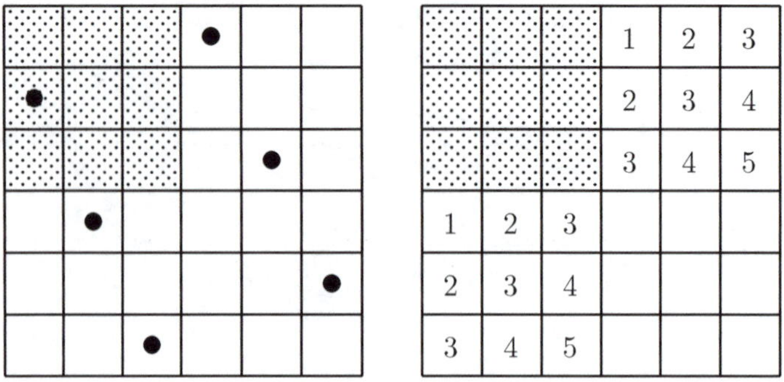

Suppose this subboard is empty. Then the three queens in the three columns on the left are all in the 3×3 subboard at the bottom left corner, while the three queens in the top three rows are all in the 3×3 subboard at the top right corner. These six queens all lie on five diagonals, as indicated in the diagram above on the right. Hence two of them must attack each other, which is not allowed.

Problem C1-3.

(a) Since the rooks do not attack one another, there is a rook in each row and a rook in each column. When two rooks attack the same square from the same distance away, they must lie on a diagonal. It follows that all six rooks are along the same diagonal, as shown in the diagram below on the left.

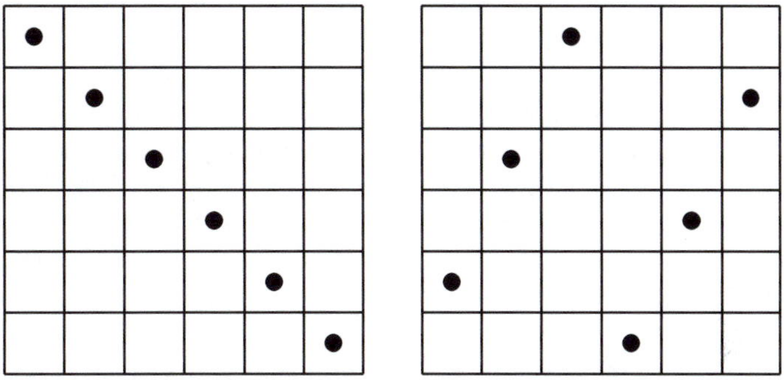

(b) When two rooks attack the same square from different distances away, they are not on the same diagonals. If we replaced all the rooks by queens, they still do not attack one another. One way of placing six non-attacking queens on a 6×6 board is shown in the diagram above on the right.

Problem C1-4.

The moves between the eight pairs of squares at the corners must be made by both rooks. Each shaded square in the diagram below is on an edge. There are exactly three moves involving it, and each rook must make exactly two of them. Hence at least one of these moves must be made by both rooks. One such move can take care of both shaded squares along each of the four edges. This brings the number of common moves to at least 8+4=12. The diagram below shows the paths of the two rooks with exactly 12 common moves.

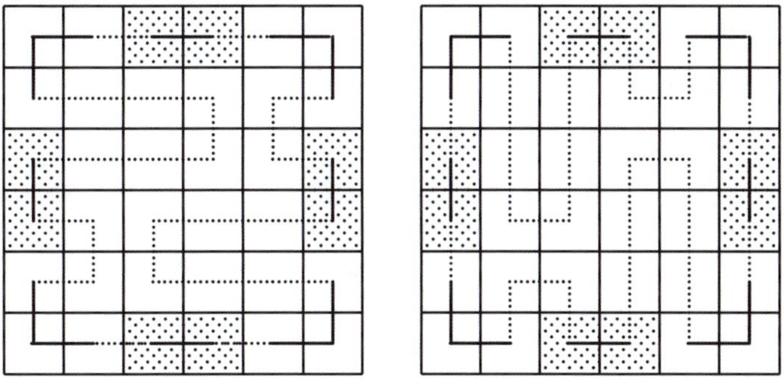

Set C2.

Problem C2-1.

The assistant passes the test by writing down the numbers 0123456, 0234561, 0345612, 0456123, 0561234 and 0612345. Since the magician's number uses seven different digits, it must use at least three of the digits 1, 2, 3, 4, 5 and 6. At most one of these these can be in the first place. Each of the other two must match a digit of some number written down by the assistant.

Problem C2-2.

The Three of Clubs should be the first or the last card. If the audience points to any other card, the magician removes it. If the audience points to the Three of Clubs, the magician removes the card from the opposite end. These are the only possible positions for the Three of Clubs to guarantee the success of the trick. Suppose it is in neither of these positions. The audience points to the Three of Clubs. The magician must remove a card which is neither the first nor the last card. These two cards will remain in place until they are the only cards left. This means that the Three of Clubs has already been removed, and the trick has failed.

Problem C2-3.

The magician and the assistant make the following arrangement. Consider the five cards arranged in a circle in the order 1, 2, 3, 4 and 5. If the audience chooses two non-adjacent cards, say 2 and 5, the assistant will choose the next card after each of them, in this case 3 and 1 since 1 comes after 5. If the audience chooses two adjacent cards, say 3 and 4, the assistant will choose the next two cards, in this case 5 and 1. Thus the magician can always deduce which two cards the audience has chosen.

Problem C2-4.

In order to be sure of knowing a particular digit of the mayor's ticket number, the magician must either check it or deduce it from that digit of every other spectator. She can succeed by asking for the first three digits of the mayor's ticket number, and the last three digits of every other spectator's ticket number. If she asks for two digits or less, she can know at most two digits of the mayor's account number and deduce two more digits.

Set C3.

Problem C3-1.

Try door 3 with key 3. If the attempt is successful, then door 4 is opened with key 4. One more attempt will settle doors and keys 1 and 2. Suppose the attempt fails. Try door 2 with key 3. If the attempt is successful, then key 1 must open door 1, key 2 must open door 3, key 3 must open door 2 and key 4 must open door 4. Suppose this attempt also fails. Then key 3 must open door 4 and key 4 must open door 3. One more attempt will settle doors and keys 1 and 2.

Problem C3-2.

Number the pies 1 to 7 in cyclic order, with the 3 banana pies following the 3 apple pies and preceding the cherry pie. Alice tastes number 4. There are three cases.
Case 1. Number 4 is the cherry pie.
Then Alice eats it.
Case 2. Number 4 is an apple pie.
Then number 7 must be a banana pie and the cherry pie is not between these two. Alice tastes number 2. If it is the cherry pie, Alice eats it. If it is another apple pie, Alice eats number 1. If it is a banana pie, Alice eats number 3.
Case 3. Number 4 is a banana pie.
Then number 1 must be an apple pie and the cherry pie is not between these two. Alice tries number 6. The analysis is analogous to Case 2.

Problem C3-3

(a) The knight starts with the bronze sword, reducing the number of heads to $36-9-3=24$. Then he uses the silver sword, reducing the number of heads to $24-8-2=14$. This is followed by three uses of the gold sword, reducing the number heads first to $14-7-1=6$, then to $6-3-1=2$ and finally to $2-1-1=0$.

(b) The knight starts with the bronze sword twice, reducing the number of heads first to $36-9-3=24$ and then to $24-6-3=15$. This is followed by using the silver sword, reducing the number of heads to $15-5-2=8$. Then he uses the bronze sword again, reducing the number of heads to $8-2-3=3$. Finally, he uses the silver sword again, reducing the number of heads to $3-1-2=0$.

(c) The knight starts with the silver sword, reducing the number of heads to $36-12-2=22$. Then he uses the gold sword twice, reducing the number of heads first to $22-11-1=10$ and then to $10-5-1=4$. This is followed by a final cut using the bronze sword, reducing the number of heads to $4-1-3=0$.

Problem C3-4.

The task is possible in nine crossings.
1. Alice, Betty and Carla go to the far shore.
2. Betty and Carla come back to the near shore.
3. Carla and Debra go to the far shore.
4. Alice and Carla come back to the near shore.
5. Alice, Betty and Carla go to the far shore.
6. Carla and Debra come back to the near shore.
7. Debra and Ellen go to the far shore.
8. Alice and Betty come back to the near shore.
9. Alice, Betty and Carla go to the far shore.

Set C4.

Problem C4-1.

The smallest number of wrestlers who win awards is at least one, since the strongest wrestler always wins an award in any arrangement of matches. We now show an arrangement of matches in which exactly one wrestler wins an award. Let each wrestler be matched with the wrestler immediately below in strength, except for the weakest wrestler who is matched with the strongest one. Then each wrestler is in two matches. Everyone wins once and loses once except that the strongest one wins both matches while the weakest one loses both matches.

Problem C4-2.

Arrange the boxes in increasing order of the numbers of pencils in them. The numbers of pencils in the respective boxes are at least 1, at least 2, at least 3 pencils and at least 4 pencils. Throw out any extra pencils so that the numbers of pencils in the respective boxes are exactly 1, exactly 2, exactly 3 and exactly 4. We must choose the only pencil in the first box. Now the two pencils in the second box have different colours, so that one of them must be of a colour different from the pencil chosen. We can then choose that pencil. The three pencils in the third box also have different colours, so that one of them must be of a colour different from the two pencils chosen. We can then choose that pencil. Similarly, we can make a suitable choice from the last box.

Problem C4-3.

The sum of the two announcements of each native, knight or knucklehead, is equal to the sum of the ages of the two neighbours. Let the natives be #1 to #10 in cyclic order. If we add up the sums of the two ages announced by each of #2, #6 and #10, we have the total age of #1, #3, #5, #7, #9 and #1 again. If we add up the sums of the two ages announced by each of #4 and #8, we have the total age of #3, #5, #7 and #9. Subtracting the second total from the first and dividing the difference by 2, we have the age of #1. The age of every native can be determined in a similar manner. Knowing the age of each native, we can tell whether any particular native is a knight or a knucklehead.

Problem C4-4.

We may have as many as five kneejerks. They answer first and all say "No". The other five, all knights, then say "Yes". We now show that there cannot be more than five knuckleheads. Let the questioner take a break when the running total of "Yes" answers and "No" answers are the same. Consider what happens between two breaks. In order to get to the next break, the number of "Yes" answers must be the same as the number of "No" answers. Moreover, all the answers by the kneejerks must be the same because the lead does cannot change. It follows that at most half of the answers are by the kneejerks. Since this is true between any two breaks, the total number of kneejerks is at most half the total populaation, namely five.

Part III

International Mathematics Tournament of the Towns

Tournament 29 to Tournament 42

Fall 2007 to Spring 2021

In the early years of the Tournament, the structure varied from time to time. Eventually it stablized into a standard form. This was maintained throughout the period covered by this book.

Each Tournament has a Fall Round and a Spring Round. Each Round has an Ordinary Level (O-Level) and an Advanced Level (A-Level). Each Level has a Junior Paper and a Senior Paper. Sometimes, the Junior Paper and the Senior Paper share common problems. The problem in the Senior Paper may also be a harder version of the corresponding problem in the Junior Paper. The time allowance is four hours for an O-Level Paper and five hours for an A-Level Paper.

Each O-Level Paper has five problems and each A-Level Paper has seven problems. The maximum credit for each problem is given. A contestant may attempt any number of questions. For each contestant, the credits obtained from individual problems are listed in descending order, and the sum of the highest three will constitute the raw score of the contestant. A student may write both the O-Level Paper and the A-Level Paper.

The Russian school system spans eleven years, six in elementary school and five in high school. The Junior Paper is attempted by students in Years 7, 8 and 9. The raw score of a Year 7 student is adjusted by a factor of $\frac{3}{2}$ while the raw score of a Year 8 student is adjusted by a factor of $\frac{4}{3}$. The Senior Paper is attempted by students in year 10 and 11. The raw score of a Year 10 student is adjusted by a factor of $\frac{5}{4}$. The raw score of a Year 9 student or a Year 11 student does not change in the adjustment.

A student's final score for the Tournament is the maximum of the adjusted scores in the O-Level and the A-Level of the Fall Round and the Spring Round.

The problems and solutions to the first 28 Tournaments can be found in the following publications.

[1] Taylor, P. J. *International Mathematics Tournament of the Towns: 1980–1984*, AMT, Canberra, 1993.

[2] Taylor, P. J. *International Mathematics Tournament of the Towns: 1984–1989*, AMT, Canberra, 1992.

[3] Taylor, P. J. *International Mathematics Tournament of the Towns: 1989–1993*, AMT, Canberra, 1994.

[4] Storozhev, A. M. and Taylor, P. J. *International Mathematics Tournament of the Towns: 1993–1997*, AMT, Canberra, 2003.

[5] Storozhev, A. M. *International Mathematics Tournament of the Towns: 1997–2002*, AMT, Canberra, 2006.

[6] Liu, A. and Taylor, P. J. *International Mathematics Tournament of the Towns : 2002–2007*, AMT, Canberra, 2009.

Tournament 29

Fall 2007

Junior O-Level Paper

1. Black and white counters are placed on an 8×8 board, with at most one counter on each square. What is the maximum number of counters that can be placed such that each row and each column contains twice as many white counters as black ones?

2. Initially, the number 1 and a non-integral number x are written on a blackboard. In each step, we can choose two numbers on the blackboard, not necessarily different, and write their sum or their difference on the blackboard. We can also choose a non-zero number on the blackboard and write its reciprocal on the blackboard. Is it possible to write x^2 on the blackboard in a finite number of moves?

3. D is the midpoint of the side BC of triangle ABC. E and F are points on CA and AB respectively, such that BE is perpendicular to CA and CF is perpendicular to AB. If DEF is an equilateral triangle, does it follow that ABC is also equilateral?

4. Each square of a 29×29 table contains one of the integers 1, 2, 3, ..., 29, and each of these integers appears 29 times. The sum of all the numbers above the main diagonal is equal to three times the sum of all the numbers below this diagonal. Determine the number in the central square of the table.

5. The audience chooses two of five cards, numbered from 1 to 5 respectively. The assistant of a magician chooses two of the remaining three cards, and asks a member of the audience to take them to the magician, who is in another room. The two cards are presented to the magician in an arbitrary order. By an arrangement with the assistant beforehand, the magician is able to deduce which two cards the audience has chosen. Explain how this may be done.

Note: The problems are worth 3, 4, 4, 5 and 5 points respectively.

Senior O-Level Paper

1. Pictures are taken of 100 adults and 100 children, with one adult and one child in each, the adult being the taller of the two. Each picture is reduced to $\frac{1}{k}$ of its original size, where k is a positive integer which may vary from picture to picture. Prove that it is possible to have the reduced image of each adult taller than the reduced image of every child.

2. Initially, the number 1 and two positive numbers x and y are written on a blackboard. In each step, we can choose two numbers on the blackboard, not necessarily different, and write their sum or their difference on the blackboard. We can also choose a non-zero number on the blackboard and write its reciprocal on the blackboard. Is it possible to write on the blackboard, in a finite number of moves, the number
   (a) x^2;
   (b) xy?

3. Give a construction by straight-edge and compass of a point C on a line ℓ parallel to a segment AB, such that the product $AC \cdot BC$ is minimum.

4. The audience chooses two of twenty-nine cards, numbered from 1 to 29 respectively. The assistant of a magician chooses two of the remaining twenty-seven cards, and asks a member of the audience to take them to the magician, who is in another room. The two cards are presented to the magician in an arbitrary order. Does there exist an arrangement with the assistant beforehand such that the magician is able to deduce which two cards the audience has chosen?

5. A square of side length 1 centimeter is cut into three convex polygons. Is it possible that the diameter of each of them does not exceed
   (a) 1 centimeter;
   (b) 1.01 centimeters;
   (c) 1.001 centimeters?

Note: The problems are worth 3, 2+2, 4, 4 and 1+2+2 points respectively.

Junior A-Level Paper

1. Let $ABCD$ be a rhombus. Let K be a point on the line CD, other than C or D, such that $AD = BK$. Let P be the point of intersection of BD with the perpendicular bisector of BC. Prove that A, K and P are collinear.

2. (a) Each of Peter and Basil thinks of three positive integers. For each pair of his numbers, Peter writes down the greatest common divisor of the two numbers. For each pair of his numbers, Basil writes down the least common multiple of the two numbers. If both Peter and Basil write down the same three numbers, must these three numbers be equal to each other?
   (b) Is the analogous result true if each of Peter and Basil thinks of four positive integers instead?

3. Michael is at the centre of a circle of radius 100 meters. Each minute, he will announce the direction in which he will be moving. Catherine can leave it as is, or change it to the opposite direction. Then Michael moves exactly 1 meter in the direction determined by Catherine. Does Michael have a strategy which guarantees that he can get out of the circle, even though Catherine will try to stop him?

4. Two players take turns entering a symbol in an empty square of a $1 \times n$ board, where n is an integer greater than 1. Aaron always enters the symbol X and Betty always enters the symbol O. Two identical symbols may not occupy adjacent squares. A player without a move loses the game. If Aaron goes first, which player has a winning strategy?

5. Attached to each of a number of objects is a tag which states the correct mass of the object. The tags have fallen off and have been replaced on the objects at random. We wish to determine if by chance all tags are in fact correct. We may use exactly once a horizontal lever which is supported at its middle. The objects can be hung from the lever at any point on either side of the support. The lever either stays horizontal or tilts to one side. Is this task always possible?

6. The audience arranges n coins in a row. The sequence of heads and tails is chosen arbitrarily. The audience also chooses a number between 1 and n inclusive. Then the assistant turns one of the coins over, and the magician is brought in to examine the resulting sequence. By an agreement with the assistant beforehand, the magician tries to determine the number chosen by the audience.

   (a) If this is possible for some n, is it also possible for $2n$?

   (b) Determine all n for which this is possible.

7. For each letter in the English alphabet, William assigns an English word which contains that letter. His first document consists only of the word assigned to the letter A. In each subsequent document, he replaces each letter of the preceding document by its assigned word. The fortieth document begins with "Till whatsoever star that guides my moving." Prove that this sentence reappears later in this document.

Note: The problems are worth 5, 3+3, 6, 7, 8, 4+5 and 9 points respectively.

Senior A-Level Paper

1. (a) Each of Peter and Basil thinks of three positive integers. For each pair of his numbers, Peter writes down the greatest common divisor of the two numbers. For each pair of his numbers, Basil writes down the least common multiple of the two numbers. If both Peter and Basil write down the same three numbers, must these three numbers be equal to one another?

   (b) Is the analogous result true if each of Peter and Basil thinks of four positive integers instead?

2. Let K, L, M and N be the midpoints of the sides AB, BC, CD and DA of a cyclic quadrilateral $ABCD$. Let P be the point of intersection of AC and BD. Prove that the circumradii of triangles PKL, PLM, PMN and PNK are equal to one another.

3. Determine all finite increasing arithmetic progressions in which each term is the reciprocal of a positive integer and the sum of all the terms is 1.

4. Attached to each of a number of objects is a tag which states the correct mass of the object. The tags have fallen off and have been replaced on the objects at random. We wish to determine if by chance all tags are in fact correct. We may use exactly once a horizontal lever which is supported at its middle. The objects can be hung from the lever at any point on either side of the support. The lever either stays horizontal or tilts to one side. Is this task always possible?

5. The audience arranges n coins in a row. The sequence of heads and tails is chosen arbitrarily. The audience also chooses a number between 1 and n inclusive. Then the assistant turns one of the coins over, and the magician is brought in to examine the resulting sequence. By an agreement with the assistant beforehand, the magician tries to determine the number chosen by the audience.

   (a) Prove that if this is possible for some n_1 and n_2, then it is also possible for $n_1 n_2$.

   (b) Determine all n for which this is possible.

6. Let P and Q be two convex polygons. Let h be the length of the projection of Q onto a line perpendicular to a side of P which is of length p. Define $f(P,Q)$ to be the sum of the products hp over all sides of P. Prove that $f(P,Q) = f(Q,P)$.

7. There are 100 boxes, each containing either a red cube or a blue cube. Alex has a sum of money initially, and places bets on the colour of the cube in each box in turn. The bet can be anywhere from 0 up to everything he has at the time. After the bet has been placed, the box is opened. If Alex loses, his bet will be taken away. If he wins, he will get his bet back, plus a sum equal to the bet. Then he moves onto the next box, until he has bet on the last one, or until he runs out of money. What is the maximum factor by which he can guarantee to increase his amount of money, if he knows that the exact number of blue cubes is

   (a) 1;

   (b) some integer k, $1 < k \leq 100$?

Note: The problems are worth 2+2, 6, 6, 6, 4+4, 8 and 3+5 points respectively.

Spring 2008

Junior O-Level Paper

1. In the convex hexagon $ABCDEF$, AB, BC and CD are respectively parallel to DE, EF and FA. If $AB = DE$, prove that $BC = EF$ and $CD = FA$.

2. There are ten congruent segments on a plane. Each point of intersection divides every segment passing through it in the ratio 3:4. Find the maximum number of points of intersection.

3. There are ten cards with the number a on each, ten with b and ten with c, where a, b and c are distinct real numbers. For every five cards, it is possible to add another five cards so that the sum of the numbers on these ten cards is 0. Prove that one of a, b and c is 0.

4. Find all positive integers n such that $(n+1)!$ is divisible by $1!+2!+\cdots+n!$.

5. Each square of a 10×10 board is painted red, blue or white, with exactly twenty of them red. No two adjacent squares are painted in the same colour. A domino consists of two adjacent squares, and it is said to be good if one square is blue and the other is white.

   (a) Prove that it is always possible to cut out 30 good dominoes from such a board.

   (b) Give an example of such a board from which it is possible to cut out 40 good dominoes.

   (c) Give an example of such a board from which it is not possible to cut out more than 30 good dominoes.

Note: The problems are worth 3, 5, 5, 6 and 2+2+2 points respectively.

Senior O-Level Paper

1. There are ten cards with the number a on each, ten with b and ten with c, where a, b and c are distinct real numbers. For every five cards, it is possible to add another five cards so that the sum of the numbers on these ten cards is 0. Prove that one of a, b and c is 0.

2. Can it happen that the least common multiple of 1, 2, ..., n is 2008 times the least common multiple of 1, 2, ..., m for some positive integers m and n?

3. In triangle ABC, $\angle A = 90°$. M is the midpoint of BC and H is the foot of the altitude from A to BC. The line passing through M and perpendicular to AC meets the circumcircle of triangle AMC again at P. If BP intersects AH at K, prove that $AK = KH$.

4. No matter how two copies of a convex polygon are placed inside a square, they always have a common point. Prove that no matter how three copies of the same polygon are placed inside this square, they also have a common point.

5. We may permute the rows and the columns of the table below. How may different tables can we generate?

1	2	3	4	5	6	7
7	1	2	3	4	5	6
6	7	1	2	3	4	5
5	6	7	1	2	3	4
4	5	6	7	1	2	3
3	4	5	6	7	1	2
2	3	4	5	6	7	1

Note: The problems are worth 4, 5, 5, 5 and 6 points respectively.

Junior A-Level Paper

1. An integer n is the product of two consecutive integers.

   (a) Prove that we can add two digits to the right of this number and obtain a perfect square.

   (b) Prove that this can be done in only one way if $n > 12$.

2. A line parallel to the side AC of triangle ABC cuts the side AB at K and the side BC at M. O is the point of intersection of AM and CK. If $AK = AO$ and $KM = MC$, prove that $AM = KB$.

3. Alice and Brian are playing a game on a $1 \times (n + 2)$ board. To start the game, Alice places a counter on any of the n interior squares. In each move, Brian chooses a positive integer k. Alice must move the counter to the kth square on the left or the right of its current position.

If the counter moves off the board, Alice wins. If it lands on either of the end squares, Brian wins. If it lands on another interior square, the game proceeds to the next move. For which values of n does Brian have a strategy which allows him to win the game in a finite number of moves?

4. Given are finitely many points in the plane, no three on a line. They are painted in four colours, with at least one point of each colour. Prove that there exist three triangles, distinct but not necessarily disjoint, such that the three vertices of each triangle have different colours, and none of them contains a coloured point in its interior.

5. Standing in a circle are 99 girls, each with a candy. In each move, each girl gives her candy to either neighbour. If a girl receives two candies in the same move, she eats one of them. What is the minimum number of moves after which only one candy remains?

6. Do there exist positive integers a, b, c and d such that $\frac{a}{b}+\frac{c}{d}=1$ and $\frac{a}{d}+\frac{c}{b}=2008$?

7. A convex quadrilateral $ABCD$ has no parallel sides. The angles between the diagonal AC and the four sides are $55°$, $55°, 19°$ and $16°$ in some order. Determine all possible values of the acute angle between AC and BD.

Note: The problems are worth 2+2, 5, 6, 6, 7, 7 and 8 points respectively.

Senior A-Level Paper

1. A triangle has an angle of measure θ. It is dissected into several triangles. Is it possible that all angles of the resulting triangles are less than θ, if

   (a) $\theta = 70°$;

   (b) $\theta = 80°$?

2. Alice and Brian are playing a game on the real line. To start the game, Alice places a counter on a number x where $0 < x < 1$. In each move, Brian chooses a positive number d. Alice must move the counter to either $x+d$ or $x-d$. If it lands on 0 or 1, Brian wins. Otherwise the game proceeds to the next move. For which values of x does Brian have a strategy which allows him to win the game in a finite number of moves?

3. A polynomial

$$x^n + a_1x^{n-1} + a_2x^{n-2} + \cdots + a_{n-2}x^2 + a_{n-1}x + a_n$$

has n distinct real roots $x_1,\ x_2,\ \ldots,\ x_n$, where $n > 1$. The polynomial

$$nx^{n-1} + (n-1)a_1x^{n-2} + (n-2)a_2x^{n-3} + \cdots + 2a_{n-2}x + a_{n-1}$$

has roots $y_1,\ y_2,\ \ldots,\ y_{n-1}$. Prove that

$$\frac{x_1^2 + x_2^2 + \cdots + x_n^2}{n} > \frac{y_1^2 + y_2^2 + \cdots + y_{n-1}^2}{n-1}.$$

4. Each of Peter and Basil draws a convex quadrilateral with no parallel sides. The angles between a diagonal and the four sides of Peter's quadrilateral are α, α, β and γ in some order. The angles between a diagonal and the four sides of Basil's quadrilateral are also α, α, β and γ in some order. Prove that the acute angle between the diagonals of Peter's quadrilateral is equal to the acute angle between the diagonals of Basil's quadrilateral.

5. The positive integers are arranged in a row in some order, each occuring exactly once. Does there always exist an adjacent block of at least two numbers somewhere in this row such that the sum of the numbers in the block is a prime number?

6. Seated in a circle are 11 wizards. A different positive integer not exceeding 1000 is pasted onto the forehead of each. A wizard can see the numbers of the other 10, but not his own. Simultaneously, each wizard puts up either his left hand or his right hand. Then each declares the number on his forehead at the same time. Is there a strategy on which the wizards can agree beforehand, which allows each of them to make the correct declaration?

7. Each of three lines cuts chords of equal lengths in two given circles. The points of intersection of these lines form a triangle. Prove that its circumcircle passes through the midpoint of the segment joining the centres of the circles.

Note: The problems are worth 3+3, 6, 6, 7, 8, 8 and 8 points respectively.

Solutions

Fall 2007

Junior O-Level Paper

1. The number of counters in each row must be a multiple of 3, since there are twice as many white counters as black ones. Since there are only 8 squares in each row, the maximum number of counters in each row in 6. Since there are 8 rows, the maximum number of counters overall is 48. This can be attained by the placement shown in the diagram below.

●	●	○	○	○	○		
●	○	○	○	○			●
○	○	○	○			●	●
○	○	○			●	●	○
○	○			●	●	○	○
○			●	●	○	○	○
		●	●	○	○	○	○
	●	●	○	○	○	○	

2. We first write down $x - 1$ and $x + 1$. Then we can write down $\frac{1}{x-1}$, $\frac{1}{x+1}$ and their difference $\frac{2}{x^2-1}$. The reciprocal of this number is $\frac{x^2-1}{2}$. Adding this number to itself yields $x^2 - 1$, and adding 1 to it yields x^2.

3. Let ABC be a triangle in which $\angle A = 60°$ and $\angle C = 30°$. Then F coincides with B. Since $\angle BEC = 90°$, E lies on the semicircle with diameter BC. Hence $DB = DE$. Since $\angle EBD = 90° - \angle C = 60°$, DEF is an equilateral triangle while ABC is not.

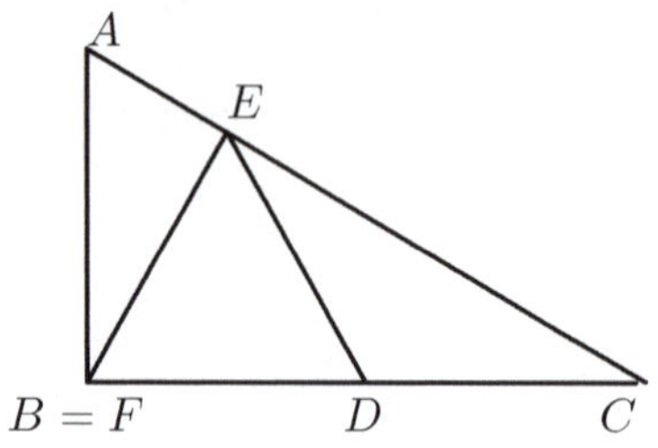

4. There are 29 numbers on the main diagonal, 29×14 numbers above it and 29×14 numbers below it. The sum of the largest 29×14 numbers is $29(16+17+\cdots+29) = 29 \times 7 \times 45$ while the the sum of the smallest 29×14 numbers is $29(1+2+\cdots+14) = 29 \times 7 \times 15$. Since the former is exactly three times as large as the latter, the largest 29×14 numbers are all above the main diagonal and the smallest 29×14 numbers are all below the main diagonal. In other words, every number on the main diagonal, including the central square, is 15.

5. The magician and her assistant can agree beforehand to arrange the numbers 1 to 5 in order on a circle, so that 1 follows 5. If the audience chooses two adjacent cards, say 3 and 4, the assistant will choose the two cards after them, which are 5 and 1. If the audience chooses two non-adjacent cards, say 3 and 5, the assistant will choose the cards after them, namely, 4 and 1. If the magician receives two adjacent cards, say 2 and 3, she will know that the audience must have chosen 5 and 1. If the magician receives two non-adjacent cards, say 2 and 5, she will know that the audience must have chosen 1 and 4.

Senior O-Level Paper

1. Let the heights of the adult and the child in the ith picture be a_i and c_i respectively, where $1 \leq i \leq 100$. For $2 \leq i \leq 100$, we have $\frac{a_i}{a_1} > \frac{c_i}{a_1}$, so that there exists a rational number $\frac{p_i}{q_i}$ such that $\frac{a_i}{a_1} > \frac{p_i}{q_i} > \frac{c_i}{a_1}$. Hence $\frac{a_i}{p_i} > \frac{a_1}{q_i} > \frac{c_i}{p_i}$. We reduce the first picture to $\frac{1}{q_1 q_2 \cdots q_{100}}$ of its size and the ith picture for $2 \leq i \leq 100$ to $\frac{1}{q_1 q_2 \cdots q_{i-1} p_i q_{i+1} q_{i+2} \cdots q_{100}}$ of its size.

2. (a) We first write down $x-1$ and $x+1$. Then we can write down $\frac{1}{x-1}$, $\frac{1}{x+1}$ and their difference $\frac{2}{x^2-1}$. The reciprocal of this number is $\frac{x^2-1}{2}$. Adding this number to itself yields x^2-1, and adding 1 to it yields x^2.

   (b) We first write down $x+y$. By (a), we can write down $(x+y)^2$, x^2 and y^2. This is followed by $2xy = (x+y)^2 - x^2 - y^2$ and $\frac{1}{2xy}$. Finally, we can write down the sum $\frac{1}{2xy} + \frac{1}{2xy} = \frac{1}{xy}$ and the reciprocal xy.

   (Compare with Problem 2 in the Junior O-Level Paper.)

3. Draw a semicircle with diameter AB towards ℓ. Suppose they have a point of intersection C. Then $\angle BCA = 90°$ so that the area of triangle ABC is $\frac{1}{2}AC \cdot BC$. For any other point P on ℓ, the area of triangle ABP is $\frac{1}{2}AP \cdot BP \sin APB \leq \frac{1}{2}AP \cdot BP$. Since ℓ is parallel to AB, triangles ABC and ABP have the same area. It follows that $AC \cdot BC \leq AP \cdot BP$.

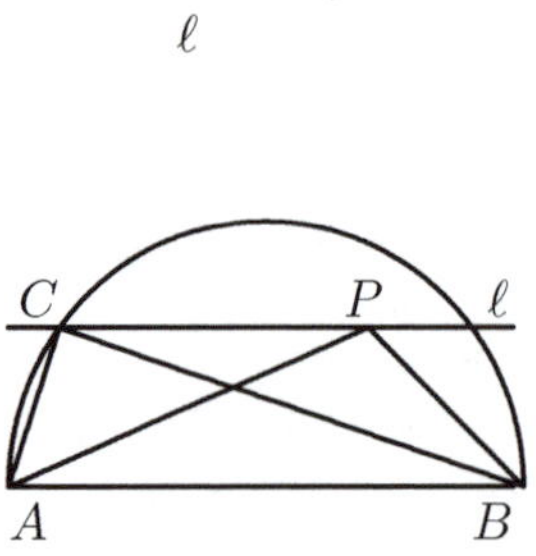

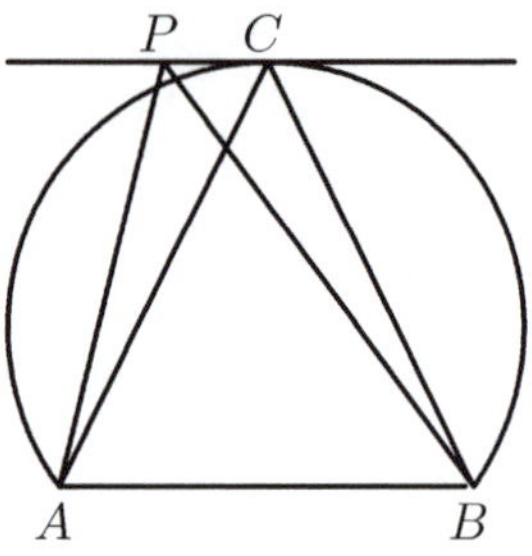

Suppose the semicircle does not intersect ℓ. Let C be the point on ℓ equidistant from A and B. Draw the circumcircle of triangle ABC. Then ℓ is one of its tangents. Hence for any point P on ℓ other than C, $\angle ACB > \angle APB$. Since triangles ABC and ABP have the same area, we have $AC \cdot BC < AP \cdot BP$ as before.

4. The magician and her assistant can agree beforehand to arrange the numbers 1 to 29 in order on a circle, so that 1 follows 29. If the audience chooses two adjacent cards, say 3 and 4, the assistant will choose the two cards after them, which are 5 and 6. If the audience chooses two non-adjacent cards, say 3 and 29, the assistant will choose the cards after them, namely, 4 and 1. If the magician receives two adjacent cards, say 2 and 3, she will know that the audience must have chosen 29 and 1. If the magician receives two non-adjacent cards, say 2 and 15, she will know that the audience must have chosen 1 and 14.
(Compare with Problem 5 of the Junior O-Level Paper.)

5. (a) Let $ABCD$ be the square. Let E, F, G and H be the respective midpoints of AB, BC, CD and DA. Let P and Q be the respective midpoints of AE and FC. (See the diagram below on the left.) Suppose $ABCD$ has been dissected into three convex polygons each of diameter at most 1. By the Pigeonhole Principle, two of E, F, G and H must belong to the same polygon. Suppose E and G belong to the same polygon. Then we must have A and D belong to a second polygon and B and C to the third. However, P is now too far from each of D, G and C. Hence we may assume that E and F belong to the same polygon. Now P is too far from F, Q is too far from E, and P and Q are too far from each other. Hence P belongs to a second polygon while Q belongs to a third. Now D is too far from each of E, P and Q, and cannot belong to any of the three polygons. This contradiction shows that the task is impossible.

   (b) We can cut a square $ABCD$ into three rectangles $ADHK$, $BFJK$ and $CFJH$, as shown in the diagram above on the right. The diameter of any rectangle is a diagonal. Now $AH^2 = 1^2 + (\frac{1}{8})^2 = \frac{65}{64}$. From (a), $FK^2 = FH^2 = \frac{65}{64}$ also. Now $\frac{1}{64} < \frac{1}{50} < \frac{201}{10000}$ so that $\frac{65}{64} < \frac{10201}{10000}$. It follows that the diameter of each of the rectangles is less than $\frac{101}{100}$.

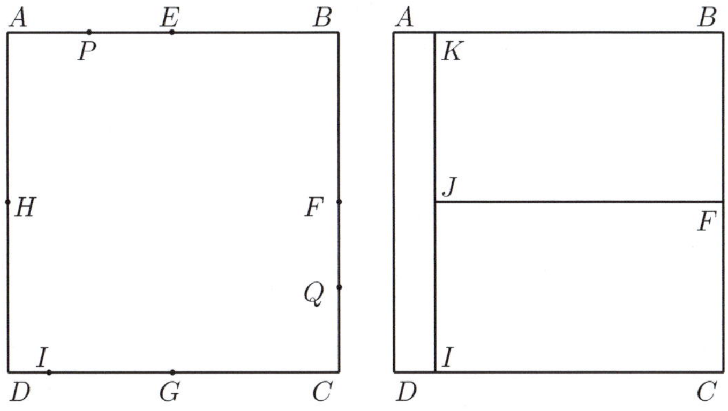

(c) Suppose $ABCD$ has been dissected into three convex polygons each of diameter at most 1. By the Pigeonhole Principle, two of the vertices must belong to the same polygon. They cannot be opposite vertices as otherwise the diameter of that polygon will be $\sqrt{2}$. Hence we may assume that A and D belong to the first polygon. Suppose B and C belong to the second. Since E is too far from C and from D, it must belong to the third polygon. Then the point I on DG which is at a distance $\frac{1}{8}$ from D is too far from each of A, E and B, and cannot belong to any of the three polygons. From this contradiction, we may assume that B belongs to the second polygon while C belongs to the third. Then E must belong to the second polygon while G must belong to the third. Since F is too far from A, we may assume that it belongs to the third polygon. However, I is too far from A, from B and also from F since $IF^2 = (\frac{1}{2})^2 + (\frac{7}{8})^2 = \frac{65}{64}$. This contradiction shows that the task is impossible since we have $AI = IF = \frac{\sqrt{65}}{8} > \frac{1001}{1000}$.

Junior A-Level Paper

1. Let AK intersect BD at E. We shall prove that $BE = CE$, so that E lies on the perpendicular bisector of BC. It will then follow that $E = P$, and that A, P and K are indeed collinear. Since $AD = BK$ and AB is parallel to DK, $ABKD$ is a cyclic quadrilateral. It follows that

$$\angle AKD = \angle ABD = \angle CBD,$$

so that $BCKE$ is also a cyclic quadrilateral. We now have

$$\angle ECB = \angle EKB = \angle ADB = \angle EBC,$$

so that $EB = EC$.

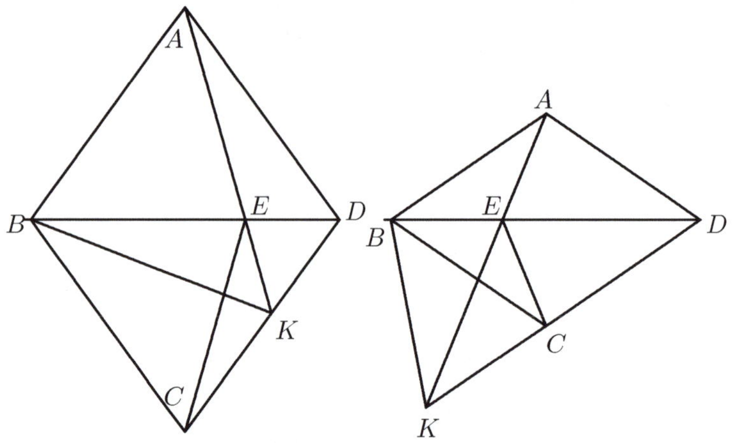

2. (a) Denote by $a \triangle b$ the greatest common divisor of a and b, and by $a \triangledown b$ the least common multiple of a and b. They are binary operations which are associative. Let the numbers Peter thinks of be p_1, p_2 and p_3, the numbers Basil thinks of be b_1, b_2 and b_3, and the numbers both write down be w_1, w_2 and w_3. Note that $w_1 \triangle w_2 = w_2 \triangle w_3 = w_3 \triangle w_1 = p_1 \triangle p_2 \triangle p_3$. Similarly, $w_1 \triangledown w_2 = w_2 \triangledown w_3 = w_3 \triangledown w_1 = b_1 \triangledown b_2 \triangledown b_3$. It follows that $w_1 w_2 = (w_1 \triangle w_2)(w_1 \triangledown w_2) = w_2 \triangle w_3)(w_2 \triangledown w_3) = w_2 w_3$, so that $w_1 = w_3$. Similarly, w_2 shares this common value.

   (b) The answer is no. Peter's numbers may be 1, 2, 2 and 2. Then his six greatest common divisors are 1, 1, 1, 2, 2 and 2. Basil's numbers may be 1, 1, 1 and 2. Then his six least common multiples are 1, 1, 1, 2, 2 and 2. The two sets of numbers are identical, but the six numbers in each set are not all the same.

3. Michael can escape. In the first move, he chooses any direction. Catherine cannot gain anything by reversing it. In each subsequent move, Michael chooses a direction which is perpendicular to the line joining his current position to the centre of circle. Again Catherine cannot gain anything by reversing it. Let d_n be the distance of Michael from the centre of the circle after the nth move. We have $d_1 = 1$ and $d_{n+1} = \sqrt{d_n^2 + 1}$. We claim that $d_n = \sqrt{n}$ for all $n \geq 1$. The basis holds, and by the induction hypothesis, $d_{n+1} = \sqrt{\sqrt{n}^2 + 1} = \sqrt{n+1}$. It follows that after 10000 moves, Michael will arrive at the circumference of the circle.

4. We claim that Betty can guarantee a win. We first prove the following auxiliary result. Suppose the first square is marked X and the last square is marked O, with n vacant squares in between. If Aaron goes first, he loses. We use induction on n. When $n = 1$, neither player has a move. Since Aaron moves first, he loses.

Assume that result holds up to $n-1$ for some $n \geq 2$. Consider a board with n vacant squares between the X and O already marked. In his first move, Aaron will partition the board into two, with i and j vacant squares respectively, where $i+j=n-1$. In the first board, the first and the last squares are marked X. Betty places an O next to either X. Then we have two boards each of which has X at one end and O at the other, with less than n vacant squares in between. Aaron will lose because by the induction hypothesis, he loses on both boards. We now return to the vacant $1 \times n$ board. Suppose Aaron marks an X on the kth square. By symmetry, we may assume that $k > 1$. Betty marks an O on the first square. It is Aaron's move, and by the auxiliary result, he will lose if $k = n-1$ or n. If not, he will at some point be forced to mark an X on the ℓth square where $\ell \geq k+2$. Then Betty will mark an O on the $(k+1)$st square, and the auxiliary result applies again. Thus Aaron is forced to open up new parts of the board again and again. Eventually he runs out of room and loses. It follows that Betty can always win if $n \geq 2$.

5. Let there be n objects, and let the mass indicated by the tag on the ith object be m_i, $1 \leq i \leq n$. We may assume that $m_1 \leq m_2 \leq \cdots \leq m_n$. Arbitrarily choose positive numbers $d_2 < d_3 < \cdots < d_n$ and choose d_1 so that $d_1m_1 = d_2m_2 + d_3m_3 + \cdots + d_nm_n$. On one side of the support, hang the 1st object at a distance d_1 from the support. On the other side of the support, hang the ith object at a distance d_i from the support for $2 \leq i \leq n$. Let the correct mass of the ith object be a_i for $1 \leq i \leq n$. We consider three cases.

   Case 1. All the tags are in fact correct.
   Then we will have equilibrium.

   Case 2. The tag on the 1st object is correct but those on some of the others are not.
   Then $\{a_2, a_3, \ldots, a_n\}$ is a permutation of $\{m_2, m_3, \ldots, m_n\}$, and by the Rearrangement Inequality,

$$d_2a_2 + d_3a_3 + \cdots + d_na_n < d_2m_2 + d_3m_3 + \cdots + d_nm_n = d_1m_1.$$

   Case 3. The tag on the 1st object is incorrect.
   Then $m_1 < a_1 = m_j$ for some j, $2 \leq j \leq n$. Hence $d_1a_1 = d_1m_j > d_1m_1$ while

$$\begin{aligned} & d_2m_2 + d_3m_3 + \cdots + d_nm_n \\ > \; & d_2m_1 + d_3m_2 + \cdots + d_jm_{j-1} + d_{j+1}m_{j+1} + \cdots + d_nm_n \\ \geq \; & d_2a_2 + d_3a_3 + \cdots + d_na_n, \end{aligned}$$

   and again we have no equilibrium.

6. (a) Given a row of n coins arbitrarily arranged heads and tails, and a number between 1 and n inclusive, the assistant can flip exactly one coin so that the magician can tell which number has been chosen. With a row of $2n$ coins and a number m between 1 and $2n$, the magician and the assistant place the numbers 1 to n in order in the first row of a $2\times n$ array, and the numbers from $n+1$ to $2n$ in order in the second row. If the row number h and the column number k of the location of m are determined, then $m=(h-1)n+k$. The magician and the assistant also consider the $2n$ coins as in a $2\times n$ array. Code each coin with heads up as 0 and each coin with tails up as 1. Compute the sum of the codes of the two coins in each column modulo 2 and regard the result as a linear array of n coded coins. By the hypothesis, the assistant can flip the qth coded coin to signal the number k to the magician. This can be achieved by flipping either of the two coins in the qth column. To signal the number h to the magician, the assistant will just use the bottom coin of the qth column, code 0 meaning $h=1$ and code 1 meaning $h=2$. If the bottom coin is not correct, flip it. Otherwise, flip the top coin.

   (b) For $n=1$, the assistant must flip the only coin. However, the chosen number can only be 1, and the magician does not require any assistance. Hence the task is possible. For $n=2$, let the coins be coded as in (a). The assistant will just use the second coin, code 0 meaning $h=1$ and code 1 meaning $h=2$. If the second coin is not correct, flip it. Otherwise, flip the first coin. Hence the task is also possible. By (a), the task is possible whenever n is a power of 2. We now show that the converse also holds. Each of the 2^n arrangement of the coins codes a specific number between 1 and n. If n is not a power of 2, then $2^n=qn+r$ where q and r are the quotient and the remainder obtained from the Division Algorithm, with $r>0$. By the Pigeonhole Principle, some number is coded by at most q arrangements. Each may be obtained by the flip of a single coin from exactly n other arrangements. This yields a total count of $qn<2^n$. On the other hand, from each of the 2^n arrangements, we must be able to obtain one of these q arrangements by the flip of a single coin. This contradiction shows that the task is impossible unless n is a power of 2.

7. Let the ith document be D_i, $0\le i\le 40$. Note that D_0 consists only of the letter A, and D_1 consists only of the word assigned to the letter A. This word does not start with A, as otherwise all documents start with A, which is not the case since D_{40} starts with T. However, the word assigned to A does contain the letter A, so that D_1 contains D_0, and not from the beginning. Similarly, D_{i+1} contains D_i for $0\le i\le 39$, again not from the beginning.

Note that no D_i can start with a single-letter word, as otherwise this word will start all subsequent documents, which is not the case since D_{40} does not start with a single-letter word. Since there are only twenty-six letters in the English alphabet, D_j and D_k must start with the same letter for some j and k where $j < k \leq 27$. Then D_{j+1} and D_{k+1} start with the same word, D_{j+2} and D_{k+2} start with the same two words, and so on. Let $t = 40 - k$. Since $k \leq 27$, $t \geq 13$ so that D_{j+t} and $D_{k+t} = D_{40}$ start with the same thirteen words, including "Till whatsoever star that guides my moving". Since D_{40} contains a copy of D_{j+t} and not from the beginning, the sentence "Till whatsoever star that guides my moving" will reappear in D_{40}.

Senior A-Level Paper

1. (See Problem 2 of the Junior A-Level Paper.)

2. Since $\angle PCD = \angle PBA$ and $\angle PDC = \angle PAB$, triangles PCD and PBA are similar. Hence $\frac{PD}{PA} = \frac{CD}{BA} = \frac{MD}{KA}$, so that triangles PMD and PKA are also similar. It follows that we have $\angle DPM = \angle APK$. Since LM is parallel to BD, $\angle DPM = \angle PML$. Since KL is parallel to AC, $\angle APK = \angle PKL$. Hence $\angle PML = \angle PKL$, which implies that the circumradii of triangles PML and PKL have a common value. In the same way, we can prove that this is also the value of the circumradii of triangles PMN and PKN.

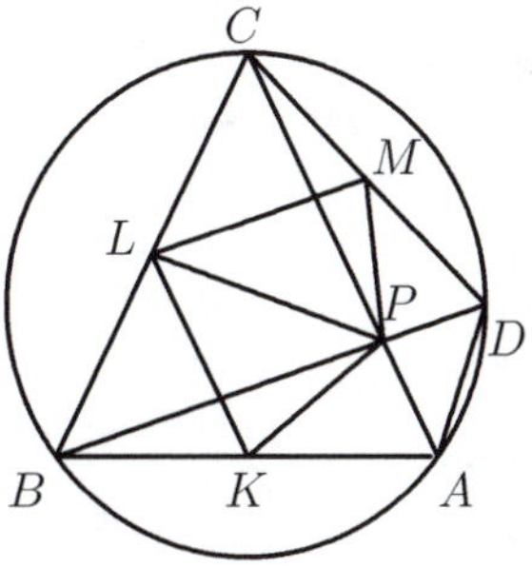

3. If the arithmetic progression has at most 2 terms, the only possibilities are $\{1\}$ and $\{\frac{1}{2}, \frac{1}{2}\}$, but neither is increasing. Henceforth, we assume that the arithmetic progression has at least 3 terms. Now suppose that $\{\frac{1}{b_i} : 1 \leq i \leq n\}$ is an arithmetic progression with the desired properties, where b_i is a positive integer for $1 \leq i \leq n$. Let m be the least common multiple of these n integers. Then $\frac{1}{b_i} = \frac{c_i}{m}$ for some positive integer c_i, $1 \leq i \leq n$. These n numbers also form an increasing arithmetic progression. Let d be its common difference.

Suppose c_{n-1} and c_n are not relatively prime. Then their greatest common divisor has a prime divisor p, and p divides $c_n - c_{n-1} = d$. It follows that d divides c_i for $1 \le i \le n$. However, this means that m cannot be the least common multiple of b_i, $1 \le i \le n$. It follows that c_{n-1} and c_n are relatively prime, so that $c_{n-1}c_n$ divide m. Hence

$$c_{n-1}c_n \le m = c_1 + c_2 + \cdots + c_n \le nc_{n-1} \le c_{n-1}c_n.$$

It follows that we have equality throughout. Now $m = nc_{n-1}$ holds if and only if $n = 3$, and $3c_2 = c_2c_3$ holds if and only if $c_3 = 3$, so that $c_2 = 2$ and $c_1 = 1$. Hence $\{\frac{1}{6}, \frac{1}{3}, \frac{1}{2}\}$ is the unique arithmetic progression with the desired properties.

4. (See Problem 5 of the Junior A-Level Paper.)

5. (a) Given a row of n_i coins, $i = 1, 2$, arbitratily arranged heads and tails, and a number between 1 and n_i inclusive, the assistant can flip exactly one coin so that the magician can tell which number has been chosen. With a row of n_1n_2 coins and a number m between 1 and n_1n_2, the magician and the assistant place the numbers 1 to n_2 in order in the first row of a $n_1 \times n_2$ array, the numbers from $n_2 + 1$ to $2n_2$ in order in the second row, and so on. If the row number h and the column number k of the location of m are determined, then $m = (h-1)n+k$. The magician and the assistant also consider the n_1n_2 coins as in a $n_1 \times n_2$ array. Code each coin with heads up as 0 and each coin with tails up as 1. Compute the sum of the codes of the coins in each row and each column modulo 2 and regard the results as two linear arrays of n_1 and n_2 coded coins respectively. By the hypothesis, the assistant can flipped the pth coded coin in the first array to signal the number h to the magician. This can be achieved by flipping any coin in the pth row. Similarly, the assistant can flipped the qth coded coin in the second array to signal the number k to the magician. This can be achieved by flipping any coin in the qth column. Both objective can be achieved by flipping the coin at the intersection of the pth row and the qth column.

   (b) For $n = 1$, the assistant must flip the only coin. However, the chosen number can only be 1, and the magician does not require any assistance. Hence the task is possible. For $n = 2$, let the coins be coded as in (a). The assistant will just use the second coin, code 0 meaning $h = 1$ and code 1 meaning $h = 2$. If the second coin is not correct, flip it. Otherwise, flip the first coin. Hence the task is also possible. By (a), the task is possible whenever n is a power of 2.

We now show that the converse also holds. Each of the 2^n arrangement of the coins codes a specific number between 1 and n. If n is not a power of 2, then $2^n = qn + r$ where q and r are the quotient and the remainder obtained from the Division Algorithm, with $r > 0$. By the Pigeonhole Principle, some number is coded by at most q arrangements. Each may be obtained by the flip of a single coin from exactly n other arrangements. This yields a total count of $qn < 2^n$. On the other hand, from each of the 2^n arrangements, we must be able to obtain one of these q arrangements by the slip of a single coin. This contradiction shows that the task is impossible unless n is a power of 2.

(Compare with Problem 6 of the Junior A-Level Paper.)

6. Let P be $P_1P_2\dots P_n$ and let p_i be the length of P_iP_{i+1} for $1 \le i \le n$, where P_{n+1} is to be interpreted as P_1. Let Q be $Q_1Q_2\dots Q_m$ and let q_j be the length of Q_jQ_{j+1} for $1 \le j \le m$, where Q_{m+1} is to be interpreted as Q_1. Let θ_{ij} be the angle formed by the lines P_iP_{i+1} and Q_jQ_{j+1}. The diagram below illustrates the case where a convex heptagon Q is projected onto a line perpendicular to P_iP_{i+1}.

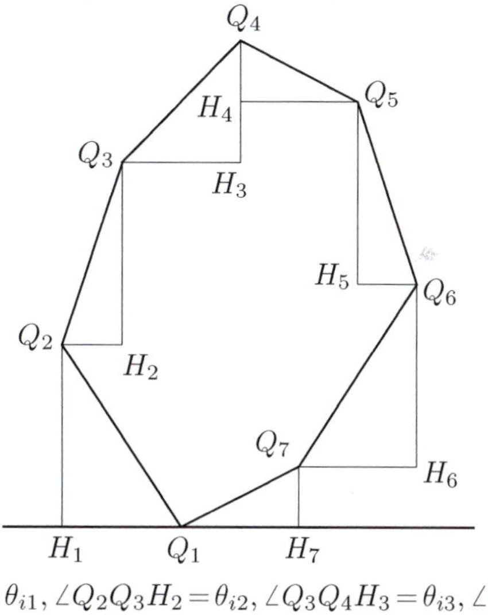

Then $\angle Q_1Q_2H_1 = \theta_{i1}$, $\angle Q_2Q_3H_2 = \theta_{i2}$, $\angle Q_3Q_4H_3 = \theta_{i3}$, $\angle Q_5Q_4H_4 = \theta_{i4}$, $\angle Q_6Q_5H_5 = \theta_{i5}$, $\angle Q_7Q_6H_6 = \theta_{i6}$ and $\angle Q_7Q_1H_7 = \theta_{i7}$. The length h_i of the projection of Q perpendicular to PiP_{i+1} is given by

$$\begin{aligned} h_i &= H_1Q_1 + Q_1H_7 + Q_7H_6 \\ &= q_1 \sin\theta_{i1} + q_7 \sin\theta_{i7} + q_6 \sin\theta_{i6} \end{aligned}$$

and also by

$$\begin{aligned} h_i &= Q_2H_2 + Q_3H_3 + H_4Q_5 + H_5Q_6 \\ &= q_2 \sin\theta_{i2} + q_3 \sin\theta_{i3} + q_4 \sin\theta_{i4} + q_5 \sin\theta_{i5}. \end{aligned}$$

Hence $h_i = \frac{1}{2}\sum_{j=1}^{7} q_j \sin\theta_{ij}$. More generally,

$$\begin{aligned} f(P,Q) &= \sum_{i=1}^{n} p_i \left(\frac{1}{2}\sum_{j=1}^{m} q_j \sin\theta_{ij}\right) \\ &= \sum_{j=1}^{m} q_j \left(\frac{1}{2}\sum_{i=1}^{n} p_i \sin\theta_{ij}\right) \\ &= f(Q,P). \end{aligned}$$

7. (a) More generally, let the number of boxes be n. Let Alex's first bet be b_n, and let the final amount of money he would end up with be a_n. Clearly $b_1 = 1$ and $a_1 = 2$, as the only box will contain a blue cube, and Alex can make a bet of 1 on blue. Let $n > 1$. Then Alex bets red. If he loses the initial bet, he would have $1 - b_n$ left. However, he can now double this amount $n-1$ times to obtain $a_n = 2^{n-1}(1-b_n)$. If he wins the initial bet instead, he would have $1 + b_n$, and there is still a blue cube in the remaining $n-1$ boxes. Hence $a_n = a_{n-1}(1+b_n)$. The values of these two expressions for a_n go in opposite directions with a change of value in b_n. Alex's best strategy is to make them equal, so that he will have a guaranteed value of a_n. From $2^{n-1}(1-b_n) = a_{n-1}(1+b_n)$, we have $b_n = \frac{2^{n-1}-a_{n-1}}{2^{n-1}+a_{n-1}}$. Simple calculations reveal that

$$\begin{aligned} b_2 &= \tfrac{2-2}{2+2} &= 0, \\ a_2 &= 2(1-0) &= 2, \\ b_3 &= \tfrac{4-2}{4+2} &= \tfrac{1}{3}, \\ a_3 &= 2^2(1-\tfrac{1}{3}) &= \tfrac{8}{3}. \end{aligned}$$

We claim that $b_n = \frac{n-2}{n}$ and $a_n = \frac{2^n}{n}$ for $n \geq 2$. By mathematical induction,

$$\begin{aligned} b_n &= \frac{2^{n-1} - \frac{2^{n-1}}{n-1}}{2^{n-1} + \frac{2^{n-1}}{n-1}} \\ &= \frac{n-2}{n}, \\ a_n &= 2^{n-1}\left(1 - \frac{n-2}{n}\right) \\ &= \frac{2^n}{n}. \end{aligned}$$

For $n = 100$, Alex can increase his amount of money by a factor of $\frac{2^{100}}{100}$.

(b) Let there be exactly k blue cubes inside the n boxes. Denote by $f(n,k)$ the maximum factor by which Alex can increase his amount of money. Note that $f(n,k)=f(n,n-k)$, so we may asume that $k\leq\frac{n}{2}$. Clearly, $f(n,0)=2^n$. For $0<k<\frac{n}{2}$, Alex bets on red. Let his first bet be x. If Alex loses, he would have $1-x$ left, and his final amount would then be $(1-x)f(n-1,k-1)$. If Alex wins, he would have $1+x$ left, and his final amount would then be $(1+x)f(n-1,k)$. As in (a), we want these two expressions to have a common value, which we take as $f(n,k)$. Solving

$$(1-x)f(n-1,k-1)=(1+x)f(n-1,k),$$

we have $x=\frac{f(n-1,k-1)-f(n-1,k)}{f(n-1,k-1)+f(n-1,k)}$, which lies between 0 and 1. This yields $f(n,k)=\frac{2f(n-1,k-1)f(n-1,k)}{f(n-1,k-1)+f(n-1,k)}$. When $k=\frac{n}{2}$ is an integer, symmetry dictates that Alex bets 0 so that $f(n,k)=f(n-1,k-1)$. Simple calculations reveal that we have $f(2,1)=f(1,0)=2$ and $f(3,1)=\frac{2\times4\times2}{4+2}=\frac{8}{3}$. We now prove by mathematical induction that $f(n,k)=\frac{2^n}{\binom{n}{k}}$. This has been checked up to $n=3$. Suppose the result holds for some $n\geq3$. We have

$$f(n+1,0)=2^{n+1}=\frac{2^{n+1}}{\binom{n+1}{0}}.$$

For $0<k<\frac{n+1}{2}$, we have

$$f(n+1,k)=\frac{2f(n,k-1)f(n,k)}{f(n,k-1)+f(n,k)}=\frac{2^{n+1}}{\binom{n}{k-1}+\binom{n}{k}}=\frac{2^{n+1}}{\binom{n+1}{k}}$$

by Pascal's Formula. When $k=\frac{n+1}{2}$ is an integer, $\binom{n}{k-1}=\binom{n}{k}$, so that

$$f(n+1,k)=f(n,k-1)=\frac{2^{n+1}}{\binom{n}{k-1}+\binom{n}{k}}=\frac{2^{n+1}}{\binom{n+1}{k}}.$$

For $n=100$, Alex can increase his amount of money by a factor of $\frac{2^{100}}{\binom{100}{k}}$.

Spring 2008

Junior O-Level Paper

1. Extend FA and CB to meet at H, and extend CD and FE to meet at K, as shown in the diagram below. Then $CHFK$ is a parallelogram. Hence triangles ABH and DEK are similar. They are in fact congruent since $AB = DE$. Now $BC = HC - HB = KF - KE = EF$ and $CD = KC - KD = HF - HA = FA$.

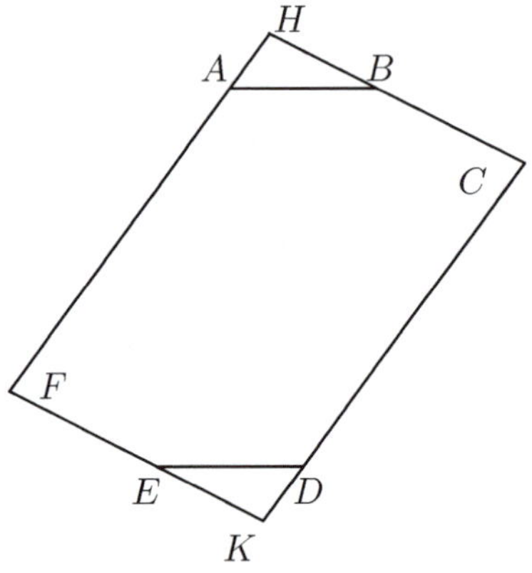

2. On each segment, there are exactly two points which divide it in the ratio 3:4. Hence the total count segment by segment is at most 20. However, it takes two segments to produce a point of intersection. Hence there are at most 10 such points. The diagram below shows how this can be attained, so that 10 is indeed the maximum.

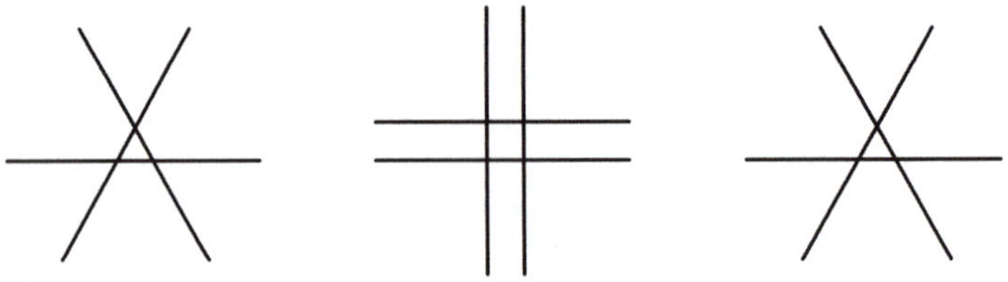

3. Suppose none of a, b and c is 0. They cannot all be positive and they cannot all be negative. By symmetry, we may assume that a and b are positive while c is negative. Since a and b are distinct, we may assume that $a > b$. If $a > -c$, we take five cards with a on each. Then it is impossible to take another five cards to bring the total down to 0. If $-c > a$, we take five cards with c on each. Then it is impossible to take another five cards to bring the total up to 0. It follows that we must have $a = -c > b$. If we now take four cards with a on each and a fifth card with b on it, it is impossible to take another five cards to bring the total down to 0.

4. For $n = 1$, $1!$ divides $2!$. For $n = 2$, $1!+2!$ divides $3!$. We claim that there are no solutions for $n \geq 3$. We have

$$\begin{aligned}(n+1)! &= n! + n(n!) \\ &= n((n-1)! + n!) \\ &< n(1! + 2! + \cdots + n!).\end{aligned}$$

For $n = 3$, $4! > 2(1! + 2! + 3!)$. Suppose for some $n \geq 3$,

$$n! > (n-2)(1! + 2! + \cdots + (n-1)!).$$

Note that $2(n-2) \geq n-1$. By mathematical induction,

$$\begin{aligned}(n+1)! &= (n-1)n! + 2n! \\ &> (n-1)n! + 2(n-2)(1! + 2! + \cdots + (n-1)!) \\ &\geq (n-1)(1! + 2! + \cdots + n!).\end{aligned}$$

It follows that $\dfrac{(n+1)!}{1! + 2! + \cdots + n!}$ lies strictly between $n-1$ and n. Hence it cannot be an integer, and the claim is justified.

5. (a) Divide the board into 10 rows each consisting of 5 dominoes. At most 20 of these 50 dominoes can contain a red square. Each of the other 30 must be a good domino since no two adjacent squares are painted in the same colour.

   (b) Paint the board blue and white in the usual pattern. Repaint into red squares 20 of the squares marked by circles, and divide the rest of the board into 40 dominoes as shown in the diagram below on the left. Each of these 40 dominoes is good.

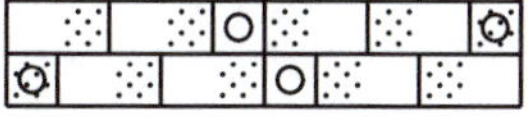

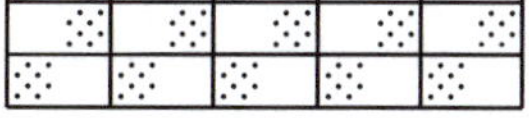

   (c) Divide the board into 50 dominoes and paint it blue and white in the usual checkerboard pattern as shown in the diagram above on the right. If we repaint any 20 blue squares into red squares, we will only have 30 blue squares left. Since we need a blue square in each good domino, we will have exactly 30 good dominoes.

Senior O-Level Paper

1. Suppose none of a, b and c is 0. They cannot all be positive and they cannot all be negative. By symmetry, we may assume that a and b are positive while c is negative. Since a and b are distinct, we may assume that $a > b$. If $a > -c$, we take five cards with a on each. Then it is impossible to take another five cards to bring the total down to 0. If $-c > a$, we take five cards with c on each. Then it is impossible to take another five cards to bring the total up to 0. It follows that we must have $a = -c > b$. If we now take four cards with a on each and a fifth card with b on it, it is impossible to take another five cards to bring the total down to 0.

2. Let the highest power of 2 less than or equal to m be 2^r. Note that $2008 = 2^3 \times 251$. The highest power of 2 less than or equal to n must be 2^{r+3}. It follows that $n > 4m$. Let the highest power of 3 less than or equal to m be 3^s. Then the highest power of 3 less than or equal to n must also be 3^s since 3 does not divide 2008. However, $n > 4m \geq 4 \times 3^s > 3^{s+1}$, which is a contradiction. Hence no such positive integers m and n exist.

3. Since both AB and MP are perpendicular to AC and $BM = MC$, MP intersects AC at its midpoint D. It follows that MP is a diameter of the circumcircle, so that MC is perpendicular to PC. It follows that triangles MCD and MPC are similar, so that $\frac{MD}{MC} = \frac{MC}{MP}$. Hence $\frac{MD}{MB} = \frac{MB}{MP}$. Since $\angle DMB = \angle BMP$, triangles DMB and BMP are also similar. It follows that $\angle CBD = \angle BPM = \angle ABK$. Now triangles BAH and BCA are also similar. Since $CD = DA$, we have $AK = KH$.

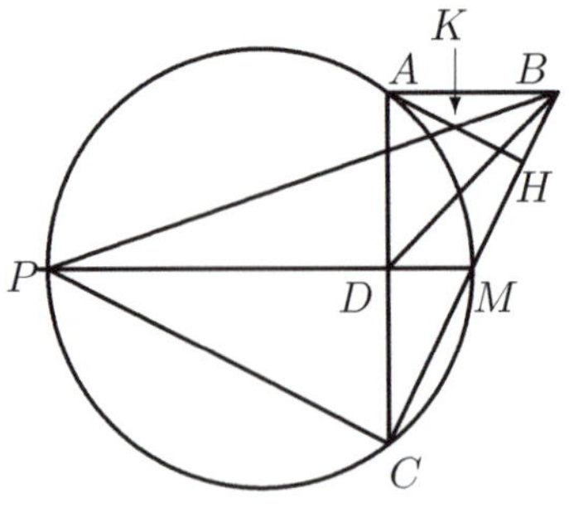

4. Let a copy F of the convex polygon be placed anywhere inside the square. Consider the copy F' obtained from F by a half-turn about the centre O of the square. By hypothesis, F and F' must have a point in common. Let it be P. Then the point P' obtained from P by a half-turn about O is also in the intersection of F and F'. Since F is convex, O is also in F.

It follows that a copy of the convex polygon placed anywhere inside the square must cover O, and if three copies are placed in the square, they will have O in common.

5. The columns may be permuted in 7! ways so that the first row is different. The remaining rows may be permuted in 6! ways so that the first column is different. Once the first row and the first column have been fixed, the remaining entries in the table are also fixed. Hence the total number of different tables we can generate is $7! \times 6!$.

Junior A-Level Paper

1. Let $n = k(k+1)$. Adding two digits to the right of 2 produces an integer m where $100k(k+1) \leq m \leq 100k(k+1) + 99$.

   (a) If we add 25, then $m = 100k(k+1) + 25 = (10k+5)^2$.

   (b) Notice that $n > 12$ means $k > 3$. It follows that

   $$100k(k+1) - (10k+4)^2 = 20k - 16 > 0$$

   and

   $$100k(k+1) + 99 - (10k+6)^2 = -20k + 63 < 0.$$

   Thus the only square within range is $(10k+5)^2$.

2. Since $AO = AK$, $\angle AKO = \angle AOK$. Since $MK = MC$, we have $\angle MCK = \angle MKC$. Since KM is parallel to AC, $\angle ACK = \angle MKC$ and $\angle BMK = \angle ACM$. Now

   $$\angle ABC = \angle AKO - \angle MCK = \angle AOK - \angle ACK = \angle MAC.$$

   Hence triangles MAC and KBM are congruent, so that $AM = BK$.

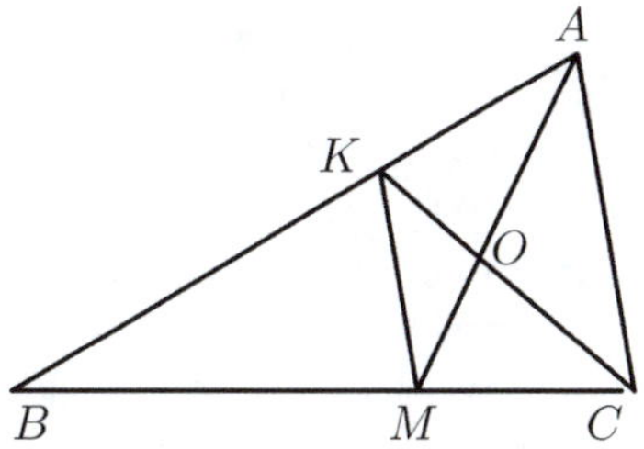

3. colour the two end squares red. Then there is a block of adjacent blank squares between them. If the block does not have a middle square, the colouring process terminates. If it does, colour the middle square red. This creates twice as many blocks of adjacent blank squares, all with the same number of squares in them.

Eventually, the colouring process stops either because all squares are red, or because the new blocks do not have middle squares. If all squares are red, then $n = 2^k - 1$ for some positive integer k, and Brian has a sure win. Wherever Alice places the counter, Brian can force it into either square which makes its current square red. Eventually, the counter will be forced into either of the two end squares. If the squares are not all red, then $n \neq 2^k - 1$ for any positive integer k, and Alice cannot lose. She simply places the counter on a blank square, and Brian can never force it onto a red square. Since both end squares are red, Brian cannot win. In summary, Brian wins if and only if $n = 2^k - 1$ for some positive integer k.

4. Consider all sets of four points of different colours. Since the number of points is finite, we can choose the set whose convex hull has the smallest area. If the convex hull is a quadrilateral, then there are no coloured points in its interior, as otherwise we have a set whose convex hull has smaller area. The four vertices of the quadrilateral determine four triangles each with vertices of different colours, and any three of these four triangles will satisfy the requirement. Suppose the convex hull is a triangle ABC, say with A red, B yellow and C blue. Then only points of the fourth colour, say green, can be inside ABC, and there is at least one such point D. If there are no green points other than D, then triangles ACD, BAD and CBD satisfy the requirement. Suppose BAD contains other green points. Choose among them a point E such that triangle BAE has the smallest area. Then it cannot contain any green points in its interior, and we can replace BAD by BAE. A similar remedy can be applied if either ACD or CBD contains green points in its interior. Hence we will get three triangles which satisfy the requirement.

5. Let the girls be labeled 1 to 99 in clockwise order. We first show that the task can be accomplished in 98 moves. In each of the first 49 moves, if she still has a candy, the kth girl gives hers to the $(k-1)$st girl for $2 \leq k \leq 50$ and to the $(k+1)$st girl for $51 \leq k \leq 99$. The 1st girl, who will always have a candy, gives hers to the 99th girl. After 49 moves, only the 1st and the 99th girl have candies. These two candies can be passed, in opposite directions, to the 50th girl in another 49 moves. We now show that the task cannot be accomplished in less than 98 moves. We will not allow the girls to eat the candies, but each must pass all she has to the same neighbour. Our target is to have all candies in the hands of one girl. Consider what happens to a candy in two consecutive moves. It either returns to the girl who has it initially, or is passed to a girl two places away.

Suppose the candies all end up with the 50th girl in at most 98 moves. The number of moves is not enough for the candy initially with the 50th girl to go once around the circle before returning to her. It follows that the number of moves must be even. However, in order for the candy initially with the 49th girl to end up in the hands of the 50th girl in an even number of moves, it must go once around the circle, and that takes 98 moves.

6. Such positive integers exist. We take $b = kd$ for some integer k. Then $a + kc = kd$ and $ka + c = 2008kd$. Solving this system of equations, we have $a = \frac{kd(2008k-1)}{k^2-1}$ and $c = \frac{kd(k-2008)}{k^2-1}$. Taking $d = k^2 - 1$, we have $b = k(k^2 - 1)$, $a = k(2008k - 1)$ and $c = k(k - 2008)$ for $k \geq 2009$. When $k = 2009$, we have $a = 8104448639$, $b = 8108484720$, $c = 2009$ and $d = 4036080$.

7. There are three cases to consider, the two equal angles are at the same vertex, they are at opposite vertices but on the same side of the diagonal, and they are at opposite vertices and on opposite sides of the diagonal. The last one can be discarded because there will be two parallel sides. In the first case, as illustrated by the diagram below on the left, let the equal angles be at the vertex C. Draw the excircle of triangle BCD opposite C with centre E. Then A lies on the line CE, and we have $\angle PBE = \angle EBD$ and $\angle QDE = \angle EDB$. Then $\angle BEC = \angle PBE - 55°$ and $\angle DEC = \angle QDE - 55°$. Hence the sum of the angles of triangle BED is $2\angle EBD + 2\angle EDB - 110°$, which is 180°. It follows that $\angle EBD + \angle EDB = 145°$ so that $\angle BED = 35°$. Hence A and E coincide. By symmetry, we can take $\angle DAC = 16°$. Then $\angle ADB = \angle QDA = 55° + 16° = 71°$. It follows that the acute angle between AC and BD is $71° + 16° = 87°$.

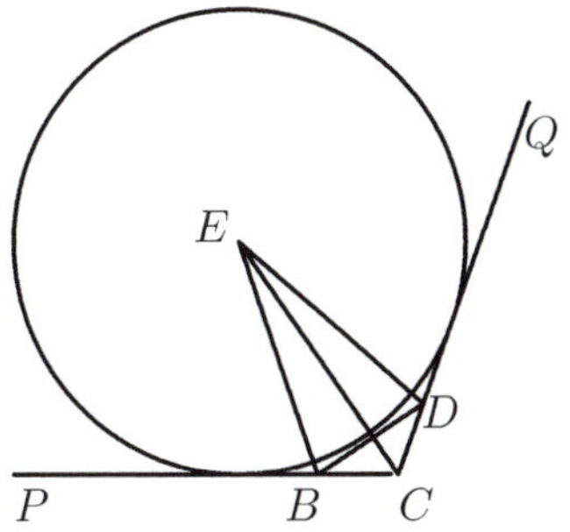

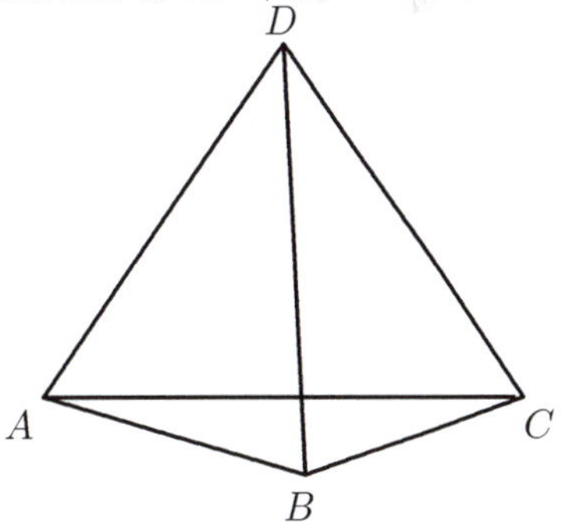

The second case is illustrated by the diagram above on the right, where $\angle DAC = \angle DCA = 55°$, $\angle BAC = 16°$ and $\angle BCA = 19°$. Then we have $\angle ABC = 180° - 16° - 19° = 145°$.

If $DB < DA = DC$, then

$$\angle ABC = \angle ABD + \angle DBC > \angle BAD + \angle BCD > 145^\circ.$$

Similarly, if $DB > DA = DC$, then $\angle ABC < 145^\circ$. It follows that $DB = DA = DC$ so that $\angle ABD = 55^\circ + 16^\circ = 71^\circ$. Hence the acute angle between AC and BD is again $71^\circ + 16^\circ = 87^\circ$. In summary, 87° is the only possible value.

Senior A-Level Paper

1. (a) Suppose the task is possible. In the resulting triangulation, the 70° angle must be subdivided into at least two angles. Hence one of these angles is at most 35°. In the triangle to which it belongs, one of the other two angles is at least $\frac{1}{2}(180^\circ - 35^\circ) = 72.5^\circ$. This is a contradiction.

   (b) In the diagram below, ABC is a triangle with $AB = AC$ and $\angle CAB = 80^\circ$. It is dissected into seven triangles where we have $AF = AH = AE$, $HF = HG = HD = HE$, HG is parallel to AB and HD to AC. Then $\angle DHG = 80^\circ$, $\angle HGD = \angle GDH = 50^\circ$, $\angle HAE = \angle HAF = 40^\circ$, $\angle BGF = \angle CDE = 75^\circ$ and

$$\angle AFH = \angle AHF = \angle FHG = \angle DHE = \angle EHA = \angle AEH = 70^\circ.$$

   The other angles all have measure 55°. If we move D and G a little closer to each other, we can make all angles to have measure less than 80°.

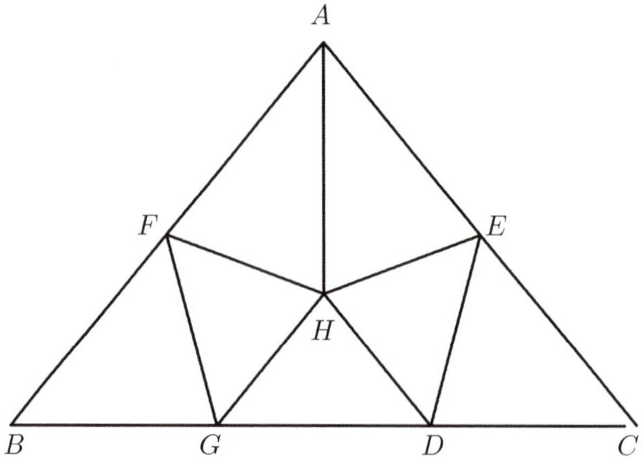

2. Call a number good if it is of the form $\frac{m}{2^n}$ for some odd integer m satisfying $0 < m < 2^n$. Suppose x is a good number. Then Brian chooses $d = \min\{\frac{m}{2^n}, 1 - \frac{m}{2^n}\}$. In order to avoid losing immediately, Alice must move the counter to $\frac{m}{2^{n-1}}$ or $1 - \frac{m}{2^{n-1}}$, which is another good number.

Repeating this procedure, the power of 2 in the denominator diminishes by 1 in each move. After n moves, the counter is forced into 0 or 1, and Brian wins. Suppose x is not a good number. Then whatever value d Brian chooses, either $x+d$ or $x-d$ is not good. This is because the sum of two good numbers is good, and half a good number is also good, but x itself is not good. It follows that Alice can never be forced to move the counter to a good number. However, since $\frac{1}{2}$ is good, and it is the only point from which Brian can force Alice to lose on the move, Brian cannot win. In summary, Brian wins if and only if x is a good number.

3. We have $x_1+x_2+\cdots+x_n=-a_1$, $x_1x_2+x_1x_3+\cdots+x_{n-1}x_n=a_2$, $y_1+y_2+\cdots+y_{n-1}=-\frac{a_1(n-1)}{n}$ and $y_1y_2+y_1y_3+\cdots+y_{n-2}y_{n-1}=\frac{a_2(n-2)}{n}$. It follows that we have $X=x_1^2+x_2^2+\cdots+x_n^2=a_1^2-2a_2$ while $Y=y_1^2+y_2^2+\cdots+y_{n-1}^2=\frac{a_1^2(n-1)^2}{n^2}-\frac{2a_2(n-2)}{n}$. Hence

$$\begin{aligned}\frac{X}{n}-\frac{Y}{n-1} &= a_1^2\left(\frac{1}{n}-\frac{n-1}{n^2}\right)-2a_2\left(\frac{1}{n}-\frac{n-2}{n(n-1)}\right)\\ &= \frac{1}{n^2(n-1)}((n-1)a_1^2-2na_2)\\ &= \frac{1}{n^2(n-1)}((n-1)X-2a_2).\end{aligned}$$

By the Rearrangement Inequality,

$$(n-1)(x_1^2+x_2^2+\cdots+x_n^2)-2(x_1x_2+x_1x_3+\cdots+x_{n-1}x_n)\geq 0,$$

with equality if and only if $x_1=x_2=\cdots=x_n$. Since the roots are distinct, we have strict inequality.

4. In the diagram below on the left is Peter's quadrilateral $ABCD$, with $\angle CAD=\angle ACD=\alpha$, $\angle BAC=\beta$ and $\angle ACB=\gamma$, the diagonals intersecting at E. In the diagram below in the middle is Basil's quadrilateral $A'B'C'D'$, with $\angle B'A'C'=\angle D'A'C'=\alpha$, $\angle B'C'A'=\beta$ and $\angle D'C'A'=\gamma$, the diagonals intersecting at E'.

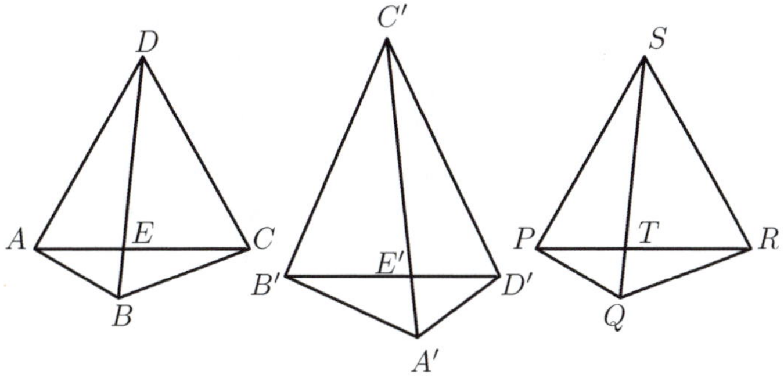

Construct triangle PQT similar to triangle $C'B'E'$. If we extend PT to R, then we have $\angle QTR = \angle D'E'C'$, and we can choose R so that triangle QRT is similar to triangle $D'C'E'$. Similarly, we can extend QT to S so that triangle RST is siimilar to triangle $A'D'E'$. Join SP to complete the quadrilateral $PQRS$, as shown in the diagram above on the right. Now $\angle STP = \angle QTR = \angle D'E'C' = \angle B'E'A'$ and

$$\frac{PT}{ST} = \frac{PT}{QT} \cdot \frac{QT}{RT} \cdot \frac{RT}{ST} = \frac{C'E'}{B'E'} \cdot \frac{D'E'}{C'E'} \cdot \frac{A'E'}{D'E'} = \frac{A'E'}{B'E'}.$$

Hence triangle SPT is similar to triangle $B'A'E'$. It follows that triangle PQR is similar to triangle ABC, and triangle RSP is similar to triangle CDA, so that the quadrilateral $PQRS$ is similar to the quadrilateral $ABCD$. It follows that $\angle C'E'B' = \angle PTQ = \angle AEB$.

5. Let the first three terms be 1, 4! and 2. So far, each of 1+4!, 4!+2 and 1+4!+2 are composite. We will lengthen the sequence by two terms at a time. The second term to be added is the smallest number not yet in the sequence, and the first term to be added is the factorial of the sum of the preceding terms plus the following term. Thus the fifth term is 3, the fourth term is (1+4!+2+3)!, the seventh term is 4 and the sixth term is (1+4!+2+(1+4!+2+3)!+3+4)!. Since the odd-numbered term added at each stage is the smallest number not yet in the sequence, every positive integer will eventually appear. No positive integer can appear twice as the even-numbered term added at each stage is larger than any number which has been chosen earlier. For any block of consecutive terms other than {1,4!}, the largest term is of the form $n!$ while the sum of the other terms is some positive integer k where $2 \le k \le n$. The sum of the terms in the block is $n! + k$, which is composite since it is divisible by k.

6. Each wizard constructs an 10×10 table, the ith row being the base-2 representation, with leading 0s, of the number on the forehead of the wizard i places away in clockwise order. Then he computes the sum of the diagonal elements modulo 2. If the sum is 0, he puts up his left hand, and if it is 1, he puts up his right hand. Consider an arbitrary wizard A. Each digit of the base-2 representation of A's number appears on the diagonal of exactly one of the other wizards. Consider the digit which is on the diagonal of wizard B. The other 9 digits on that diagonal are known to A. From B's show of hand, A has the sum of the 10 digits on that diagonal, and he can determine the missing digit. Recovering the other digits in a similar manner, A can reconstruct his own number.

7. Let the centres of the two circles be L and N, with M the midpoint of LN. Let a line α cut the circles at A_1, A_2, A_3 and A_4 as shown in the diagram below, with $A_1A_2 = A_3A_4$. Let A_5 and A_6 be the respective midpoints of A_1A_2 and A_3A_4. Then LA_5 and NA_6 are both perpendicular to α. Let A_7 be the foot of the perpendicular from M to α. Then A_7 is the midpoint of A_5A_6, and hence of A_2A_3 and of A_1A_4. Now $A_7A_1 \cdot A_7A_2 = A_7A_4 \cdot A_7A_3$. Hence A_7 has equal power with respect to both circles, and lies on their radical axis ℓ. Let β and γ be two other lines which cut the circle in equal segments. Then the feet of the perpendicular from M to both lines also lie on ℓ. It follows that ℓ is the Simson line with respect to M of the triangle formed by α, β and γ. Hence M lies on the circumcircle of this triangle.

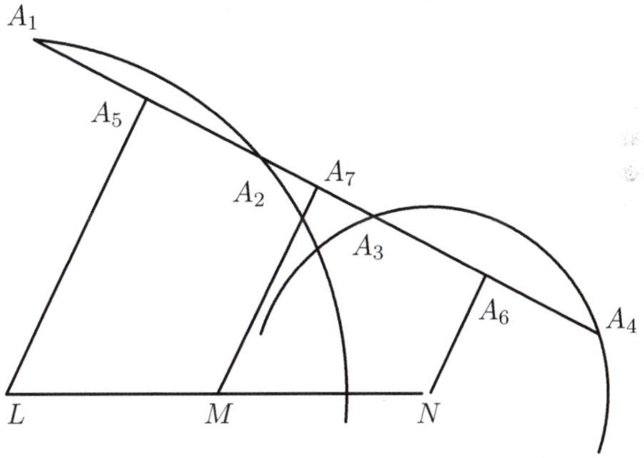

Tournament 30

Fall 2008

Junior O-Level Paper

1. Each of ten boxes contains a different number of pencils. No two pencils in the same box are of the same colour. Prove that one can choose one pencil from each box so that no two chosen pencils are of the same colour.

2. Twenty-five different numbers are chosen from 1, 2, ..., 50. Twenty-five different numbers are also chosen from 51, 52, ..., 100. No two chosen numbers differ by 50. Find the sum of all 50 chosen numbers.

3. Triangle ACE is inscribed in a circle of radius 2. Prove that one can choose a point D on the arc CE, a point F on the arc EA and a point B on the arc AC, such that the numerical value of the area of the hexagon $ABCDEF$ is equal to the numerical value of the perimeter of triangle ACE.

4. The average of two distinct positive integers is calculated. Can the product of all three numbers be equal to a^{2008} for some positive integer a?

5. On a straight track, several people are running at different constant speeds. They start at one end of the track at the same time. When they reach either end of the track, they turn around and continue to run indefinitely. Some time after the start, all runners meet at the same point. Prove that this will happen again.

Note: The problems are worth 3, 3, 4, 4 and 4 points respectively.

Senior O-Level Paper

1. Alex distributes some cookies into several boxes and records the number of cookies in each box. If the same number appears more than once, it is recorded only once. Serge takes one cookie from each box and puts them on the first plate. Then he takes one cookie from each box that is still non-empty and puts them on the second plate. He continues until all the boxes are empty. Then Serge records the number of cookies on each plate. As with Alex, if the same number appears more than once, it is recorded only once. Prove that Alex's record contains the same number of numbers as Serge's record.

2. Let n be an integer such that $n > 2$. Find all non-negative numbers $x_1, x_2, \ldots, x_n$ such that $x_1 - x_2 = 1$ and

$$\begin{aligned} & \sqrt{x_1} + \sqrt{x_2 + x_3 + \cdots + x_n} \\ = & \sqrt{x_2} + \sqrt{x_3 + x_4 + \cdots + x_1} \\ = & \cdots \\ = & \sqrt{x_n} + \sqrt{x_1 + x_2 + \cdots + x_{n-1}}. \end{aligned}$$

3. A 30-gon $A_1A_2 \ldots A_{30}$ is inscribed in a circle of radius 2. Prove that one can choose a point B_k on the arc A_kA_{k+1} for $1 \le k \le 29$ and a point B_{30} on the arc $A_{30}A_1$, such that the numerical value of the area of the 60-gon $A_1B_1A_2B_2 \ldots A_{30}B_{30}$ is equal to the numerical value of the perimeter of the original 30-gon.

4. Five distinct positive integers form an arithmetic progression. Can their product be equal to a^{2008} for some positive integer a?

5. On the infinite board are several rectangles whose sides run along the grid lines. They have no interior points in common, and each consists of an odd number of the squares. Prove that these rectangles can be painted in four colours such that two rectangles painted in the same colour do not have any boundary points in common.

Note: The problems are worth 3, 3, 4, 4 and 4 points respectively.

Junior A-Level Paper

1. On a 100×100 board, 100 queens are placed so that no two attack each other. Prove that if the board is divided into four 50×50 subboards, then there is at least one queen in each subboard.

2. Each of four coins weighs an integral number of grams. Available for use is a balance which shows the difference of the weights between the objects in the left pan and those in the right pan. Is it possible to determine the weight of each coin by using this balance four times, if it may make a mistake of 1 gram either way in at most one weighing?

3. Serge has drawn triangle ABC and one of its medians AD. When informed of the ratio $\frac{AD}{AC}$, Elias is able to prove that $\angle CAB$ is obtuse and $\angle BAD$ is acute. Determine the ratio $\frac{AD}{AC}$ and justify your result.

4. Baron Münchausen asserts that he has a map of Oz showing five towns and ten roads, each road connecting exactly two cities. A road may intersect at most one other road once. The four roads connected to each town are alternately red and yellow. Can this assertion be true?

5. Let a_1, a_2, ..., a_n be positive numbers such that $a_1+a_2+\cdots+a_n \le \frac{1}{2}$. Prove that $(1+a_1)(1+a_2)\cdots(1+a_n) < 2$.

6. ABC is a scalene triangle. E and F are points outside triangle ABC such that $\angle ECA = \angle EAC = \angle FAB = \angle FBA = \theta$. The line through A perpendicular to EF intersects the perpendicular bisector of BC at D. Determine $\angle BDC$.

7. In the infinite sequence $\{a_n\}$, $a_0 = 0$. For $n \ge 1$, if the greatest odd divisor of n is congruent modulo 4 to 1, then $a_n = a_{n-1} + 1$, but if the greatest odd divisor of n is congruent modulo 4 to 3, then $a_n = a_{n-1} - 1$. The initial terms are 0, 1, 2, 1, 2, 3, 2, 1, 2, 3, 4, 3, 2, 3, 2 and 1.

   (a) Prove that the number 1 appears infinitely many times in this sequence.

   (b) Prove that every positive integer appears infinitely many times in this sequence.

Note: The problems are worth 4, 6, 6, 6, 8, 9 and 5+5 points respectively.

Senior A-Level Paper

1. A standard 8×8 board is modified by varying the distances between parallel grid lines, so that the squares are rectangles which are not necessarily squares, and do not necessarily have constant area. The ratio between the area of any white square and the area of any black square is at most 2. Determine the maximum possible ratio of the total area of the white squares to the total area of the black squares.

2. Space is dissected into non-overlapping unit cubes. Is it necessarily true that for each of these cubes, there exists another one sharing a common face with it?

3. Anna and Boris play a game with $n > 2$ piles each initially consisting of a single counter. The players take turns, Anna going first. In each move, a player chooses two piles containing numbers of counters relatively prime to each other, and merging the two piles into one. The player who cannot make a move loses the game. For each n, determine the player with a winning strategy.

4. In the quadrilateral $ABCD$, AD is parallel to BC but $AB \ne CD$. The diagonal AC meets the circumcircle of triangle BCD again at A' and the circumcircle of triangle BAD again at C'. The diagonal BD meets the circumcircle of triangle ABC again at D' and the circumcircle of triangle ADC again at B'. Prove that the quadrilateral $A'B'C'D'$ also has a pair of parallel sides.

5. In the infinite sequence $\{a_n\}$, $a_0 = 0$. For $n \geq 1$, if the greatest odd divisor of n is congruent modulo 4 to 1, then $a_n = a_{n-1} + 1$, but if the greatest odd divisor of n is congruent modulo 4 to 3, then $a_n = a_{n-1} - 1$. The initial terms are 0, 1, 2, 1, 2, 3, 2, 1, 2, 3, 4, 3, 2, 3, 2 and 1. Prove that every positive integer appears infinitely many times in this sequence.

6. $P(x)$ is a polynomial with real coefficients such that there exist infinitely many pairs (m, n) of integers satisfying $P(m) + P(n) = 0$. Prove that the graph $y = P(x)$ has a centre of symmetry.

7. A contest consists of 30 true or false questions. Victor knows nothing about the subject matter. He may write the contest several times, with exactly the same questions, and is told how many questions he has answered correctly each time. Can he be sure that he will answer all 30 questions correctly

   (a) on his 30th attempt;

   (b) on his 25th attempt?

Note: The problems are worth 4, 6, 6, 6, 8, 9 and 5+5 points respectively.

Spring 2009

Junior O-Level Paper

1. In a convex 2009-gon, all diagonals are drawn. A line intersects the 2009-gon but does not pass through any of its vertices. Prove that the line intersects an even number of diagonals.

2. Let $a \wedge b$ denote the number a^b. The order of operations in the expresson $7 \wedge 7 \wedge 7 \wedge 7 \wedge 7 \wedge 7 \wedge 7$ must be determined by inserting five pairs of brackets. Is it possible to put brackets in two distinct ways so that the expressions have the same value?

3. Vlad is going to print a digit on each face of several unit cubes, in such a way that a 6 does not turn into a 9. If it is possible to form every 30-digit number with these blocks, what is the minimum number of blocks?

4. When a positive integer is increased by 10%, the result is another positive integer whose digit-sum has decreased by 10%. Is this possible?

5. In the rhombus $ABCD$, $\angle A = 120°$. M is a point on the side BC and N is a point on the side CD such that $\angle MAN = 30°$. Prove that the circumcentre of triangle MAN lies on a diagonal of $ABCD$.

Note: The problems are worth 3, 4, 4, 4 and 5 points respectively.

Senior O-Level Paper

1. Let $a \wedge b$ denote the number a^b. The order of operations in the expresson $7 \wedge 7 \wedge 7 \wedge 7 \wedge 7 \wedge 7 \wedge 7$ must be determined by inserting five pairs of brackets. Is it possible to put brackets in two distinct ways so that the expressions have the same value?

2. On the plane are finitely many points, no three on a straight line. Some pairs of points are connected by segments. If any line which does not pass through any of these points intersects an even number of these segments, prove that each of these points is connected to an even number of the other points.

3. Let a and b be arbitrary positive integers. The sequence $\{x_k\}$ is defined by $x_1 = a$, $x_2 = b$ and for $k \geq 3$, x_k is the greatest odd divisor of $x_{k-1} + x_{k-2}$.

   (a) Prove that the sequence is eventually constant.

   (b) How can this constant value be determined from a and b?

4. In an arbitrary binary number, consider any digit 1 and any digit 0 which follows it, not necessarily immediately. They form an odd pair if the number of other digits in between is odd, and an even pair if this number is even. Prove that the number of even pairs is greater than or equal to the number of odd pairs.

5. X is an arbitrary point inside a tetrahedron. Through each of the vertices of the tetrahedron, draw a line parallel to the line joining X to the centroid of the opposite face. Prove that these four lines are concurrent.

Note: The problems are worth 3, 4, 2+2, 4 and 4 points respectively.

Junior A-Level Paper

1. Basil and Peter play the following game. Initially, there are two numbers on the blackboard, $\frac{1}{2009}$ and $\frac{1}{2008}$. At each move, Basil chooses an arbitrary positive number x, and Peter selects one of the two numbers on the blackboard and increases it by x. Basil wins if one of the numbers on the blackboard increases to 1. Does Basil have a winning strategy, regardless of what Peter does?

2. (a) Does there exist a non-convex polygon which can be dissected into two congruent parts by a line segment which cuts one side of the original polygon in half and another side in the ratio 1:2?
   (b) Can such a polygon be convex?

3. The central square of an 101×101 board is the bank. Every other square is marked S or T. A bank robber who enters a square marked S must go straight ahead in the same direction. A bank robber who enters a square marked T must make a right turn or a left turn. Is it possible to mark the squares in such a way that no bank robber can get to the bank?

4. In a sequence of distinct positive integers, each term except the first is either the arithmetic mean or the geometric mean of the term immediately before and the term immediately after. Is it necessarily true that from a certain point on, the means are either all arithmetic means or all geometric means?

5. A castle is surrounded by a circular wall with 9 towers. Some knights stand on guard on these towers. After every hour, each knight moves to a adjacent tower. A knight always moves in the same direction, whether clockwise or counter-clockwise. At some hour, there are at least two knights on each tower. At another hour, there are exactly 5 towers each of which has exactly one knight on it. Prove that at some other hour, there is a tower with no knights on it.

6. In triangle ABC, $AB = AC$ and $\angle CAB = 120°$. D and E are points on BC, with D closer to B, such that $\angle DAE = 60°$. F and G are points on AB and AC, respectively, such that $\angle FDB = \angle ADE$ and $\angle GEC = \angle AED$. Prove that the area of triangle ADE is equal to the sum of the areas of triangles FBD and GCE.

7. Let $\binom{n}{k}$ be the number of ways of choosing a subset of k objects from a set of n objects. Prove that if k and ℓ are positive integers less than n, then $\binom{n}{k}$ and $\binom{n}{\ell}$ have a common divisor greater than 1.

Note: The problems are worth 3, 2+3, 5, 5, 6, 7 and 9 points respectively.

Senior A-Level Paper

1. A rectangle is divided into several smaller rectangles. Is it possible that for each pair of rectangles so obtained, the line segment joining their centres intersects some other rectangle?

2. In a sequence of distinct positive integers, each term except the first is either the arithmetic mean or the geometric mean of the term immediately before and the term immediately after. Is it necessarily true that from a certain point on, the means are either all arithmetic means or all geometric means?

3. There is a counter in each square of a 10×10 board. We may choose a diagonal containing an even number of counters and remove any counter from it. What is the maximum number of counters which can be removed from the board by these operations?

4. Three planes dissect a parallelepiped into eight hexahedra such that all of their faces are quadrilaterals. One of the hexahedra has a circumsphere. Prove that each of these hexahedra has a circumsphere.

5. Let $\binom{n}{k}$ be the number of ways of choosing a subset of k objects from a set of n objects. Prove that if k and ℓ are positive integers less than n, then $\binom{n}{k}$ and $\binom{n}{\ell}$ have a common divisor greater than 1.

6. A positive integer n is given. Anna and Boris take turns marking points on a circle. Anna goes first and uses the red colour while Boris uses the blue colour. When n points of each colour have been marked, the game is over, and the circle has been divided into $2n$ arcs. The winner is the player who has the longest arc both endpoints of which are of this player's colour. Which player can always win, regardless of any action of the opponent?

7. At step 1, the computer has the number 6 in a memory square. In step n, it computes the greatest common divisor d of n and the number m currently in that square, and replaces m with $m+d$. Prove that if $d > 1$, then d must be prime.

Note: The problems are worth 4, 4, 6, 6, 8, 9 and 9 points respectively.

Solutions

Fall 2008

Junior O-Level Paper

1. Label the boxes from 1 to 10 in increasing order of the number of pencils inside. Then there are at least k pencils in the kth box for $1 \leq k \leq 10$. We will choose the pencils starting from box 1, and going on to box 10. From box 1, we choose any pencil. Suppose we have chosen a pencil from each of boxes 1, 2, ..., k for some k, $1 \leq k \leq 9$, with no two pencils the same colour. Now there are at least $k+1$ pencils in box $k+1$. Since these pencils are of different colours, there is one which does not match in colour any of the k pencils chosen so far, and we can add it to our collection. The desired conclusion follows from mathematical induction.

2. We can choose at most one number from each pair k and $50 + k$ for $1 \leq k \leq 50$. Since fifty numbers are chosen, we must choose exactly one number from each of these pairs. Twenty-five of the numbers are of the form k, and the other twenty-five are of the form $50 + k$. Hence their sum is $25 \times 50 + 1 + 2 + \cdots + 50 = 1250 + 1275 = 2525$.

3. From the centre O of the circle, drop perpendiculars to CE, EA and AC, and extend them to meet the circle at D, F and B respectively. The area of $ABCDEF$ is the sum of the areas of $OCDE$, $OEFA$ and $OABC$, which are equal respectively to $\frac{1}{2}CE \cdot OD$, $\frac{1}{2}EA \cdot OF$ and $\frac{1}{2}AC \cdot OB$. Since $OD = OF = OB = 2$, the area of $ABCDEF$ is numerically equal to $CE + EA + AC$.

4. Let the two positive integers be b and $3b$. Then their average is $2b$, and the product is $6b^3$. Let $b = 6^{669}$. Then $6b^3 = 6^{2008}$.

5. Consider two runners meeting each other at time t, either at an end of the track or somewhere in between and coming from opposite directions. Between them, they have covered the length of the track an even number of times. This will be true again at time $2t$, and they will therefore meet again, either at an end of the track, or somewhere in between and coming from opposite directions. Consider now two runners meeting each other at time t, somewhere between the ends of the track and going in the same direction. Create a hypothetical runner who meets both of them at time t, but coming from the opposite direction. So at time $2t$, both runners will meet the hypothetical runner again, and hence meet each other.

Senior O-Level Paper

1. The diagram below represents a typical inventory of cookies. Alex's numbers will be 2, 3, 5 and 7 while Serge's numbers will be 1, 4, 5 and 6. In the general situation, the number of numbers on Alex's record is equal to the number of vertical segments in the polygonal path while the number of numbers on Serge's record is equal to the number of horizontal segments in the polygonal path. Since the path starts with a horizontal segment and ends with a vertical segment, we have a one-to-one correspondence which yields the desired conclusion.

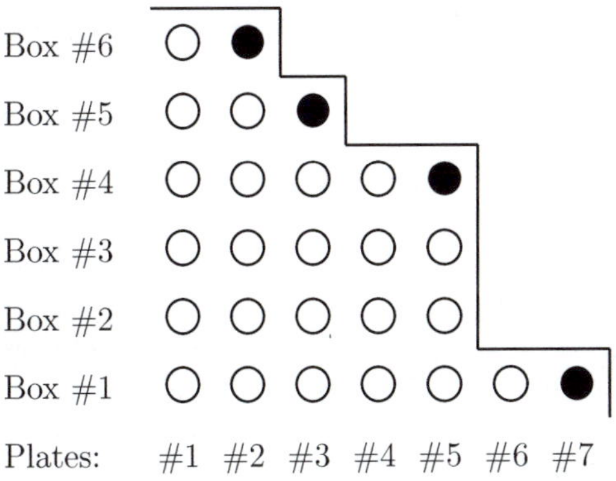

2. Let $S = x_1 + x_2 + \cdots + x_n$. Squaring $\sqrt{x_1} + \sqrt{S - x_1} = \sqrt{x_2} + \sqrt{S - x_2}$, we have

$$S + 2\sqrt{x_1x_2 + x_1(S - x_1 - x_2)} = S + 2\sqrt{x_1x_2 + x_2(S - x_1 - x_2)},$$

which yields either $x_1 = x_2$ or $x_3 + x_4 + \cdots x_n = 0$. Since $x_1 - x_2 = 1$, the latter is the case. Now $\sqrt{x_1} + \sqrt{x_2} = \sqrt{x_1 + x_2}$ which leads to $x_1x_2 = 0$. Hence $x_1 = 1$ and $x_2 = x_3 = \cdots = x_n = 0$.

3. Identify A_{31} with A_1. From the centre O of the circle, drop perpendiculars to A_kA_{k+1} and extend them to meet the circle at B_k for $1 \leq k \leq 30$. The area of the 60-gon is the sum of the areas of $OA_kB_kA_{k+1}$ for all k, $1 \leq k \leq 30$. These are equal respectively to $\frac{1}{2}A_kA_{k+1} \cdot OB_k$. Since $OB_k = 2$, the area of the 60-gon is numerically equal to $A_1A_2 + A_2A_3 + \cdots + A_{30}A_1$.

4. Let the five positive integers in arithmetic progression be b, $2b$, $3b, 4b$ and $5b$. Let $a = c^2$ for some positive integer c. Then $120b^5 = c^{4016}$. We can choose $b = 120^{803}$ so that we have $120b^5 = 120^{4016} = 14400^{2008}$.

5. Label the rows and columns $\ldots, -3, -2, -1, 0, 1, 2, 3, \ldots$. Each square whose row and column numbers are both odd is marked 1. Each square whose row and column numbers are both even is marked 2. Each square whose row number is odd and whose column number is even is marked 3. Each square whose row number is even and whose column number is odd is marked 4. Consider a rectangle consisting of an odd number of squares. Then both dimensions are odd, so that all four corner squares are marked with the same number n. Paint this rectangle in the nth colour. Two rectangles sharing a common horizontal border have corner squares on rows of opposite parity. The number used to mark the corner squares of one rectangle is different from that used to mark the corner squares of the other rectangle. Hence these two rectangles have different colours.

Junior A-Level Paper

1. Clearly, the western half of the board has 50 queens, as does the northern half of the board. Assume by symmetry that the northwestern quadrant is empty. Then 50 queens must be in the southwestern quadrant and another 50 in the northeastern quadrant. Hence the southeastern quadrant is also empty. However, the squares in the southwestern and the northeastern quadrants all lie on 99 diagonals going between the southwest and the northeast. By the Pigeonhole Principle, two of the queens will be on squares of the same diagonal, and hence attack each other. This is a contradiction. Hence no quadrants may be empty.

2. Label the coins A, B, C and D, with respective weights a, b, c and d. In the four weighings, weigh B, C and D, A and B against C and D, A and C against B and D, and A and D against B and C. Let the results be $b+c+d=w$, $a+b-c-d=x$, $a-b+c-d=y$ and $a-b-c+d=z$. For now, assume that no mistakes are possible. We have $w+x+y+z=3a$ so that $a=\frac{w+x+y+z}{3}$. Since $y+z=2a-2b$, we have $b=\frac{2a-(y+z)}{2}$. Similarly, $c=\frac{2a-(z+x)}{2}$ and $d=\frac{2a-(x+y)}{2}$. Suppose now a mistake of 1 gram is possible. If $w+x+y+z$ is a multiple of 3, no mistakes have been made. If it is one more or one less, we know the direction of the mistake. In any case, we can round the total to the nearest multiple of 3 and use it to determine a. Now each of $a-w$, x, y and z has the same parity as $a+b+c+d$, and hence as one another. Whichever has the opposite parity to the other three is where the mistake has been made.

3. Note that $\angle CAB$ is obtuse if and only if A lies inside the circle with diameter BC, and $\angle BAD$ is acute if and only if A lies outside the circle with diameter BD. If $\frac{AD}{AC}$ is constant, A lies on a circle of Apollonius of C and D, with a diameter on the line CD.

The only such circle which lies between the circles with diameters BC and BD must also have B as one end of the diameter on CD. Since $\frac{BD}{BC} = \frac{1}{2}$, the other end of this diameter must be the point E between C and D such that $\frac{ED}{EC} = \frac{1}{2}$. It follows that the ratio Serge gives Elias must be $\frac{1}{2}$. Since ABC is a triangle, A does not lie on BC. Then $\angle CAB$ is obtuse since A lies inside the circle with diameter BC, and $\angle BAD$ is acute since A lies outside the circle with diameter BD.

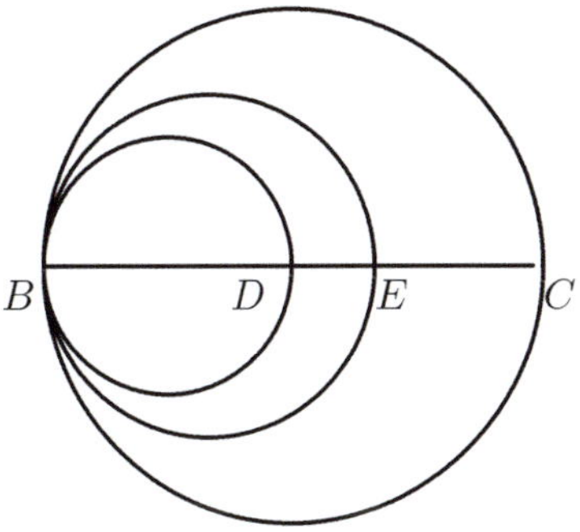

4. The Baron may have the following map. The double lines indicate the yellow brick road while the single lines indicate the red sidestreets.

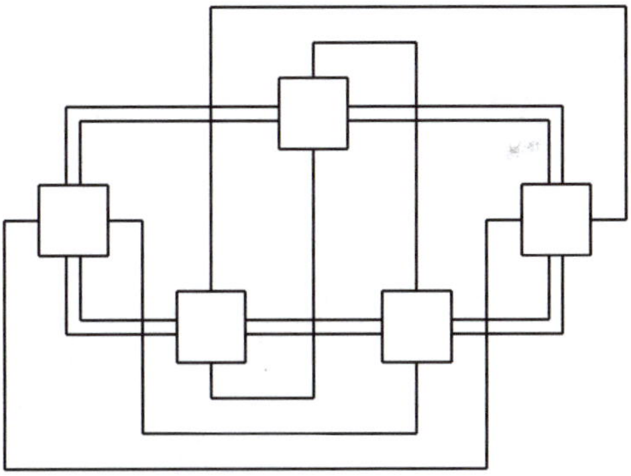

5. By the Arithmetic-Geometric Means Inequality, we have

$$
\begin{aligned}
& (1+a_1)(1+a_2)\cdots(1+a_n) \\
\leq & \left(\frac{(1+a_1)+(1+a_2)+\cdots+(1+a_n)}{n}\right)^n \\
\leq & \left(1+\frac{1}{2n}\right)^n.
\end{aligned}
$$

By the Binomial Theorem,

$$\begin{aligned}\left(1+\frac{1}{m}\right)^m &= \sum_{k=0}^{m}\binom{m}{k}\frac{1}{m^k}\\ &< \sum_{k=0}^{n}\frac{1}{m!}\\ &\leq \frac{1}{0!}+\frac{1}{1!}+\sum_{k=2}^{m}\left(\frac{1}{k-1}-\frac{1}{k}\right)\\ &= 3-\frac{1}{m}.\end{aligned}$$

Hence $(1+\frac{1}{2n})^n = \sqrt{(1+\frac{1}{2n})^{2n}} < \sqrt{3} < 2$.

6. Let D be the point outside triangle ABC such that $DB = DC$ and $\angle BDC = 2\theta$. Draw the circles through A with respective centres E and F. The powers of the point D with respect to these circles are $DE^2 - CE^2$ and $DF^2 - BF^2$. We claim that they are equal, so that D lies on the common chord of these circles, which passes through A and is perpendicular to the line EF of centres.

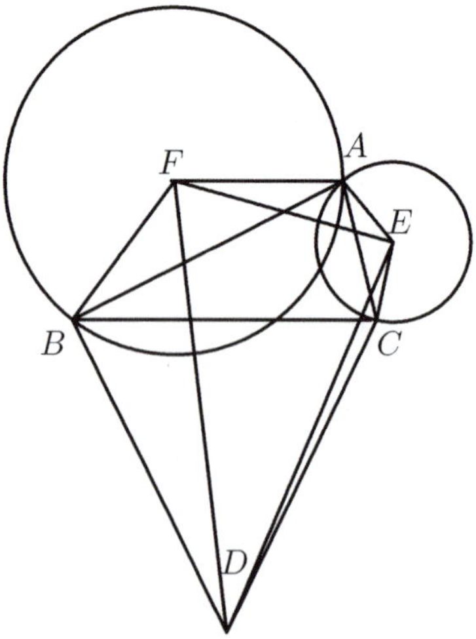

By the Cosine Law,

$$DE^2 - CE^2 = DC^2 - 2DC\cdot CE\cos DCE = DC^2 + 2DC\cdot CE\sin BCA$$

since $\angle DCB + \angle ACE = 90°$. Similarly,

$$DF^2 - BF^2 = DB^2 - 2DB\cdot BF\cos DBF = DB^2 + 2DB\cdot BF\sin ABC.$$

Since triangles ACE and ABF are similar, $\frac{CE}{BF} = \frac{AC}{AB}$. By the Sine Law, $\frac{AC}{AB} = \frac{\sin BCA}{\sin ABC}$. Hence $CE \sin BCA = BF \sin ABC$. Since $DC = DB$, we indeed have $DE^2 - CE^2 = DF^2 - BF^2$ and the claim is justified. Since the position of D is uniquely determined, $\angle BDC = 2\theta$.

7. (a) We will prove by induction on m that $a_n = 1$ whenever $n = 2^m - 1$. It is easy to verify that $a_1 = 1$, $a_2 = 2$, $a_3 = 1$, $a_4 = 2$, $a_5 = 3$, $a_6 = 2$ and $a_7 = 1$. Suppose that the result holds for some $m \geq 3$. For $1 \leq k \leq 2^m - 1$, $k \neq 2^{m-1}$, let $k = 2^s g$ where $s \leq m - 2$ and g is odd. Then $2^m + k = 2^s(2^{m-s} + g)$ and $2^{m-s} + g \equiv g \pmod 4$ since $m - s \geq 2$. On the other hand, the greatest odd divisor of 2^{m-1} is 1 while that of $2^m + 2^{m-1} = 2^{m-1}3$ is 3. Hence $a_{2^{m+1}-1} - a_{2^m}$ is 2 less than $a_{2^m-1} - a_0$. By the induction hypothesis, $a_{2^m-1} = 1$. Since the greastest odd divisor of 2^m is 1, $a_{2^m} = 2$. It follows that $a_{2^{m+1}-1} = 2 + 1 - 0 - 2 = 1$.

   (b) We claim that if a value h appears in the sequence, then the value $h + 2$ also appears in the sequence. Since 1 and 2 appear in the sequence, every positive integer appears in the sequence. Suppose $a_k = h$ for some k and let m be such that $k < 2^m$. By the argument in (a), $a_k - a_0 = a_{2^{m+1}+k} - a_{2^{m+1}}$, so that $a_{2^{m+1}+k} = h+2$. Thus the claim is justified and the sequence is unbounded. Suppose there is a value h which appears only a finite number of times. Every time the sequence hits a new high, it has to return to 1 at some point by the result in (a). After the last appearance of h, the sequence either cannot return to 1 or cannot get to a new high. This is a contradiction.

Senior A-Level Paper

1. We may have rows and columns of alternating widths $\frac{1}{3}$ and $\frac{2}{3}$. Let the white squares remain squares while the black squares become non-squares. Then the area of each white square is either $\frac{1}{9}$ or $\frac{4}{9}$ while the area of each black square is $\frac{2}{9}$. Thus the ratio between the area of any white square and the area of any black square is at most 2. The total area of the white squares is $16(\frac{1}{9} + \frac{4}{9}) = \frac{80}{9}$ while the total area of the black squares is $32(\frac{2}{9}) = \frac{64}{9}$. Here, the ratio of the total area of the white squares to the total area of the black squares is $\frac{80}{9} : \frac{64}{9} = 5 : 4$. To show that this is the maximum possible, divide the modified board into 16 subboards each consisting of four squares in a 2×2 configuration. Let the dimensions of one of the subboards be $a \times b$. Let the vertical grid line divide it into two rectangles of widths x and $a - x$, and we may assume that $x > \frac{a}{2}$. Let the horizontal grid line divide the subboard into two rectangles of heights y and $b - y$, and we may assume that $y > \frac{b}{2}$.

The condition that the ratio between the area of any white square and the area of any black square is at most 2 applies here also, and this is satisfied if and only if $x \leq \frac{2a}{3}$ and $y \leq \frac{2b}{3}$. Let the white squares be $x \times y$ and $(a-x) \times (b-y)$. Their total area is

$$T = 2xy + ab - bx - ay = x(2y - b) + a(b - y).$$

Since $2y > b$, T increases as x increases to its maximum value of $\frac{2a}{3}$. Similarly, $T = y(2x-a)+b(a-x)$ increases as y increases to its maximum value of $\frac{2b}{3}$. Hence $T \leq \frac{8ab}{9} + ab - 2\frac{2ab}{3} = \frac{5ab}{9}$. It follows the ratio of the total area of the white squares to the total area of the black squares is at most 5:4. Since this is true in each of the 16 subboards, it is true on the entire board.

2. It is not necessarily true. We tile space in the standard way with unit cubes. Choose one of them and call it C. There are two cubes sharing the front and back faces of C. They belong to two infinite columns of cubes parallel to the x-axis. There are two cubes sharing the top and bottom faces of C. They belong to two infinite columns of cubes parallel to the y-axis. There are two cubes sharing the left and right faces of C. They belong to two infinite columns of cubes parallel to the z-axis. These six columns do not intersect. Now we shift all the cubes in each column by half a unit. Then C does not share a complete face with any other cube.

3. Call a pile *even* if it has an even number of counters. If it has an odd number of counters, call it *small* if it has only one, and *big* otherwise. We first consider the case when n is odd. Anna is forced to form an even pile with two counters on the first move, and Boris merges it with a small pile into a big pile. The strategy of Boris is to always leave one big pile and an even number of small piles for Anna. In each move, Anna is forced to form an even pile, and it will be the only even pile at the time. If this merger involves the big pile, Boris merges the even pile with a small pile. If this merger does not involve the big pile, then the even pile has two counters, and 2 is relatively prime to any odd number, in particular, the number of counters in the big pile. Thus Boris can merge the big pile with the even pile. It follows that Boris always has a move, and hence wins the game. We now consider the case where n is even. Boris uses exactly the same strategy until the game is down to one big pile and three small piles (or four small piles at the beginning of the game for $n = 4$). After Anna creates once again an even pile, Boris merges the other two piles into a second even pile. Now Anna has no move, and Boris wins.

4. We have $\angle CAD' = \angle CBD' = \angle CA'D$ and $\angle ACB' = \angle ADB' = \angle AC'B$ from the cyclic quadrilaterals. Since AD is parallel to BC, all six angles are equal. Hence BC', $B'C, AD'$ and $A'D$ are all parallel. Let O be the point of intersection of AC and BD. From similar triangles, we have $\frac{OB'}{OC} = \frac{OD'}{OA}$, $\frac{OC}{OB} = \frac{OA}{OD}$ and $\frac{OB}{OC'} = \frac{OD}{OA'}$. Multiplication yields $\frac{OB'}{OC'} = \frac{OD'}{OA'}$, so that triangles $OB'C'$ and $OD'A'$ are similar. It follows that $A'D'$ is parallel to $B'C'$.

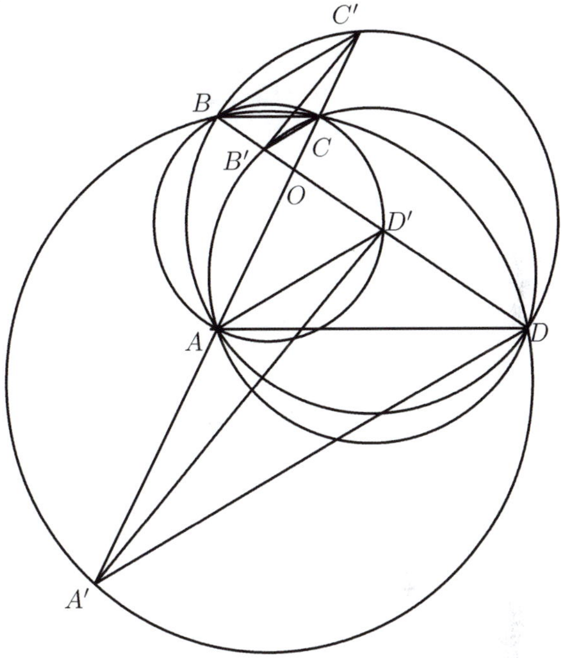

5. We first prove by induction on m that $a_n = 1$ whenever $n = 2^m - 1$. It is easy to verify that $a_1 = 1$, $a_2 = 2$, $a_3 = 1$, $a_4 = 2$, $a_5 = 3$, $a_6 = 2$ and $a_7 = 1$. Suppose that the result holds for some $m \geq 3$. Then for $1 \leq k \leq 2^m - 1, k \neq 2^{m-1}$, let $k = 2^s g$ where g is odd and $s \leq m - 2$. Now $2^m + k = 2^s(2^{m-s} + g)$ and $2^{m-s} + g \equiv g \pmod 4$ since $m - s \geq 2$. On the other hand, the greatest odd divisor of 2^{m-1} is 1 while that of $2^m + 2^{m-1} = 2^{m-1}3$ is 3. Hence $a_{2^{m+1}-1} - a_{2^m}$ is 2 less than $a_{2^m-1} - a_0$. By the induction hypothesis, $a_{2^m-1} = 1$. Since the greastest odd divisor of 2^m is 1, $a_{2^m} = 2$. It follows that $a_{2^{m+1}-1} = 2 + 1 - 0 - 2 = 1$. We claim that if a value h appears in the sequence, then the value $h+2$ also appears in the sequence. Since 1 and 2 appear in the sequence, every positive integer appears in the sequence. Let $a_k = h$ for some k and let m be such that $k < 2^m$. As before, $a_k - a_0 = a_{2^{m+1}+k} - a_{2^{m+1}}$, so that $a_{2^{m+1}+k} = h + 2$. Thus the claim is justified and the sequence is unbounded.

Suppose there is a value h which appears only a finite number of times. Every time the sequence hits a new high, it has to return to 1 at some point. After the last appearance of h, the sequence either cannot return to 1 or cannot get to a new high. This is a contradiction.
(Compare with Problem 7 in the Junior A-Level Paper.)

6. Let $P(x) = a_0x^k + a_1x^{k-1} + \cdots + a_{k-1}x + a_k$. We may assume that $a_0 > 0$. Suppose m and n are of the same sign, say positive. Note that $P(x) > 0$ whenever $x > \alpha$ for some positive number α. In order for $P(m) + P(n) = 0$ to hold, we must have $P(m) \leq 0$ or $P(n) \leq 0$. By symmetry, we may assume the former is the case. Then $m < \alpha$, so that we have finitely many values of m. For each of these values of m, we have finitely many values of n for which $P(n) = -P(m)$. This contradicts the condition that there exist infinitely many pairs (m, n) of integers satisfying $P(m)+P(n) = 0$. It follows that m and n must have opposite signs. If k is even, then $P(x) > 0$ whenever $|x| > \alpha$ for some positive number α, and we have the same contradiction as before. It follows that k must be odd. Then $P(m) + P(n) = (m + n)Q_1(m, n) + Q_2(m, n)$, where $Q_1(m, n) = m^{k-1} - m^{k-2}n + \cdots - mn^{k-2} + n^{k-1}$ and $Q_2(m, n)$ is of degree at most $k - 1$. All terms in $Q_1(m, n)$ are positive. Hence $|\frac{Q_2(m,n)}{Q_1(m,n)}| < \beta$ for some positive number β. When $P(m) + P(n) = 0$, we have $m + n = \frac{Q_2(m,n)}{Q_1(m,n)}$ so that $|m + n| < \beta$. Since there exist infinitely many pairs (m, n) of integers satisfying $P(m) + P(n) = 0$, some value of $m + n$ must occur infinitely often. Let this value be c. Define $R(x) = P(x) + P(c - x)$. Then $R(x)$ has infinitely many roots. Since it is a polynomial, it is identically zero. Hence $P(x) + P(c - x) = 0$ for all real numbers x, meaning that the graph $y = P(x)$ has a centre of symmetry at $(\frac{c}{2}, 0)$.

7. (a) In the first test, Victor answers True for all 30 questions. Suppose he gets 15 questions correct. In the second test, Victor changes the answers in test 1 to Questions 2, 3 and 4. The number of correct answers must be 12, 18, 14 or 16. In the first two cases, Victor knows the correct answers to Questions 2, 3 and 4, and has enough tests left to sort out the remaining questions. Hence we may assume by symmetry that the number of correct answers is 14. This means that the correct answers to two of Questions 2, 3 and 4 are True, and the other one False. Victor then changes the answers from test 1 to Questions $2k - 1$ and $2k$ in the kth test, $3 \leq k \leq 15$. If in the kth test, he gets either 13 or 17 questions correct, then he knows the correct answers to Questions $2k - 1$ and $2k$. Thus we may assume that he gets 15 correct answers in each test. Thus one correct answer is True and other False in each of these 13 pairs of questions. Moreover, Victor now knows that the correct answer to Question 1 is False.

So far, he has used 15 tests. In test 16, Victor changes the answers from test 1 to Questions 2, 3 and 5, and in test 17 to Questions 2, 4 and 5. The following chart shows he can deduce the correct answers to Questions 2 to 6. He has just enough tests left to sort out the remaining pairs.

Correct Answer to Question					Numbers of Correct Answers in Test	
2	3	4	5	6	16	17
T	T	F	T	F	12	14
T	F	T	T	F	14	12
F	T	T	T	F	14	14
T	T	F	F	T	14	16
T	F	T	F	T	16	14
F	T	T	F	T	16	16

Suppose that in the first test, Victor gets a correct answers where $a \neq 15$. He changes the answers from test 1 to Questions $2k-1$ and $2k$ in the kth test, $2 \leq k \leq 15$. In each of these 14 tests, he will get either a questions correct again, or $a \pm 2$ questions correct. In the latter case, he will know the correct answers to Questions $2k-1$ and $2k$. In the former case, he will know that one of these two answers is True and the other is False. Victor will also have similar knowledge about Questions 1 and 2 since he knows the total number of answers that should be True. Because $a \neq 15$, Victor must know the correct answers to one pair of questions. Hence he only needs at most 14 more questions to sort out the remaining pairs.

(b) In the first test, Victor answers True for all 30 questions. Suppose he gets a questions correct. In the second test, he changes the answers to the first two questions to False. He will get either a questions correct again, or $a \pm 2$ questions correct. In the latter case, he will know the correct answers to the first two questions. In the former case, Victor changes the first four answers to True, False, False and False in the third test, and to False, True, False and True in the fourth test. In the third test, the number of correct answers may be $a \pm 1$ or $a \pm 3$, while in the fourth test, the number of correct answers may be a or $a \pm 2$. From these data, he can deduce the correct answers to the first four questions, as shown in the chart below. Victor now handles each of the six subsequent groups of four questions in the same manner in 3 more questions, because Question 1 is relevant throughout.

Number of Correct Answers in Test		Correct Answer to Question			
3	4	1	2	3	4
$a-3$	a	F	T	T	T
	$a-2$	T	F	T	T
$a-1$	a	F	T	T	F
	$a+2$	F	T	F	T
	$a-2$	T	F	T	F
$a+1$	a	T	F	F	T
	$a+2$	F	T	F	F
$a+3$	a	T	F	F	F

After 22 tests, he knows all the correct answers except to the last two questions. He can use the 23rd test to determine the correct answer to the second to last question. Then he also knows the answer to the last question because he knows the total number of answers that should be True. Thus in the 24th test, Victor can answer all 30 questions correctly.

Spring 2009

Junior O-Level Paper

1. Let there be m vertices of the convex 2009-gon on one side of the line and n vertices on the other side. Since $m + n = 2009$, one of m and n is odd and the other is even. Hence mn is even. The line intersects mn segments which join two points on opposite sides. Two of them are sides of the 2009-gon, but the remaining $mn - 2$ are diagonals, and this number is even.

2. More generally, $(n\wedge(n\wedge n))\wedge n = (n^{n\wedge n})^n = (n^n)^{n\wedge n} = (n\wedge n)\wedge(n\wedge n)$. Adding three more terms to both sides the same way maintains the equal value.

3. We need at least 30 copies of each non-zero digits because there is a 30-digit number consisting only of that digit. We need at least 29 copies of zero because there is a 30-digit number whose last 29 digits are zeros. Since $30 \times 10 = 50 \times 6$, Vlad needs at least 50 cubes. His first five cubes may consist of the numbers (0,1,2,3,4,5), (6,7,8,9,0,1), (2,3,4,5,6,7), (8,9,0,1,2,3) and (4,5,6,7,8,9). Since each digit appears three times and no two copies of the same number appear on the same cube, Vlad can use this set to form any 3-digit number. If he makes nine more copies of this set, he can use the 50 cubes to form any 30-digit number.

4. Let the number consist of m 9s, preceded by 90 and followed by 0. Its digit-sum is $9(m+1)$. The new number consists of 1000, $m-2$ 9s and 89 in that order. Its digit-sum is $9m$. From $9(9m + 1) = 10(9m)$, we have $9m = 81$ so that $m = 9$.

5. Let O be the circumcentre of MAN. Then $\angle MON = 2\angle MAN = 60°$. Hence OMN is an equilateral triangle. Since $\angle MON + \angle MCN = 180°$, $CMON$ is cyclic. Now $\angle OCM = \angle ONM = 60° = \angle OMN = \angle OCN$. Hence O lies on the bisector of $\angle MCN$, which is the diagonal AC.

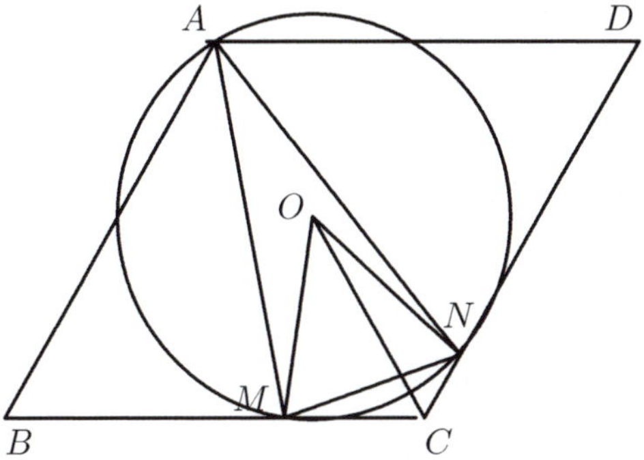

Senior O-Level Paper

1. (See Problem 2 of the Junior O-Level Paper.)

2. Suppose there is a point A connected to an odd number of other points. Then there must be a second such point B, because each connection involves two points. Take a line very close to AB, so that it does not pass through any given point. This line cuts a segments connected to A, b segments connected to B and c segments not connected to A or B, where $a+b+c$ is an even number. We now rotate this line slightly so that A remains on the same side but B moves to the opposite side of this line. Apart from possibly AB, this line cuts a segments connected to A, d segments connected to B and c segments not connected to A or B. If A is connected to B, then $d-b$ is even, and the total count $1+a+d+c=1+a+b+c+(d-b)$ is odd. If A is not connected to B, then $d-b$ is odd, and the total count $a+d+c=a+b+c+(d-b)$ is still odd. In either case, we have a contradiction.

3. (a) Note that x_k is odd for all $k \geq 3$. Hence $x_{k+2} = \frac{x_{k+1}+x_k}{2^t}$ for some positive integer t. If $x_{k+1} = x_k$, then $t = 1$ and the sequence in constant from this point on. Otherwise, $x_{k+2} < \max\{x_{k+1}, x_k\}$. Similarly, $x_{k+3} < \max\{x_{k+2}, x_{k+1}\}$. If $x_k < x_{k+1}$, then we have $x_{k+2} < x_{k+1}$ and $x_{k+3} < x_{k+1}$. If $x_k > x_{k+1}$, then $x_{k+2} < x_k$ and $x_{k+3} < x_k$. Thus $\max\{x_{k+3}, x_{k+2}\} < \max\{x_{k+1}, x_k\}$, so that the sequence is essentially decreasing, though not monotonically. Since the terms are positive integers, an infinite descent is impossible. Hence the sequence must eventually be constant.

   (b) Let g be the greatest common odd divisor of a and b. Then g is an odd divisor of x_k for $k \geq 3$. Hence it is the greatest common odd divisor of x_{k+1} and x_k. When the sequence becomes constant, g is the greatest common odd divisor of two equal terms both of which are odd. Hence this constant term must be equal to g.

4. We claim that if there is a pair of adjacent 0s, then they may be removed. This affects two kinds of pairs, those in which the 0 is one of the digits removed, and those in which the 0 comes after the digits removed. Among the first kind, whenever one of the removed 0 had formed part of an odd pair, the other removed 0 had formed part of an even pair with the same digit 1, and vice versa. Among the second kind, odd pairs remain odd and even pairs remain even with the removal of the two 0s. This justifies our claim. Similarly, pairs of adjacent 1s may be removed. When no pairs of adjacent digits are identical, the digits of the number are alternately 0 and 1.

If we are left with a 0-digit or 1-digit number, then the numbers of odd and even pairs are both 0. Suppose we are left with a number with at least 2 digits. Since leading 0s and trailing 1s do not count, we may assume that our number has the form $1010\cdots10$. Clearly all pairs are even.

5. Let O be the centroid of the tetrahedron $ABCD$ and G be the centroid of the face BCD. Then O lies on AG, with $AO = 3OG$. Let P be the point on the extension of XO such that $PO = 3OX$. Then triangles GOX and AOP are similar, so that XG is parallel to AP. By symmetry, the fixed point P lies on each of the four lines.

Junior A-Level Paper

1. Basil starts by choosing the number $\frac{2007}{2008}$. If Peter adds it to $\frac{1}{2008}$, Basil wins immediately. Hence he must add it to $\frac{1}{2009}$. The sum is $\frac{4034071}{4034072}$. Basil now chooses $\frac{1}{4034072}$, and Peter can only add it to the number which is not $\frac{4034071}{4034072}$. However, when this is done 4032062 times, the other number will also become $\frac{4034071}{4034072}$. Basil now wins by choosing $\frac{1}{4034072}$ once more.

2. (a) The diagram below shows such a polygon along with the dissecting line.

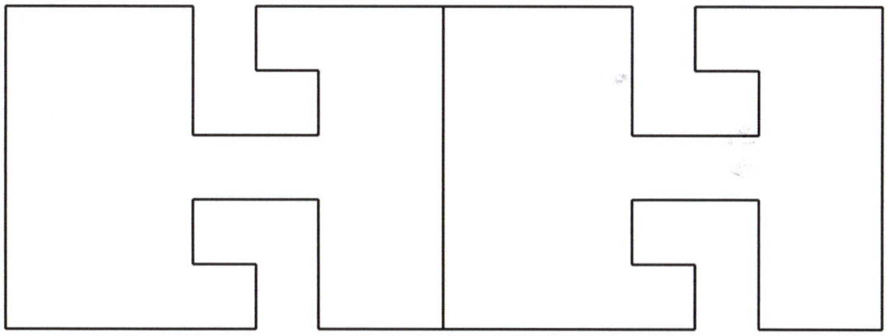

   (b) Divide the sides of a square in counter-clockwise order in the ratio 1:2. If we connect both pairs of points of division on opposite sides, the square is dissected into four congruent parts. If we connect only one pair, we have two congruent convex quadrilaterals. Disregard one of them, and the line connecting the other pair of points of division will dissect the remaining convex quadrilateral into two congruent parts.

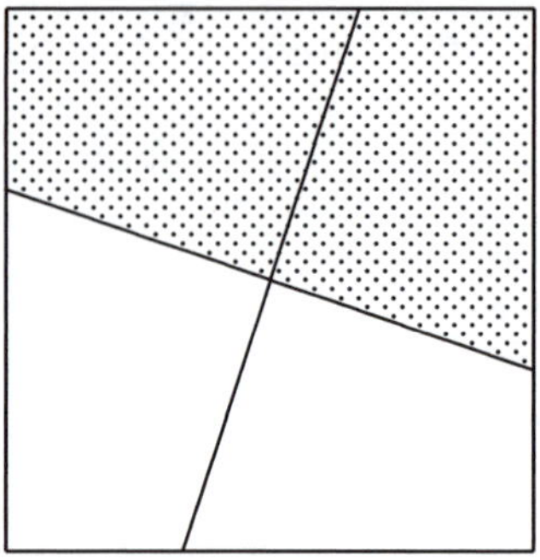

3. The robber plans ahead for his escape route. Divide the square into 51 layers of concentric squares. The bank is the sole square of the 0th layer. The eight surrounding squares constitute the 1st layer, and so on. With no movement restriction while in the bank, the robber can get to the 1st layer. Suppose the robber gets to the nth layer. If the square of entry is marked S, he can go straight into the $(n+1)$st layer. If the square is marked T, he turns either way and heads for a corner. If he passes on his way a square marked T, he can turn and get to the $(n+1)st$ layer. If this does not happen by the time he gets to a corner of the nth layer, he can go to the $(n+1)st$ layer regardless of whether the corner square is marked S or T. It follows that he can leave the 101×101 board. The bank robber then realizes that the escape route can be traversed in the opposite direction and leads him from outside to the bank!

4. In the sequence defined by $a_{2k-1} = k^2$ and $a_{2k} = k(k+1)$ for all $k \geq 1$, we have
$$\sqrt{a_{2k-1}a_{2k+1}} = \sqrt{k^2(k+1)^2} = k(k+1) = a_{2k}$$
while
$$\frac{1}{2}(a_{2k} + a_{2k+2}) = \frac{1}{2}(k(k+1) + (k+1)(k+2)) = (k+1)^2 = a_{2k+1}.$$
Hence the means are alternately geometric and arithmetic.

5. Let us call the knights who move clockwise black knights and the knights who move counterclockwise white knights. We may take the hour when there are at least two knights on each tower to be midnight. Suppose that there is a knight of the same colour in every tower at that time, say black. There will always be a black knight in every tower, because these nine just rotate around the wall. At midnight, there is another knight on each tower. Among these nine, there must be five of the same colour. They will always be on different towers, meaning that at any time, there will be five towers with at least two knights. This contradicts the given condition that there is a moment when exactly five towers have exactly one knight each.

It follows that at midnight, there is a tower A with no black knights, as well as a tower B with no white knights. Let the black knights be stationary while the white knights move counterclockwise every hour but skipping over exactly one tower. The distribution of knights remains the same. Tower A will never have black knights. There will always be a tower without white knights, shifting counterclockwise two towers per hour. Since nine is an odd number, it will shift onto tower A within nine hours. At that point, tower A will have no knights.

6. We use the symbol [ ] to denote area. Reflect the diagram about BC so that A', F' and G' are the respective images of A, F and G. Then D lies on AF' and E lies on AG', and both ABA' and ACA' are equilateral triangles. Now

$$\angle A'AG' = \angle DAE - \angle F'AG' = \angle BAA' - \angle F'AG' = \angle BAF'.$$

It follows that triangles BAF' and $A'AG'$ are congruent.

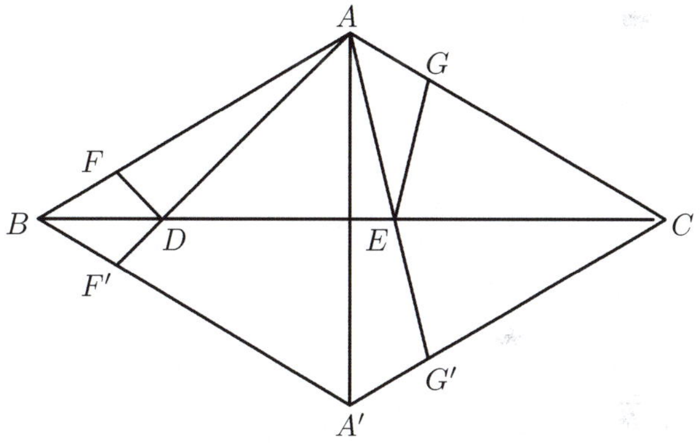

We have

$$\begin{aligned}[ADE] + [DEG'A'F'] &= [AF'A'] + [AA'G'] \\ &= [AF'A'] + [BAF'] \\ &= [BAA'] = \frac{1}{2}[BACA'].\end{aligned}$$

On the other hand, triangles BDF and CEG are congruent respectively to triangles BDF' and CEG'. Hence

$$\begin{aligned}[BDF] + [CEG] + [DEG'A'F'] &= [BDF'] + [CEG'] + [DEG'A'F'] \\ &= [BCA'] \\ &= \frac{1}{2}[BACA'].\end{aligned}$$

It follows that $[ADE] = [BDF] + [CEG]$.

7. Let $0 < k < \ell < n$. Then $\binom{\ell}{k} < \binom{n}{k}$. Suppose we have n players from which we wish to choose a team of size ℓ, and to choose k captains among the team players. The team can be chosen in $\binom{n}{\ell}$ ways and the captains can be chosen in $\binom{\ell}{k}$ ways. On the other hand, if we choose the captains first among all the players, the number of ways is $\binom{n}{k}$. From the remaining $n-k$ players, there are $\binom{n-k}{\ell-k}$ ways of choosing the $\ell-k$ non-captain players. Hence $\binom{n}{\ell}\binom{\ell}{k} = \binom{n}{k}\binom{n-k}{\ell-k}$. Now $\binom{n}{k}$ divides $\binom{n}{\ell}\binom{\ell}{k}$. If it is relatively prime to $\binom{n}{\ell}$, then it must divide $\binom{\ell}{k}$. This is a contradiction since $\binom{\ell}{k} < \binom{n}{k}$.

Senior A-Level Paper

1. Place the original rectangle in the first quadrant so that its southwest corner coincides with the origin of the coordinate plane. The small rectangle whose southwest corner also coincides with the origin is called the main rectangle. Let its centre be at (x, y). Consider the rectangles which have parts of the northern edge of the main rectangle as their southern edges. Let there be n of them and let their centres be at (x_i, y_i), with $x_1 < x_2 < \cdots < x_n$. Let k be such that $x_k < x < x_{k+1}$. The segment joining (x_k, y_k) and (x, y) must therefore pass through the $(k+1)$st rectangle, which means that the segment joining (x, y) and (x_{k+1}, y_{k+1}) cannot pass through the kth rectangle. It follows that it must intersect the eastern edge of the main rectangle rather than its northern edge, so that $n = k + 1$. Similarly, if there are m rectangles which have parts of the eastern edge of the main rectangle as their western edges, the line segment joining (x, y) to the centre of the last of these rectangles must intersect the northern edge of the main rectangle rather than its eastern edge. However, the nth northern neighbour and the mth eastern neighbour of the main rectangle share common interior points, which is a contradiction. Thus it is not possible that for each pair of rectangles so obtained, the line segment joining their centres intersects some other rectangle.

2. (See Problem 4 of the Junior A-Level Paper.)

3. The status of a diagonal is the parity of the number of counters currently on it. Initially, twenty of them are odd. Whenever a counter is removed, it affects the status of the two diagonals on which it lies. They cannot both be odd. If one is odd and the other is even, the total number of odd diagonals remains the same. If both are even, that number increases by two. Hence it cannot fall below its initial value of twenty. It follows that at least ten counters must remain on the board.

Label the squares (i,j) where $0 \leq i,j \leq 9$. We can remove all but five counters on the squares (i,j) where $i+j$ is odd, namely (1,0), (3,0), (5,0), (7,0) and (9,0). This is accomplished in ten stages by removing the counters on the squares, using even diagonals in alternating directions:
0. (0,1), (0,3), (0,5), (0,7) and (0,9), 1. (1,2), (1,4), (1,6), (1,8).
2. (2,1), (2,3), (2,5), (2,7) and (2,9), 3. (3,2), (3,4), (3,6), (3,8).
4. (4,1), (4,3), (4,5), (4,7) and (4,9), 5. (5,2), (5,4), (5,6), (5,8).
6. (6,1), (6,3), (6,5), (6,7) and (6,9), 7. (7,2), (7,4), (7,6), (7,8).
8. (8,1), (8,3), (8,5), (8,7) and (8,9), 9. (9,2), (9,4), (9,6), (9,8).
Similarly, we can remove all but five counters on the squares (i,j) where $i+j$ is even, namely (0,0), (2,0), (4,0), (6,0) and (8,0). Thus only the ten counters on the first row are left.

4. Our solution makes use of the following two auxillary results.

Lemma 1. Let $ABCD$ be a parallelogram. Let a line intersect AB at X and CD at Z. Let another line intersect BC at Y and DA at W. Let WY intersect XZ at P. If the quadrilateral $AXPW$ is cyclic, then so are the quadrilaterals $BYPX$, $CZPY$ and $DWPZ$.

Proof:

Let $\angle WAX = \theta$. Then $\angle YCZ = \theta$ also since $ABCD$ is a parallelogram. We also have $\angle WPZ = \angle XPY = \theta$ since $AXPW$ is a cyclic quadrilateral. It follows that so is $CZPY$. Now

$$\angle XBY = \angle ZDW = \angle WPX = \angle YPZ = 180^\circ - \theta.$$

Hence $BYPX$ and $DWPZ$ are cyclic quadrilaterals also.

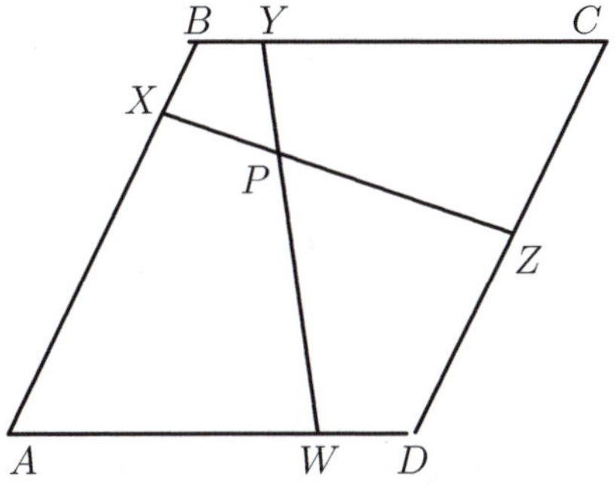

Lemma 2. Let $AXPW$ be a face of a hexahedron. Let $A'X'P'W'$ be the opposite face such that AA', XX', PP' and WW' are the lateral edges. If all six faces are cyclic quadrilaterals, then the hexahedron itself is cyclic.

Proof:

Since $AXPW$ is cyclic, there are many spheres which contains its circumcircle, and we can find one which passes through A'. Now A, X and

A' determines a unique circle. It must be the circumcircle of $AXA'X'$, and it must lie on this sphere. It follows that X' lies on this sphere. The same argument shows that P' and W' also lie on this sphere, so that the hexahedron is cyclic. We now return to the original problem. Each of the three planes has a cross-section with the parallelepiped in the form of a parallelogram. This cross-section does not meet two opposite faces of the parallelepiped, which are also parallelograms. All three parallelograms are divided into four quadrilaterals. In two of these parallolgrams, one of the four quadrilaterals is cyclic. By Lemma 1, the others are also cyclic. In the third parallelogram, which is a face of the parallelepiped, the dividing lines form the same angles with the sides of the parallelogram as those in the opposite face. Hence the four quadrilaterals here are cyclic too. It follows that all faces of the eight hexahedra are cyclic. By Lemma 2, the hexahedra are all cyclic.

5. (See Problem 7 of the Junior A-Level Paper.)

6. Boris has a winning strategy divided into four stages.

    1. After Anna has marked the initial red point, Boris defines as principal points the remaining vertices of a regular n-gon inscribed in the circle, with this red point as one of the vertices.
    2. Boris marks principal points whenever possible, until all have been marked. Since he has n moves and there are only $n-1$ unmarked principal points initially, this stage ends before his last move.
    3. Once all the principal points have been marked, Boris find pairs of adjacent red principal points. For each such pair, he marks a blue point between the two red points. Suppose Anna has marked k principal points red while Boris has marked the remaining $n-k$ principal points blue. There are at most $k-1$ pairs of adjacent red principal points. Hence this stage also ends before the last move of Boris.
    4. When Boris is ready to make his last move, all n principal points have been marked. There are $n-1$ other marked points. Hence there exist two adjacent principal points with no other points in between. At least one of them is blue since he has ensured there is a blue point between any two adjacent red principal points. His final marked point is on this arc, arbitrarily close to a principal point where the other principal point is blue.

The longest arc Boris can claim may be made arbitrarily close to $\frac{1}{n}$ of the circle, while all arcs Anna can claim are shorter than $\frac{1}{n}$ of the circle. Hence Boris can be assured of a win regardless of any action by Anna.

7. We denote by $a \triangle b$ the greatest common divisor of a and b. Let $f(1) = 6$. For $n \geq 2$, let $g(n) = n \triangle f(n-1)$ and $f(n) = f(n-1) + g(n)$. Then $g(2) = 2\triangle 6 = 2$, $f(2) = 6+2 = 8$, $g(3) = 3\triangle 8 = 1$ and $f(3) = 8+1 = 9$. Suppose $f(n) = 3n$ for some $n \geq 3$. Let p be the smallest prime divisor of $2n-1$. We claim that
(1) $f(n+k) = 3n + k$ for $0 \leq k \leq \frac{p-3}{2}$;
(2) $f(n + \frac{p-1}{2}) = 3n + 3(\frac{p-1}{2})$.
To establish (1), we use induction on k. The result holds for $k = 0$ as we are given that $f(n) = 3n$. Suppose it holds for some $k \geq 0$ such that $k + 1 \leq \frac{p-3}{2}$. Then

$$\begin{aligned} g(n+k+1) &= (n+k+1) \triangle f(n) \\ &= (n+k+1) \triangle (3n+k) \\ &= (n+k+1) \triangle (2n-1) \\ &= (2n+2k+2) \triangle (2n-1) \\ &= (2k+3) \triangle (2n-1) \\ &= 1, \end{aligned}$$

because $2k+3 = 2(k+1)+1 \leq p-3+1 < p$. It follows from the induction hypothesis that $f(n+k+1) = f(n+k) + 1 = 3n+k+1$. This completes the inductive proof of (1). Now

$$\begin{aligned} g\left(n + \frac{p-1}{2}\right) &= \left(n + \frac{p-1}{2}\right) \triangle f\left(n + \frac{p-3}{2}\right) \\ &= \left(n + \frac{p-1}{2}\right) \triangle \left(3n + \frac{p-3}{2}\right) \\ &= \left(n + \frac{p-1}{2}\right) \triangle (2n-1) \\ &= (2n+p-1) \triangle (2n-1) \\ &= p \triangle (2n-1) \\ &= p. \end{aligned}$$

Hence $f(n+\frac{p-1}{2}) = g(n+\frac{p-1}{2}) + f(n+\frac{p-3}{2}) = p+3n+\frac{p-3}{2} = 3n+3(\frac{p-1}{2})$. Thus (2) is justified. We now return to the original problem. Since $f(3) = 3(3)$, we have $f(n) = 3n$ infinitely often. Between this and the next occurence, all values of g are either 1 or the smallest prime divisor of $2n-1$, which is the desired result.

Tournament 31

Fall 2009

Junior O-Level Paper

1. Is it possible to dissect a square into nine squares, with five of them of one size, three of them of another size and one of them of a third size?

2. There are forty weights: 1, 2, ..., 40 grams. Ten weights with even masses were put on the left pan of a balance. Ten weights with odd masses were put on the right pan of the balance. The left and the right pans are balanced. Prove that one pan contains two weights whose masses differ by exactly 20 grams.

3. A cardboard circular disk of radius 5 centimetres is placed on the table. While it is possible, Peter puts cardboard squares with side 5 centimetres outside the disk so that:
   (1) one vertex of each square lies on the boundary of the disk;
   (2) the squares do not overlap;
   (3) each square has a common vertex with the preceding one.
   Find how many squares Peter can put on the table, and prove that the first and the last of them must also have a common vertex.

4. On a lottery ticket, a number consisting of seven different digits is to be written. On the draw date, an official number with seven different digits is revealed. A ticket wins a prize if it matches the official number in at least one digit. Is it possible to guarantee winning a prize by buying at most six tickets?

5. A new website registered 2000 people. Each of them invited 1000 other registered people to be their friends. Two people are considered to be friends if and only if they have invited each other. What is the minimum number of pairs of friends on this website?

Note: The problems are worth 3, 4, 4, 5 and 5 points respectively.

Senior O-Level Paper

1. We only know that the password of a safe consists of 7 different digits. The safe will open if we enter 7 different digits, and one of them matches the corresponding digit of the password. Can we open this safe in less than 7 attempts?

2. A, B, C, D, E and F are points in space such that AB is parallel to DE, BC is parallel to EF, CD is parallel to FA, but $AB \neq DE$. Prove that all six points lie in the same plane.

3. Are there positive integers a, b, c and d such that

$$a^3 + b^3 + c^3 + d^3 = 100^{100}?$$

4. A point is chosen on each side of a regular 2009-gon. Let S be the area of the 2009-gon with vertices at these points. For each of the chosen points, reflect it across the midpoint of its side. Prove that the 2009-gon with vertices at the images of these reflections also has area S.

5. A country has two capitals and several towns. Some of them are connected by roads. Some of the roads are toll roads where a fee is charged for driving along them. It is known that any route from the south capital to the north capital contains at least ten toll roads. Prove that all toll roads can be distributed among ten companies so that anybody driving from the south capital to the north capital must pay each of these companies.

Junior A-Level Paper

1. Ten jars contain varying amounts of milk. Each is large enough to hold all the milk. At any time, we can tell the precise amount of milk in each jar. In a move, we may pour out an exact amount of milk from one jar into each of the other 9 jars, the same amount in each case. Prove that we can have the same amount of milk in each jar after at most ten moves.

2. Mike has 1000 unit cubes. Each has 2 opposite red faces, 2 opposite blue faces and 2 opposite white faces. Mike assembles them into a $10 \times 10 \times 10$ cube. Whenever two unit cubes meet face to face, these two faces have the same colour. Prove that an entire face of the $10 \times 10 \times 10$ cube has the same colour.

3. Find all positive integers a and b such that $(a + b^2)(b + a^2) = 2^m$ for some integer m.

4. Let $ABCD$ be a rhombus. P is a point on side BC and Q is a point on side CD such that $BP = CQ$. Prove that the centroid of triangle APQ lies on the segment BD.

5. We have n objects with weights $1, 2, \ldots, n$ grams. We wish to choose two or more of these objects so that the total weight of the chosen objects is equal to the average weight of the remaining objects. Prove that
(a) if $n + 1$ is a perfect square, then the task is possible;
(b) if the task is possible, then $n + 1$ is a perfect square.

6. On an infinite board are placed 2009 $n \times n$ cardboard pieces such that each of them covers exactly n^2 squares of the infinite board. The cardboard pieces may overlap. Prove that the number of squares of the infinite board which are covered by an odd number of cardboard pieces is at least n^2.

7. Olga and Max visit a certain archipelago with 2009 islands. Some pairs of islands are connected by boats which run both ways. Olga chooses the first island on which they land. Then Max chooses the next island which they can visit. Thereafter, the two take turns choosing an accessible island which they have not yet visited. When they arrive at an island which is connected only to islands they have already visited, whoever's turn to choose next will be the loser. Prove that Olga can always win, regardless of how Max plays and regardless of the way the islands are connected.

Note: The problems are worth 4, 6, 6, 6, 2+7, 10 and 14 points respectively.

Senior A-Level Paper

1. After a gambling session, each of one hundred pirates calculates the amount he has won or lost. Money can only change hands in the following way. Either one pirate pays an equal amount to every other pirate, or one pirate receives the same amount from every other pirate. Each pirate has enough money to make any payment. Is it always possible, after several such steps, for all the winners to receive exactly what they have won and for all losers to pay exactly what they have lost?

2. A non-square rectangle is cut into N rectangles of various shapes and sizes. Prove that one can always cut each of these rectangles into two rectangles so that one can construct a square and rectangle, each figure consisting of one piece from each of the N rectangles.

3. Every edge of a tetrahedron is tangent to a given sphere. Prove that the three line segments joining the points of tangency of the three pairs of opposite edges of the tetrahedron are concurrent.

4. Denote by $[n]!$ the product of $1, 11, \ldots, 11\ldots1$, where the last factor has n ones. Prove that $[n+m]!$ is divisible by $[n]![m]!$.

5. Let XYZ be a triangle. The convex hexagon $ABCDEF$ is such that AB, CD and EF are parallel and equal to XY, YZ and ZX, respectively. XYZ and $ABCDEF$ are in the same orientation. Prove that the area of the triangle with vertices at the midpoints of BC, DE and FA is not less than the area of triangle XYZ.

6. Olga and Max visit a certain archipelago with 2009 islands. Some pairs of islands are connected by boats which run both ways. Olga chooses the first island on which they land. Then Max chooses the next island which they can visit. Thereafter, the two take turns choosing an accessible island which they have not yet visited. When they arrive at an island which is connected only to islands they have already visited, whoever's turn to choose next will be the loser. Prove that Olga can always win, regardless of how Max plays and regardless of the way the islands are connected.

7. At the entrance to a cave is a rotating round table. On top of the table are n identical barrels, evenly spaced along its circumference. Inside each barrel is a herring either with its head up or its head down. In a move, Ali Baba chooses from 1 to n of the barrels and turns them upside down. Then the table spins around. When it stops, it is impossible to tell which barrels have been turned over. The cave will open if the heads of the herrings in all n barrels are all up or are all down. Determine all values of n for which Ali Baba can open the cave in a finite number of moves.

Note: The problems are worth 4, 6, 7, 9, 9, 12 and 14 points respectively.

Spring 2010

Junior O-Level Paper

1. Each of six baskets contains some pears, plums and apples. The number of plums in each basket is equal to the total number of apples in the other five baskets, and the number of apples in each basket is equal to the total number of pears in the other five baskets. Prove that the total number of fruit in the six baskets is a multiple of 31.

2. Karlsson and Lillebror are dividing a square cake. Karlsson chooses a point P of the cake which is not on the boundary. Lillebror makes a straight cut from P to the boundary of the cake, in any direction he chooses. Then Karlsson makes a straight cut from P to the boundary, at a right angle to the first cut. Lillebror will get the smaller of the two pieces. Can Karlsson prevent Lillebror from getting at least one quarter of the cake?

3. An angle is given in the plane, and a compass is the only available tool.

   (a) Use the compass the minimum number of times to determine if the angle is acute or obtuse.

   (b) Use the compass any number of times to determine if the angle is exactly 31°.

4. At a party, each person knows at least three other people. Prove that an even number of them, at least four, can sit at a round table such that each knows both neighbours.

5. On the blackboard are the squares of the first 101 positive integers. In each move, we can replace two of them by the absolute value of their difference. After 100 moves, only one number remains. What is the minimum value of this number?

Note: The problems are worth 3, 3, 2+2, 5 and 5 points respectively.

Senior O-Level Paper

1. Bananas, lemons and pineapples are being delivered by 2010 ships. The number of bananas in each ship is equal to the total number of lemons in the other 2009 ships, and the number of lemons in each ship is equal to the total number of pineapples in the other 2009 ships. Prove that the total number of fruit being delivered is a multiple of 31.

2. Each line in the coordinate plane has the same number of common points with the parabola $y = x^2$ and with the graph $y = f(x)$. Prove that $f(x) = x^2$.

3. Is it possible to cover the surface of a regular octahedron by several regular hexagons, without gaps or overlaps?

4. Baron Münchausen claims that a non-constant polynomial $P(x)$ with non-negative integers as coefficients is uniquely determined by the values of $P(2)$ and $P(P(2))$. Surely the Baron is wrong, isn't he?

5. A segment is given on the plane. In each move, it may be rotated about either of its endpoints in a $45°$ angle clockwise or counterclockwise. Is it possible that after a finite number of moves, the segment returns to its original position except that its endpoints are interchanged?

Note: The problems are worth 3, 4, 5, 5 and 6 points respectively.

Junior A-Level Paper

1. Alex cuts a piece of cheese in the ratio $1 : \alpha$, where α is a positive real number not equal to 1. He then chooses any piece and cuts it in the same way. Is it possible for him to choose α so that after a finite number of cuts, he can obtain two piles of pieces each containing half the original amount of cheese?

2. M is the midpoint of the side CA of triangle ABC. P is some point on the side BC. AP and BM intersect at the point O. If $BO = BP$, determine $\frac{OM}{PC}$.

3. Along a circle are placed 999 numbers, each 1 or -1, and there is at least one of each. The product of each block of 10 adjacent numbers along the circle is computed. Let S denote the sum of these 999 products.
   (a) What is the minimum value of S?
   (b) What is the maximum value of S?

4. Is it possible that the sum of the digits of a positive integer n is 100 while the sum of the digits of the number n^3 is 100^3?

5. On a circular road are N horsemen, riding in the same direction, each at a different constant speed. There is only one point along the road at which a horseman is allowed to pass another horseman. Can they continue to ride for an arbitrarily long period if
   (a) $N = 3$;
   (b) $N = 10$?

6. A broken line consists of 31 segments joined end to end. It does not intersect itself, and has distinct end points. Adjacent segments are not on the same straight line. What is the smallest number of straight lines which can contain all segments of such a broken line?

7. A number of ants are on a 10×10 board, each in a different square. Every minute, each ant crawls to the adjacent square either to the east, to the south, to the west or to the north. It continues to crawl in the same direction as long as this is possible, but reverses direction if it has reached the edge of the board. In one hour, no two ants ever occupy the same square. What is the maximum number of ants on the board?

Note: The problems are worth 3, 4, 3+3, 6, 3+5, 8 and 11 points respectively.

Senior A-Level Paper

1. Is it possible to divide all the lines in the plane into pairs of perpendicular lines so that every line belongs to exactly one pair?

2. Alex cuts a piece of cheese in the ratio $1 : \alpha$, where

   (a) α is a positive irrational number;

   (b) α is a positive rational number not equal to 1.

   He then chooses any piece and cuts it in the same way. Is it possible for him to choose α so that after a finite number of cuts, he can obtain two piles of pieces each containing half the original amount of cheese?

3. In the first step, we apply one of the functions sin, cos, tan, cot, arcsin, arccos, arctan and arccot to the number 1. In each subsequent step, we apply one of these functions to the number obtained in the preceding step. Can we obtain the number 2010 after a finite number of steps?

4. At a convention, each of the 5000 participants watches at least one movie. Several participants can form a discussion group if either they have all watched the same movie, or each has watched a movie nobody else in the group has. A single participant may also form a group, and every participant must belong to exactly one group. Prove that the number of groups can always be exactly 100.

5. On a circular road are 33 horsemen, riding in the same direction, each at a different constant speed. There is only one point along the road at which a horseman is allowed to pass another horseman. Can they continue to ride for an arbitrarily long period?

6. A circle with centre I is tangent to all four sides of a convex quadrilateral $ABCD$. M and N are the midpoints of AB and CD respectively. If $\frac{IM}{AB} = \frac{IN}{CD}$, prove that $ABCD$ has a pair of parallel sides.

7. A multi-digit number is written on the blackboard. Susan puts in a number of plus signs between some pairs of adjacent digits. The addition is performed and the process is repeated with the sum. Prove that regardless of what number was initially on the blackboard, Susan can always obtain a single-digit number in at most ten steps.

Note: The problems are worth 3, 2+2, 6, 6, 7, 8 and 9 points respectively.

Solutions

Fall 2009

Junior O-Level Paper

1. The diagram below shows that a 6×6 square can be cut into one 2×2 square, three 3×3 squares and five 1×1 squares.

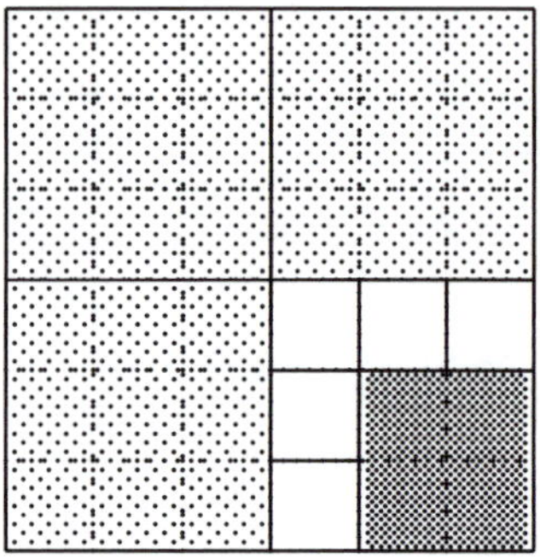

2. Suppose to the contrary that no two weights in the same pan differ in mass by exactly 20 grams. Then in the right pan, we must have put in exactly one weight from each of the following ten pairs: (1,21), (3,23), ..., (19,39). The total mass in the right pan is

$$1 + 3 + \cdots + 19 + 20k = 100 + 20k,$$

where k is the number of times we chose the heavier weight from a pair. This is a multiple of 4. Similarly, the total mass in the left pan is $2 + 4 + \cdots + 20 + 20h = 110 + 20h$, where h is the number of times we chose the heavier weight from a pair. This is not a multiple of 4. We have a contradiction as the two pans cannot possibly balance.

3. Let O be the centre of the circle, A, B and C be the points of contact with the circle of three squares in order, and P and Q be the common vertices of these squares. Call OA, OB and OC the root canals of the respective squares. Then $OAPB$ and $OBQC$ are rhombi. Moreover, $\angle PBQ = 90°$. Hence $\angle AOC = 90°$. This means that every two alternate root canals are perpendicular. It follows that there must be 8 root canals, and the last square must have a common vertex with the first.

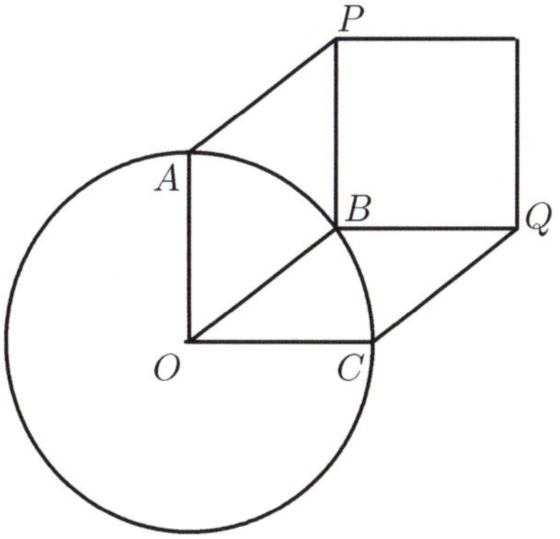

4. One can buy six tickets with the numbers 0123456, 0234561, 0345612, 0456123, 0561234 and 0612345. Since the official number uses seven different digits, it must use at least three of the digits 1, 2, 3, 4, 5 and 6. At most one of these these can be in the first place. Each of the other two must match a digit on some ticket.

5. Pretend that the 2000 people are seated at a round table, evenly spaced. Each invites the next 1000 people in clockwise order. Then only two people who are diametrically opposite to each other become friends. This shows that the number of pairs of friends may be as low as 1000. Construct a directed graph with 2000 vertices representing the people. Each vertex is incident to 1000 outgoing arcs representing the invitations. The total number of arcs is 2000×1000. The total number of pairs of vertices is $2000 \times 1999 \div 2 = 1999 \times 1000$. Even if every pair of vertices is connected by an arc, we still have $2000 \times 1000 - 1999 = 1000$ extra arcs. These can only appear as arcs going in the opposite direction to existing arcs. It follows that there must be at least 1000 reciprocal invitations, and therefore at least 1000 pairs of friends.

Senior O-Level Paper

1. In six attempts, we enter 0123456, 0234561, 0345612, 0456123, 0561234 and 0612345. Since the password uses 7 different digits, it must use at least 3 of the digits 1, 2, 3, 4, 5 and 6. At most one of these 3 can be in the first place. The other 2 must match one of our attempts.

2. Suppose to the contrary the six points do not all lie in the same plane. Now B, C and D determine a plane, which we may assume to be horizontal. Suppose that E does not lie in this plane. Since AB is parallel to DE , A does not lie in this plane either. Since $AB \neq DE$, A and E do not lie in the same horizontal plane. Since BC is parallel to EF, F lies on the same horizontal plane as E. Since CD is parallel to FA, A lies on the same horizontal plane as F. This is a contradiction.

It follows that E also lies on the horizontal plane determined by B, C and D. Since BC is parallel to EF, F also lies in this plane, and since FA is parallel to CD, A does also.

3. For $a = 10^{66}$, $b = 2a$, $c = 3a$ and $d = 4a$, we have

$$a^3 + b^3 + c^3 + d^3 = (1^3 + 2^3 + 3^3 + 4^3)(100^{33})^3 = 100^{100}.$$

4. Let 1 be the side length of the regular 2009-gon $A_1A_2 \dots A_{2009}$. For indexing purposes, we treat 2010 as 1. For $1 \leq k \leq 2009$, let B_k be the chosen point on A_kA_{k+1} with $A_{2010} = A_1$, C_k be the image of reflection of B_k, and $d_k = A_kB_k$. Let

$$\begin{aligned} S &= d_1 + d_2 + \cdots + d_{2009}, \\ T &= d_1d_2 + d_2d_3 + \cdots + d_{2009}d_1. \end{aligned}$$

Now $B_1B_2 \dots B_{2009}$ may be obtained from the regular 2009-gon by removing 2009 triangles, each with an angle equal to the interior θ angle of the regular 2009-gon, flanked by two sides of lengths $1 - d_k$ and d_{k+1}. Hence its area is equal to that of the regular 2009-gon minus $\frac{1}{2}\sin\theta$ times $(1 - d_1)d_2 + (1 - d_2)d_3 + \cdots + (1 - d_{2009})d_1 = S - T$. Similarly, the area of $C_1C_2 \dots C_{2009}$ is equal to that of the regular 2009-gon minus $\frac{1}{2}\sin\theta$ times $d_1(1-d_2)+d_2(1-d_3)+\cdots+d_{2009}(1-d_1) = S-T$. Hence these two 2009-gons have the same area.

5. Label each town with the minimum number of toll roads one has to drive from the south capital to that town. In particular, the south capital itself is labeled with 0. Then any two towns connected by a road either have the same label, or have labels that differ by 1. If a toll road connects two towns with the same label, distribute it to any company; if it connects towns with labels $n - 1$ and n, distribute it to company n if $n \leq 10$. Otherwise, distribute it to any company. Anybody driving from the south capital to the north capital must cross towns with labels 1, 2, ..., and 10, and they will pay all 10 companies.

Junior A-Level Paper

1. Pour from each jar exactly one tenth of what it initially contains into each of the other nine jars. At the end of these ten operations, each jar will contain one tenth of what is inside each jar initially. Since the total amount of milk remains unchanged, each jar will contain one tenth of the total amount of milk.

2. Assign spatial coordinates to the unit cubes, each dimension ranging from 1 to 10. If all cubes are in the same colour orientation, there is nothing to prove. Hence we may assume that (i, j, k) and $(i+1, j, k)$ do not. Since they share a left-right face, let the common colour be red. We may assign blue to the front-back faces of (i, j, k). Then its top-bottom faces are white, the front-back faces of $(i + 1, j, k)$ are white and the top-bottom faces of $(i + 1, j, k)$ are blue. Now $(i, j + 1, k)$ share a white face with (i, j, k) while $(i+1, j+1, k)$ share a blue face with $(i+1, j, k)$. Since $(i, j + 1, k)$ and $(i + 1, j + 1, k)$ share a left-right face, the only available colour is red. It follows that the $1 \times 2 \times 10$ block with $(i, 1, k)$ and $(i + 1, 1, k)$ at one end and $(i, 10, k)$ and $(i + 1, 10, k)$ at the other has 1×10 faces left and right which are all red. Similarly, if we carry out the expansion vertically, we obtain a $2 \times 10 \times 10$ black with 10×10 faces left and right which are all red. Finally, if we carry out the expansion sideways, we will have the left and right faces of the large cube all red.

3. Suppose $a = b$. Then $a + a^2 = a(a + 1)$ is a power of 2, so that each of a and $a + 1$ is a power of 2. This is only possible if $a = 1$. Suppose $a \neq b$. By symmetry, we may assume that $a > b$, so that $a^2 + b > a + b^2$. Since their product is a power of 2, each is a power of 2. Let $a^2 + b = 2^r$ and $a + b^2 = 2^s$ with $r > s$. Then $2^s(2^{r-s} - 1) = 2^r - 2^s = (a - b)(a + b - 1)$. Now $a - b$ and $a + b - 1$ have opposite parity. Hence one of them is equal to 2^s and the other to $2^{r-s} - 1$. If $a - b = 2^s = a + b^2$, then $-b = b^2$. If $a + b - 1 = 2^s = a + b^2$, then $b - 1 = b^2$. Both are contradictions. Hence there is a unique solution $a = b = 1$.

4. Extend AB to P' so that $BP' = BP = CQ$. Then $BP'CQ$ is a parallelogram so that $P'Q$ and BC bisect each other at a point K. Let AK intersect BD at G' and let QG' intersect AB at R'.

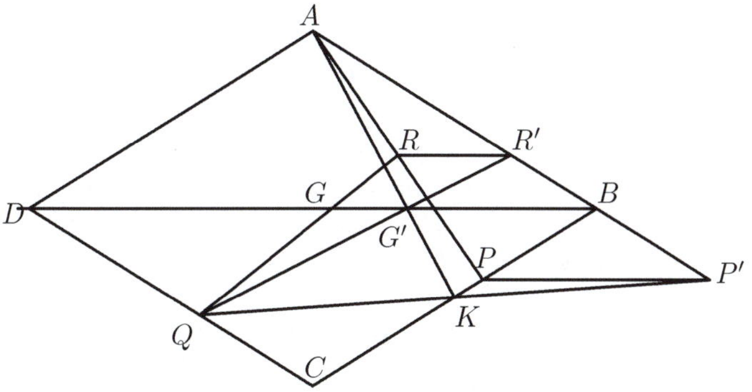

Since K is the midpoint of BC, its distance from BD is half the distance of C from BD, which is equal to the distance of A from BD. It follows that $AG' = 2KG'$. Since K is the midpoint of $P'Q$, G' is the centroid of triangle $AP'Q$. Hence $QG' = 2R'G'$ and R' is the midpoint of AP'.

Let R be the midpoint of AP and let QR intersect BD at G. Then RR' is parallel to PP', which is in turn parallel to BD. Hence $QG = 2RG$ so that G is the centroid of triangle APQ.

5. (a) Suppose $n+1 = k^2$ for some positive integer k. We take the lightest k objects with total weight $1 + 2 + \cdots + k = \frac{k(k+1)}{2}$ grams. The average weight of the remaining objects is $\frac{(k+1)+(k^2-1)}{2} = \frac{k(k+1)}{2}$ grams also.

   (b) The total weight of the n objects is $1+2+\cdots+n = \frac{n(n+1)}{2}$ grams. Let T grams be the total weight of the k chosen objects. This is also the average weight of the remaining $n-k$ objects. Hence $\frac{n(n+1)}{2} = T(n-k+1)$. Now

   $$2T(n-k+1) = n(n+1) > n^2 + n - k^2 + k = (n+k)(n-k+1),$$

   so that $2T > n+k$. If we choose the lightest k objects, then T attains its maximum value $\frac{(k+1)+n}{2}$, so that $2T \le n+k+1$. It follows that we must have $2T = n+k+1$, and we must take the lightest k objects. Then $\frac{n+k+1}{2} = T = 1+2+\cdots+k = \frac{k^2+k}{2}$, so that $n+1 = k^2$.

6. Partition the infinite board into $n \times n$ subboards by horizontal and vertical lines n units apart. Within each subboard, assign the coordinates (i, j) to the square at the ith row and the jth column, where $1 \le i, j \le n$. Whenever an $n \times n$ cardboard is placed on the infinite board, it covers n^2 squares all with different coordinates. The total number of times squares with coordinates (1,1) are covered is 2009. Since 2009 is odd, at least one of the squares with coordinates (1,1) is covered by an odd number of cardboards. The same goes for the other $n^2 - 1$ coordinates. Hence the total number of squares which are covered an odd number of times is at least n^2.

7. We construct a graph, with the vertices representing the islands and the edges representing connecting routes. The graph may have one or more connected components. Since the total number of vertices is odd, there must be a connected component with an odd number of vertices. Olga chooses from this component the largest set of independent edges, that is, edges no two of which have a common endpoint. She will colour these edges red. Since the number of vertices is odd, there is at least one vertex which is not incident with a red vertex. Olga will start the tour there. Suppose Max has a move. It must take the tour to a vertex incident with a red edge. Otherwise, Olga could have colour one more edge red. Olga simply continues the tour by following that red edge. If Max continues to go to vertices incident with red edges, Olga will always have a ready response.

Suppose somehow Max manages to get to a vertex not incident with a red edge. Consider the tour so far. Both the starting and the finishing vertices are not incident with red edges. In between, the edges are alternately red and uncoloured. If Olga interchanges the red and uncoloured edges on this tour, she could have obtained a larger independent set of edges. This contradiction shows that Max could never get to a vertex not incident with red edges, so that Olga always wins if she follows the above strategy.

Senior A-Level Paper

1. A pirate who owes money is put in group A, and the others are put in group B. Each pirate in group A puts the full amount of money he owes into a pot, and the pot is shared equally among all 100 pirates. For each pirate in group B, each of the 100 pirates puts $\frac{1}{100}$th of the amount owed to him in a pot, and this pirate takes the pot. We claim that all debts are then settled.

   Let a be the total amount of money the pirates in group A owe, and let b be the total amount of money owed to the pirates in group B. Clearly, $a = b$. Each pirate in group A pays off his debt, takes back $\frac{a}{100}$ and then pays out another $\frac{b}{100}$. Hence he has paid off his debt exactly. Each pirate in group B takes in $\frac{a}{100}$, pays out $\frac{b}{100}$ and then takes in what is owed him. Hence the debts to him have been settled too.

2. Let the given rectangle R have length m and width n with $m > n$. Contract the length of R by a factor of $\frac{n}{m}$, resulting in an $n \times n$ square. For each of the N rectangles in R, the corresponding rectangle in S has the same width but shorter length. Thus we can cut the former into a primary piece congruent to the latter, plus a secondary piece. Using S as a model, the N primary pieces may be assembled into an $n \times n$ square while the N secondary pieces may be assembled into an $(m-n) \times n$ rectangle.

3. Let the points of tangency to the sphere of AB, AC, DB and DC be K, L, M and N respectively. The line KL intersects the line BC at some point P not between B and C. By the converse of the undirected version of Menelaus' Theorem, $1 = \frac{BP}{PC} \cdot \frac{CL}{LA} \cdot \frac{AK}{KB} = \frac{BP}{PC} \cdot \frac{CL}{KB}$ since $LA = AK$. Since $CL = CN$, $KB = MB$ and $ND = DM$, we have $1 = \frac{BP}{PC} \cdot \frac{CN}{MB} = \frac{BP}{PC} \cdot \frac{CN}{ND} \cdot \frac{DM}{MB}$. By the undirected version of Menelaus' Theorem, P, M and N are collinear. It follows that K, L, M and N are coplanar, so that KN intersects LM. Similarly, the line joining the points of tangency to the sphere of AD and BC also intersects KN and LM. Since the three lines are not coplanar, they must intersect one another at a single point.

4. Define $f(n) = 111\ldots1$ with n 1s and $f(0) = 1$ so that $[0]!=1$. Define $\begin{bmatrix} n \\ k \end{bmatrix} = \frac{[n]!}{[k]![n-k]!}$ for $0 \le k \le n$. We use induction on n to prove that $\begin{bmatrix} n \\ k \end{bmatrix}$ is always a positive integer for all $n \ge 1$. For $n = 0$, $\begin{bmatrix} 0 \\ 0 \end{bmatrix} = \frac{[0]!}{[0]![0]!} = 1$. Suppose the result holds for some $n \ge 0$. Consider the next case.

$$\begin{aligned} \begin{bmatrix} n+1 \\ k \end{bmatrix} &= \frac{[n+1]!}{[k]![n+1-k]!} \\ &= \frac{[n]!f(n+1)}{[k]![n+1-k]!} \\ &= \frac{[n]!f(n+1-k)10^k}{[k]![n-k]!f(n+1-k)} + \frac{[n]!f(k)}{[k-1]![n+1-k]!} \\ &= 10^k \begin{bmatrix} n \\ k \end{bmatrix} + \begin{bmatrix} n \\ k-1 \end{bmatrix}. \end{aligned}$$

Since both terms in the last line are positive integers, the induction argument is complete. In particular, for any positive integers m and n, $\begin{bmatrix} m+n \\ m \end{bmatrix} = \frac{[m+n]!}{[m]![n]!}$ is a positive integer, so that $[m+n]!$ is divisible by $[m]![n]!$.

5. Denote the area of a polygon P by $[P]$. We first establish a preliminary result.

Lemma.

Let M be the midpoint of a segment AB which does not intersect another segment CD. Then $[CMD] = \frac{[CAD]+[CBD]}{2}$.

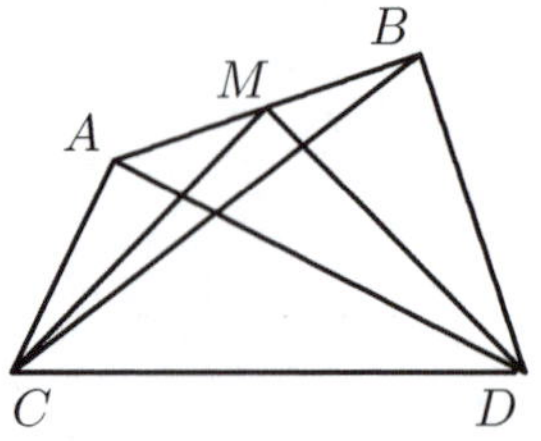

Proof:

Since M is the midpoint of AB, we have

$$[CAD] = [ABDC] - [BAD] = [ABDC] - 2[BMD]$$

and

$$[CBD] = [ABDC] - [ABC] = [ABDC] - 2[AMC].$$

Hence

$$2[CMD] = 2([ABDC] - [AMC] - [BMD]) = [CAD] + [CBD].$$

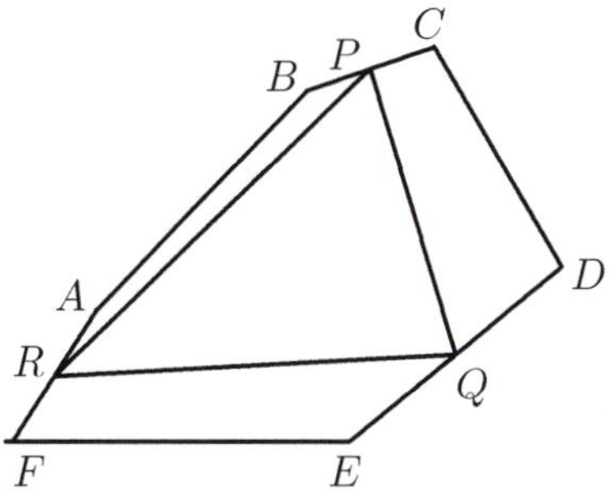

Returning to the problem, let P, Q and R be the respective midpoints of BC, DE and FA. By the Lemma, we have

$$\begin{aligned}[PQR] &= \frac{1}{2}([BQR]+[CQR]) \\ &= \frac{1}{4}([BDR]+[BER]+[CDR]+[CER]) \\ &= \frac{1}{8}([BAD]+[BFD]+[BAE]+[BFE] \\ &\quad +[CAD]+[CFD]+[CAE]+[CFE]).\end{aligned}$$

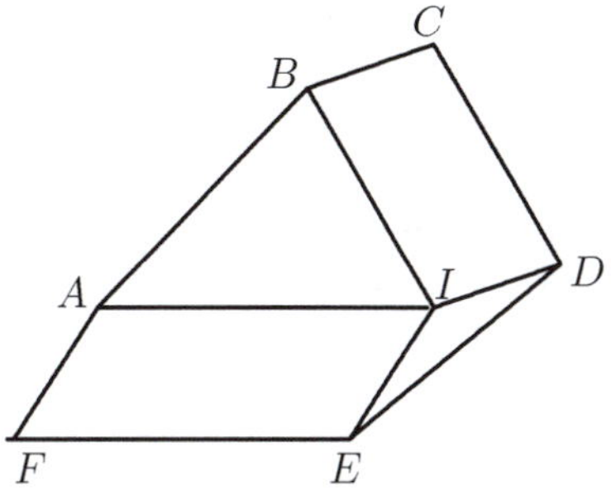

Let I be the point such that ABI is congruent to XYZ. Then $BCDI$ and $EFAI$ are parallelograms. Since $ABCDEF$ is convex, I is inside the hexagon. Hence $[XYZ] < [ABCDEF]$. Note that the distance of D from AB is equal to the sum of the distances from C and I to AB, Hence $[BAD] = [BAC] + [BAI] = [BAC] + [XYZ]$. Similarly, $[BAE] = [BAF] + [XYZ]$.

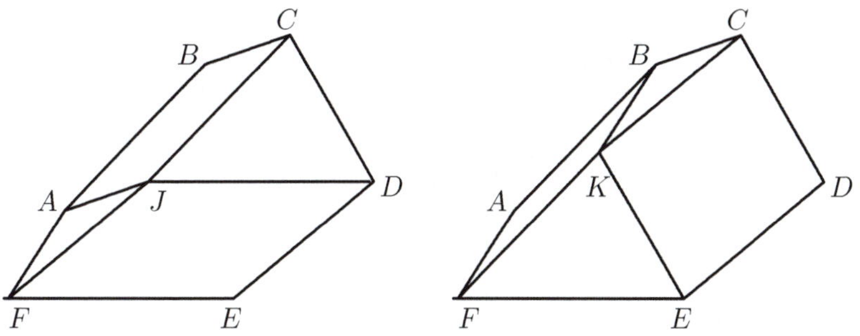

Let J and K be the points such that JCD and FKE are congruent to XYZ. Then $[ACD] = [BCD] + [XYZ]$, $[FCD] = [ECD] + [XYZ]$, $[BFE] = [AFE] + [XYZ]$ and $[CFE] = [DFE] + [XYZ]$. It follows that $[PQR] = \frac{1}{8}(2[ABCDEF] + 6[XYZ]) > [XYZ]$.

Note that if XYZ and $ABCDEF$ are in opposite orientations, the first of the three diagrams above may look like the one below, and minor modifications to the argument are necessary.

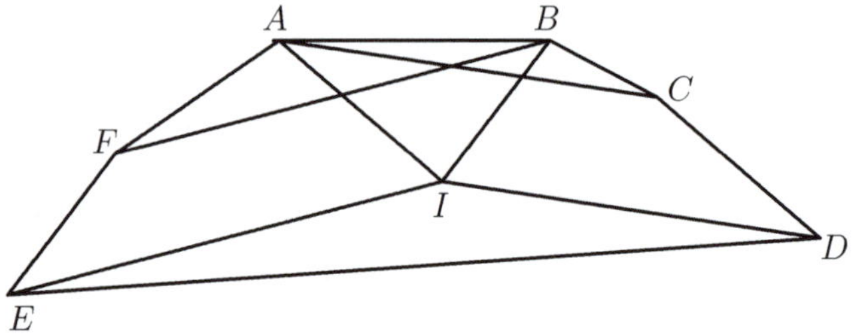

6. (See Problem 7 of the Junior A-Level Paper.)

7. The task is guaranteed to succeed if and only if n is a power of 2. Suppose n is not a power of 2. Then it has an odd prime factor p. Choose p evenly spaced barrels and make sure that the herrings inside are not all pointing the same way. Ignore all other barrels. At any point, let the herrings in r barrels be pointing up while the herrings in the other s barrels are pointing down. Since $r + s = p$ is odd, $r \neq s$. We may assume that $r > s$. In order for Ali Baba to succeed, he must turn over all r barrels of the first kind or all s barrels of the second kind. A pagan god who is having fun with Ali Baba can spin the table so that if Ali Baba plans to turn over r barrels, the herring in at least one of them is pointing down; and if Ali Baba plans to turn over s barrels, the herring in all of them are pointing up. This way, Ali Baba will never be able to open the cave.

   If $n = 2^k$ for some non-negative integer k, we will prove by induction on k that Ali Baba can open the cave. The case $k = 0$ is trivial as the cave opens automatically. The case $k = 1$ is easy. If the cave is not already open, turning one barrel over will do. For $k = 2$, let 0 or 1 indicate whether the herring is heads up or heads down. The diagram below represents the four possible states the table may be in, as well as the transition between states by the following operations. Operation **A:** Turn over any two opposite barrels. Operation **B:** Turn over any two adjacent barrels. Operation **C:** Turn over any one barrel. By performing the sequence **ABACABA**, the cave will open.

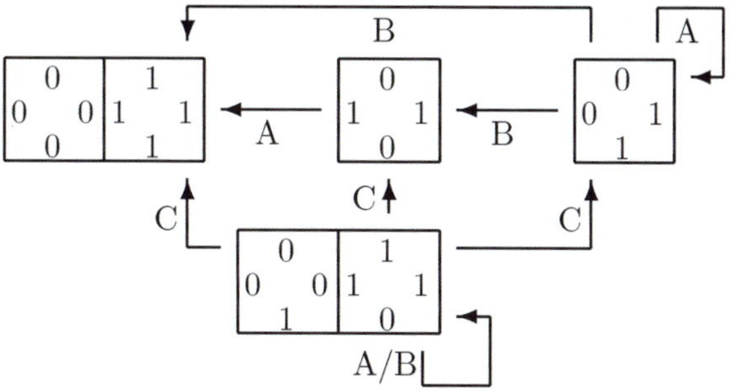

The first state is called an *absorbing* state, in that once there, no further transition takes place as the cave will open immediately. The second state becomes the first state upon the first operation A. The third state remains in place during the first operation A, but will become either the first state or the second state upon the first operation B. In the latter case, it will become the first state upon the second operation A. The fourth state remains in place during the first three operations, but will become any of the other three states upon the operation C. It will become the first state at the latest after three more operations. The success of the case $k = 2$ paves the way for the case $k = 3$. The process is typical of the general inductive argument so that we give a detailed analysis. The idea is to treat each pair of diametrically opposite barrels as a single entity.

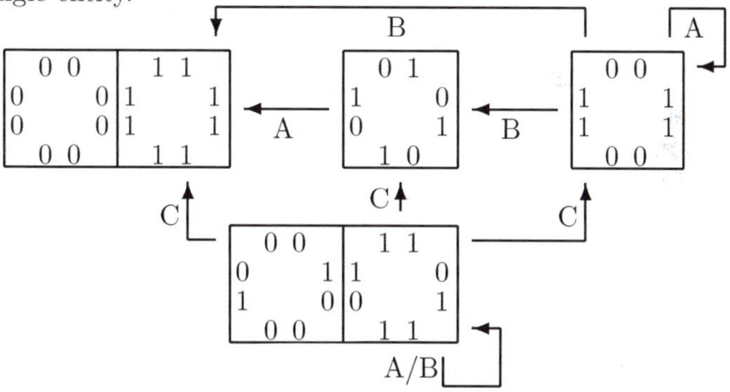

The above diagram, which is essentially copied from that for $k = 2$, is part of a much bigger state-transition diagram for $k = 3$. Here, all the states have the property that opposite pairs of barrels are all matching, that is, both are 0 or both are 1. The operations are modified from those in the case $k = 2$ as follows.Operation **A:** Turn over every other pair of opposite barrels; in other words, turn over every other barrel. Operation **B:** Turn over any two adjacent pairs of opposite barrels. Operation **C:** Turn over any pair of opposite barrels. By performing the sequence **ABACABA**, the cave will open.

These states together form an expanded absorbing state in the overall diagram below.

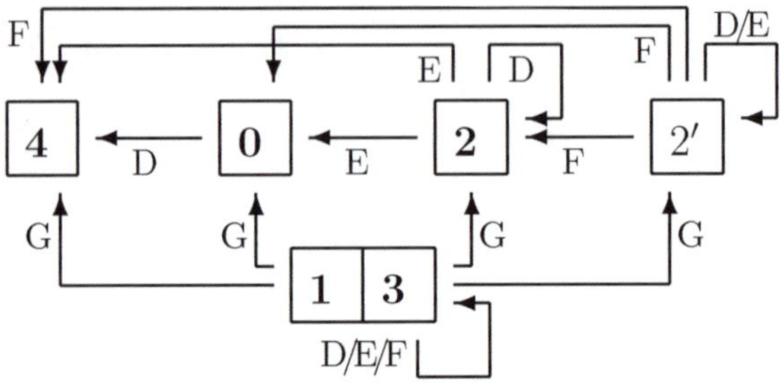

Here, the box marked m contains all states with m matching opposite pairs, where $0 \leq m \leq 4$. The box marked 4 is the expanded absorbing state mentioned above. The states with 2 matching pairs are classified according to whether these matching are alternating or adjacent. The former states are contained in the box marked 2 while the latter states are contained in the box marked $2'$. We have four new operations. Operation **D:** Turn over any 4 adjacent barrels. Operation **E:** Turn over any 2 barrels separated by one other barrel. Operation **F:** Turn over any two adjacent barrels. Operation **G:** Turn over any barrel. Let **X** denote the sequence ABACABA. Then the sequence for the case $k = 3$ is

XDXEXDXFXDXEXDXGXDXEXDXFXDXEXDX.

We keep repeating X to clear any state that has entered the box marked 4, to prevent them from returning to another box. Whatever the state the table is in, the cave will open by the end of this sequence. The general procedure is now clear. We treat each opposite pair as a single entity, thereby reducing to the preceding case. Then we moving progressively all states into the expanded absorbing state. Thus the task is possible whenever n is a power of 2.

Spring 2010

Junior O-Level Paper

1. In counting the total number of apples, we have counted each pear five times. Hence the total number of apples is five times the total number of pears. Similarly, in counting the total number of plums, we have counted each apple five times, so the total number of plums is five times the total number of apples, and twenty-five times the total number of pears. It follows that the total number of fruit is equal to the total number of pears times 1+5+25=31, and is therefore a multiple of 31.

2. Lillebror can always get at least one quarter of the cake, making the first cut from P past the centre O and to the boundary of the cake. If Karlsson had chosen P to coincide with O, then Lillebror just makes any cut. Now draw the line ℓ through O perpendicular to the first cut. Its cut off two quadrants of the cake. Karlsson has only two directions to choose for the second cut, which must be parallel to ℓ. In either case, the smaller of the two pieces contains at least one of these quadrants.

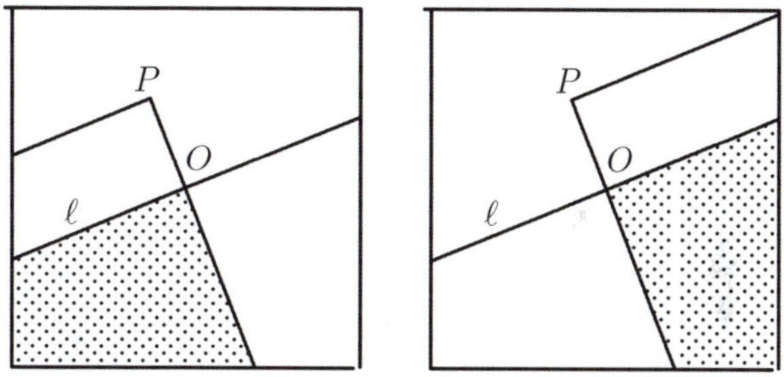

3. (a) Let O be the vertex of the given angle. Let P be any point on one arm of the angle other than O. Draw a circle with centre P and radius OP. If the other arm is tangent to the circle, then the given angle is a right angle. If the other arm intersects the circle in two points, then the given angle is acute. If the other arm misses the circle, then the given angle is obtuse. Hence the task can be accomplished using the compass only once.

   (b) Let O be the vertex of the given angle. Draw a circle Ω with centre O and arbitrary radius, cutting the two arms of the angle at A_0 and A_1 respectively. Using A_0A_1 as radius, mark off on Ω successive points A_2, A_3, $\ldots$ so that $A_0A_1 = A_1A_2 = A_2A_3 = \cdots$. Then $\angle A_0OA_1 = 31^\circ$ if and only if $A_{360} = A_0$ but $A_k \neq A_0$ for $1 \leq k \leq 359$, and we have gone around Ω exactly 31 times.

4. Form the longest possible line of the people at the party so that each knows the next one. Let the first person be A. Then all the acquaintances of A must be in the line, as otherwise any missing one could be put in front of A to form a longer line. Let B, C and D be the first three acquaintances of A down the line. Suppose there are an odd number of people between B and C. Then we can put the segment of the line from B to C at a round table and insert A between B and C. This will meet the condition of the problem. Similarly, if there are an odd number of people between C and D, the condition can also be met. Finally, if there is an even number, possibly 0, of people both between B and C and between C and D, then there are an odd number of people betweeen B and D.

5. There are 51 odd numbers and 50 even numbers on the blackboard. Each move either keeps the number of odd numbers unchanged, or reduces it by 2. It follows that the last number must be odd, and its minimum value is 1. The squares of four consecutive integers can be replaced by a 4 because $(n+2)^2+(n-1)^2-(n+1)^2-n^2=4$. Hence the squares of eight consecutive integers can be replaced by 0. Taking the squares off from the end eight at a time, we may be left with 1, 4, 9, 16 and 25. However, the best we can get out of these five numbers is 3. Hence we must include 36, 49, 64, 81, 100, 121, 144 and 169. The sequence of combinations may be $169-144=25$, $25-25=0$, $100-0=100$, $100-64=36$, $36-36=0$, $121-0=121$, $121-81=40$, $49-40=9$, $16-9=7$, $9-7=2$, $4-2=2$ and $2-1=1$.

Senior O-Level Paper

1. In counting the total number of lemons, we have counted each pineapple 2009 times. Hence the total number of lemons is 2009 times the total number of pineapples. Similarly, in counting the total number of bananas, we have counted each lemon 2009 times. Hence the total number of bananas is 2009 times the total number of lemons, and $2009^2=4036081$ times the total number of pineapples. It follows that the total number of fruit is equal to the total number of pineapples times $1+2009+4036081=4038091=31\times 130261$, a multiple of 31.

2. Note that $f(x)$ is uniquely defined for all x since it is given to be a function. In any case, since $y=x^2$ intersects each vertical line in exactly one point, so does $y=f(x)$. Let S be the region of the plane below $y=x^2$. Every point in S lies on a line which does not intersect $y=x^2$. Hence no point of $y=f(x)$ can belong to S. For any real number r, consider the line tangent to $y=x^2$ at the point (r,r^2). Except for this point, the line lies entirely in S. Since this line intersects $y=f(x)$ at exactly one point, we must have $f(r)=r^2$. Since r is an arbitrary real number, $f(x)=x^2$.

3. It is possible to accomplish the task with twelve hexagons, as shown in the diagram below.

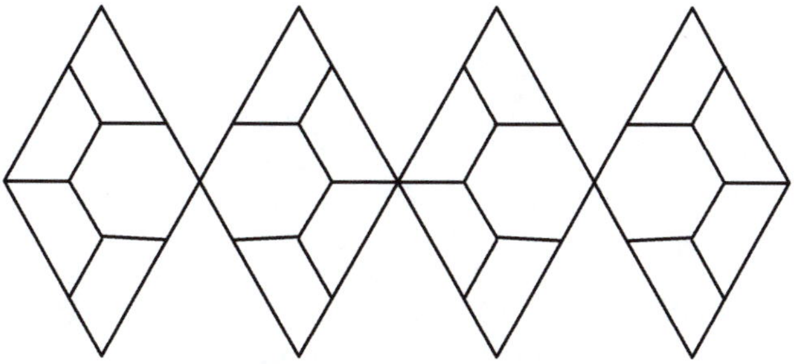

4. Let $P(x)=a_0x^n+a_1x^{n-1}+\cdots+a_n$. Suppose $P(2)=b$. Then

$$b=a_02^n+a_12^{n-1}+\cdots+a_n>a_0+a_1+\cdots+a_n.$$

It follows that we have

$$\begin{aligned} b^n &> b^{n-1}(a_0+a_1+\cdots+a_n)\\ &> a_1b^{n-1}+\cdots+a_{n-1}b+a_n.\end{aligned}$$

Now $\frac{P(b)}{b^n}=a_0+\frac{a_1b^{n-1}+\cdots+a_{n-1}b+a_n}{b^n}$. Hence the degree n of the polynomial is the largest integer for which $P(b)\geq b^n$, and we have $a_0=\lfloor\frac{P(b)}{b^n}\rfloor$. In an analogous manner, $a_1=\lfloor\frac{P(b)-a_0b^n}{b^{n-1}}\rfloor$, and so on. In other words, a_k is the kth digit of $P(b)$ expressed in base b. It follows that $P(x)$ is uniquely determined, and the Baron is right!

Note that the values of $P(2)$ and $P(P(2))$ cannot be assigned arbitrarily. Suppose we have $P(2)=13$ and $P(13)=2224$, the above algorithm yields $n=3$, $a_0=\lfloor\frac{2224}{13^3}\rfloor=1$, $a_1=\lfloor\frac{2224-13^3}{13^2}\rfloor=0$, $a_2=\lfloor\frac{2224-2197}{13}\rfloor=2$ and $a_3=27-2\times13=1$. On the other hand, if $P(2)=3$ and $P(3)=5$, we get $n=1$, $a_0=\lfloor\frac{5}{3}\rfloor=1$ and $a_1=5-3=2$, but $P(x)=x+2$ yields $P(2)=4$. This trouble arises because the correct polynomial $P(x)=2x-1$ does not satisfy the hypothesis of the problem, and the above algorithm cannot be applied.

5. Suppose the task is possible. Let the segment be of length 1. Label one of its endpoints A and the other B. We combine consecutive moves making rotations about the same point into one, so that the new moves alternately rotate about A and B through an angle which is a multiple of $45°$. Denote the initial positions of A and B by A_0 and B_0 respectively. By symmetry, we may assume that the first rotation is about B. Denote the new position of A by A_1.

The next rotation is about A_1. Denote the new position of B by B_1. Continue until $A_k = B_0$ or $B_k = A_0$ for some k. We may assume that the former is the case. Then we have a $(2k-1)$-gon $A_1B_1 \ldots A_k$ whose edges are all of length 1 and may intersect one another. Each horizontal edge represents a horizontal displacement of 1 unit, while each slanting edge represents a horizontal displacement of $\frac{1}{\sqrt{2}}$ unit. These are incommensurable. In going around the perimeter of the polygon once, the net horizontal displacement is 0. Hence we must have an even number of horizontal edges and an even number of slanting edges. Similarly, we must also have an even number of vertical edges. Hence the total number of edges of this polygon must be even, but a $(2k-1)$-gon has an odd number of edges. This is a contradiction.

Junior A-Level Paper

1. Let $\alpha < 1$. The first cut creates the piece $\frac{1}{1+\alpha}$ and $\frac{\alpha}{1+\alpha}$. Then cut the larger piece into $\frac{1}{(1+\alpha)^2}$ and $\frac{\alpha}{(1+\alpha)^2}$. We want $\frac{1}{(1+\alpha)^2} = \frac{\alpha}{(1+\alpha)^2} + \frac{\alpha}{1+\alpha}$ or $1 = \alpha + \alpha(1+\alpha)$. From $\alpha^2 + 2\alpha - 1 = 0$, $\alpha = \frac{-2\pm\sqrt{4+4}}{2} = -1 \pm \sqrt{2}$. Since $\alpha > 0$, $\alpha = \sqrt{2} - 1$.

2. Through M, draw a line parallel to AP, intersecting BC at N. Since triangles CMN and CAP are similar and $AM = MC$, $PN = NC$. Since triangles BOP and BMN are similar and $BO = BP$, $OM = PN$. Hence $\frac{OM}{PC} = \frac{OM}{2PN} = \frac{1}{2}$.

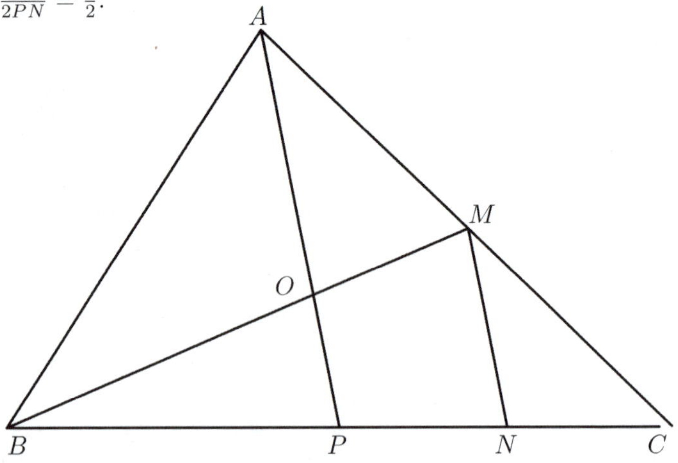

3. Since each product is equal to 1 or -1, the value of S is always odd. Let the numbers be $a_1, a_2, \ldots, a_{999}$ in cyclic order, with $a_n = a_{n-999}$ whenever $n > 999$.

(a) The minimum value of S is -997. This can be attained by having 100 copies of -1, each adjacent pair separated by 9 copies of 1, except for one pair which is separated by just 8 copies of 1. Every block of 10 adjacent numbers contains exactly 1 copy of -1, the sole exception being the block with 1 copy of -1 at either end and 8 copies of 1 in between. If -997 is not the minimum value, it would have to be -999, meaning that all 999 products are equal to -1. We must have $a_1 = a_{11}$ since $a_1a_2\cdots a_{10} = -1 = a_2a_3\cdots a_{11}$. Similarly, we also have $a_{11} = a_{21} = \cdots = a_{9981}$. Since 10 and 999 are relatively prime, all these 999 subscripts are different. This means that we have either 999 copies of 1 or 999 copies of -1. This is forbidden.

(b) The maximum value of S is 995. This can be attained by having 2 adjacent copies of -1 and 997 copies of 1. There are only two blocks of 10 adjacent numbers which contain exactly one copy of -1 and have -1 as products. All other blocks have 1 as products. If 995 is not the maximum value, it would have to be 997 or 999. We cannot have 999 since this means all 999 numbers are copies of 1, or all are copies of -1, which is forbidden. Suppose it is 997, which means that exactly one block of 10 adjacent numbers has product -1. Let $a_1a_2\cdots a_{10} = -1$. Then $a_1 = -a_{11}$ but $a_{11} = a_{21} = \cdots = a_{9981}$. Since 10 and 999 are relatively prime, all these 999 subscripts are different. Hence all 999 numbers are equal except one, and there are exactly ten blocks of 10 adjacent numbers with product -1, so that $S = 979$.

4. Let $n = 10^{4^1} + 10^{4^2} + \cdots + 10^{4^{100}}$. Then the sum of the digits of n is 100. Consider n^3. It is the sum of 100^3 terms each a product of three powers of 10. We claim that if two such terms are equal, they must be products of the same three powers of 10. If $4^a + 4^b + 4^c = 4^x + 4^y + 4^z$, where $a \le b \le c$ and $x \le y \le z \le c$, we must have $z = c$. Otherwise, even if $x = y = z = c - 1$, we still have $3(4^{c-1}) < 4^c$. Similarly, we must have $y = b$ and $x = a$, justifying the claim. Now a product of the same three powers of 10 can occur at most 3!=6 times. Hence there is no carrying in adding these 100^3 terms, so that the sum of the digits of n^3 is exactly 100^3.

5. We will give a construction to show that the answer to each part is affirmative. The constant speed of each rider is an integral number of laps per time period. The time period may be chosen as the hour, though some of the speeds may be unreasonably high.

(a) Let the constant speeds of riders C, B and A be 2, 3 and 4 laps per hour respectively. The diagram below represents what happens in an hour, and this can be repeated indefinitely.

The horizontal axis is time, and the hour is divided into 12 five-minute blocks. The vertical axis is distance, and the top point is to be identified with the bottom point to represent the starting position, which also happens to be the only point at which passing is allowed. As can be seen, no two of the three riders are ever together except at the starting point.

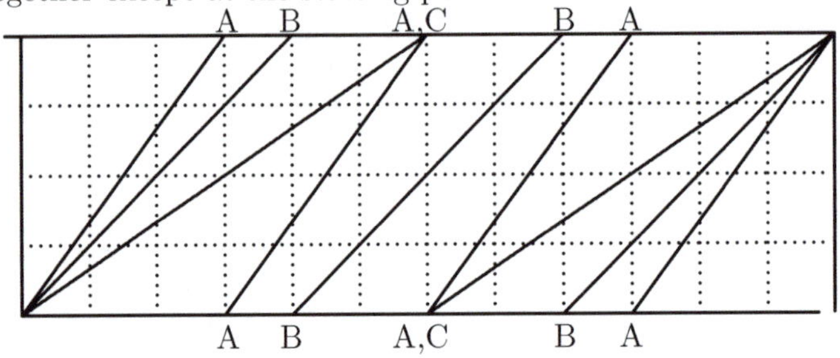

(b) Our plan is to add new riders one at a time. First, if the new rider D has constant speed 0 laps per hour, and is stuck at the starting point, then obviously everything is fine. However, no two riders may have the same constant speed. We shall increase the constant speeds of the four riders we have so far by the same amount. Currently, they are 0, 2, 3 and 4 laps per hour. Their pairwise differences are $4-3=3-2=1$, $4-2=2-0=2$, $3-0=3$ and $4-0=4$. The least common multiple of 1, 2, 3 and 4 is 12. So the new constant speeds are 12, 14, 15 and 16 laps per hour. We claim that if the constant speeds of two riders S and F are kv and $k(v+1)$ laps per hour respectively, where k and v are positive integers, then they only pass each other at the starting point. Divide the hour into $v(v+1)$ equal intervals. Then each lap by F takes v units while each lap by S takes $v+1$ units. Suppose they pass each other somewhere other than the starting point, during the lap of F from time a to time d, and the lap of S from time b to time c. Then $a+1 \leq b$, $c+1 \leq d$, $d-a=v+1$ while $c-b=v$. Adding the two inequalities, we have $a+c+2 \leq b+d$. Subtracting the second equation from the first, we have $b+d=a+c+1$. This is a contradiction. Now 12:14=2(6:7), 12:15=3(4:5), 12:16=4(3:4), 14:16=2(6:7), 14:15=1(14:15) and 15:16=1(15:16). Hence the plan works for 4 riders. In the same manner, we can have additional riders up to 10 and beyond.

6. We first show that 9 lines are necessary. If we only have 8 lines, they generate at most 28 points of intersection. Since the broken line can only change direction at these points, it can have at most 29 segments. The diagram below shows a broken line with 32 segments all lying on 9 lines. Hence 9 lines are also sufficient in this case.

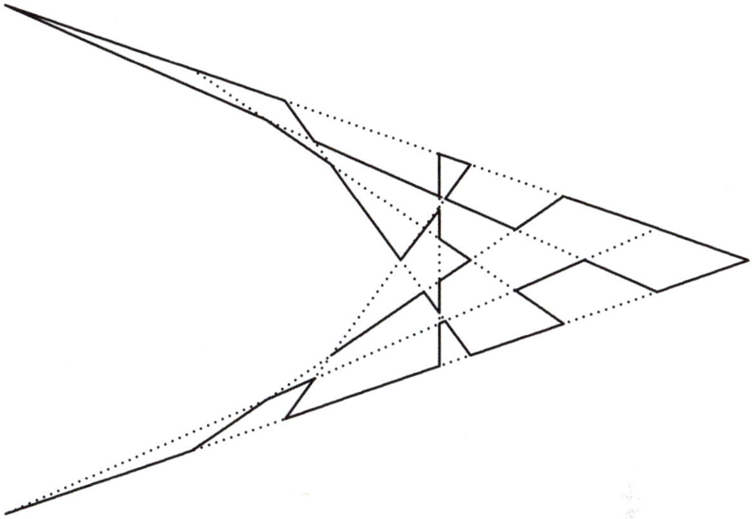

7. We claim that on each row or column, there are at most 2 ants crawling along that row or column. Hence the total number of ants on the board is at most $2 \times 10 + 2 \times 10 = 40$. Suppose there are 3 ants crawling along a row or column. By the Pigeonhole Principle, two of them must occupy squares of the same colour in the standard colouring. These 2 ants must occupy the same square well before an hour has elapsed. This justifies the claim. We now show a construction whereby there can be as many as 40 ants on the board.

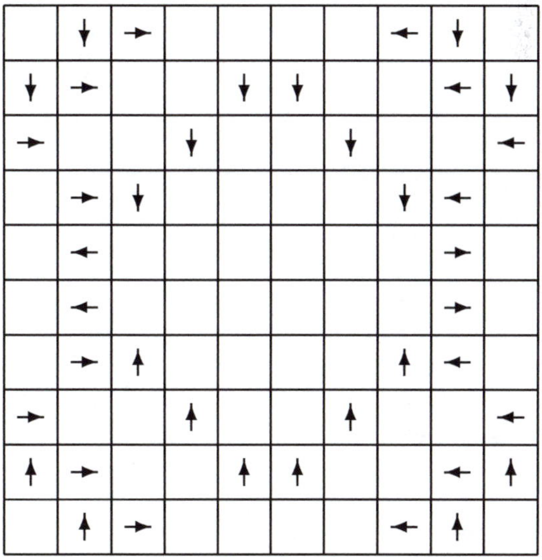

Senior A-Level Paper

1. Form families consisting of all mutually parallel lines. Put into a group two families whose lines are perpendicular. For each group, choose an arbitrary line ℓ not parallel to either family. Each line in a family intersects exactly one point of ℓ, and each point of ℓ lies on exactly one line in the family. Thus each point of ℓ defined one line from each family, and these two lines form a pair. This procedure may be applied to all groups, so that every line in the plane is in exactly one pair.
2. (a) Let $\alpha > 1$. The first cut creates the piece $\frac{\alpha}{\alpha+1}$ and $\frac{1}{\alpha+1}$. Then cut the larger piece into $\frac{\alpha^2}{(\alpha+1)^2}$ and $\frac{\alpha}{(\alpha+1)^2}$. Note that we want $\frac{\alpha^2}{(\alpha+1)^2} = \frac{\alpha}{(\alpha+1)^2} + \frac{1}{\alpha+1}$ or $\alpha^2 = \alpha + \alpha + 1$. Solving $\alpha^2 - 2\alpha - 1 = 0$, we have $\alpha = \frac{2\pm\sqrt{4+4}}{2} = 1 \pm \sqrt{2}$. Since $\alpha > 0$, $\alpha = \sqrt{2} + 1$.

   (b) Let $\alpha = \frac{m}{n} > 1$ where m and n are relatively prime positive integers. In the first step, we have the pieces $\frac{m}{m+n}$ and $\frac{n}{m+n}$. In all subsequent steps, we will cut all pieces. There is no harm in assuming this since the two parts of a piece which is not to be cut can just stay together. Suppose the task is accomplished after k steps. Each of the 2^k pieces is $\frac{m^i n^{k-i}}{(m+n)^k}$ for $0 \le i \le k$, with $i = 0$ occurring only once. Each numerator is a multiple of m except for n^k. Thus the division into two piles of equal amount is not possible.

   (Compare with Problem 1 of the Junior A-Level Paper.)

3. If $x = \frac{1}{\sqrt{n}}$, then $\arctan x$ is an angle θ in a right triangle with opposite side 1 and adjacent side $\sqrt{n}$. By Pythagoras' Theorem, the length of the hypotenuse is $\sqrt{n+1}$, so that $\sin\theta = \frac{1}{\sqrt{n+1}}$. Define $f(x) = \sin(\arctan x)$. Starting with $1 = \frac{1}{\sqrt{1}}$, we can apply $f(x)$ repeatedly and obtain $\frac{1}{\sqrt{2010^2}} = \frac{1}{2010}$. Now $\cot(\arctan \frac{1}{2010}) = 2010$.

4. We construct united groups in the first stage and diverse groups in the second stage. In the kth step of the first stage, we create a united group of size at least $101-k$. This stage terminates, perhaps even immediately, when no such groups can be formed. If this is as a result of having nobody left, then we have formed at most 100 groups since $1 + 2 + \cdots + 100 > 5000$. We can take individuals out of existing groups to form groups of one until we have exactly 100 groups. Suppose after the nth step, we cannot form a united group of size at least $101 - (n + 1)$ from the remaining participants.

We proceed to the second stage. We have created n united groups, and now we create $100 - n$ diverse groups. Start with any movie watched by at least one of the remaining participants. There are less than $100 - n$ of the remaining participants who have watched this movie, and they can be put into separate groups. This will be the movie each of them has watched that nobody else in their group would have. Take another movie watched by at least one of the remaining participants. There are less than $100 - n$ of such participants, possibly including some who are already in the groups. Those that are not already assigned to a group can now join groups not including anyone who has watched this movie. The remaining participants can be added to the groups in an analogous manner. If some of these $100 - n$ groups happen to be empty, we can take individuals out of existing groups to form groups of one until we have exactly 100 groups.

5. We first prove an auxiliary result.

Lemma.

If the constant speeds of two riders S and F are kv and $k(v+1)$ laps per hour respectively, where k and v are positive integers, then they only pass each other at the starting point.

Proof:

Divide the hour into $v(v+1)$ equal intervals. Then each lap by F takes v units while each lap by S takes $v+1$ units. Suppose they pass each other somewhere other than the starting point, during the lap of F from time a to time d, and the lap of S from time b to time c. Then $a+1 \leq b$, $c+1 \leq d$, $d-a = v+1$ while $c-b = v$. Adding the two inequalities, we have $a+c+2 \leq b+d$. Subtracting the second equation from the first, we have $b+d = a+c+1$. This is a contradiction. In our construction, the passing point is the same as the starting point, the constant speed of each rider is a different number of laps per hour and for any two riders, their constant speeds are respectively kv and $k(v+1)$ laps per hour for some positive integers k and v. We use induction on the number n of runners. For $n = 1$, there is nothing to prove. Suppose the result holds for some $n \geq 1$. If we add an $(n+1)$st runner with speed 0 at the starting point, the result still holds. Let M be the least common multiple of the pairwise differences of these $n+1$ constant speeds. Increase the speed of each rider by M. Let the constant speeds of two riders before the increment be kv and $k(v+1)$ respectively. Their difference k is a divisor of M, so that $M = kh$ for some positive integer h. Now $(M + kv) : (M + k(v+1)) = k(h+v) : k(h+v+1)$. This completes the inductive argument, and the result is true in particular for $n = 33$.

(Compare with Problem 5 of the Junior A-Level Paper.)

6. Let P, Q, R and S be the points of tangency of the circle with AB, BC, CD and DA respectively. Let $\angle AIS = \angle AIP = \alpha$, $\angle BIP = \angle BIQ = \beta$, $\angle CIQ = \angle CIR = \gamma$ and $\angle DIR = \angle DIS = \delta$. Then

$$\angle AIB + \angle CID = \alpha + \beta + \gamma + \delta = 180^\circ.$$

If $\angle AIB > 90^\circ$, then $\angle CID < 90^\circ$. The point I will be inside the circle with AB as diameter but outside the circle with CD as diameter. Hence $\frac{IM}{AB} < \frac{1}{2} < \frac{IN}{CD}$. Similarly, if $\angle AIB < 90^\circ$, then $\frac{IM}{AB} > \frac{1}{2} > \frac{IN}{CD}$. Both contradict the hypothesis that $\frac{IM}{AB} = \frac{IN}{CD}$. Hence $\alpha+\beta = \angle AIB = 90^\circ$ so that Q, I and S are collinear. Since both BC and DA are perpendicular to QS, they are parallel to each other.

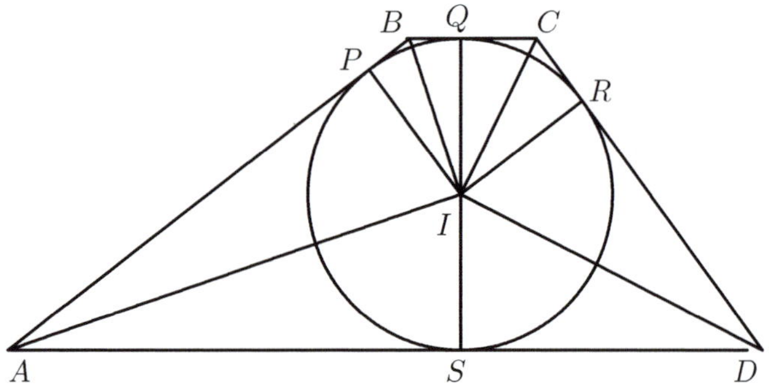

7. We use an overline to denote the concatenation of digits. Let the given number be $\overline{a_0a_1a_2\dots a_n}$ and the sum of the digits be S. Suppose we have $S \le 10^{10}$. By putting a plus sign between every pair of adjacent digits in each step, Susan obtains a number with at most 11 digits in the first step, a number at most 99 in the second step, a number at most 18 in the third step and a single-digit number in the fourth step. Suppose $S > 10^{10}$. Define

$$\begin{aligned} A &= \overline{a_0a_1a_2} + \overline{a_3a_4a_5} + \overline{a_6a_7a_8} + \cdots, \\ B &= a_0 + \overline{a_1a_2a_3} + \overline{a_4a_5a_6} + \overline{a_7a_8a_9} + \cdots, \\ C &= \overline{a_0a_1} + \overline{a_2a_3a_4} + \overline{a_5a_6a_7} + \cdots. \end{aligned}$$

Note that $A+B+C > 100S$ so that one of A, B and C is greater than $10S$. By symmetry, we may assume it is B. Then there exists a positive integer t such that $S < 10^t < B$. Consider now the following sequence of numbers.

$$\begin{aligned} S &= a_0 + a_1 + a_2 + a_3 + \cdots, \\ & \quad a_0 + \overline{a_1a_2a_3} + a_4 + a_5 + \cdots, \\ & \quad a_0 + \overline{a_1a_2a_3} + \overline{a_4a_5a_6} + a_7 + \cdots, \\ & \quad \cdots \\ B &= a_0 + \overline{a_1a_2a_3} + \overline{a_4a_5a_6} + \overline{a_7a_8a_9} + \cdots. \end{aligned}$$

These numbers are increasing by steps of less than 1000. Hence one of them will be at least 10^t and at most $\min\{10^t+999, B\}$. This will be the first step for Susan, arriving at a number with at most four non-zero digits. By putting a plus sign between every pair of adjacent digits in each subsequent step, Susan obtains a number at most 36 in the second step, a number at most 11 in the third step and a single-digit number on the fourth step.

Tournament 32

Fall 2010

Junior O-Level Paper

1. In a multiplication table, the entry in the ith row and the jth column is the product ij. From an $m \times n$ subtable with both m and n odd, the interior $(m-2) \times (n-2)$ rectangle is removed, leaving behind a frame of width 1. The squares of the frame are painted alternately black and white. Prove that the sum of the numbers in the black squares is equal to the sum of the numbers in the white squares.

2. In a quadrilateral $ABCD$ with an incircle, $AB = CD$, $BC < AD$ and BC is parallel to AD. Prove that the bisector of $\angle C$ bisects the area of $ABCD$.

3. A $1 \times 1 \times 1$ cube is placed on an 8×8 board so that its bottom face coincides with a square of the board. The cube rolls over a bottom edge so that the adjacent face now lands on the board. In this way, the cube rolls around the board, landing on each square at least once. Is it possible that a particular face of the cube never lands on the board?

4. In a school, more than 90% of the students know both English and German, and more than 90% of the students know both English and French. Prove that more than 90% of the students who know both German and French also know English.

5. There are $2n$ points on a circle. They are joined in pairs to form n chords. For each chord,

6. there is an odd number of the other $2n-2$ points on each side. Prove that n is even.

Note: The problems are worth 4, 4, 4, 4 and 4 points respectively.

Senior O-Level Paper

1. The exchange rate in a Funny-Money machine is s McLoonies for a Loonie or $\frac{1}{s}$ Loonies for a McLoonie, where s is a positive real number. The number of coins returned is rounded off to the nearest integer. If it is exactly in between two integers, then it is rounded up to the greater integer.

   (a) Is it possible to achieve a one-time gain by changing some Loonies into McLoonies and changing all the McLoonies back to Loonies?

(b) Assuming that the answer to (a) is "yes", is it possible to achieve multiple gains by repeating this procedure, changing all the coins in hand and back again each time?

2. The diagonals of a convex quadrilateral $ABCD$ are perpendicular to each other and intersect at the point O. The sum of the inradii of triangles AOB and COD is equal to the sum of the inradii of triangles BOC and DOA.

   (a) Prove that $ABCD$ has an incircle.

   (b) Prove that $ABCD$ is symmetric about one of its diagonals.

3. From a police station situated on a straight road infinite in both directions, a thief has stolen a police car. Its maximal speed equals 90% of the maximal speed of a police cruiser. When the theft is discovered some time later, a policeman starts to pursue the thief on a cruiser. However, he does not know in which direction along the road the thief has gone, nor does he know how long ago the car has been stolen. Is it possible for the policeman to catch the thief?

4. A square board is dissected into n^2 rectangular squares by $n-1$ horizontal and $n-1$ vertical lines. The squares are painted alternately black and white in a board pattern. One diagonal consists of n black squares which are squares. Prove that the total area of all black squares is not less than the total area of all white squares.

5. In a tournament with 55 participants, one match is played at a time, with the loser dropping out. In each match, the numbers of wins so far of the two participants differ by not more than 1. What is the maximal number of matches for the winner of the tournament?

Note: The problems are worth 2+3, 2+3, 5, 5 and 5 points respectively.

Junior A-Level Paper

1. A round coin may be used to construct a circle passing through one or two given points on the plane. Given a line on the plane, show how to use this coin to construct two points such that they define a line perpendicular to the given line. Note that the coin may not be used to construct a circle tangent to the given line.

2. Pete has an instrument which can locate the midpoint of a line segment, and also the point which divides the line segment into two segments whose lengths are in a ratio of $n:(n+1)$, where n is any positive integer. Pete claims that with this instrument, he can locate the point which divides a line segment into two segments whose lengths are at any given rational ratio. Is Pete right?

3. At a circular track, 10 cyclists started from some point at the same time in the same direction with different constant speeds. If any two cyclists are at some point at the same time again, we say that they meet. No three or more of them have met at the same time. Prove that by the time every two cyclists have met at least once, each cyclist has had at least 25 meetings.

4. A rectangle is divided into 2×1 and 1×2 dominoes. In each domino, a diagonal is drawn, and no two diagonals have common endpoints. Prove that exactly two corners of the rectangle are endpoints of these diagonals.

5. For each side of a given pentagon, divide its length by the total length of all other sides. Prove that the sum of all the fractions obtained is less than 2.

6. In acute triangle ABC, an arbitrary point P is chosen on altitude AH. Points E and F are the midpoints of sides CA and AB respectively. The perpendiculars from E to CP and from F to BP meet at point K. Prove that $KB = KC$.

7. Merlin summons the n knights of Camelot for a conference. Each day, he assigns them to the n seats at the Round Table. From the second day on, any two neighbours may interchange their seats if they were not neighbours on the first day. The knights try to sit in some cyclic order which has already occurred before on an earlier day. If they succeed, then the conference comes to an end when the day is over. What is the maximum number of days for which Merlin can guarantee that the conference will last?

Note: The problems are worth 4, 5, 8, 8, 8, 8 and 12 points respectively.

Senior A-Level Paper

1. There are 100 points on the plane. All 4950 pairwise distances between two points have been recorded.

   (a) A single record has been erased. Is it always possible to restore it using the remaining records?

   (b) Suppose no three points are on a line, and k records were erased. What is the maximum value of k such that restoration of all the erased records is always possible?

2. At a circular track, $2n$ cyclists started from some point at the same time in the same direction with different constant speeds. If any two cyclists are at some point at the same time again, we say that they meet. No three or more of them have met at the same time. Prove that by the time every two cyclists have met at least once, each cyclist has had at least n^2 meetings.

3. For each side of a given polygon, divide its length by the total length of all other sides. Prove that the sum of all the fractions obtained is less than 2.

4. Two dueling wizards are at an altitude of 100 metres above the sea. They cast spells in turn, and each spell is of the form "decrease the altitude by a metres for me and by b metres for my rival," where a and b are real numbers such that $0 < a < b$. Different spells have different values for a and b. The set of spells is the same for both wizards, the spells may be cast in any order, and the same spell may be cast many times. A wizard wins if after some spell, he is still above water but his rival is not. Does there exist a set of spells such that the second wizard has a guaranteed win, if the number of spells is

   (a) finite;

   (b) infinite?

5. The quadrilateral $ABCD$ is inscribed in a circle with centre O. The diagonals AC and BD do not pass through O. If the circumcentre of triangle AOC lies on the line BD, prove that the circumcentre of triangle BOD lies on the line AC.

6. Each square of a 1000×1000 table contains 0 or 1. Prove that one can either cut out 990 rows so that at least one 1 remains in each column, or cut out 990 columns so that at least one 0 remains in each row.

7. A square board is divided into congruent rectangles with integral side lengths. A rectangle is important if it has at least one point in common with a given diagonal of the board. Prove that this diagonal bisects the total area of the important rectangles.

Note: The problems are worth 2+3, 6, 6, 2+5, 8, 12 and 14 points respectively.

Spring 2011

Junior O-Level Paper

1. The numbers from 1 to 2010 inclusive are placed along a circle so that if we move along the circle in clockwise order, they increase and decrease alternately. Prove that the difference between some two adjacent integers is even.

2. A rectangle is divided by 10 horizontal and 10 vertical lines into 121 rectangules. If 111 of them have integer perimeters, prove that they all have integer perimeters.

3. Worms grow at the rate of 1 metre per hour. When they reach their maximum length of 1 metre, they stop growing. A full-grown worm may be dissected into two new worms of arbitrary lengths totalling 1 metre. Starting with 1 full-grown worm, can one obtain 10 full-grown worms in less than 1 hour?

4. Each diagonal of a convex quadrilateral divides it into two isosceles triangles. The two diagonals of the same quadrilateral divide it into four isosceles triangles. Must this quadrilateral be a square?

5. A dragon gives a captured knight 100 coins. Half of them are magical, but only the dragon knows which are. Each day, the knight divides the coins into two piles which are not necessarily equal in size. If each pile contains the same number of magic coins, or the same number of non-magic coins, the knight will be set free. Can the knight guarantee himself freedom in at most

   (a) 50 days;

   (b) 25 days?

Note: The problems are worth 3, 4, 5, 5 and 2+3 points respectively.

Senior O-Level Paper

1. The faces of a convex polyhedron are similar triangles. Prove that this polyhedron has two pairs of congruent faces.

2. Worms grow at the rate of 1 metre per hour. When they reach their maximum length of 1 metre, they stop growing. A full-grown worm may be dissected into two new worms of arbitrary lengths totalling 1 metre. Starting with 1 full-grown worm, can one obtain 10 full-grown worms in less than 1 hour?

3. An integer k is given, where $2 \leq k \leq 50$. Along a circle are 100 white points. In each move, we choose a block of k adjacent points such that the first and the last are white, and we paint both of them black. For which values of k is it possible for us to paint all 100 points black after 50 moves?

4. Four perpendiculars are drawn from four vertices of a convex pentagon to the opposite sides. If these four lines pass through the same point, prove that the perpendicular from the fifth vertex to the opposite side also passes through this point.

5. In a country, there are 100 towns. Some pairs of towns are joined by roads. The roads do not intersect one another except at towns. It is possible to go from any town to any other town by road. Prove that it is possible to pave some of the roads so that the number of paved roads at each town is odd.

Note: The problems are worth 3, 4, 4, 5 and 5 points respectively.

Junior A-Level Paper

1. Does there exist a hexagon that can be dissected into four congruent triangles by a straight cut?

2. Passing through the origin of the coordinate plane are 180 lines, including the coordinate axes, which form $1°$ angles with one another at the origin. Determine the sum of the x-coordinates of the points of intersection of these lines with the line $y = -x + 100$.

3. Baron Münchausen has a set of 50 coins. The mass of each is a distinct positive integer not exceeding 100, and the total mass is even. The Baron claims that it is not possible to divide the coins into two piles with equal total mass. Can the Baron be right?

4. Given an integer $n > 1$, prove that there exist distinct positive integers a, b, c and d such that $a + b = c + d$ and $\frac{a}{b} = \frac{nc}{d}$.

5. AD and BE are altitudes of an acute triangle ABC. From D, perpendiculars are dropped to AB at G and AC at K. From E, perpendiculars are dropped to AB at F and BC at H. Prove that FG is parallel to HK and $FK = GH$.

6. Two ants crawl along the sides of the 49 squares of a 7×7 board. Each ant passes through all 64 vertices exactly once and returns to its starting point. What is the smallest possible number of sides covered by both ants?

7. In a square table of numbers, the sum of the largest two numbers in each row is a and the sum of the largest two numbers in each column is b. Prove that $a = b$.

Note: The problems are worth 4, 4, 5, 6, 7, 10 and 10 points respectively.

Senior A-Level Paper

1. Baron Münchausen has a set of 50 coins. The mass of each is a distinct positive integer not exceeding 100, and the total mass is even. The Baron claims that it is not possible to divide the coins into two piles with equal total mass. Can the Baron be right?

2. In the coordinate space, each of the eight vertices of a rectangular box has integer coordinates. If the volume of the solid is 2011, prove that the sides of the rectangular box are parallel to the coordinate axes.

3. (a) Does there exist an infinite triangular prism such that two of its cross-sections are similar but not congruent triangles?

   (b) Does there exist an infinite triangular prism such that two of its cross-sections are equilateral triangles of sides 1 and 2 respectively?

4. There are n red sticks and n blue sticks. The sticks of each colour have the same total length, and can be used to construct an n-gon. We wish to repaint one stick of each colour in the other colour so that the sticks of each colour can still be used to construct an n-gon. Is this always possible if

   (a) $n = 3$;

   (b) $n > 3$?

5. In the convex quadrilateral $ABCD$, BC is parallel to AD. Two circular arcs ω_1 and ω_3 pass through A and B and are on the same side of AB. Two circular arcs ω_2 and ω_4 pass through C and D and are on the same side of CD. The measures of ω_1, ω_2, ω_3 and ω_4 are α, β, β and α respectively. If ω_1 and ω_2 are tangent to each other externally, prove that so are ω_3 and ω_4.

6. In every square of a square table is a number. The sum of the largest two numbers in each row is a and the sum of the largest two numbers in each column is b. Prove that $a = b$.

7. Among a group of programmers, every two either know each other or do not know each other. Eleven of them are geniuses. Two companies hire them one at a time, alternately, and may not hire someone already hired by the other company. There are no conditions on which programmer a company may hire in the first round. Thereafter, a company may only hire a programmer who knows another programmer already hired by that company. Is it possible for the company which hires second to hire ten of the geniuses, no matter what the hiring strategy of the other company may be?

Note: The problems are worth 4, 6, 3+4, 4+4, 8, 8 and 11 points respectively.

Solutions

Fall 2010

Junior O-Level Paper

1. Let the top row of the frame be row a, the bottom row be row b, the left column be column c and the right column be column d. From the given condition, the frame can be partitioned into dominoes as shown in the diagram below. We may assume that one of the corner squares is black. Then all corner squares are black. Consider all numbers in black squares positive and all numbers in white squares negative. In row a, there are $\frac{d-c}{2}$ dominoes each with sum $-a$. In row b, there are $\frac{d-c}{2}$ dominoes each with sum b. In column c, there are $\frac{b-a}{2}$ dominoes each with sum c, and in column d, there are $\frac{b-a}{2}$ dominoes each with sum $-d$. Hence the grand total is $\frac{1}{2}(-a(d-c)+b(d-c)+c(b-a)-d(b-a))=0$, meaning that the sum of all numbers in black squares is equal to the sum of all numbers in white squares.

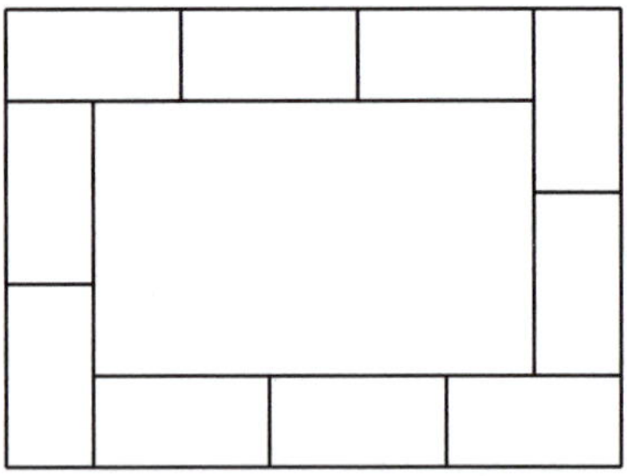

2. The bisectors of $\angle B$ and $\angle C$ both pass through the centre O of the circle. Let them intersect AD at P and Q respectively. By symmetry, $OCDP$ and $OBAQ$ are congruent. Triangles OBC and OPQ are also congruent by symmetry. It follows that CQ indeed bisects the area of $ABCD$

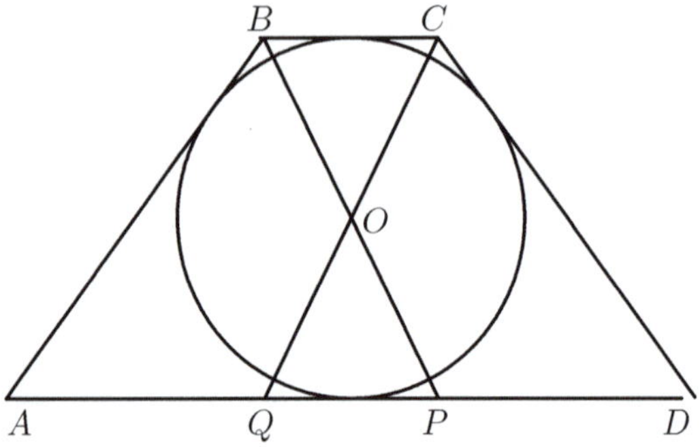

3. It is possible, and the path is indicated in the diagram below. For convenience in description, we use a cubical die starting on the top left corner with the face 2 at the bottom, the face 3 to the left, the face 5 on top, the face 4 to the right, the face 6 to the front and the face 1 to the back. The faces that land on the board are (4,5,3,2,4,5,3), (6,4), (5), (6), (3), (5,4), (6,3), (2), (6,5), (4,2), (6), (4), (5), (6,2), (4,5,3,2,4,5,3), (6), (5), (4), (6), (2), (4,5,3), (6), (5), (4), (6,3), (5,4,2,3,5), (6,2), (3), (6), (5), (3,2,4), (6), (2), (3), (6) and (5). The face 1 never lands on the board.

4. Let the total number of students be T, the number of those who know English, French and German w, the number of those who know English and French but not German x, the number of those who know English and German but not French y, and the number of those who know French and German but not English z. We are given that $\frac{w+x}{T} > \frac{9}{10}$ and $\frac{w+y}{T} > \frac{9}{10}$. From $\frac{w+x}{w+x+y+z} \geq \frac{w+x}{T} > \frac{9}{10}$, we have $w+y > 9(x+z)$. Simlarly, we have $w+x > 9(y+z)$. Hence

$$\begin{aligned} 2w + 9(x+y) &> (w+x) + (w+y) \\ &> 9(x+z) + 9(y+z) \\ &= 9(x+y) + 18z. \end{aligned}$$

It follows that $w > 9z$ and $10w > 9(w+z)$, so that $\frac{w}{w+z} > \frac{9}{10}$.

5. Paint the $2n$ endpoints of the n chords alternately yellow and blue around the circle. For a chord with the desired property, its two endpoints must have the same colour. Since there are n endpoints of each colour, n must be even.

Senior O-Level Paper

1. (a) For $s = 1$, there is obviously no chances of any gain. If $s > 1$ and we trade in n Loonies, we get $ns + \alpha$ McLoonies, where the real number α satisfies $0 \le \alpha \le \frac{1}{2}$. Then we trade these $ns + \alpha$ McLoonies back. The exchange rate is $\frac{ns+\alpha}{s} = n + \frac{\alpha}{s}$. Since $s > 1$, $\frac{\alpha}{s} < \frac{1}{2}$, so that there will not be any rounding up. Thus a one-time gain can only be possible if $s < 1$. Let $s = \frac{1}{2}$. We can trade in 1 Loonie for $\frac{1}{2}$ McLoonie, rounded up to 1. When we trade this back, we get 2 Loonies.

   (b) For an affirmative answer in (a), we must have $s < 1$. This means that $\frac{1}{s} > 1$, so that the argument in (a) shows that there will never be any increase in the number of McLoonies. So after the initial one-time gain in Loonies, no further gain is possible.

2. (a) The sum of the inradii of right triangles AOB and COD is given by
$$\frac{1}{2}(OA + OB - AB) + \frac{1}{2}(OC + OD - CD).$$
The sum of the inradii of right triangles OCB and DOA is given by
$$\frac{1}{2}(OB + OC - BC) + \frac{1}{2}(OD + OA - DA).$$
Hence the given condition is equivalent to $AB + CD = BC + DA$, which is the necessary and sufficient condition for $ABCD$ to have an incircle.

   (b) We may assume that among AB, BC, CD and DA, the longest one is DA. By Pythagoras' Theorem,
$$AB^2 + CD^2 = OA^2 + OB^2 + OC^2 + OD^2 = BC^2 + DA^2.$$
Combined with $(AB+CD)^2 = (BC+DA)^2$, $AB \cdot CD = BC \cdot DA$. Hence $(AB - CD)^2 = (BC - DA)^2$. If $DA - BC = AB - CD$, then $DA = AB$ and $ABCD$ is symmetric about AC. If instead $DA - BC = CD - AB$, then $DA = CD$ and $ABCD$ is symmetric about BD.

3. Let the speed of the cruiser be 1. The policeman's strategy is to go in one direction for a time period q, then go in the opposite direction for a time period q^2, and then go in the original direction for a time period q^3, and so on. At the end of the time period q^n, the total time elapsed is $q^n + q^{n-1} + \cdots + q = \frac{q^{n+1}-q}{q-1}$ and the net distance covered in the current direction is $q^n - q^{n-1} + \cdots + (-1)^n q = \frac{q^{n+1}-(-1)^n q}{q+1}$. Thus the net speed in this direction is $\frac{q^{n+1}-(-1)^n q}{q+1} \div \frac{q^{n+1}-q}{q-1} > \frac{q-1}{q+1}$. Solving for $\frac{q-1}{q+1} > \frac{9}{10}$, we have $q > 19$. Then the net speed of the cruiser still exceeds the raw speed of the car, and capture is inevitable.

4. Let the $n-1$ vertical lines divide the board into n vertical strips of respective width $a_1, a_2, \ldots, a_n$, and let the $n-1$ horizontal lines divide the board into n horizontal strips of respective height $b_1, b_2, \ldots, b_n$. We may assume that the bottom left square is a black square, and from the given conditions, we have $a_i = b_i$ for all $1 \le i \le n$. Consider the area of any black square positive and the area of any white square negative. Then the area of the square of height a_i and width a_j is $(-1)^{i+j}a_i a_j$. The total area of the board is

$$\sum_{i=1}^{n}\sum_{j=1}^{n}(-1)^{i+j}a_i a_j = \sum_{i=1}^{n}(-1)^i a_i \sum_{j=1}^{n}(-1)^j a_j = \left(\sum_{i=1}^{n}(-1)^i a_i\right)^2 \ge 0.$$

Hence the sum of the areas of the black squares cannot be less than the sum of the areas of the white squares.

5. Let a_n denote the total number of participants needed to produce a winner with n victories. We have $a_1 = 2$ and $a_2 = 3$. In order to have a winner with n victories, we must have a candidate with $n-1$ victories so far, and an also-ran with $n-2$ victories so far. Since no participant can lose to both of them, the total number of participants needed is $a_n = a_{n-1} + a_{n-2}$. Thus we have a shifted Fibonacci sequence. Iteration yields $a_3 = 5$, $a_4 = 8$, $a_5 = 13$, $a_6 = 21$, $a_7 = 34$ and $a_8 = 55$. Since we have 55 participants, the winner can have at most 8 victories. It is easy to construct recursively a tournament with 55 participants in which the winner has 8 victories.

Junior A-Level Paper

1. Take two points on the given line at a distance less than the diameter of the round coin. Draw two circles passing through these two points. They are situated symmetrically about the given line. Repeat this operation so that a circle from the second operation intersects a circle from the first operation. The point symmetric to this point about the given line is also a point of intersection of two constructed circles, one from each operation. These two points satisfy the requirement of the problem.

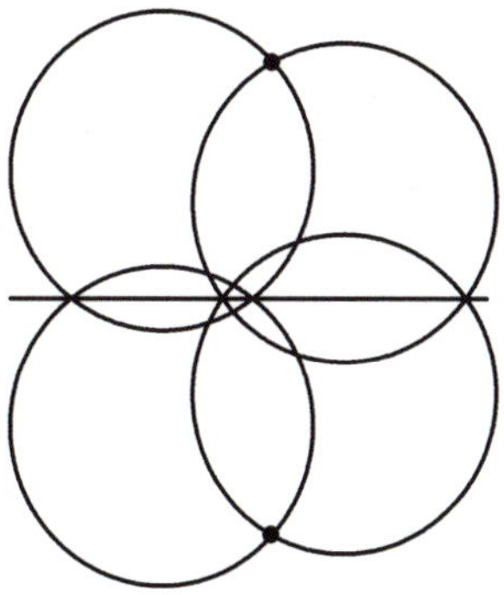

2. Pete is right. Suppose he is asked to divide a line segment into two whose lengths are in the ratio $p : q$, where p and q are relatively prime integers. We may assume that the line segment has length $p + q$. Then Pete can in fact divide the line segment into $p + q$ unit segments. For any segment of length greater than 1, if it has even length, Pete divides it by its midpoint. If it has odd length $2n + 1$, Pete divides it into two whose length are in the ratio $n : (n + 1)$. Eventually, all segments are of length 1. It is then easy to pick out the point which divides the segment into two whose lengths are in the ratio $p : q$.

3. For $1 \leq i \leq 10$, let the constant speed of the cyclist C_i be v_i, where $v_1 < v_2 < \cdots < v_{10}$. Let $u = \min\{v_2 - v_1, v_3 - v_2, \ldots, v_{10} - v_9\}$. Then $v_j - v_i \geq (j - i)u$ for all $j > i$. Let d be the length of the track. Then the meeting between the last pair of cyclists occurs at time $\frac{d}{u}$. Now C_i and C_j meet once in each time interval of length $\frac{d}{v_j - v_i}$. They would have met at least $j - i$ times by the time of the meeting of the last pair, because $(j - i)\frac{d}{v_j - v_i} \leq \frac{d}{u}$. For C_i, this means at least $1 + 2 + \cdots + (i - 1)$ meetings with $C_1, C_2, \ldots, C_{i-1}$ and at least $1 + 2 + \cdots + (10 - i)$ meetings with $C_{i+1}, C_{i+2}, \ldots, C_{10}$. The total is at least

$$\frac{i(i-1) + (10-i)(10-i+1)}{2} = (i-5)(i-6) + 25 \geq 25.$$

4. We first prove that at least one corner of the rectangle is the endpoint of a diagonal. Consider the domino at the bottom right corner. If its diagonal is in the down direction (from the left), then the bottom right corner is the endpoint of a diagonal. Suppose its diagonal is in the up direction. Consider all the dominoes touching the bottom edge of the rectangle. All of their diagonals must be in the up direction, which means that the bottom left corner of the rectangle is the endpoint of a diagonal. Henceforth, we assume that the bottom left corner is the endpoint A_1 of an up-diagonal A_1B_1 of a domino. The next domino has B_1 as one of its vertices but not the one at the bottom left corner. The diagram below illustrates some of the possible choices. The diagonal A_2B_2 in this domino must also be in the up direction. Note that B_2 is above B_1, to the right of B_1 or both.

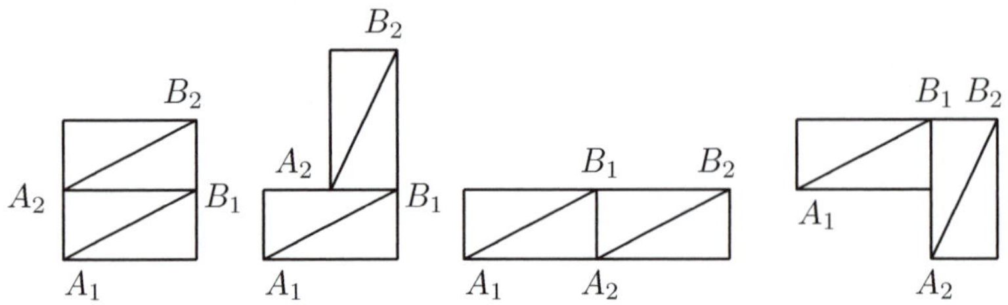

Continuing in this manner, we can build a connected chain of dominoes with up-diagonals going from the bottom left vertex to the upper right vertex of the rectangle. It is impossible to build simultaneously another connected chain of dominoes with down-diagonals going from the bottom right vertex to the upper left vertex of the rectangle. This is because two such chains must share a common domino, and only one diagonal of that domino is drawn.

5. Let the side lengths be $a_1 \le a_2 \le \cdots \le a_5 < p - a_5$ where

$$p = a_1 + a_2 + \cdots + a_5.$$

Since $p > 2a_5$, we have

$$\begin{aligned} & \frac{a_1}{p-a_1} + \frac{a_2}{p-a_2} + \cdots + \frac{a_5}{p-a_5} \\ \le\; & \frac{a_1}{p-a_5} + \frac{a_2}{p-a_5} + \cdots + \frac{a_5}{p-a_5} \\ =\; & \frac{p}{p-a_5} \\ <\; & 2. \end{aligned}$$

6. We first prove an auxiliary result. Let the line through the midpoint D of BC and perpendicular to BC cut EF at G. Then $BH \cdot FG = CH \cdot EG$. Drop perpendiculars FX and EY from F and E to BC respectively. Then X is the midpoint of BH and Y is the midpoint of CD. Since triangles FXD and EYC are congruent, we have $XD = YC = YH$ so that $XH = YD$. Now $BH \cdot FG = 2XH \cdot XD = 2YH \cdot YD = CH \cdot EG$.

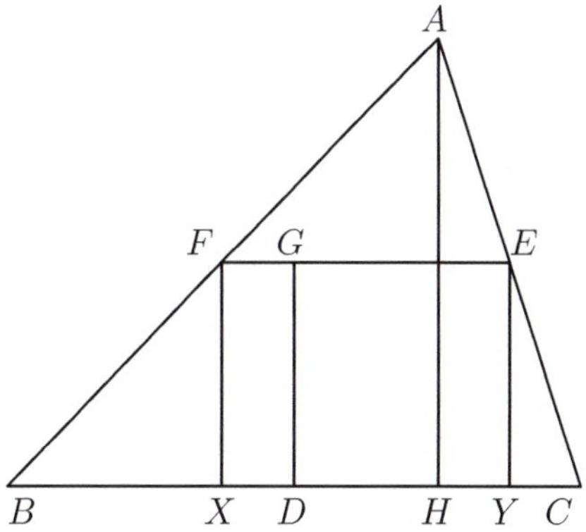

Returning to the main problem, let the line through F perpendicular to BP intersect the line DG at K_1. Triangles FGK_1 and PHB are similar. Hence $GK_1 = \frac{FG \cdot BH}{PH}$. Let the line through E perpendicular to CP intersect the line DG at K_2. We can prove in an analogous manner that $GK_2 = \frac{EG \cdot CH}{PH}$.

By the auxiliary result, $GK_1 = GK_2$. Hence K_1 and K_2 are the same point K. Since K lies on the perpendicular bisector of BC, we have $KB = KC$.

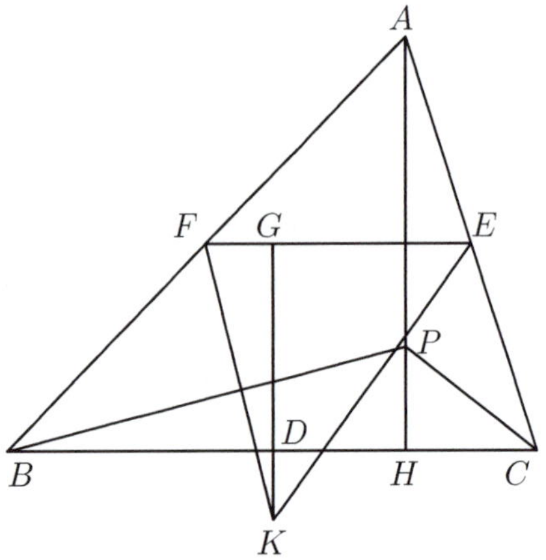

7. We may assume that the knights are seated from 1 to n in clockwise order on day 1. Then seat exchanges are not permitted between knights with consecutive numbers (1 and n are considered consecutive). We construct an invariant for a cyclic order called the winding number as follows. Merlin has n hats numbered from 1 to n from top to bottom. He starts by giving hat 1 to knight 1. Then he continues in the clockwise order round the table until he gets to knight 2, when he will give him hat 2. After he has handed out all the hats, Merlin returns to knight 1 which is his starting point. The number of times he has gone round the table is called the winding number of the cyclic order. For instance, the winding number of the cyclic order 1 4 7 2 3 6 5 is 4: (1,2,3)(4,5)(6)(7). Suppose two adjacent knights change seats. If neither is knight 1, the hats handed out in each round remain the same, so that the winding number remains constant. Suppose the seat exchange is between knight 1 and knight h, where $h \neq 2$ or n. Then knight h either becomes the first knight to get a hat in the next cycle instead of the last knight to get a hat in the preceding cycle, or vice versa. Still, the winding number remains constant. On the kth day, $1 \leq k \leq n-1$, of the conference, Merlin can start by having the knights sit in the cyclic order $k, k-1, \ldots, 2, 1, k+1, k+2, \ldots, n$. It is easy to verify that the winding number of the starting cyclic order on the kth day is k. It follows that no cyclic order can repeat on two different days within the first $n-1$ days. Therefore, Merlin can make the conference last at least n days. We claim that given any cyclic order, it can be transformed into one of those with which Merlin starts a day. Then the knights can make the conference end when the nth day is over.

Suppose the cyclic order is not one with which Merlin starts a day. We will push knight 2 forward in the clockwise direction until he is adjacent to knight 1. This can be accomplished by a sequence of exchanges if he does not encounter knight 3 along the way. If he does, we will push both of them forward towards knight 1. Eventually, we will have knights 2, 3, $\dots$, h and 1 in a block. Now $h < n$ as otherwise the initial cyclic order is indeed one of those with which Merlin starts a day. Hence we can push knight 1 counter-clockwise so that he is adjacent to knight 2. We now attempt to put knight 3 on the other side of knight 2. As before, we have knights 3, 4, $\dots$, ℓ, 2, 1 in a block. If $\ell < n$, we can push knights 2 and 1 counter-clockwise towards knight 3. If $\ell = n$, knight 1 cannot get past, but we notice that we have arrived at one of the cyclic orders with which Merlin starts a day. This justifies the claim.

Senior A-Level Paper

1. (a) This is not always possible. Suppose the record of the distance AB is lost. If the other 98 points all lie on a line ℓ, we cannot tell whether A and B are on the same side of ℓ or on opposite sides of ℓ. Thus the lost record cannot be restored from the remaining ones.

   (b) The answer is 96. Suppose 97 records are erased. All of them may be associated with a point A so that we only know the distances AB and AC, where B and C are 2 of the other 99 points. A does not lie on BC as no three of the 100 points lie on a line. Now we cannot determine whether A is on one side or the other side of the line BC. Suppose at most 96 records are erased. Construct a graph with 100 vertices representing the 100 points. Two vertices are joined by an edge if the record of the distance between the two points they represent is erased. The graph has at most 96 edges, and therefore at least 4 components. Take four vertices A, B, C and D, one from each component. The pairwise distances between the points A, B, C and D are on record, so that their relative position can be determined. For any other vertex P, it is in the same component with only one of these four vertices. Hence the distance between the point P and three of the points A, B, C and D are on record. This is enough to determine the position of the point relative to the points A, B, C and D. It follows that all erased records may be restored.

2. For $1 \leq i \leq 2n$, let the constant speed of the cyclist C_i be v_i, where $v_1 < v_2 < \cdots < v_{2n}$. Let $u = \min\{v_2 - v_1, v_3 - v_2, \dots, v_{2n} - v_{2n-1}\}$. Then $v_j - v_i \geq (j - i)u$ for all $j > i$. Let d be the length of the track. Then the meeting between the last pair of cyclists occurs at time $\frac{d}{u}$.

Now C_i and C_j meet once in each time interval of length $\frac{d}{v_j - v_i}$. They would have met at least $j - i$ times by the time of the meeting of the last pair, because $(j-i)\frac{d}{v_j - v_i} \le \frac{d}{u}$. For C_i, this means at least

$$1 + 2 + \cdots + (i-1)$$

meetings with $C_1, C_2, \ldots, C_{i-1}$ and at least

$$1 + 2 + \cdots + (2n-i)$$

meetings with $C_{i+1}, C_{i+2}, \ldots, C_{2n}$. The total is at least

$$\frac{i(i-1) + (2n-i)(2n-i+1)}{2} = (i-n)(i-(n+1)) + n^2 \ge n^2.$$

(Compare with Problem 3 of the Junior A-Level Paper.)

3. (See Problem 5 of the Junior A-Level Paper.)

4. (a) The answer is no. With a finite number of spells, there is one for which $b - a$ is maximum. If the first wizard keeps casting this spell, the best that the second wizard can do is to maintain status quo by casting the same spell. Hence the second wizard will hit the water first, giving the first wizard a win.

   (b) The answer is yes. In the nth spell, let $a = \frac{1}{n}$ and $b = 100 - \frac{1}{n}$. By symmetry, we may assume that the first wizard casts the nth spell. He is then $100 - \frac{1}{n}$ above water while the second wizard is $\frac{1}{n}$ above water. However, the second wizard wins immediately by casting the $(n+1)$st spell. He will still be $\frac{1}{n} - \frac{1}{n+1} = \frac{1}{n(n+1)}$ above water while the first wizard is submerged in water since $(100 - \frac{1}{n}) - (100 - \frac{1}{n+1}) = -\frac{1}{n(n+1)}$.

5. Let P be the circumcentre of triangle OAC. Then PO is perpendicular to AC, intersecting AC at X. Let the line through O perpendicular to BD intersect BD at Y and AC at Q.

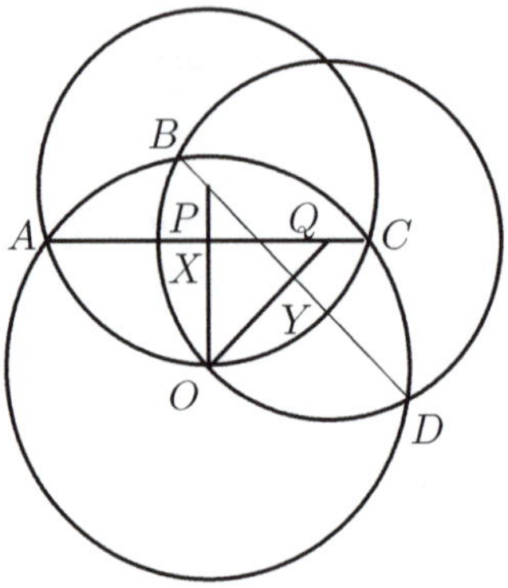

We claim that Q is the circumcentre of triangle OBD. By Pythagoras' Theorem,

$$\begin{aligned} QD^2 &= QY^2 + DY^2 \\ &= QY^2 + (OD^2 - OY^2) \\ &= QY^2 + OA^2 - (OP^2 - PY^2) \\ &= OA^2 - AP^2 + PQ^2 \\ &= QX^2 + OA^2 - (AP^2 - PX^2) \\ &= QX^2 + (OA^2 - AX^2) \\ &= QX^2 + OX^2 \\ &= QO^2. \end{aligned}$$

6. Let $S(p,q)$ denote the following statement.
"In any binary $a \times b$ table with $ab \le p$, one of the following is true.
(A) There exists an $a \times q$ subtable with at least one 0 in each row.
(B) There exists a $q \times b$ subtable with at least one 1 in each column.
The q rows or columns may be chosen arbitrarily."
We wish to prove that $S(1000000, 10)$ is true. We first examine $S(4,1)$. The relevant tables are 1×4, 1×3, 1×2, 1×1, 2×1, 3×1, 4×1 and 2×2. In the first four, if there is at least one 0, then (A) is true. Otherwise, (B) is true. In the next three, if there is at least one 1, then (B) is true. Otherwise, (A) is true. In the last one, if there are no 0s in the first row, then (B) is true. Suppose there is at least one 0 in the first row. If there is at least one 0 in the second row as well, then (A) is true. Otherwise, (B) is true. It follows that $S(4,1)$ is true. We claim that $S(p,q)$ implies $S(4p, q+1)$. Consider any $a \times b$ table with $ab \le 4p$. Let x be the minimum number of 1s in any row and y be the minimum number of 0s in any column. Then the total number of 1s is at least ax and the total number of 0s is at least by. It follows that $ax + by \le ab$. By the Arithmetic-Geometric Means Inequality, $\sqrt{(ax)(by)} \le \frac{ax+by}{2} \le \frac{ab}{2}$. Hence $(ax)(by) \le \frac{(ab)^2}{4}$ so that $xy \le p$. Let R be a row with exactly x 1s and C be a column with exactly y 0s. Consider the $y \times x$ table whose rows have 0s at the intersections with C and whose columns have 1s at the intersections with R. We are assuming that $S(p,q)$ is true, so that either (A) or (B) holds for this table. If (A) holds, adding C would make (A) hold for the $a \times b$ table. If (B) holds, adding R would make (B) hold for the $a \times b$ table. This justifies the claim. From $P(4,1)$, we can deduce in turns $S(16,2)$, $S(64,3)$ and so on, up to $S(1048576, 10)$. This clearly implies $S(1000000, 10)$.

7. Let the rectangles be of dimensions $m \times n$ or $n \times m$. Divide the whole board into unit squares. Put the label 0 on each square on the chosen diagonal. Put the label 1 on each square of the next $m+n-1$ diagonals above and parallel to the given diagonal, and put the label -1 on each square of the next $m+n-1$ diagonals below and parallel to the given diagonal. For each important rectangle, all squares have been labeled, and their sum is equal to the area of the part of the rectangle above the given diagonal minus the area of the part below. For each unimportant rectangle, at least one square is unlabeled. We now proceed to complete the labelling, diagonal by diagonal away from and parallel to the given one. Place an $m \times n$ or $n \times m$ rectangle on the board so that the square we are trying to label is the only unlabeled square in that rectangle. We choose a label so that the sum of all the labels in this rectangle is 0. Because labels on diagonals parallel to the given one are the same, the choice of position or orientation of this rectangle is immaterial.

0	1	1	1	1	−5	7	−5	1	1	1	−5
−1	0	1	1	1	1	−5	7	−5	1	1	1
−1	−1	0	1	1	1	1	−5	7	−5	1	1
−1	−1	−1	0	1	1	1	1	−5	7	−5	1
−1	−1	−1	−1	0	1	1	1	1	−5	7	−5
5	−1	−1	−1	−1	0	1	1	1	1	−5	7
−7	5	−1	−1	−1	−1	0	1	1	1	1	−5
5	−7	5	−1	−1	−1	−1	0	1	1	1	1
−1	5	−7	5	−1	−1	−1	−1	0	1	1	1
−1	−1	5	−7	5	−1	−1	−1	−1	0	1	1
−1	−1	−1	5	−7	5	−1	−1	−1	−1	0	1
5	−1	−1	−1	5	−7	5	−1	−1	−1	−1	0

The completed labelling of a 12×12 board, with $m = 2$ and $n = 3$, is shown in the diagram above. Each label off the given diagonal is the negative of the label symmetric to it about the given diagonal. It follows that the sum of all the labels in the whole square is 0. The sum of the labels in each unimportant rectangle is chosen to be 0. Hence the sum of the labels in all the important rectangles is also 0. This means that the given diagonal bisects their total area.

Spring 2011

Junior O-Level Paper

1. If all differences between two adjacent numbers are odd, then the numbers must be alternately odd and even. Also, the numbers are given to be alternately higher than both neighbours and lower than both neighbours. Now 2010 must be a high number. Hence all the even numbers are high numbers. However, 2 clearly cannot be a high number. This is a contradiction.

2. Let the widths of the columns be x_i and the heights of the rows be y_i, $1 \leq i \leq 11$. A square is said to be good if its perimeter is an integer. Since there are at least 111 good squares, there are at most 10 bad squares, so that an entire row and an entire column is free of bad squares. Let these be the first row and the first column. In particular, $2x_1 + 2y_1$ is an integer. Consider the square on the ith row and the jth column, $2 \leq i, j \leq 11$. Its perimeter is $2x_j + 2y_i$. Both the squares on the ith row and the first column and on the jth column and the first row are good. Hence $2x_1 + 2y_i$ and $2x_j + 2y_1$ are integers. It follows that $2x_j + 2y_i$ is also an integer, meaning that every square is good.

3. We divide 1 metre into 512 sillimetres and 1 hour into 512 sillihours. Then an under-sized worm grows at the rate of 1 sillimetre per sillihour. At the start, we cut the full-grown worm into two, one of length 1 sillimetre and the other 511 sillimetres. After 1 sillihour, the shorter worm has length 2 sillimetres and the longer worm is full-grown. It is then dissected into two, so that the shorter one is also of length 2 sillimetres. After another 2 sillihours, the shorter worms have length 4 sillimetres and the longer worm is full-grown. It is then dissected into two so that the shorter one is also of length 4 sillimetres. Continuing this way, we will have 10 full-grown worms after $1 + 2 + 4 + \cdots + 256 = 511$ sillihours, just beating the deadline of 1 hour.

4. The answer is no. In the convex quadrilateral $ABCD$ in the diagram below, $\angle ADB = \angle BDC = \angle DCA = \angle ACB = \angle BAC = \angle ABD = 36°$, where E is the point of intersection of the diagonals. Then we have $\angle ADC = \angle DAE = \angle DEA = \angle CEB = \angle CBE = \angle BCD = 72°$. Hence all of the triangles ABC, BAD, ACD, BDC, ABE, BCE, CDE and DAE are isosceles, and yet $ABCD$ is not a square.

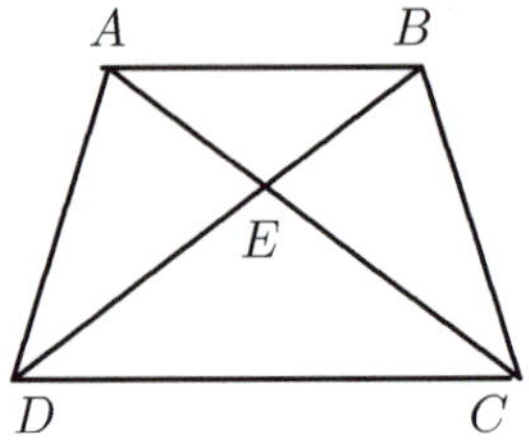

5. We just solve (b), and (a) follows. On the first day, the knight divides the 100 coins into a small pile containing 25 coins and a large pile containing 75 coins. Each day for which he is still in confinement, he transfers 1 coin from the large pile to the small pile. If he escapes, the dragon continues the transfer. If he is still confined by the 25th day, there will be 49 coins in the small pile and 51 coins in the large pile. On the 1st day, if the small pile contains 25 magic coins or 25 non-magic coins, the knight will be free. If not, the pile contains less than 25 magic coins and less than 25 non-magic coins. On the 25th day, the small pile contains either at least 25 magic coins or at least 25 non-magic coins. By symmetry, we may assume it contains at least 25 magic coins. Since the coins are transferred one at a time to the small pile, the number of magic coins must be exactly 25 at some point, on or before the 25th day. Hence the knight can guarantee freedom by the 25th day at the latest.

Senior O-Level Paper

1. If each face is an equilateral triangle, then all faces must be congruent. Since a polyhedron must have at least four faces, we easily have two pairs of congruent faces. If the faces are not equilateral triangles, consider the longest and shortest edge of the polyhedron. Then the two triangles sharing the longest edge of the polyhedron must be congruent in order to be similar. Otherwise, blowing one of them up to be larger will produce an edge of greater length. Similarly, the two triangles sharing the shortest edge of the polyhedron are also congruent. If the longest edge and the shortest edge appear in the same face, then all faces must again be congruent in order to be similar.

2. (See Problem 3 of the Junior O-Level Paper.)

3. For each k, $2 \leq k \leq 50$, construct a graph with the 100 points as vertices. Two vertices are joined by an edge if and only if they are the end vertices of a block of k adjacent vertices. In each move, we paint the two vertices of some edge.

Now the graph may be a cycle or a union of disjoint cycles of the same length. If the length is even, choose every other edge and paint the vertices of the chosen edges. Then all 100 points are painted. If the length is odd, then there is an unpaintable vertex on each cycle. The number of cycles is given by the greatest common divisor d of $k-1$ and 100, which may be any of 1, 2, 4, 5, 10, 20 and 25. The respective lengths are 100, 50, 25, 20, 10, 5 and 4. So the only cases for which the task fails is when $d = 4$ or 20. The former yields $k = 5, 9, 13, 17, 25, 29, 33, 37, 45$ and 49. and the latter yields $k = 21$ or 41. For all other values of k, $2 \leq k \leq 50$, the task is possible.

4. Let $ABCDE$ be the pentagon. Let BG, CH, DI and EJ be the altitudes concurrent at O. Let F be the point of intersection of AO with CD.

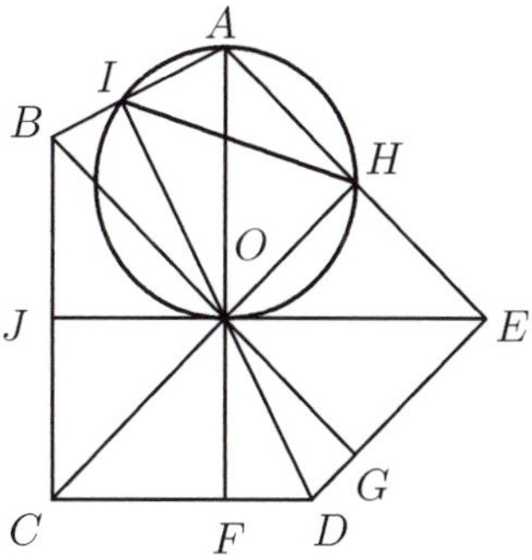

Since the right triangles COJ, JOB and BOI are similarly to the right triangles EOH, HOD and DOG, we have

$$OC \cdot OH = OE \cdot OJ = OB \cdot OG = OD \cdot OI.$$

It follows that triangles HOI and DOC are also similar, so that we have $\angle OHI = \angle ODC$. Since $\angle OHA = 90° = \angle OIA$, $AHOI$ is a cyclic quadrilateral. Hence $\angle AHI = \angle AOI = \angle DOF$. Now

$$\angle OFD = \angle 180° - \angle DOF - \angle ODF = 180° - \angle OHA - \angle OHI = 90°.$$

5. Let $\mathcal{F}$ be the set of towns with an odd number of paved roads and $\mathcal{G}$ be the set of towns with an even number of paved roads. Note that $|\mathcal{F}|$ is even at any time. Initially, $|\mathcal{F}| = 0$. If we have $|\mathcal{F}| = 100$ at some point, the task is accomplished. Suppose $|\mathcal{F}| < 100$. Then there are at least 2 towns A and B in $\mathcal{G}$. Since the graph is connected, there exists a tour from A to B, going along the roads without visiting any town more than once. Interchange the status of each road on this tour, from paved to unpaved and vice versa. (This is of course done on the planning map, before any actual paving is carried out.) Then A and B move from $\mathcal{G}$ to $\mathcal{F}$ while all other towns stay in $\mathcal{F}$ or $\mathcal{G}$ as before. Hence we can make $|\mathcal{F}|$ increase by 2 at a time, until it reaches 100.

Junior A-Level Paper

1. The diagram below shows such a hexagon and how it is dissected.

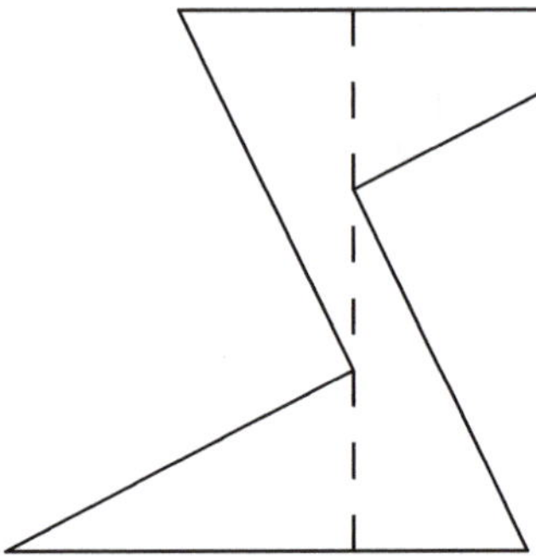

2. The line $x + y = 100$ is parallel to the line making a $135°$ angle with the positive x-axis. Hence there are only 179 points of intersection. The point of intersection of $x + y = 100$ with $y = x$ is (50,50), and the other points of intersections are arranged in symmetric pairs with respect to this point. Hence the desired sum is $179 \times 50 = 8950$.

3. The Baron is right, as usual. The masses of the 50 coins in his collection may just be the 50 even numbers up to 100. Their total mass is 50 times 51, so that each of two piles would have total mass 25 times 51, which is odd. This is impossible since all the coins have even weights.

4. To satisfy $\frac{a}{b} = \frac{nc}{d}$ only, we can take $a = 2n$, $b = 1$, $c = 2$ and $d = 1$. Note that $a + b = 2n + 1$ while $c + d = 3$. To satisfy $a + b = c + d$ as well, we now change these values to $a = 3(2n) = 6n$, $b = 3(1) = 3$, $c = (2n+1)2 = 4n+2$ and $d = (2n+1)1 = 2n+1$. These four numbers are distinct when $n > 1$.

5. Since $\angle EHD = 90° = \angle EKD$, $EDHK$ is a cyclic quadrilateral.

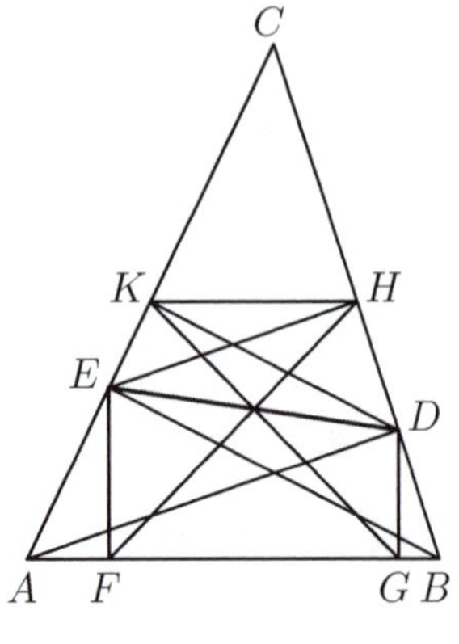

Hence $\angle EHK = \angle EDK$. Now DK and BE are parallel since both are perpendicular to AC. Hence $\angle EDK = \angle DEB$. Finally, we have $\angle DEB = \angle DAB$ since $ABDE$ is also a cyclic quadrilateral. Hence $\angle EHK = \angle DAB$. Now HK and AB make equal angles with the parallel lines EH and DA. Hence HK is parallel to AB and therefore to FG. Since $\angle EFB = 90° = \angle EHB$, $EFBH$ is a cyclic quadrilateral. Hence $\angle EHF = \angle EBA$. Similarly, $\angle DKG = \angle DAB$. It follows that $\angle FHK = \angle FHE + \angle EHK = \angle EBA + \angle DAB$. By symmetry, $\angle GKH = \angle GKD + \angle DKH = \angle DAB + \angle EBA$ also. It follows easily that triangles FHK and GKH are congruent, so that $FK = GH$.

6. The eight sides at the four corner vertices must be traversed by both ants. Along each of the four edges of the board, the middle four vertices all have degree 3, and must lie on a side traversed by both ants. To minimize the number of such sides, they must cover these sixteen vertices in pairs. Hence their number cannot be less than 8+8=16. The following diagram shows the paths of the two ants with exactly 16 sides covered by both, every other side along the four edges of the board.

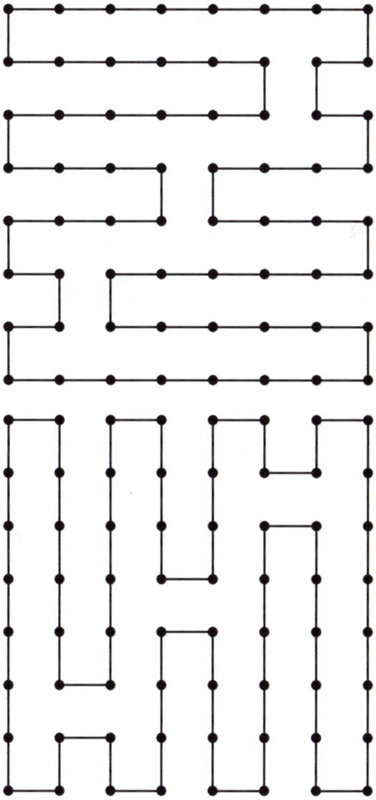

7. We only need to prove that $a \geq b$ since we will then have $b \geq a$ by symmetry. Circle the largest number x_j in column j, $1 \leq j \leq n$, where n is the number of rows and therefore of columns. By relabelling if necessary, we may assume that x_1 is the smallest of these n numbers. We consider two cases.
   Case 1. Two circled numbers x_j and x_k are on the same row.
   Then $a \geq x_j + x_k \geq 2x_1 \geq b$ since the sum of the largest two numbers in column 1 is b.
   Case 2. Each circled number is in a different row.
   Let the second largest number y_1 in column 1 be in row j, and let the circled number in row j be x_k. Then $b = x_1 + y_1 \leq x_k + y_1 \leq a$ since the sum of the largest two numbers in row k is a.

Senior A-Level Paper

1. The Baron is right, as usual. The masses of the 50 coins in his collection may just be the 50 even numbers up to 100. Their total mass is 50 times 51, so that each of two piles would have total mass 25 times 51, which is odd. This is impossible since all the coins have even weights.

2. Let the side lengths of the rectangular box be $a \leq b \leq c$. Because the vertices are lattice points, each of a^2, b^2 and c^2 is an integer. Since 2011 is a prime, we must have either $a = b = 1$ and $c = 2011$, or $a = 1$ and $b = c = \sqrt{2011}$. In the former case, the sides of lengths 1 must be parallel to two of the coordinate axes. It follows that the side of length 2001 must also be parallel to the third coordinate axis. In the latter case, the side of length 1 is parallel to a coordinate axis, and the sides of length $\sqrt{2011}$ define a square in a plane parallel to a coordinate plane. Again, because the vertices are lattice points, we must be able to express 2011 as a sum of two squares. One of them must be the square of an even integer and the other the square of an odd integer, say $(2m)^2 + (2n + 1)^2 = 2011$. This simplifies to $4(m^2 + n^2 + n) = 2010$, which is impossible since 2010 is not a multiple of 4.
3. (a) Take a prism with a cross-section ABC where $AB = \sqrt{3}$, $AC = 2$ and $BC = 1$, so that $\angle ABC = 90°$.

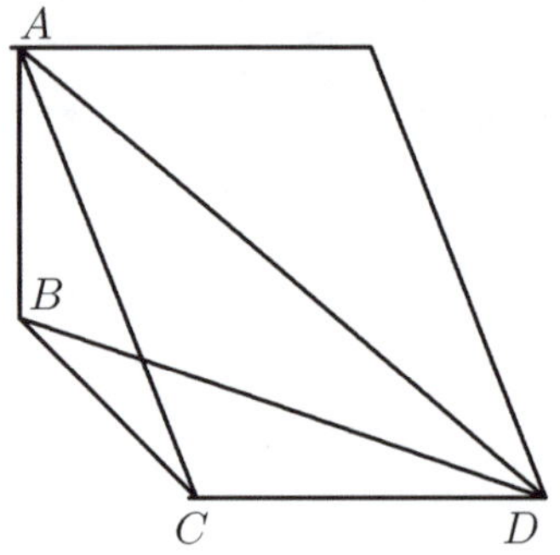

Let D be the point on the line through C perpendicular to ABC such that $BD = 3$. Then $AD = 2\sqrt{3}$ and triangle DBA is similar to triangle ABC.

(b) Take a prism $ABCA'B'C'$ as shown in the diagram below, with $AB = c$, $AC = b$ and $BC = a$. If we can inscribe an equilateral triangle of side 1 in this prism, then each of a, b and c is less than 1. Take the cross-section XBC' where X is a point on AA' to be chosen. Let $AX = x$ and $XA' = y$. Note that we have $BC' = \sqrt{a^2 + (x+y)^2}$, $BX = \sqrt{b^2 + x^2}$ and $XC' = \sqrt{c^2 + y^2}$. If XBC' is an equilateral triangle of side 2, then $x = \sqrt{4 - b^2} \geq \sqrt{3}$ and $y = \sqrt{4 - c^2} \geq \sqrt{3}$. Hence $4 \geq a^2 + (\sqrt{3} + \sqrt{3})^2 > 12$, which is a contradiction.

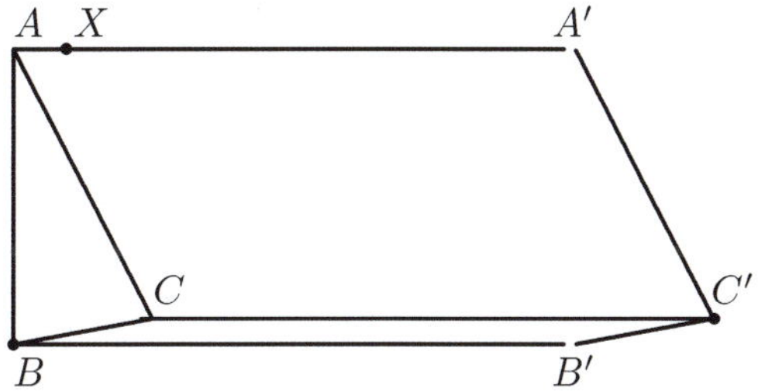

4. (a) The red sticks may be of lengths 16, 16 and 1, and the blue sticks may be of lengths 13, 11 and 9. Note that 16+16+1=13+11+9, $16 < 16 + 1$ and $13 < 11 + 9$. The conditions of the problem are satisfied. If we swap the red stick of length 1 with any blue stick, it will not form a triangle with the other two blue sticks. If we swap any blue stick with a red stick of length 16, it will not form a triangle with the other two red sticks.

(b) There is a red stick of length $5 - \frac{1}{n(n+1)}$ and another of length $4 + \frac{1}{n(n+1)}$. The other $n - 2$ are of length $\frac{1}{n-2}$. There is a blue stick of length $5 - \frac{1}{n^2(n+1)}$ and another of length $5 - \frac{1}{n^3(n+1)}$. The other $n - 2$ are of length $\frac{1}{n^3(n-2)}$. The total length of the sticks of each colour is 10. We consider four cases.

Case 1. A short red stick is swapped for a short blue stick.
We will have trouble on the predominantly red side, because

$$\begin{aligned}
& 4 + \frac{1}{n(n+1)} + \frac{n-3}{n-2} + \frac{1}{n^3(n-2)} - 5 + \frac{1}{n(n+1)} \\
< \quad & \frac{3}{n(n+1)} + \frac{n-3}{n-2} - 1 \\
< \quad & \frac{1}{n-2} + \frac{n-3}{n-2} - 1 \\
= \quad & 0.
\end{aligned}$$

Case 2. A long red stick is swapped for a short blue stick.
We will have trouble on the predominantly red side because we have $4+\frac{1}{n(n+1)}>1+\frac{1}{n^3(n-2)}$.
Case 3. A long red stick is swapped for a long blue stick.
We will have trouble on the predominantly blue side because we have $5-\frac{1}{n^2(n+1)}>5-\frac{1}{n(n+1)}+\frac{1}{n^3}$.
Case 4. A short red stick is swapped for a long blue stick.
We will have even greater trouble on the predominantly blue side.

5. Let X and Y be the respective midpoints of AB and CD. Let O, P, Q' and P' be the respective centres of ω_1, ω_2, ω_3 and ω_4. Let E, F, M, N, M' and N' be points on AD such that YE, XF, OM, PN, $O'M'$ and $P'N'$ are perpendicular to AD. Let K be the point on XF such that OK is perpendicular to XF, L be the points on PN such that LY is perpendicular to PN, K' be the point on XF such that $O'K'$ is perpendicular to XF, and L' be the point on $P'N'$ such that $L'Y$ is perperndicular to $P'N'$. Let $AX=a$, $DY=c$, $DE=e$, $AF=f$ and $XF=YE=h$.

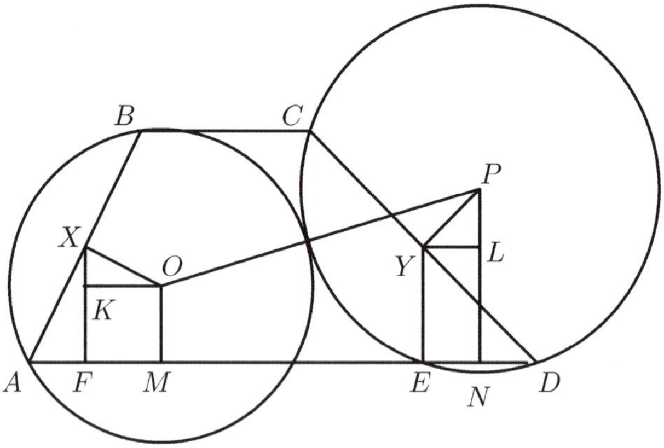

We first prove three preliminary results.
(1) $OB\cdot PC=O'B\cdot P'C$.
By the Law of Sines, we have $2OB\sin\alpha=2a=2O'B\sin\beta$ as well as $2PC\sin\beta=2c=2P'C\sin\alpha$. Hence $cOB=aP'C$ and $aPC=cO'B$, so that $OB\cdot PC=O'B\cdot P'C$.
(2) $OX\cdot PY=O'X\cdot P'Y$.
We have $c^2OX^2=c^2(OB^2-a^2)=a^2(P'C^2-c^2)=a^2P'Y^2$ so that $cOX=aP'Y$. Similarly, $cO'X=aPY$. Hence $OX\cdot PY=O'X\cdot P'Y$.
(3) $KX\cdot LP=K'O'\cdot L'Y$.
We have $KX=\frac{f}{a}OX$, $LP=\frac{e}{c}PY$, $K'O'=\frac{f}{a}O'X$ and $L'Y=\frac{e}{c}P'Y$ It follows from (2) that $KX\cdot LP=K'O'\cdot L'Y$.
The horizontal distance between O and P is $MN=EF-MF+EN$.
The vertical distance between O and P is $PN-OM=KX+LP$.

We are given that ω_1 and ω_2 are tangent, so that

$$MN^2 + (KX + LP)^2 = (OB + PC)^2.$$

The horizontal distance between O' and P' is $M'N' = EF - E'N + F'M$, and the vertical distance is $O'M' - P'N' = K'O' + L'Y$. To have ω_3 tangent to ω_4, we need to prove that

$$(M'N')^2 + (K'O' + L'Y)^2 = (O'B + P'C)^2.$$

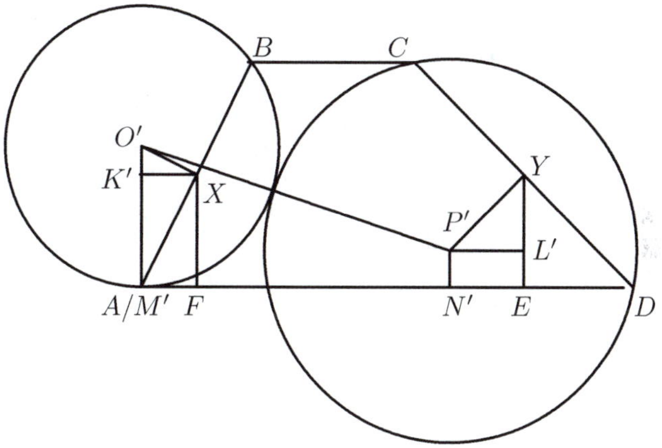

Note that

$$FM = \frac{h}{a}OX = \frac{h}{c}P'Y = EN'$$

and

$$FM' = \frac{h}{a}O'X = \frac{h}{c}PY = EN.$$

Hence $MN = M'N'$. In view of (1) and (3), the desired result now follows from

$$\begin{aligned}
& OB^2 - KX^2 + PC^2 - LP^2 \\
= \; & \frac{a^2}{c^2}P'C^2 - \frac{f^2}{a^2}OX^2 + \frac{c^2}{a^2}O'B^2 - \frac{e^2}{c^2}PY^2 \\
= \; & \frac{a^2}{c^2}P'C^2 - \frac{f^2}{c^2}P'Y^2 + \frac{c^2}{a^2}O'B^2 - \frac{e^2}{a^2}O'X^2 \\
= \; & \frac{h^2}{c^2}P'C^2 - f^2 + \frac{h^2}{a^2}O'B^2 - e^2 \\
= \; & \frac{h^2}{a^2}OB^2 - e^2 + \frac{h^2}{c^2}PC^2 - f^2 \\
= \; & \frac{a^2}{c^2}PC^2 - \frac{f^2}{c^2}PY^2 + \frac{c^2}{a^2}OB^2 - \frac{e^2}{a^2}OX^2 \\
= \; & \frac{a^2}{c^2}PC^2 - \frac{f^2}{a^2}O'X^2 + \frac{c^2}{a^2}OB^2 - \frac{e^2}{c^2}P'Y^2 \\
= \; & O'B^2 - (K'O')^2 + P'C^2 - L'Y^2.
\end{aligned}$$

6. We only need to prove that $a \geq b$ since we will then have $b \geq a$ by symmetry. Circle the largest number x_j in column j, $1 \leq j \leq n$, where n is the number of rows and therefore of columns. By relabelling if necessary, we may assume that x_1 is the smallest of these n numbers. We consider two cases.
Case 1. Two circled numbers x_j and x_k are on the same row.
Then $a \geq x_j + x_k \geq 2x_1 \geq b$ since the sum of the largest two numbers in column 1 is b.
Case 2. Each circled number is in a different row.
Let the second largest number y_1 in column 1 be in row j, and let the circled number in row j be x_k. Then $b = x_1 + y_1 \leq x_k + y_1 \leq a$ since the sum of the largest two numbers in row k is a.

7. Let there be eleven attributes on which the companies rank the candidates. The ranking of each attribute for each candidate is a non-negative integer. It turns out that for each candidate, the sum of the eleven rankings is exactly 100. Moreover, no two candidates have exactly the same set of rankings, and for each possible set of rankings, there is such a candidate. The eleven geniuses are those with a ranking of 100 in one attribute and a ranking of 1 in every other attribute. Two candidates know each other if their sets of rankings differ only in two attributes, and those two rankings differ by 1. Consider candidate A who is the first hired by the first company. By the Pigeonhole Principle, the ranking of at least one attribute for A is at least 10, and we may assume that this is the first attribute. The second company hires the candidate whose ranking in the first attribute is exactly 10 lower than that of A, but exactly 1 higher in each of the other ten attributes. At this point, the first company has a big edge in hiring the genius of the first attribute, but the second company has a small edge in hiring the genius of each of the other ten attributes. The second company concedes the genius of the first attribute to the first company, but aims to hire the other ten geniuses by maintaining these small advantages. Note that among the candidates hired by each company, the highest ranking in any attribute can only increase by 1 with each new hiring. Whenever the first company makes a hiring, the second company will respond by hiring a candidate whose rankings change in the same attributes and in the same directions.

Tournament 33

Fall 2011

Junior O-Level Paper

1. P and Q are points on the longest side AB of triangle ABC such that $AQ = AC$ and $BP = BC$. Prove that the circumcentre of triangle CPQ coincides with the incentre of triangle ABC.

2. Several guests at a round table are eating from a basket containing 2011 berries. Going in clockwise direction, each guest has eaten either twice as many berries as or six fewer berries than the next guest. Prove that not all the berries have been eaten.

3. From the 9×9 board, all 16 squares whose row numbers and column numbers are both even have been removed. The punctured board is dissected into rectangular pieces. What is the minimum number of square pieces?

4. Along a circle are the integers from 1 to 33 in some order. The sum of every pair of adjacent numbers is computed. Is it possible for these sums to consist of 33 consecutive numbers?

5. On a highway, a pedestrian and a cyclist are going in the same direction, while a cart and a car are coming from the opposite direction. All are travelling at constant speeds, not necessarily equal to one another. The cyclist catches up with the pedestrian at 10 o'clock. After a time interval, the cyclist meets the cart, and after another time interval equal to the first, she meets the car. After a third time interval, the car meets the pedestrian, and after another time interval equal to the third, the car catches up with the cart. If the pedestrian meets the car at 11 o'clock, when does he meet the cart?

Note: The problems are worth 3, 4, 4, 4 and 5 points respectively.

Senior O-Level Paper

1. Several guests at a round table are eating from a basket containing 2011 berries. Going in clockwise direction, each guest has eaten either twice as many berries as or six fewer berries than the next guest. Prove that not all the berries have been eaten.

2. Peter buys a lottery ticket on which he enters an n-digit number, none of the digits being 0. On the draw date, the lottery administrators will reveal an $n \times n$ table, each square containing one of the digits from 1 to 9. A ticket wins a prize if it does *not* match any row or column of this table, read in either direction. Peter wants to bribe the administrators to reveal the digits on some squares chosen by Peter, so that Peter can guarantee to have a winning ticket. What is the minimum number of digits Peter has to know?

3. In a convex quadrilateral $ABCD$, $AB = 10$, $BC = 14$, $CD = 11$ and $DA = 5$. Determine the angle between its diagonals.

4. Positive integers $a < b < c$ are such that $b + a$ is a multiple of $b - a$ and $c + b$ is a multiple of $c - b$. If a is a 2011-digit number and b is a 2012-digit number, exactly how many digits does c have?

5. In the plane are 10 lines in general position, which means that no 2 are parallel and no 3 are concurrent. Where 2 lines intersect, we measure the smaller of the two angles formed between them. What is the maximum value of the sum of the measures of these 45 angles?

Note: The problems are worth 3, 4, 4, 4 and 5 points respectively.

Junior A-Level Paper

1. An integer $n > 1$ is written on the board. Alex replaces it by $n + d$ or $n - d$, where d is any divisor of n greater than 1. This is repeated with the new value of n. Is it possible for Alex to write on the board the number 2011 at some point, regardless of the initial value of n?

2. P is a point on the side AB of triangle ABC such that $AP = 2PB$. If $CP = 2PQ$, where Q is the midpoint of AC, prove that ABC is a right triangle.

3. A set of at least two objects with pairwise different weights has the property that for any pair of objects from this set, we can choose a subset of the remaining objects so that their total weight is equal to the total weight of the given pair. What is the minimum number of objects in this set?

4. A game is played on a board with 2012 horizontal rows and $k > 2$ vertical columns. A marker is placed in an arbitrarily chosen square of the left-most column. Two players move the marker in turns. During each move, the player moves the marker one square to the right, or one square up or down to a square that has never been occupied by the marker before. The game is over when either player moves the marker to the right-most column.

There are two versions of this game. In Version A, the player who gets the marker to the right-most column wins. In Version B, this player loses. However, it is only when the marker reaches the second column from the right that the players learn whether they are playing Version A or Version B. Does either player have a winning strategy?

5. Let $abcd = (1-a)(1-b)(1-c)(1-d)$, where $0 < a, b, c, d < 1$ are real numbers. Prove that $(a+b+c+d) - (a+c)(b+d) \geq 1$.

6. A car goes along a straight highway at the speed of 60 kilometres per hour. A 100 metre long fence is standing parallel to the highway. Every second, the passenger of the car measures the angle subtended by the fence. Prove that the sum of all angles measured by him is less than $1100°$.

7. Each vertex of a regular 45-gon is red, yellow or green, and there are 15 vertices of each colour. Prove that we can choose three vertices of each colour so that the three triangles formed by the chosen vertices of the same colour are congruent to one another.

Note: The problems are worth 3, 4, 5, 6, 6, 7 and 9 points respectively.

Senior A-Level Paper

1. Pete has marked at least 3 points in the plane such that all distances between them are different. A pair of marked points A and B will be called *unusual* if A is the furthest marked point from B, and B is the nearest marked point to A (apart from A itself). What is the largest possible number of unusual pairs that Pete can obtain?

2. Let $abcd = (1-a)(1-b)(1-c)(1-d)$, where $0 < a, b, c, d < 1$ are real numbers. Prove that $(a+b+c+d) - (a+c)(b+d) \geq 1$.

3. In triangle ABC, points D, E and F are bases of altitudes from vertices A, B and C respectively. Points P and Q are the projections of F to AC and BC respectively. Prove that the line PQ bisects the segments DF and EF.

4. Does there exist a convex n-gon such that all its sides are equal and all vertices lie on the parabola $y = x^2$, where
   (a) $n = 2011$;
   (b) $n = 2012$?

5. We will call a positive integer *good* if all its digits are nonzero. A good integer will be called *special* if it has at least k digits and their values are strictly increasing from left to right. Let a good integer be given. In each move, one may insert a special integer into the digital expression of the current number, on the left, on the right or in between any two of the digits. Alternatively, one may also delete a special number from the digital expression of the current number. What is the largest k such that any good integer can be turned into any other good integer by a finite number of such moves?

6. Prove that for $n > 1$, the integer $1^1 + 3^3 + 5^5 + \ldots + (2^n - 1)^{2^n - 1}$ is a multiple of 2^n but not a multiple of 2^{n+1}.

7. A blue circle is divided into 100 arcs by 100 red points such that the lengths of the arcs are the positive integers from 1 to 100 in an arbitrary order. Prove that there exist two perpendicular chords with red endpoints.

Note: The problems are worth 4, 4, 5, 3+4, 7, 7 and 9 points respectively.

Spring 2012

Junior O-Level Paper

1. Buried under each square of a 8×8 board is either a treasure or a message. There is only one square with a treasure. A message indicates the minimum number of steps needed to go from its square to the treasure square. Each step takes one from a square to another square sharing a common side. What is the minimum number of squares one must dig up in order to bring up the treasure for sure?

2. The number 4 has an odd number of odd positive divisors, namely 1, and an even number of even positive divisors, namely 2 and 4. Is there a number with an odd number of even positive divisors and an even number of odd positive divisors?

3. In the parallelogram $ABCD$, the diagonal AC touches the incircles of triangles ABC and ADC at W and Y respectively, and the diagonal BD touches the incircles of triangles BAD and BCD at X and Z respectively. Prove that either W, X, Y and Z coincide, or $WXYZ$ is a rectangle.

4. Brackets are to be inserted into the expression $10 \div 9 \div 8 \div 7 \div 6 \div 5 \div 4 \div 3 \div 2$ so that the resulting number is an integer.

   (a) Determine the maximum value of this integer.

   (b) Determine the minimum value of this integer.

5. Ryno, a little rhinoceros, has 17 scratch marks on its body. Some are horizontal and the rest are vertical. Some are on the left side and the rest are on the right side. If Ryno rubs one side of its body against a tree, two scratch marks, either both horizontal or both vertical, will disappear from that side. However, at the same time, two new scratch marks, one horizontal and one vertical, will appear on the other side. If there are less than two horizontal and less than two vertical scratch marks on the side being rubbed, then nothing happens. If Ryno continues to rub its body against trees, is it possible that at some point in time, the numbers of horizontal and vertical scratch marks have interchanged on each side of its body?

Note: The problems are worth 3, 4, 4, 2+3 and 5 points respectively.

Senior O-Level Paper

1. Each vertex of a convex polyhedron lies on exactly three edges, at least two of which have the same length. Prove that the polyhedron has three edges of the same length.

2. The squares of a $1\times 2n$ board are labeled $1, 2, \ldots, n, -n, \ldots, -2, -1$ from left to right. A marker is placed on an arbitrary square. If the label of the square is positive, the marker moves to the right a number of squares equal to the value of the label. If the label is negative, the marker moves to the left a number of squares equal to the absolute value of the label. Prove that if the marker can always visit all squares of the board, then $2n+1$ is prime.

3. Consider the points of intersection of the graphs of $y = \cos x$ and of $x = 100\cos(100y)$ for which both coordinates are positive. Let a be the sum of their x-coordinates and b be the sum of their y-coordinates. Determine the value of $\frac{a}{b}$.

4. A quadrilateral $ABCD$ with no parallel sides is inscribed in a circle. Two circles, one passing through A and B, and the other through C and D, are tangent to each other at X. Prove that the locus of X is a circle.

5. In an 8×8 board, the rows are numbered from 1 to 8 and the columns are labeled from a to h. In a two-player game on this board, Alice has a white rook which starts on the square b2, and Bob has a black rook which starts on the square c4. They take turns moving their rooks, Alice going first. In each move, a rook lands on another square in the same row or the same column as its starting square. However, that square cannot be under attack by the other rook, and cannot have been landed on before by either rook. The player without a move loses the game. Which player has a winning strategy?

Note: The problems are worth 4, 4, 5, 5 and 5 points respectively.

Junior A-Level Paper

1. It is possible to place an even number of pears in a row such that the masses of any two adjacent pears differ by at most 1 gram. Prove that it is then possible to put the pears two in a bag and place the bags in a row such that the masses of any two adjacent bags differ by at most 1 gram.

2. One hundred points are marked in the plane, with no three in a line. Is it always possible to connect the points in pairs such that all fifty segments intersect one another?

3. In a team of guards, each is assigned a different positive integer. For any two guards, the ratio of the two numbers assigned to them is at least 3:1. A guard assigned the number n is on duty for n days in a row, off duty for n days in a row, back on duty for n days in a row, and so on. The guards need not start their duties on the same day. Is it possible that on any day, at least one in such a team of guards is on duty?

4. Each entry in an $n \times n$ table is either $+$ or $-$. At each step, one can choose a row or a column and reverse all signs in it. From the initial position, it is possible to obtain the table in which all signs are $+$. Prove that this can be accomplished in at most n steps.

5. Let p be a prime number. A set of $p+2$ positive integers, not necessarily distinct, is called *interesting* if the sum of any p of them is divisible by each of the other two. Determine all interesting sets.

6. A bank has one million clients, one of whom is Inspector Gadget. Each client has a unique PIN number consisting of six digits. Dr. Claw has a list of all the clients. He is able to break into the account of any client, choose any n digits of the PIN number and copy them. The n digits he copies from different clients need not be in the same n positions. He can break into the account of each client, but only once. What is the smallest value of n which allows Dr. Claw to determine the complete PIN number of Inspector Gadget?

7. Let AH be an altitude of an equilateral triangle ABC. Let I be the incentre of triangle ABH, and let L, K and J be the incentres of triangles ABI, BCI and CAI respectively. Determine $\angle KJL$.

Note: The problems are worth 4, 4, 6, 6, 8, 8 and 8 points respectively.

Senior A-Level Paper

1. In a team of guards, each is assigned a different positive integer. For any two guards, the ratio of the two numbers assigned to them is at least 3:1. A guard assigned the number n is on duty for n days in a row, off duty for n days in a row, back on duty for n days in a row, and so on. The guards need not start their duties on the same day. Is it possible that on any day, at least one in such a team of guards is on duty?

2. One hundred points are marked inside a circle, with no three in a line. Prove that it is possible to connect the points in pairs such that all fifty lines intersect one another inside the circle.

3. Let n be a positive integer. Prove that there exist integers $a_1, a_2, \ldots, a_n$ such that for any integer x, the number

$$(\cdots(((x^2+a_1)^2+a_2)^2+\cdots)^2+a_{n-1})^2+a_n$$

is divisible by $2n-1$.

4. Alex marked one point on each of the six interior faces of a hollow unit cube. Then he connected by strings all pairs of marked points on adjacent faces. Prove that the total length of these strings is at least $6\sqrt{2}$.

5. Let ℓ be a tangent to the incircle of triangle ABC. Let ℓ_a, ℓ_b and ℓ_c be the respective images of ℓ under reflection across the exterior bisector of $\angle A$, $\angle B$ and $\angle C$. Prove that the triangle formed by these lines is congruent to ABC.

6. We attempt to cover the plane with an infinite sequence of rectangles, overlapping allowed.

   (a) Is the task always possible if the area of the nth rectangle is n^2 for each n?

   (b) Is the task always possible if each rectangle is a square, and for any number N, there exist squares with total area greater than N?

7. Konstantin has a pile of 100 pebbles. In each move, he chooses a pile and splits it into two smaller ones until he gets 100 piles each with a single pebble.

   (a) Prove that at some point, there are 30 piles containing a total of exactly 60 pebbles.

   (b) Prove that at some point, there are 20 piles containing a total of exactly 60 pebbles.

   (c) Prove that Konstantin may proceed in such a way that at no point, there are 19 piles containing a total of exactly 60 pebbles.

Note: The problems are worth 4, 5, 6, 6, 8, 3+6 and 6+3+3 points respectively.

Solutions

Fall 2011

Junior O-Level Paper

1. The bisector of $\angle A$ is also the perpendicular bisector of CQ, and the bisector of $\angle B$ is also the perpendicular bisector of CP. The incentre of triangle ABC is the point of intersection of the bisectors of $\angle A$ and $\angle B$. The circumcentre of triangle CPQ is the point of intersection of the perpendicular bisectors of CQ and CP. Hence the incentre of triangle ABC is also the circumcentre of triangle CPQ.

2. It is not possible for each guest to eat six fewer berries than the next guest. Hence one of them has to eat twice as many, and therefore an even number of berries. Going now in the counter-clockwise direction, the next guest eats either twice as many as or six fewer than the preceding guest. It follows that every guest has eaten an even number of berries. Since 2011 is odd, not all the berries have been eaten.

3. The diagram below shows a dissection of the punctured board into rectangular pieces, none of them being squares.

4. The task is possible. Let the numbers along the circle be 17, 1, 18, 2, 19, 3, ..., 15, 32, 16 and 33. Then the sums are 18, 19, 20, 21, 22, ..., 47, 48, 49 and 50.

5. The diagram below shows five snapshots of the highway. Since all speeds are constant, the motions can be represented by straight lines, AD for the pedestrian, AC for the cyclist, BE for the cart and CE for the car. The equality of time intervals yield $AB = BC$ and $CD = DE$. Hence F, which represents the moment the pedestrian met the cart, is the centroid of triangle ACE, so that $AF = \frac{2}{3}AD$. Since A is at 10 o'clock and D is at 11 o'clock, F is at 10:40.

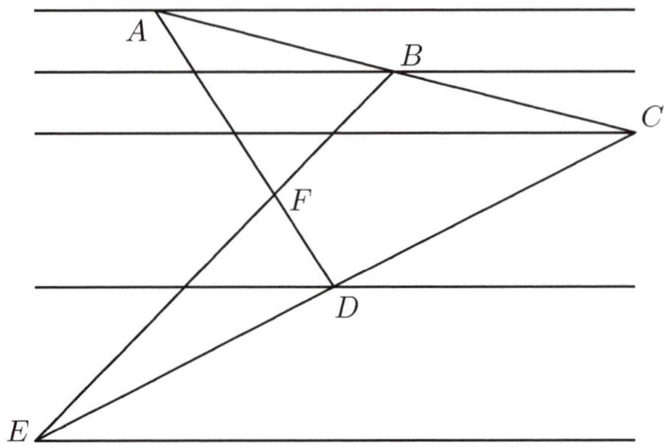

Senior O-Level Paper

1. (See Problem 2 of the Junior O-Level Paper.)

2. The minimum number is n. If Peter knows at most $n-1$ of the digits, he will not know any digit on one of the rows, and his ticket may match that row. On the other hand, if Peter knows every digit on a diagonal, he can guarantee to have a winning ticket. Let the digits on this diagonal be $d_1, d_2, \ldots, d_n$. Peter can enter the digits $t_1, t_2, \ldots, t_n$ on his ticket such that neither t_k nor t_{n+1-k} matches d_k or d_{n+1-k} for any k, $1 \le k \le n$. Then his ticket cannot match the kth row or the kth column for any k in either direction.

3. Let AC and BD intersect at O. Suppose the diagonals are not perpendicular to each other. We may assume that $\angle AOB = \angle COD < 90°$ by symmetry. Then $(OA^2 + OB^2) + (OC^2 + OD^2) > AB^2 + CD^2 = 221$. We also have $(OD^2 + OA^2) + (OB^2 + OC^2) < DA^2 + BC^2 = 221$. This is a contradiction. Hence both angles between the diagonals are 90°.

4. Since $c > b$, c has at least 2012 digits. We have $b + a = k(b-a)$ for some integer $k > 1$. Hence $a(k+1) = b(k-1)$, so that $\frac{b}{a} = \frac{k+1}{k-1} = 1 + \frac{2}{k-1} \le 3$, with equality if and only if $k = 2$. Similarly, $\frac{c}{b} \le 3$, so that $\frac{c}{a} = \frac{c}{b} \cdot \frac{b}{a} \le 9$. Hence $c < 10a$. Since a has 2011 digits, c has at most 2012 digits. It follows that c has exactly 2012 digits.

5. Despite the statement of the problem, whether the lines are concurrent or not is totally irrelevant. In fact, it facilitates our argument to have them all pass through the same point. Let the lines be $\ell_0, \ell_1, \ldots, \ell_9$, forming the angles $\theta_0, \theta_1, \ldots, \theta_9$ as shown in the diagram below.

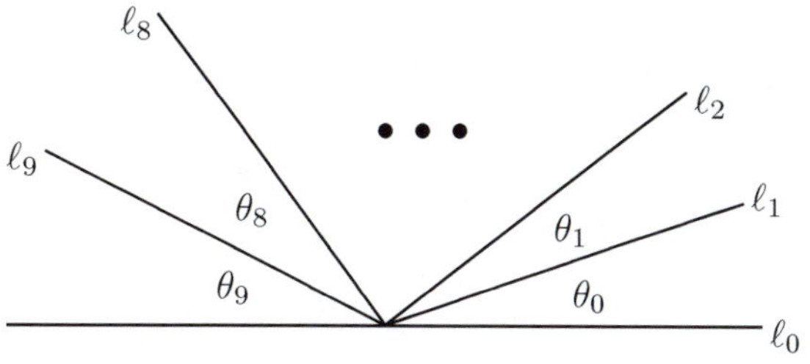

Define $\phi(i,j)$ to be the smaller angle formed between ℓ_i and ℓ_j. Then $\phi(i,j) \le \theta_i + \theta_{i+1} + \cdots + \theta_{j-1}$. We have

$$\begin{aligned}
\phi(0,1)+\phi(1,2)+\cdots+\phi(9,0) &\le \theta_0+\theta_1+\cdots+\theta_9 \\
&= 180^\circ; \\
\phi(0,2)+\phi(1,3)+\cdots+\phi(9,1) &\le (\theta_0+\theta_1)+(\theta_1+\theta_2) \\
&\quad +\cdots+(\theta_9+\theta_0) \\
&= 2(\theta_0+\theta_1+\cdots+\theta_9) \\
&= 360^\circ; \\
\phi(0,3)+\phi(1,4)+\cdots+\phi(9,2) &\le (\theta_0+\theta_1+\theta_2)+(\theta_1+\theta_2+\theta_3) \\
&\quad +\cdots+(\theta_9+\theta_0+\theta_1) \\
&= 540^\circ; \\
\phi(0,4)+\phi(1,5)+\cdots+\phi(9,3) &\le 4(\theta_0+\theta_1+\cdots+\theta_9) \\
&= 720^\circ; \\
\phi(0,5)+\phi(1,6)+\cdots+\phi(9,4) &\le 900^\circ.
\end{aligned}$$

The last expression yields

$$\phi(0,5)+\phi(1,6)+\phi(2,7)+\phi(3,8)+\phi(4,9) \le 450^\circ.$$

It follows that the overall sum cannot exceed

$$180^\circ + 360^\circ + 540^\circ + 720^\circ + 450^\circ = 2250^\circ.$$

Equality holds if $\theta_0 = \theta_1 = \cdots = \theta_9 = 18^\circ$, but the maximum value 2250° can be attained in many other ways.

Junior A-Level Paper

1. Starting from any positive integer n, Alex adds n a total of 2010 times to get $2011n$. Then he subtracts 2011 a total of $n-1$ times to get 2011.

2. Extend QP to R so that $RP = 2PQ$. Then P is the centroid of triangle ARC. Since $AP = 2PB$, the extension of AP intersects RC at its midpoint B. Since $RP = CP$, triangles PRB and PCB are congruent, so that $\angle ABR = \angle ABC$. Since their sum is $180°$, each is $90°$.

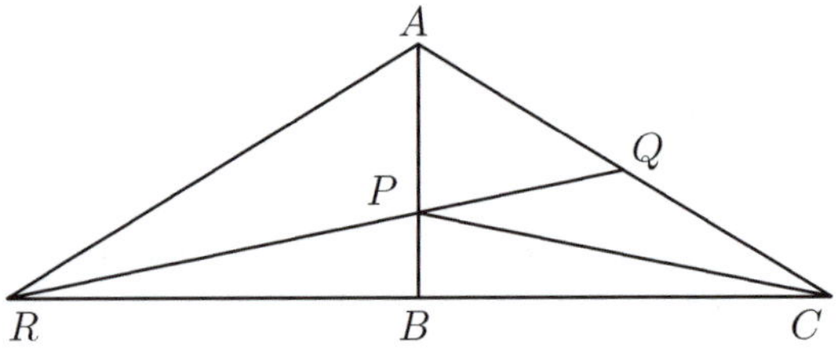

3. Clearly, the set cannot have 2, 3 or 4 objects, as it would not be possible to balance the heaviest two objects. Suppose it has only 5 objects, of respective weights $a > b > c > d > e$. Clearly, we must have $a + b = c + d + e$. Since $a + c > b + d$, we must also have $a + c = b + d + e$, which implies $b = c$, a contradiction. The set may have 6 objects, of respective weights 8, 7, 6, 5, 4 and 3. Then 8+7=6+5+4, 8+6=7+4+3, 8+5=7+6, 8+4=7+5, 8+3=7+4=6+5, 7+3=6+4, 6+3=5+4, 5+3=8 and 4+3=7.

4. The first player has a winning strategy. She will only move the counter up or down until she learns whether Version A or Version B of the game is being played. Since 2012 is even, she can choose a direction (up or down) so that the marker stays in the same column for an odd number of moves. If the marker starts at the end of the column, the only possible direction allows her to keep it in the same column for an odd number of moves. Thus she can ensure that the second player is always the one to move the marker to the right, whether by choice earlier or being forced to do so when the marker reaches the end of the column.

   When the marker reaches the second column from the right, if Version A is being played, the first player can win by simply moving the marker to the right. If Version B is being played, she can keep the marker in this column as before, and wait for the second player to lose.

5. From $abcd = (1-a)(1-b)(1-c)(1-d)$, we have

$$\begin{aligned}
& a+b+c+d-(a+c)(b+d) \\
= \ & 1+ac(1-b-d)+bd(1-a-c) \\
= \ & 1+ac(1-b)(1-d)+bd(1-a)(1-c)-2abcd \\
\geq \ & 1+2\sqrt{ac(1-b)(1-d)bd(1-a)(1-c)}-2abcd \\
= \ & 1+2abcd-2abcd \\
= \ & 1.
\end{aligned}$$

6. Divide the points of observation into six groups cyclically, so that the points in each group are 100 metres apart, the same as the length of the fence. The diagram below shows the angles subtended by the fence at the points of a group.

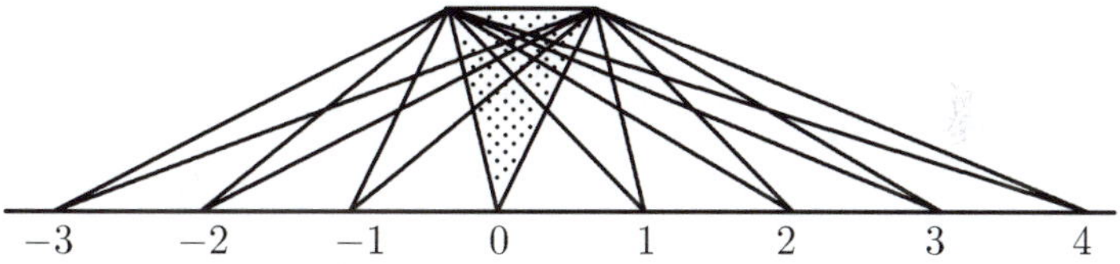

We now parallel translate all these points to a single point, along with their fences and subtended angles, as shown in the diagram below.

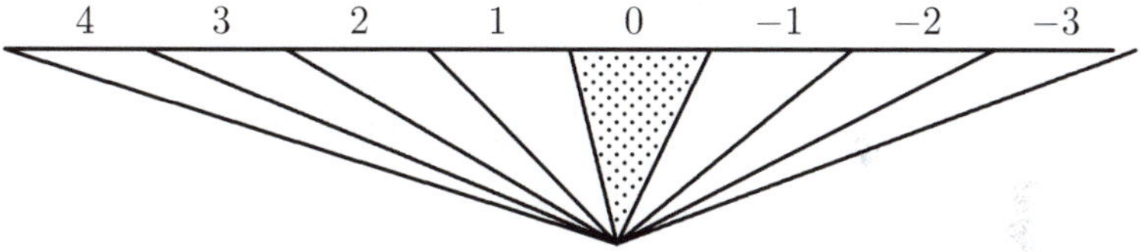

The sum of all these angles is clearly at most 180°. Since there are six groups of points of observations, the sum of all subtended angles is at most $6 \times 180° < 1100°$.

7. Copy the regular 45-gon onto a piece of transparency and mark on it the 15 red points. Call this the Red position, and rotate the piece of transparency about the centre of the 45-gon 8° at a time. For each of the 45 positions, count the number of matches of yellow points with the 15 marked points. Since each of the 15 yellow points may match up with any of the 15 marked points, the total number of matches is $15 \times 15 = 225$, so that the average number of matches per position is 5. However, in the Red position, the number of matches is 0. Hence there is a position with at least 6 matches.

Call this the Yellow position, and choose any 6 of the matched marked points and erase the other 9. Repeat the rotation process, but this time counting the number of matches of green points with the 6 marked points. The total number of matches is $6 \times 15 = 90$, so that the average number of matches per position is 2. As before, there is a position with at least 3 matches. Call this the Green position, choose any 3 of the matched marked points and erase the other 3. The 3 remaining marked points define three congruent triangles, a red one in the Red position, a yellow one in the Yellow position and a green one at the Green position.

Senior A-Level Paper

1. First, we show by example that we may have one unusual pair when there are at least 3 points. Let A and B be chosen arbitrarily. Add some points within the circle with centre B and radius AB but outside the circle with centre A and radius AB. Then B is the point nearest to A while A is the point furthest from B. We now prove that if there is another pair of unusual points, we will have a contradiction. We consider two cases.

   Case 1.
   The additional unusual pair consists of C and D such that D is the point nearest to C while C is the point furthest from D. Now $DA > AB$ since B is the point nearest to A, $AB > BC$ because A is the point furthest from B, $BC > CD$ since D is the point nearest to C, and $CD > DA$ because C is the point furthest from D. Hence $DA > DA$, which is a contradiction.

   Case 2.
   The additional unusual pair consists of B and C such that C is the point nearest to B and B is the point furthest from C. Now $CA > AB$ since B is the point nearest to A, $AB > BC$ since A is the point furthest from B, and $BC > CA$ because B is the point furthest from C. Hence $CA > CA$, which is a contradiction.

2. From $abcd = (1-a)(1-b)(1-c)(1-d)$, we have $\frac{ac}{(1-a)(1-c)} = \frac{(1-b)(1-d)}{bd}$. It follows that

$$\frac{a+c-1}{(1-a)(1-c)} = \frac{ac}{(1-a)(1-c)} - 1 = \frac{(1-b)(1-d)}{bd} - 1 = \frac{1-b-d}{bd}.$$

   Now $\frac{(a+c-1)(1-b-d)}{(1-a)(1-c)bd} \geq 0$ since it is the product of two equal terms. From $(1-a)(1-c)bd > 0$, we have $(a+c-1)(1-b-d) \geq 0$. Expansion yields $a - ab - ad + c - cb - cd - 1 + b + d \geq 0$, which is equivalent to $a + b + c + d - (a+c)(b+d) \geq 1$.

3. Let H be the orthocentre of triangle ABC. Then H is the incentre of triangle DEF. Note that $CDHE$ and $CQFP$ are cyclic quadrilaterals. Hence $\angle ADE = \angle FCP = \angle FQP$. Since AD and FQ are parallel to each other, so are ED and PQ. Thus $\angle PQD = \angle EDC = \angle FDQ$. Let M and N be the points of intersection of PQ with EF and DF respectively. Then $ND = NQ$. Since $\angle FQD = 90°$, N is the circumcentre of triangle FQD so that $FN = ND$. Similarly, we have $FM = ME$.

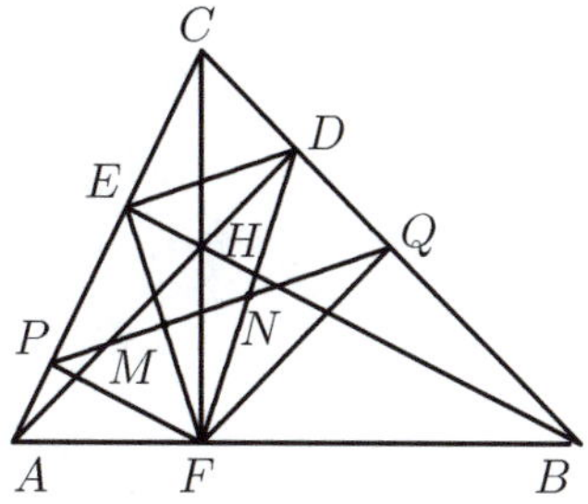

4. (a) Such a 2011-gon exists. Let V be the vertex of the parabola. On one side, mark on the parabola points A_1, A_2, $\dots$, A_{1005} such that $VA_1 = A_1A_2 = \cdots = A_{1004}A_{1005} = t$, and on the other side, points B_1, $B_2, \dots, B_{1005}$ such that $VB_1 = B_1B_2 = \cdots = B_{1004}B_{1005} = t$. The length ℓ of $A_{1005}B_{1005}$ varies continuously with t. When t is very small, we have $\ell > t$. When t is very large, ℓ is less than a constant times $\sqrt{t}$, which is in turn less than t. Hence at some point in between, we have $\ell = t$.

   (b) We first prove a geometric result.

   Lemma.

   In the convex quadrilateral $ABCD$, if we have $AB = CD$ and $\angle ABC + \angle BCD > 180°$, then $AD > BC$.

   Proof:

   Complete the parallelogram $BCDE$. Then

   $$\angle EBC + \angle BCD = 180° < \angle ABC + \angle BCD.$$

   Hence $\angle ABC > \angle EBC$. Since $BD = BD$ and $AB = CD = EB$, we have $AD > ED + BC$ by the Side-angle-side Inequality.

   Corollary.

   If A, B, C and D are four points in order on a parabola and $AB = CD$, then $AD > BC$.

   Proof:

   Since the parabola is a convex curve, the extension of AB and DC meet at some point L. Then $AD > BC$ since

   $$\begin{aligned} \angle ABC + \angle BCD &= (\angle ALD + \angle BCL) + (\angle ALD + \angle CBL) \\ &= 180° + \angle ALD \\ &> 180°. \end{aligned}$$

Returning to our problem, suppose P_1, P_2, $\ldots$, P_{2012} are points in order on a parabola, such that $P_1P_2 = P_2P_3 = \cdots = P_{2011}P_{2012}$. We now obtain $P_1P_{2012} > P_2P_{2011} > \cdots > P_{1006}P_{1007}$ by applying the Corollary 1005 times. Hence the 2012-gon cannot be equilateral.

5. We cannot have $k = 9$ as the only special number would be 123456789. Adding or deleting it does not change anything. We may have $k = 8$. We can convert any good number into any other good number by adding or deleting one digit at a time. We give below the procedures for adding digits. Reversing the steps allows us to delete digits.

   Adding 1 or 9 anywhere.
   Add 123456789 and delete 23456789 or 12345678.

   Adding 2 anywhere.
   Add 23456789 and then add 1 between 2 and 3. Now delete 13456789.

   Adding 8 anywhere.
   Add 12345678 and then add 9 between 7 and 8. Now delete 12345679.

   Adding 3 anywhere.
   Add 23456789 and delete 2. Now add 1 and 2 between 3 and 4 and delete 12456789.

   Adding 7 answhere.
   Add 12345678 and delete 8. Now add 8 and 9 between 6 and 7 and delete 12345689.

   Adding 4 anywhere.
   Add 23456789 and delete 2 and 3, Now add 1, 2 and 3 between 4 and 5 and delete 12356789.

   Adding 6 anywhere.
   Add 12345678 and delete 7 and 8. Now add 7, 8 and 9 between 5 and 6 and delete 12345789.

   Adding 5 anywhere.
   Add 23456789, delete 2, 3 and 4 and add 1, 2, 3 and 4 betwen 5 and 6. Alternately, add 12345678, delete 6, 7 and 8 and add 6, 7, 8 and 9 between 4 and 5. Now delete 12346789.

6. We first prove two preliminary results.

   Lemma 1.

   For any positive odd integer k, $k^{2^n} \equiv 1 \pmod{2^{n+2}}$.

   Proof:

   We have $k^{2^n} - 1 = (k-1)(k+1)(k^2+1)(k^4+1)\cdots(k^{2^{n-1}}+1)$. This is a product of $n+1$ even factors. Moreover, one of $k-1$ and $k+1$ is divisible by 4. The desired result follows.

   Lemma 2.

   For any integer $n \geq 2$, $(2^n + k)^k \equiv k^k(2^n + 1) \pmod{2^{n+2}}$.

   Proof:

   Expanding $(2^n + k)^k$, all the terms are divisible by 2^{n+2} except for $\binom{k}{1}k^{k-1}2^n$ and k^k. The desired result follows.

Let the given sum be denoted by S_n. We now use induction on n to prove that S_n is divisible by 2^n but not 2^{n+1} for all $n \geq 2$. Note that $S_2 = 1^1 + 3^3 = 28$ is divisible by 2^2 but not by 2^3. Using Lemmas 1 and 2, we have

$$\begin{aligned} S_{n+1} - S_n &= \sum_{i=1}^{2^{n-1}} (2^n + 2i - 1)^{2^n+2i-1} \\ &\equiv \sum_{i=1}^{2^{n-1}} (2^n + 2i - 1)^{2i-1} \pmod{2^{n+2}} \\ &\equiv \sum_{i=1}^{2^{n-1}} (2i-1)^{2i-1}(2^n + 1) \pmod{2^{n+2}} \\ &= S_n(2^n + 1). \end{aligned}$$

Now $S_{n+1} = 2S_n(2^{n-1}+1)$. By the induction hypothesis, S_n is divisible by 2^n but not by 2^{n+1}. It follows that S_{n+1} is divisible by 2^{n+1} but not by 2^{n+2}.

7. We first prove a geometric result.

Lemma.

Let P, Q, R and S be four points on a circle in cyclic order. If the arcs PQ and RS add up to a semicircle, then PR and QS are perpendicular chords.

Proof:

Let O be the centre of the circle. From the given condition, we have $\angle POQ+\angle ROS = 180°$ and $\angle PSQ+\angle RPS = \frac{1}{2}(\angle POQ+\angle ROS) = 90°$. It follows that PR and QS are perpendicular chords.

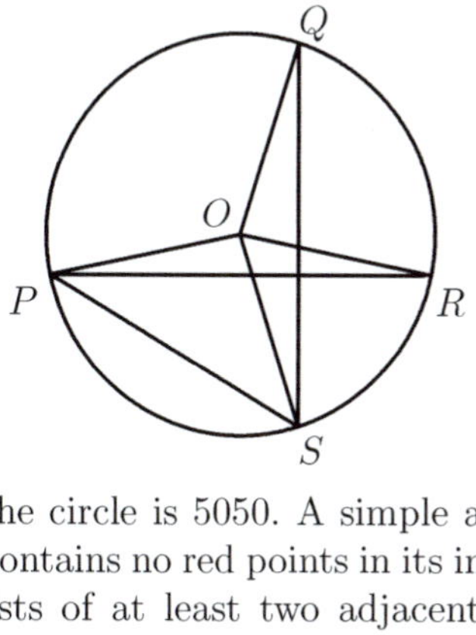

The circumference of the circle is 5050. A simple arc is one which has two red endpoints but contains no red points in its interior. A compound arc is one which consists of at least two adjacent simple arcs whose total length is less than 2525. For each red point P, choose the longest compound arc with P as one of its endpoints. We consider two cases.

Case 1.
Suppose for some P there are two equal choices in opposite directions. Then they must be of length 2475, and the part of the circle not in either of them is the simple arc of length 100. It follows that one of these two compound arcs, say PQ, does not contain the simple arc RS of length 50. We may assume that P, Q, R and S are in cyclic order. If $Q = R$, then PS is a diameter of the circle. If $S = P$, then QR is a diameter of the circle. In either case, PQ and RS are perpendicular chords. If these four points are distinct, then the arcs PQ and RS add up to a semicircle. By the Lemma, PR and QS are perpendicular chords.
Case 2.
The choice is unique for every P. Then we have chosen at least 50 such arcs since each may serve as the maximal arc for at most two red points. Each of these arcs has length at least 2476 but less than 2525, a range of 49 possible values. By the Pigeonhole Principle, two of them have the same length $2525 - k$ for some k where $1 \leq k \leq 49$. If one of them does not contain the simple arc $\mathcal{C}$ of length k, we can then argue as in Case 1. Suppose both of them contain $\mathcal{C}$. This is only possible if one of them ends with $\mathcal{C}$ and the other starts with it. By removing $\mathcal{C}$ from both, we obtain two disjoint compound arcs both of length $2525 - 2k$, and note that $2 \leq 2k \leq 98$. Now one of them does not contain the simple arc of length $2k$, and we can argue as in Case 1.

Spring 2012

Junior O-Level Paper

1. Whichever square we dig up first, there is no guarantee that the treasure is there. If the message we get says that the treasure is one square away, we cannot determine its location uniquely. Thus we have to dig up at least three squares. Let the first two squares we dig up be at the lower left corner and the lower right corner of the board. We may as well suppose that we do not find the treasure under either of them. Now we know the diagonal which runs from north-west to south-east, as well as the diagonal which runs from north-east to south-west, which contains the square with the treasure. The treasure can be brought up by digging up just one more square, the one at the intersection of these two diagonals.

2. All integers under discussion are taken to be positive. The divisors of an integer n can be divided into pairs such that the product of the two numbers in each pair is n, except when n is a square, with $\sqrt{n}$ having no partner. Suppose there exists an integer n with an even number of odd divisors and an odd number of even divisors. Then it has an odd number of divisors in total, and must be a square, say $n = x^2$. Let $x = 2^k y$ where y is odd. Then $n = 2^{2k}y^2$, and the odd divisors of n are precisely the divisors of y^2. This number of divisors cannot be even.

3. Suppose $ABCD$ is a rhombus. Then both W and Y coincide with the midpoint of AC, and both X and Z coincide with the midpoint of BD. Since AC and BD bisect each other, all four points coincide. Suppose $ABCD$ is not a rhombus.

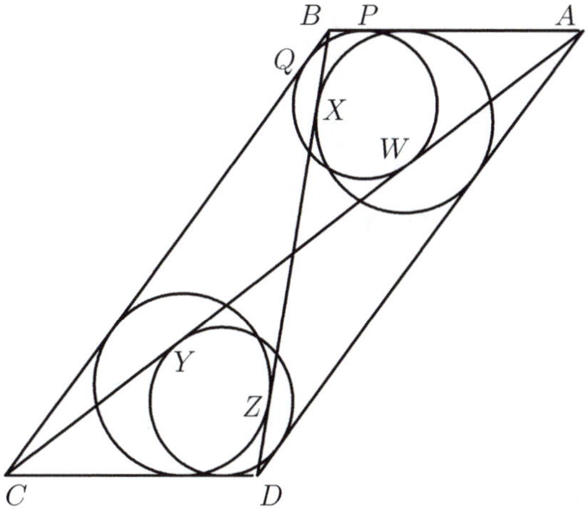

None of the four points coincides with the common midpoint of AC and BD. Hence they are distinct. By symmetry, $AW = CY$ and $BX = DZ$. Hence WY and XZ also bisect each other, so that $WXYZ$ is a parallelogram. Let the incircle of ABC touch AB at P and BC at Q. We may assume that W is closer to A and Y is closer to C. Then

$$WY = CW - CY = CW - AW = CQ - AP = CB - AB.$$

Similarly, $XZ = AD - AB = WY$. Being a parallelogram with equal diagonals, $WXYZ$ is a rectangle.

4. Bracketing simply separates the factors 10, 9, ..., 2 into the numerator and the denominator of the overall expression.

   (a) We have
$$\begin{aligned} & 10 \div (((((((9 \div 8) \div 7) \div 6) \div 5) \div 4) \div 3) \div 2) \\ = \; & 10 \div \frac{9}{8!} \\ = \; & \frac{10!}{9^2} \\ = \; & 44800. \end{aligned}$$

   Since 9 is the second number in the sequence, it must be in the denominator. Hence the maximum value cannot be higher than 44800.

   (b) Since 7 is the only number in the sequence divisible by 7, it must be in the numerator. Hence the minimum value cannot be lower than 7. We have
$$\begin{aligned} & (10 \div 9) \div ((8 \div 7) \div (6 \div ((5 \div 4) \div (3 \div 2)))) \\ = \; & \frac{10}{9} \div \left(\frac{8}{7} \div \left(6 \div \left(\frac{5}{4} \div \frac{3}{2}\right)\right)\right) \\ = \; & \frac{10}{9} \div \left(\frac{8}{7} \div \left(6 \div \frac{6}{5}\right)\right) \\ = \; & \frac{10}{9} \div \left(\frac{8}{7} \div \frac{36}{5}\right) \\ = \; & \frac{10}{9} \div \frac{10}{63} \\ = \; & 7. \end{aligned}$$

5. Let a, b, c and d be the numbers of scratch marks which are horizontal and on the left side, vertical and on the left side, horizontal and on the right side, and vertical and on the right side. Suppose the initial values of a and b have been interchanged, and so are those of c and d, then $a + b$ and $c + d$ are unchanged. Since each of these two sums changes by 2 after a rubbing, the total number of rubbings must be even.

If we allow negative values temporarily, the order of the rubbings is immaterial, and we can assume that they occur alternately on the left side and on the right side. After each pair of rubbings, the parity of each of a, b, c and d has changed. Suppose initially $a + b$ is odd so that $c + d$ is even. After an odd number of pairs of rubbings, the final values of a and b may have interchanged from their initial values, the odd one becomes even and the even one becomes odd. However, this is not possible for c and d, as they either change from both even to both odd, or from both odd to both even. Similarly, after an even number of pairs of rubbings, the final values of c and d may have interchanged from their initial values, but this is not possible for a and b. Thus the desired scenario cannot occur.

Senior O-Level Paper

1. Let the number of vertices be v and the number of edges be e. Since each vertex lies on exactly 3 edges and each edge joins exactly 2 vertices, $3v = 2e$. Suppose to the contrary that the polyhedron does not have three edges of equal length. Then each vertex has a pair of equal edges of length different from any other edge. Hence the total number of edges is at least $2v$, but $3v = 2e \geq 4v$ is a contradiction.

2. Suppose $2n + 1$ is not prime. Then it has a prime divisor $p < 2n + 1$. If the marker starts on a square whose label is divisible by p, it must move in either direction by a number of spaces equal to a multiple of p. We claim that the squares whose labels are divisible by p are evenly spaced. Then the marker must stay on squares whose labels are divisible by p, and cannot visit all squares. The claim certainly holds among the squares with positive labels and among those with negative labels. To see that it also holds across the two sides, simply add $2n + 1$ to each of the negative labels. Then a label is divisible by p after the modification if and only if it is divisible by p before the modification. The desired result follows since the modified labels are $1, 2, \ldots, 2n$ from left to right.

3. Define $X = \frac{x}{10}$ and $Y = 10y$. Then the graphs become the symmetric pair $Y = 10\cos(10X)$ and $X = 10\cos(10Y)$. Now X and Y are both positive if and only if both x and y are positive. Let A be the sum of the X-coordinates and B be the sum of the Y-coordinates.for which both X and Y are positive. Then $A = \frac{a}{10}$ and $B = 10b$ by definition, and $A = B$ by symmetry. Hence $\frac{a}{b} = \frac{10A}{\frac{B}{10}} = 100$.

4. Draw the circumcircle of $ABCD$, a circle ω_1 through A and B and a circle ω_2 through C and D such that ω_1 and ω_2 are tangent to each other. Let the lines AB and CD omtersect at P, which is necessarily outside all three circle. Let the line PX intersect ω_1 again at Y_1 and ω_2 again at Y_2. We have $PA \cdot PB = PC \cdot PD = k$, $PA \cdot PB = PX \cdot PY_1$ and $PC \cdot PD = PX \cdot PY_2$. It follows that Y_1 and Y_2 coincide. Since this point lies on both ω_1 and ω_2, it must also coincide with X. Hence PX^2 is equal to the constant k, so that the locus of X is a circle with centre P.

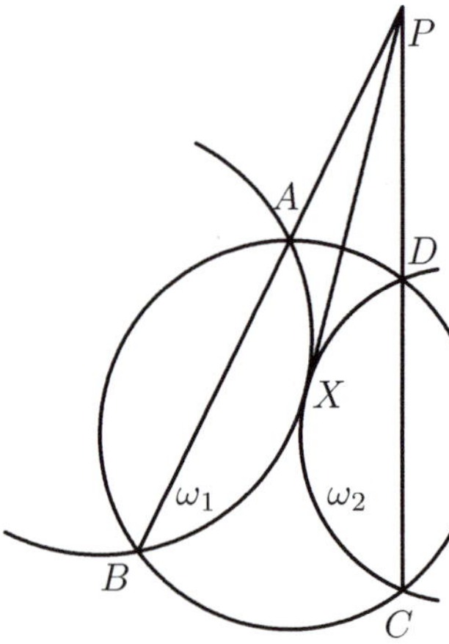

5. Bob has a winning strategy. Divide the eight rows into four pairs (1,3), (2,4), (5,7) and (6,8), and the eight columns also into four pairs (b, c), (d, e), (f, g) and (h, a). Then divide the sixty-four squares into thirty-two pairs. Two squares are in the same pair if and only if they are on two different rows which form a pair, and on two different columns which also form a pair. Thus the starting squares of the two rooks form a pair. Bob's strategy is to move the black rook to the square which forms a pair with the square where the white rook has just landed. First, this can always be done, because if the white rook stays on its current row, the black rook will do the same, and if the white rook stays on its current column, the black rook will do the same. Second, since the square on which the white rook has just landed cannot have been landed on before, the square to which the black rook is moving has never been landed on before, since the squares are occupied by the two rooks in pairs. Third, the black rook will not be under attack by the white rook since the two squares in the same pair are on different rows and on different columns. Hence Bob always has a move, and can simply wait for the Alice to run out of moves.

Junior A-Level Paper

1. Let the pears be P_1, $P_2, \ldots, P_{2n}$, with masses $a_1 \leq a_2 \leq \cdots \leq a_{2n}$ respectively. For $1 \leq k \leq n$, put P_k and P_{2n-k+1} in bag B_k and place the bags in a row in numerical order. We claim that the difference in the masses of B_k and B_{k+1} is at most 1 gram for $1 \leq k < n$. Now the mass of B_k is $a_k + a_{2n-k+1}$ and the mass of B_{k+1} is $a_{k+1} + a_{2n-k}$. Both $a_{k+1} - a_k$ and $a_{2n-k+1} - a_{2n-k}$ are non-negative, and we claim that each is at most 1 gram. The desired result will then follow. Consider P_k and P_{k+1} in the original line-up. If they are in fact neighbours, the claim follows from the given condition. Suppose there are other pears in between. Moving from P_k to P_{k+1}, let P_j be the first pear not lighter than P_k and P_i be the pear before P_j. Then $a_i \leq a_k \leq a_j$ and $a_j - a_i \leq 1$. Hence $a_j - a_k \leq 1$. However, we must have $a_{k+1} \leq a_j$. It follows that $a_{k+1} - a_k \leq 1$. Similarly, we can prove that $a_{2n-k+1} - a_{2n-k} \leq 1$, justifying the claim.

2. This is not always possible. Let P_1, P_2, $\ldots, P_{99}$ be evenly spaced points on a circle with centre P_{100}. We may assume that when the points are connected in pairs, P_{100} is connected to P_1. Let P_{50} be connected to P_k where $k \neq 1, 50, 100$. If $2 \leq k \leq 49$, $P_{50}P_k$ and P_1P_{100} are in opposite semicircles divided by the diameter passing through P_{50}. If $51 \leq k \leq 99$, $P_{50}P_k$ and P_1P_{100} are in opposite semicircles divided by the diameter passing through P_{49}. In either case, $P_{50}P_k$ and P_1P_{100} do not intersect.

3. Let the guards be G_1, G_2, $\ldots$, G_k and let $n_1 > n_2 > \cdots > n_k \geq 1$ be the numbers assigned to them. In fact, $n_i \geq 3n_{i+1}$ for $1 \leq i < k$. There is an interval of $3n_2$ days during which G_1 is not on duty. Within this interval, there is a subinterval of $n_2 \geq 3n_3$ days during which G_2 is not on duty either. Repeating this argument until we reach G_k, we will have an interval of n_k days in which none of the guards are on duty.

4. Note that the order in which the steps are taken is irrelevant, and that reversing the signs of the same row or column twice cancels out. Hence each row and column features in at most one step. Let A be the set of rows and columns that are reversed in order to obtain the table with all +s. Let B be the set of the other rows and columns. For each entry, its row and column may be both in A, both in B or one in each. It follows that reversing the signs of the rows in columns in B instead of those in A also results in a table with all +s. Between A and B, there is a total of $2n$ rows and columns. Hence one of them has at most n rows and columns, so that n steps are indeed sufficient.

5. We may assume that the greatest common divisor of the $p+2$ numbers in an interesting set is 1. Let $S = a_1 + a_2 + \cdots + a_{p+2}$.

For $1 \le k \le p+2$, a_k divides $S - a_j$ for $j \ne k$. Hence a_k divides $(p+1)S - (S - a_k)$, so that a_k divides pS. We consider two cases.
Case 1. None of a_k is divisible by p.
Then a_k divides S. For any $j \ne k$, a_k divides $S - a_j$, so that it divides a_j. It follows that all the numbers are equal. Since their greatest common divisor is 1, they are all equal to 1.
Case 2. At least one a_k is divisible by p.
Not all of them can be divisible by p since their greatest common divisor is 1. We may assume that a_k is not divisible by p for $1 \le k \le n$ and divisible by p for $n+1 \le k \le p+2$, where $n \le p+1$. Suppose $n \le p$. Let $T = a_1 + a_2 + \cdots + a_n$. If T is not divisible by p, then $a_1 + a_2 + \cdots + a_p$ is not divisible by a_{p+2}. If T is divisible by p, then $a_2 + a_3 + \cdots + a_{p+1}$ is not divisible by a_{p+2}. It follows that $n = p+1$, so that a_{p+2} is the only number divisible by p. As in Case 1, $a_1 = a_2 = \cdots = a_{p+1} = 1$ so that $a_{p+2} = p$.
In summary, all interesting sets are of the form $a_1 = a_2 = \cdots = a_{p+1} = d$ and $a_{p+2} = d$ or pd for an arbitrary positive integer d.

6. In order to be sure of knowing a digit of Inspector Gadget's PIN number, Dr. Claw either must check it or check that digit of every other client. For $n = 3$, Dr. Claw can find out the first three digits of Inspector Gadget's PIN number, and deduce the last three digits by checking those of every other client. There is no solution for $n \le 2$ since Dr. Claw can know at most 2 digits of Inspector Gadget's PIN number and deduce 2 more by checking those of every other client.

7. Since K is the incentre of triangle BCI, $\angle BKI = 90^\circ + \frac{1}{2}\angle BCI = 97.5^\circ$. Let O be the centre of triangle ABC and let M be the point symmetric to I about OA. Note that K lies on BM.

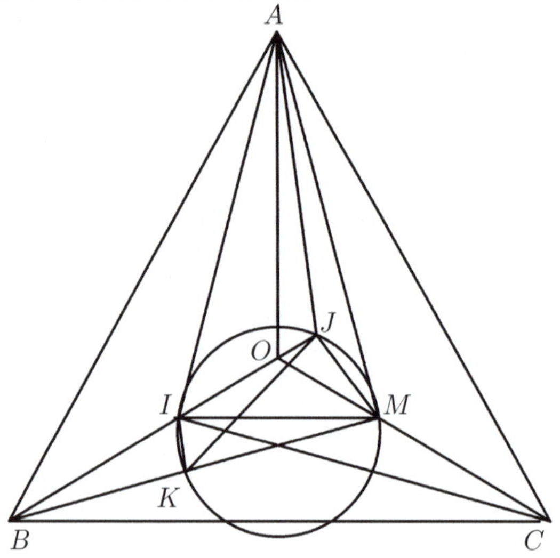

We have $\angle MAJ = 7.5^\circ = \angle OAJ$. Since $\angle AOJ = 60^\circ = \angle MOJ$, J is the incentre of triangle MAO. It follows that

$$\angle MJO = 90^\circ + \frac{1}{2}\angle MAO = 97.5^\circ = \angle BKI.$$

Hence $IJMK$ is a cyclic quadrilateral and $\angle IJK = \angle IMK = 15^\circ$. Since L is symmetric to K about BO, $\angle KJL = 2\angle IJK = 30^\circ$.

Senior A-Level Paper

1. (See Problem 3 of the Junior A-Level Paper.)

2. Among all the ways of connecting the one hundred points in pairs, consider the one for which the total length of the fifty segments is maximum, We claim that this connection has the desired property. Suppose to the contrary that two lines, AB and CD, intersect outside the circle. Then these four points form a convex quadrilateral, and we may assume that it is $ABCD$. Let AC intersect BD at E. Then $AC+BD = AE+BE+CE+DE > AB+CD$. Replacing AB and CD by AC and BD increases the total length of the fifty segments. This contradiction justifies our claim.

3. Note that $1^2 \equiv (2n-2)^2$, $2^2 \equiv (2n-3)^2$, $\ldots$, $(n-1)^2 \equiv n^2 \pmod{2n-1}$. We claim that for any i and j, $1 \le i < j \le n-1$, we can find k such that $(i+k)^2 \equiv (j+k)^2 \pmod{2n-1}$. Suppose $j-i = 2m-1$ for some m. Choose k so that $j+k \equiv n+m-1$ and $i+k \equiv n-m$. Suppose $j-i$ is even. Then $(2n-1)+i-j$ is odd and we can make a similar choice for k. This justifies the claim. Now x^2 takes on n different values modulo $2n-1$. By a suitable choice of a_1, we can make $(x^2+a_1)^2$ take on at most $n-1$ different values modulo $2n-1$. By a suitable choice of a_2, we can make $((x^2+a_1)^2+a_2)^2$ take on at most $n-2$ different values modulo $2n-1$. Continuing in this manner, we can eventually choose a_{n-1} so that $(\cdots(((x^2+a_1)^2+a_2)^2+\cdots)^2+a_{n-1})^2$ takes on only one value. By a suitable choice of a_n, we can make the final expression divisible by $2n-1$.

4. Let the points Alex marked be F on the front, B at the back, R to the right, L to the left, U on the up face and D on the down face. The twelve strings formed three closed loops $FRBL$, $FUBD$ and $RULD$. We claim that the total length of each loop is at least $2\sqrt{2}$.

Let $FRBL$ be projected onto the down face. Then each point lies on one side of a unit square, as shown in the diagram below on the left. We now fold the loop out as shown in the diagram below on the right. Since $FXYF'$ is a parallelogram, the total length of the strings FR, RB, BL and LF' is at least XY. This is twice the diagonal of a unit square, which is $2\sqrt{2}$.

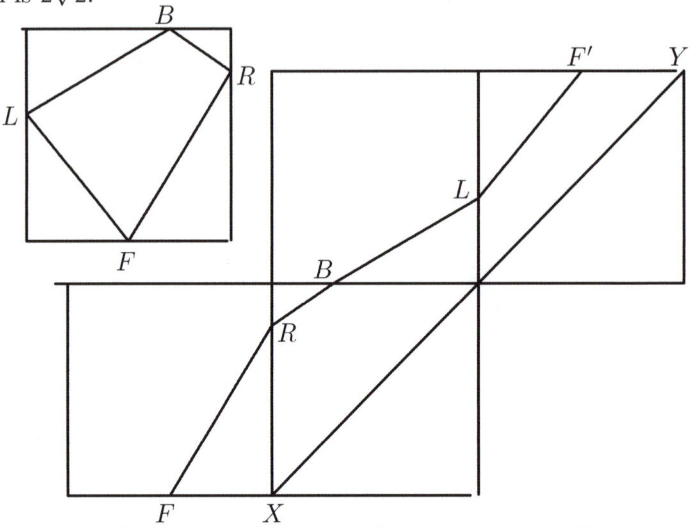

5. Let ℓ intersect the exterior bisectors of $\angle A$, $\angle B$ and $\angle C$ at P, Q and R respectively. Let ℓ_b intersect ℓ_c at D, ℓ_c intersect ℓ_a at E, and ℓ_a interssect ℓ_b at F. We first show that DEF is similar to ABC. Let the exterior bisectors of $\angle A$ and $\angle B$ intersect at M. Then

$$\begin{aligned}\angle AMB &= 180^\circ - \left(90^\circ - \frac{1}{2}\angle A\right) - \left(90^\circ - \frac{1}{2}\angle B\right)\\ &= \frac{1}{2}(\angle A + \angle B)\\ &= 90^\circ - \frac{1}{2}\angle C.\end{aligned}$$

It follows that

$$\begin{aligned}\angle EFD &= \angle PFQ\\ &= 180^\circ - \angle FPQ - \angle FQP\\ &= 180^\circ - 2(\angle MPQ + \angle MQP)\\ &= 180^\circ - 2\angle AMB\\ &= 180^\circ - 2\left(90^\circ - \frac{1}{2}\angle C\right)\\ &= \angle C.\end{aligned}$$

Similarly, $\angle FDE = \angle A$ and $\angle DEF = \angle B$. Hence we have similarity.

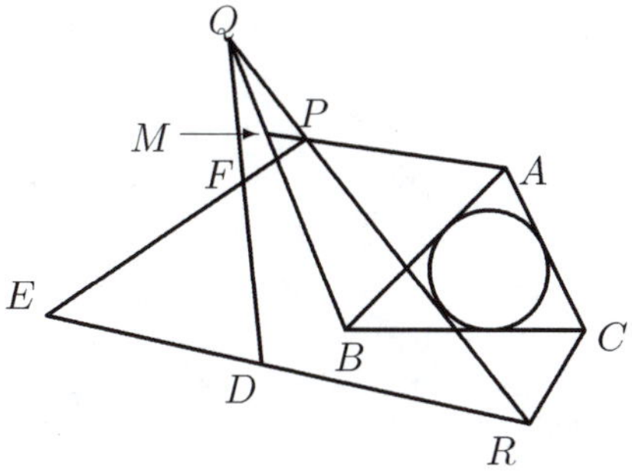

We now establish congruence. We may assume that the incircle of ABC is a unit circle. Reflect it across the exterior bisectors of $\angle A$, $\angle B$ and $\angle C$, resulting in three unit circles with centres X, Y and Z respectively. Then XYZ is similar to ABC with all side lengths doubled. Draw the line through X parallel to ℓ_a, intersecting the circumcircle of XYZ again at a point T which we may assume lies on the arc XY. Then

$$\angle XTZ = \angle XYZ = \angle B = \angle DEF,$$

so that TZ is parallel to ℓ_c. Similarly, TY is parallel to ℓ_b. Consider the unit circle ω with centre T. Since ℓ_a is tangent to the unit circle with centre X, it is also tangent to ω. Similarly, ℓ_b and ℓ_c are also tangent to ω. We must still show that ω is not an excircle of DEF. Note that the positions of D, E, F and T obviously depend on the position of ℓ. As ℓ moves continuously around the incircle of ABC, T does not cross the perimeter of DEF. If we let ℓ coincide with a side of ABC, it is easy to see that T must be inside DEF. It follows that ω is the incricle of DEF. Since ABC and DEF are similar triangles with equal incircles, they are congruent.

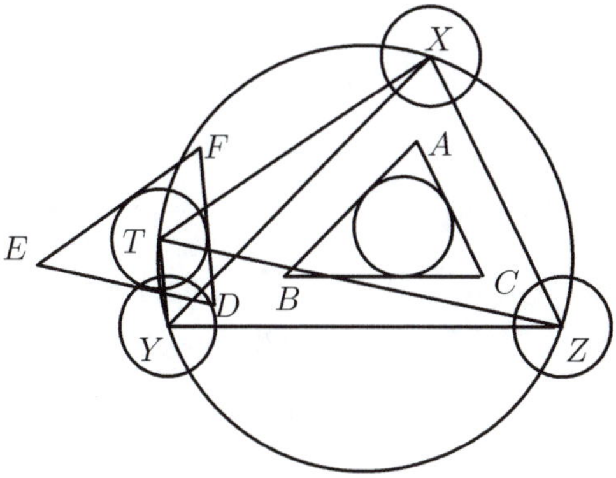

6. (a) Let the dimensions of the nth rectangle be $n^2 2^n \times \frac{1}{2^n}$. We claim that this sequence of rectangles cannot even cover a disk with radius 1. The intersection of the nth rectangle with the disk is contained in a $2 \times \frac{1}{2^n}$ rectangle and has area less than $\frac{1}{2^{n-1}}$. The total area of these intersections is less than $1 + \frac{1}{2} + \frac{1}{4} + \cdots < 2 < \pi$.

   (b) Suppose there exists a positive number a such that the side length of infinitely many of the squares in the sequence is at least a. Then we divide the plane into a sequence of $a \times a$ squares in an outward spiral, and these squares can be covered one at a time. Henceforth, we assume that for each positive real number a, the number of squares in the sequence with side length at least a is finite. This induces a well-ordering on the squares of the sequence in non-ascending order of side lengths $a_1 \geq a_2 \geq a_3 \geq \cdots$. We may assume that $a_1 < 1$. We divide the plane into a sequence of unit squares in an outward spiral, and try to cover these squares one at a time. Place the $a_1 \times a_1$ square at the bottom left corner of the first unit square. Place the $a_2 \times a_2$ square on the bottom edge of this unit square to the right of the $a_1 \times a_1$ square. In this manner, we can cover the bottom edge of the unit square because $a_1 + a_2 + \cdots + a_{k_1} > a_1^2 + a_2^2 + \cdots + a_{k_1}^2 \geq 1$ for some k_1. Let $b_1 = a_{k_1}$. Then we have covered the bottom strip of the unit square of height b_1. We now focus on the $1 \times (1 - b_1)$ rectangle, and apply the same process to cover the bottom strip of height b_2 with squares of side lengths $a_{k_1+1}, a_{k_1+2}, \ldots, a_{k_2} = b_2$.Continuing in this manner, we cover strips of height $b_3, b_4, \ldots$. We claim that $b_1 + b_2 + \cdots + b_h \geq 1$ for some h. Suppose this is not so. Then for all h,

$$\begin{aligned}
1 &> (b_1 + b_2 + \cdots + b_h)^2 \\
&\geq b_1(a_{k_1+1} + a_{k_1+2} + \cdots + a_{k_2}) \\
&\quad + b_2(a_{k_2+1} + a_{k_2+2} + \cdots + a_{k_3}) + \cdots \\
&\quad + b_h(a_{k_h+1} + a_{k_h+2} + \cdots + a_{k_{h+1}}) \\
&\geq a_{k_1+1}^2 + a_{k_1+2}^2 + \cdots + a_{k_{h+1}}^2.
\end{aligned}$$

   This is a contradiction since the last expression is not bounded above.

7. (a) At some point in time, we must have exactly 70 piles. At least 40 of them contain exactly 1 pebble each, as otherwise the total number of pebbles is at least $39 + 2 \times 31 = 101$. Removing these 40 piles leave behind exactly 30 piles containing exactly 60 pebbles among them.

(b) We call k piles containing a total of exactly $2k + 20$ pebbles a good collection. We claim that if $k \geq 23$, a good collection contains either 1 pile with exactly 2 pebbles or 2 piles each with exactly 1 pebble. Otherwise, the total number of pebbles in the collection is at least $1 = 3(k - 1) = 3k - 2$, which is strictly greater than $2k + 20$ when $k \geq 23$. Now any partition of the original pile into 40 piles results in a good collection with $k = 40$. From this, we can obtain a good collection with $k = 39$ by either removing 1 pile with exactly 2 pebbles or removing 2 piles each with exactly 1 pebble and then subdividing any other pile with at least 2 pebbles. In the same way, we can obtain good collections down to $k = 22$, with a total of exactly 64 pebbles. We claim that there exist 2 or more piles containing a total of exactly 4 pebbles. Suppose this is not the case. If there are no piles with exactly 1 pebble, then the total number of pebbles in the collection is at least $2+3\times 21 > 64$. If there are piles with exactly 1 pebble, then the total is at least $3+4\times 19 > 64$. Thus the claim is justified. We now remove these 4 pebbles, obtaining 60 pebbles in at most 20 piles. Eventual subdivision of these piles will bring the number of piles to 20 while keeping the total number of pebbles at 60.

(c) Separate out piles of 3 until we have 32 piles of 3 and are left with 1 pile of 4. Throughout this process, exactly one pile contains a number of pebbles not divisible by 3. If we include this pile, the total cannot be 60. If we exclude this pile, the total of 19 piles of 3 is only 57. We now separate the pile of 4 into 2 piles of 2. Now every pile contains at most 3 pebbles, so that 19 piles can contain at most 57 pebbles. Further separation will not change this situation.

Tournament 34

Fall 2012

Junior O-Level Paper

1. The family names of Clark, Donald, Jack, Robin and Steven are Clarkson, Donaldson, Jackson, Robinson and Stevenson, but not in that order. Clark is 1 year older than Clarkson, Donald is 2 years older than Donaldson, Jack is 3 years older than Jackson and Robin is 4 years older than Robinson. Who is older, Steven or Stevenson, and what is the difference in their ages?

2. Let $C(n)$ be the number of distinct prime divisors of a positive integer n. For instance, $C(10) = 2$, $C(11) = 1$ and $C(12) = 2$. Is the number of pairs of positive integers (a, b) such that $a \neq b$ and $C(a+b) = C(a) + C(b)$ finite or infinite?

3. The game Minesweeper is played on a 10×10 board. Each square either contains a bomb or is vacant. On each vacant square is recorded the number of bombs among the eight adjacent squares. Then all the bombs are removed, and new bombs are placed in all squares which were previously vacant. Then numbers are recorded on vacant squares as before. Can the sum of all numbers on the board now be greater than the sum of all numbers on the board before?

4. A circle touches sides AB, BC and CD of a parallelogram $ABCD$ at points K, L and M respectively. Prove that the line KL bisects the perpendicular dropped from C to AB.

5. For a class of 20 students, several field trips were organized. For each field trip, at least one student participated in it. Prove that there was a field trip such that each student who participated in it took part in at least $\frac{1}{20}$th of all field trips.

Note: The problems are worth 3, 4, 5, 5 and 5 points respectively.

Senior O-Level Paper

1. The game Minesweeper is played on a 10×10 board. Each square either contains a bomb or is vacant. On each vacant square is recorded the number of bombs among the eight adjacent squares. Then all the bombs are removed, and new bombs are placed in all squares which were previously vacant. Then numbers are recorded on vacant squares as before. Can the sum of all numbers on the board now be greater than the sum of all numbers on the board before?

2. A sphere trisects each edge of a convex polyhedron at two points. Is it necessarily true that all faces of the polyhedron are
   (a) congruent to one another;
   (b) regular polygons?

3. For a class of 20 students, several field trips were organized. For each field trip, at least four students participated in it. Prove that there was a field trip such that each student who participated in it took part in at least $\frac{1}{17}$th of all field trips.

4. (a) Let $C(n)$ be the number of distinct prime divisors of a positive integer n. For instance, $C(10) = 2$, $C(11) = 1$ and $C(12) = 2$. Is the number of pairs of positive integers (a, b) such that $a \neq b$ and $C(a + b) = C(a) + C(b)$ finite or infinite?
   (b) What is the answer if it is also required that $C(a + b) > 1000$?

5. Among 239 coins, there are two fake coins of the same weight, and 237 real coins of the same weight but different from that of the fake coins. Is it possible, in three weighings on a balance, to determine whether the fake coins are heavier or lighter than the real coins? It is not necessary to identify the fake coins.

Note: The problems are worth 4, 2+3, 5, 2+3 and 5 points respectively.

Junior A-Level Paper

1. The base ten representation of an integer consists of two different digits appearing alternately. The number has at least 10 digits. What is the highest power of 2 that can divide this number?

2. With Dale watching, Chip divides 222 nuts between two boxes. Dale chooses a number n between 1 and 222 inclusive. Then Chip moves some nuts, possibly none, from one of the boxes to a third box, so that there will be exactly n nuts in some box, or in two of the boxes. Chip wants to minimize the number of nuts in the third box while Dale wants to maximize this number. What is the optimal value of this number?

3. Markers are placed in some squares of an 11×11 board. The number of markers on the whole board is even. The number of markers on any four squares with a common corner is also even. Prove that the number of markers on a longest diagonal of the board is also even.

4. The incentre of triangle ABC is I. The incentres of triangles BIC, CIA and AIB are X, Y and Z respectively. If the incentre of triangle XYZ is also I must triangle ABC be equilateral?

5. A car goes along a circular track in the clockwise direction. At noon, Peter and Paul take their positions at two different points of the track. Some moment later, they simultaneously end their duties and compare their notes. The car passes each of them at exactly 30 times. Peter notices that each circle is passed by the car 1 second faster than the preceding one while Paul's observation is the opposite: each circle is passed 1 second slower than the preceding one. Prove that their duty last at least one and a half hours.

6. A is a fixed point inside a given circle.

   (a) Two perpendicular lines are drawn through A, intersecting the circle at four points. Prove that the centre of mass of these four points does not depend on the choice of the two lines.

   (b) A regular $2n$-gon with centre A is drawn, and the rays from A to its vertices intersect the circle.at $2n$ points. Prove that the centre of mass of these $2n$ points does not depend on the choice of the regular $2n$-gon.

7. Peter chooses some positive integer a and Paul wants to determine it. Paul only knows that the sum of the digits of Peter's number is 2012. In each moves, Paul chooses a positive integer x and Peter tells him the sum of the digits of $|x - a|$. What is the minimal number of moves Paul needs to determine Peter's number for sure?

Note: The problems are worth 5, 5, 6, 7, 8, 4+4 and 10 points respectively.

Senior A-Level Paper

1. An infinite sequence of numbers $a_1, a_2, a_3, \ldots$ is given. For any positive integer k, there exists a positive integer $t = t(k)$ such that

$$a_k = a_{k+t} = a_{k+2t} = \cdots.$$

   Is this sequence necessarily periodic?

2. With Dale watching, Chip divides 1001 nuts among three boxes. Dale chooses a number n between 1 and 1001 inclusive. Then Chip moves some nuts, possibly none, from one of the boxes to a fourth box, so that there will be exactly n nuts in some box, in two of the boxes or in three of the boxes. Chip wants to minimize the number of nuts in the fourth box while Dale wants to maximize this number. What is the optimal value of this number?

3. A car goes along a circular track in the clockwise direction. At noon, Peter and Paul take their positions at two different points of the track. Some moment later, they simultaneously end their duties and compare their notes. The car passes each of them exactly 30 times. Peter notices that each circle is passed by the car 1 second faster than the preceding one while Paul's observation is the opposite: each circle is passed 1 second slower than the preceding one. Prove that their duty last at least one and a half hours.

4. I is the incentre of triangle ABC. A circle through B and I intersects AB at F and BC at D. K is the midpoint of DF. Prove that $\angle AKC$ is obtuse.

5. Peter chooses some positive integer a and Paul wants to determine it. Paul only knows that the sum of the digits of Peter's number is 2012. In each moves, Paul chooses a positive integer x and Peter tells him the sum of the digits of $|x - a|$. What is the minimal number of moves Paul needs to determine Peter's number for sure?

6. A is a fixed point inside a given sphere.

   (a) Three mutually perpendicular lines are drawn through A, intersecting the sphere at six points. Prove that the centre of mass of these six points does not depend on the choice of the three lines.

   (b) A regular icosahedron with centre A is drawn, and the rays from A to its vertices intersect the sphere.at twelve points. Prove that the centre of mass of these twelve points does not depend on the choice of the regular icosahedron.

7. There are 1000000 soldiers in a line. The sergeant divides the line into 100 segments of varying lengths. He then form a new line by permuting the segments without changing the order of soldiers within each segment. This is repeated several times, with the segments having the same lengths in the same order as before, and permuted in exactly the same way. Every soldier originally from the first segment records the number of performed procedures that takes him back to the first segment for the first time. Prove that at most 100 of these numbers are different.

Note: The problems are worth 4, 5, 6, 8, 8, 5+5 and 10 points respectively.

Spring 2013

Junior O-Level Paper

1. ABC and DEF are two triangles. Is it always possible to form two triangles with A, B, C, D, E and F as vertices, such that the triangles have no common points in the interiors or on the boundaries?

2. Start with an non-negative integer n. In each move, we may add 9 to the current number or, if the number contains a digit 1, we may delete it. If there are leading 0s as a result, they are also deleted. Is it always possible to obtain the number $n + 1$ in a finite number of steps?

3. The weight of each of 11 objects is a distinct integral number of grams. Whenever two subsets of these objects of unequal sizes are weighed against each other, the subset with the larger number of objects is always the heavier one. Prove that the weight of one of the objects exceeds 35 grams.

4. Eight rooks are placed on an 8×8 board such that no two attack each other. Each rook claims the square on which it stands. An vacant square attacked by two rooks is claimed by the rook which is nearer. If both rooks are at the same distance away from the vacant square, each claims half of it. Prove that each rook claims the same number of squares.

5. In the quadrilateral $ABCD$, $AB = CD$, $\angle B = 150°$ and $\angle C = 90°$. Determine the angle between BC and the line joining the midpoints of AD and BC.

Note: The problems are worth 3, 4, 4, 5 and 5 points respectively.

Senior O-Level Paper

1. Start with an non-negative integer n. In each move, we may add 9 to the current number or, if the number contains a digit 1, we may delete it. If there are leading 0s as a result, they are also deleted. Is it always possible to obtain the number $n + 1$ in a finite number of steps?

2. In triangle ABC, $\angle C = 90°$ and CE is an altitude. Squares $ACKL$ and $BCMN$ are constructed outside ABC. Prove that $\angle LEM = 90°$.

3. Eight rooks are placed on an 8×8 board such that no two attack each other. Each rook claims the square on which it stands. An vacant square attacked by two rooks is claimed by the rook which is nearer. If both rooks are at the same distance away from the vacant square, each claims half of it. Prove that each rook claims the same number of squares.

4. Each of 100 stones with distinct weights has a label showing its weight. Is it always possible to rearrange the labels so that the total weight of any group of less than 100 stones is not equal to the sum of the corresponding labels?

5. A polynomial of the form $x^2 + ax + b$ is said to be admissible if the absolute values of a and b are both at most 2013, and both roots are integers. Prove that the sum of all admissible polynomials has no real roots.

Note: The problems are worth 3, 4, 4, 4 and 5 points respectively.

Junior A-Level Paper

1. The sum of any two of n distinct numbers is a positive integer power of 2. What is the maximum value of n?

2. Ten boys and ten girls are standing in a row. Each boy counts the number of girls to his left, and each girl counts the number of boys to her right. Prove that the total number counted by the boys is equal to the total numbers counted by the girls.

3. Is it possible to mark some of the squares of a 19×19 board so that each 10×10 board contains a different number of marked squares?

4. On a circle are 1000 non-zero numbers painted alternately black and white. Each black number is the sum of its two neighbours while each white number is the product of its two neighbours. What are the possible values of the sum of these 1000 numbers?

5. A lattice point in the plane is one both coordinates of which are integers. If there are exactly two lattice points in the interior of a triange whose vertices are all lattice points, prove that the line joining these two points either passes through a vertex or is parallel to one side of the triangle.

6. In triangle ABC, $\angle C = 90°$. The incircle is tangent to BC at D and to CA at E. The line through D perpendicular to the bisector of $\angle A$ intersects the line through E perpendicular to the bisector of $\angle B$ at P. Prove that CP is perpendicular to AB.

7. In a school with more girls than boys, there is only one ping-pong table. A boy is playing against a girl while all remaining students form a single line. At the end of a game, the student at the head of the line replaces the student of the same gender who has just finished playing, and the replaced student goes to the end of the line. Prove that eventually, every boy has played every girl.

Note: The problems are worth 4, 4, 5, 5, 6, 8 and 9 points respectively.

Senior A-Level Paper

1. A set of positive integers is such that the sum of any two of them is a positive integral power of 2. What is the maximum size of this set?

2. A boy and a girl are sitting at opposite ends of a long bench. One at a time, twenty other children take seats on the bench. If a boy takes a seat between two girls or if a girl takes a seat between two boys, he or she is said to be brave. At the end, the boys and girls are sitting alternately. What is the number of brave children?

3. If both coordinates of a point in the plane are integers, we call it a *lattice* point. If there are at least two lattice points in the interior of a triangle whose vertices are all lattice points, prove that there exist two of these points such that the line joining them either passes through a vertex or is parallel to one side of the triangle.

4. Is it possible to arrange the numbers $1, 2, \ldots, 100$ on a circle in some order so that the absolute value of the difference between any two adjacent numbers is at least 30 and at most 50?

5. Initially, only three points on the plane are painted, one red, one yellow and one blue. In each step, we choose two points of different colours. A point is painted in the third colour so that an equilateral triangle with vertices painted red, yellow and blue in clockwise order is formed with the two chosen points. Note that a painted point may be painted again, and it retains all its colours. Prove that after any number of moves, all points of the same colour lie on a straight line.

6. The sum of the squares of five distinct positive numbers is equal to the sum of the ten pairwise products of the five numbers.

   (a) Prove that there exists a subset of size three of these five numbers such that no triangle can have side lengths equal to the numbers in the subset.

   (b) Prove that the number of such subsets is at least six.

7. One thousand wizards are standing in a column. Each is wearing one of the hats numbered from 1 to 1001 in some order, one hat not being used. Each wizard can see the number of the hat of any wizard in front of him, but not that of any wizard behind. Starting from the back, each wizard in turn calls out a number from 1 to 1001 so that every other wizard can hear it. Each number can be called out at most once. At the end, a wizard who fails to call out the number on his hat is removed from the Council of Wizards. This procedure is known to the wizards in advance, and they have a chance to discuss strategy. Is there a strategy which can keep in the Council of Wizards

   (a) more than 500 of these wizards;

   (b) at least 999 of these wizards?

Note: The problems are worth 3, 4, 6, 6, 7, 4+5 and 5+7 points respectively.

Solutions

Fall 2012

Junior O-Level Paper

1. The total age of Clark, Donald, Jack, Robin and Steven must be the same as the total age of Clarkson, Donaldson, Jackson, Robinson and Stevenson. Since Clark, Donald, Jack and Robin are 1+2+3+4=10 years older than Clarkson, Donaldson, Jackson and Robinson, Stevenson must be 10 years older than Steven.

2. For any positive integer n, let $a = 2^n$ and $b = 2^{n+1}$. Then $a+b = 3\times 2^n$. We have $C(a) + C(b) = 1 + 1 = 2 = C(a + b)$. Since n is arbitrary, the number of pairs (a, b) is infinite.

3. Represent each square by a vertex and join two adjacent squares by an edge. An edge is called a scoring edge if it joins a bomb square and vacant square, since it will contribute 1 to the number recorded on the vacant square. An edge which joins two bomb squares or two vacant squares is non-scoring. Hence the sum of the recorded numbers is equal to the number of scoring edges. After the transformation of the board, a scoring edge remains a scoring edge, and a non-scoring edge remains a non-scoring edge. Hence the sum of all the numbers on the board before must be equal to the sum of all the numbers on the board after.

4. Let KL intersect the perpendicular CH dropped from C to AB at P. Let the extensions of KL and DC intersect at Q. Now $\angle KLM = 90°$ since KM is a diameter of the circle. Hence QLM is a right triangle, so that its circumcentre lies on QM as well as on the perpendicular bisector of LM. Now $CL = CM$ since both are tangents from C to the circle. Hence C lies on the perpendicular bisector of LM. Being on QM, C must be the circumcentre of triangle QLM. This means that $QC = CM$. Now PC is parallel to KM. By the Midpoint Theorem, $PC = \frac{1}{2}KM = \frac{1}{2}HC$, which is equivalent to the desired result.

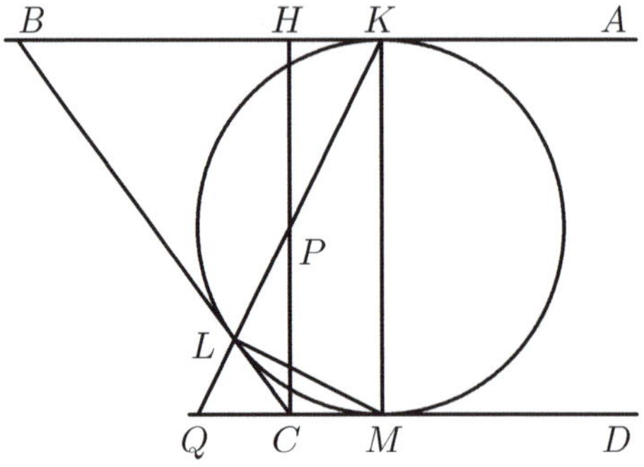

5. Let n be the number of field trips. Let us call a student "enthusiastic" if he or she takes part in no fewer than $\frac{n}{20}$ of the trips, and "lazy" otherwise. Obviously the number of lazy students does not exceed 20. Each of them attends fewer than $\frac{n}{20}$ of the trips. Therefore the total number of trips attended by lazy students is fewer than $20 \times \frac{n}{20} = n$. Since each trip has at least one participant, there is a trip attended only by enthusiastic students.

Senior O-Level Paper

1. (See Problem 3 of the Junior O-Level Paper.)

2. (a) Consider a triangular prism where both bases are equilateral triangles and all three lateral faces are squares. We can draw a circle passing through all the points of trisection of the edges of the top triangle, and another circle passing through all the points of trisection of the edges of a lateral square. There is a unique sphere on which both circles lie. Its centre is in the plane of bisection of the edge common to the top triangle and the lateral square. By symmetry, all edges of this triangular prism are trisected, but not all faces are congruent to one another.

   (b) Consider any face of the polyhedron and let AB and BC be two of its edges. Let the points of trisection for AB be P and Q, and those for BC be R and S. Now the cross-section of the sphere in the plane of this face is a circle passing through P, Q, R and S. By the Power-of-a-point Theorem, we have $BP \cdot BQ = BR \cdot BS$. Now $BP \cdot BQ = \frac{2}{9}AB^2$ while $BE \cdot BS = \frac{2}{9}BC$. Hence $AB = BC$, and it follows that this face is an equilateral polygon. Moreover, $PQ = RS$, so that AB and BC are equidistant from the centre of this circular cross-section. It follows that the polygon has an incircle. Being equilateral as well, it must be a regular polygon. Since this is an arbitrary face, all faces are regular polygons.

3. Let n be the number of field trips. Let us call a student "enthusiastic" if she or he takes part in no fewer than $\frac{n}{17}$ of the trips, and "lazy" otherwise. Let each student fill in a registration form for each trip in which she or he participates. Then there are at least $4n$ registration forms. Suppose there are at most 3 enthusiastic students. Then there are at least 17 lazy ones. The enthusiastic students fill in at most $3n$ registration forms. Hence the lazy students fill in at least n of them. Thus at least one of them must have participated in at least $\frac{n}{17}$ trips, and not called lazy. It follows that there are at least 4 enthusiastic students and at most 16 lazy ones. The total number of trips taken by any lazy students is less than $16 \times \frac{n}{17} < n$, so that there is a trip attended only by enthusiastic students.

4. (a) For any prime number $p > 23$, let $a = 7p$ and $b = 23p$. Then $a + b = 30p$. We have $C(a) + C(b) = 2 + 2 = 4 = C(a + b)$. Since there are infinitely many primes $p > 23$, the number of pairs (a, b) is infinite.

   (b) Let $p_1, p_2, \ldots, p_k$ be the first k prime numbers, where $k > 1000$. Then we can write $p_1 p_2 \cdots p_k - 1 = q_1^{t_1} q_2^{t_2} \cdots q_s^{t_s}$, where the qs are prime number disjoint from the ps and $0 < s < k$. Let $m = k - s$ and choose prime numbers $r_1, r_2, \ldots, r_m$ such that the r are disjoint from the qs and distinct from p_l. Now let $a = r_1 r_2 \cdots r_m$ and $b = r_1 r_2 \cdots r_m q_1^{t_1} q_2^{t_2} \cdots q_s^{t_s}$. Then $a + b = r_1 r_2 \cdots r_m p_1 p_2 \cdots p_k$. We have $C(a) = m$, $C(b) = m + s$ and $C(a + b) = m + k$. Since $m + s = k$, we have $C(a + b) = C(a) + C(b)$. There being an infinite number of prime numbers, we can always choose different $r_1, r_2, \ldots, r_m$ and therefore create an infinite number of different pairs (a, b) satisfying the given conditions.

5. Put 80 coins into group A, 79 coins into each of groups B and C, and we have one coin left. If we add this coin to group B, we call the expanded group B$^+$. C$^+$ is similarly obtained from C. In the first two weighings, we try to balance group A against group B$^+$ and group A against group C$^+$. We consider three cases.
   Case 1. We have equilibrium both times.
   This means that each group has one fake coin, so that the extra coin must be fake. The coins in both groups B and C are real. Weigh one of them against the known fake coin, and that will tell us the desired answer.
   Case 2. We have equilibrium only once, say between A and B$^+$.
   Either each has a fake or neither has a fake coin. Divide the coins in A into two subgroups of 40 and weigh them against each other. If we have equilibrium, A does not have any fake coins, and neither does B, which means C has both of them. If we do not have equilibrium, A has a fake coin, and so does B$^+$. This means that the extra coin is real and C$^+$ does not have any fake coins. In either situation, the weighing between A and C$^+$ tells us the desired answer.
   Case 3. We have no equilibrium.
   We claim that either A is heavy both times or A is light both times. Suppose to the contrary that B$^+$ is heavier than A and A is heavier than C$^+$. Then they must have two, one and zero fake coins in either order. However, if A has one fake coin, it is impossible for either B or C to have two of them. This justifies our claim. So either both B$^+$ and C$^+$ have no fake coins or both have one. Divide the coins in B$^+$ into two subgroups of 40 and weigh them against each other. This will tell us the desired answer.

Junior A-Level Paper

1. Let the two digits be x and y. If the number has an even number of digits, then it is the product of $10x + y$ with a number of the form $1010\ldots101$, which is odd. Hence the highest power of 2 which divides this number is also the highest power of 2 which divides $10x + y$. The highest power 2 which is a two-digit number is $2^6 = 64$. If the number has an odd number of digits, then it is the sum of $y \times 10^k$ for some integer $k > 10$ and the same number in the even case. The former is divisible by 2^6 while the latter is divisible by 2^6 but not by 2^7. It follows that the highest power of 2 which divides this number is 6.

2. Chip puts 74 nuts into the first box and 148 nuts into the second one. By acting according to the chart below, he can keep the number nuts in the third box to at most 37.

Dale's Number	Chip's Action
$1 \le n \le 37$	Move $n \le 37$ from First
$38 \le n \le 74$	Move $74 - n \le 36$ from First
$75 \le n \le 111$	Move $n - 74 \le 37$ from Second
$112 \le n \le 148$	Move $148 - n \le 36$ from Second
$149 \le n \le 185$	Move $n - 148 \le 37$ from First
$186 \le n \le 222$	Move $222 - n \le 36$ from First

Conversely, suppose Chip puts $m \le 111$ nuts in the first box and $222-m$ nuts in the second one. Suppose $m \le 74$. By choosing $n = 111$, Dale can force Chip to move at least 37 nuts. Suppose $74 < m$. By choosing 37, Dale can force Chip to move 37 nuts. Hence the optimal value is 37.

3. The 11 squares of the main diagonal of the 11×11 board are shaded in the diagram below. The rest of the board is divided into regions consisting of 4 or 6 unshaded squares. By the given condition, the number of markers in each region is even. Since the number of markers on the whole board is even, the number of markers on the shaded squares is also even.

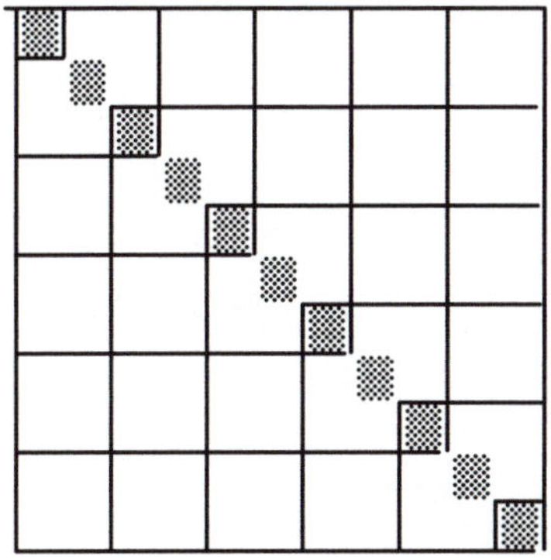

4. We claim that XYZ is an equilateral triangle. Let AI, BI and CI intersect YZ, ZX and XY at D, E and F respectively. Since I is the incentre of XYZ, $\angle EXI = \angle FXI$. Since X is the incentre of BIC, $\angle EIX = \angle FIX$. Hence EIX and FIX are congruent triangles, so that $\angle IEX = \angle IFX$. Similarly, $\angle IFY = \angle IDY$ and $\angle IDZ = \angle IEZ$. Now the sum of any two of $\angle IEX$, $\angle IFY$ and $\angle IDZ$ is $180°$. Hence each is equal to $90°$. Now AI bisects $\angle CAB$, AY bisects $\angle CAI$ and AZ bisects $\angle BAI$. Hence AI bisects $\angle YAZ$, so that triangles YAD and ZAD are congruent. It follows that D is the midpoint of YZ. Similarly, E is the midpoint of ZX and F is the midpoint of XY. Hence I is also the circumcentre of XYZ, which justifies our claim. So $\angle EIF = \angle FID = \angle DIE = 120°$. It follows that the sum of any two of $\frac{1}{2}\angle CAB$, $\frac{1}{2}\angle ABC$ and $\frac{1}{2}\angle BCA$ is $60°$. Hence each is equal to $30°$, and ABC is indeed an equilateral triangle.

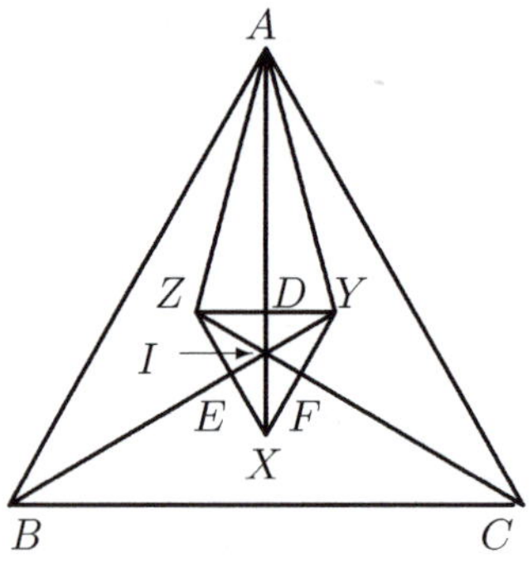

5. Let the duty period start when the car passes by Paul for the first time, and end when it passes by Peter for the thirtieth time. For $1 \leq n \leq 30$, let x_n be the time the car takes to go from Paul to Peter, and for $1 \leq n \leq 29$, let y_n be the time it takes to get back to Paul. Then the length of the duty period is

$$T = x_1 + y_1 + x_2 + y_2 + x_3 + \cdots + y_{29} + x_{30}.$$

Let $1 \leq n \leq 29$. From Paul's observations,

$$x_n + y_n = A + n$$

for some constant A, and from Peter's observations,

$$y_n + x_{n+1} = B - n$$

for some constant B. Subtracting this from $x_{n+1} + y_{n+1} = A + (n+1)$, we have

$$y_{n+1} - y_n = 2n + 1 + A - B.$$

Since the difference is a linear function, y_n is a quadratic function which we can express in the form

$$y_n = (n-15)^2 + C(n-15) + y_{15},$$

where C is some constant. If $C \geq 0$, then $y_{29} = 14^2 + 14C + y_{15} > 196$. If $C < 0$, then $y_1 = 14^2 - 14C + y_{15} > 196$. Taking $n = 1$ or 29, whichever yields $y_n > 196$, we have

$$A + B = x_n + 2y_n + x_{n+1} > 2 \times 196.$$

Summing this from $n = 1$ to $n = 29$, we have

$$2\times 196\times 29 < 29(A+B) = x_1+2y_1+2x_2+2y_2+2x_3+\cdots+2y_{29}+x_{30} < 2T.$$

It follows that $T > 196 \times 29 = 5684$. The conclusion follows since there are only 5400 seconds in one and a half hours.

6. Let O be the centre of the given circle. Let C be the midpoint of OA. For points D_k and E_k defined later, let B_k be the midpoint of D_kE_k. Then the centre of mass of D_k and E_k is B_k, with a weight of 2.

   (a) Let the two perpendicular chords be D_1E_1 and D_2E_2 as shown in the diagram below. Then $\angle OB_1A = 90° = \angle OB_2A$ so that B_1 and B_2 both lie on the circle with diameter OA. Since $\angle B_1AB_2 = 90°$, B_1B_2 is also a diameter, so that C is the midpoint of B_1B_2. Since B_1 and B_2 have equal weight, the overall centre of mass is at the fixed point C.

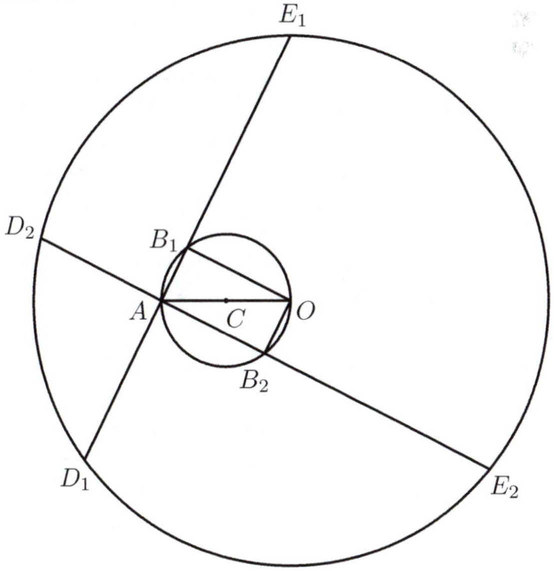

(b) Let the regular $2n$-gon be $D_1D_2 \ldots D_nE_1E_2 \ldots E_n$. Then we have $\angle OB_kA = 90°$ for $1 \leq k \leq n$. Hence B_1, B_2, ..., B_n all lie on the circle with diameter OA. We may assume that A is between B_1 and B_n. The diagram below illustrates the case $n = 5$. Now $\angle B_kAB_{k+1} = \frac{1}{n}180°$ for $1 \leq k \leq n-1$ and $B_nAB_1 = (1 - \frac{1}{n})180°$. Hence $B_1B_2 \ldots B_n$ is a regular n-gon whose centre is the fixed point C. Since each B_k has the same weight, the overall centre of mass is C.

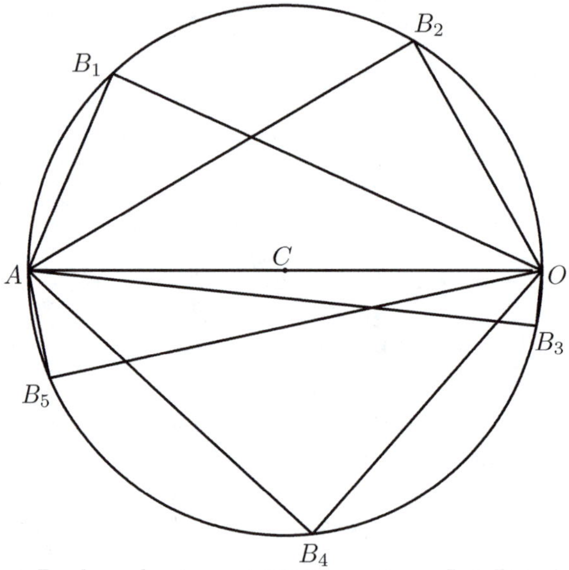

7. We claim that Paul needs at most 2012 questions. Let Peter's number have n non-zero digits a_1, a_2, ..., a_n from right to left, with $n > 1$ and $a_1 + a_2 + \cdots + a_n = 2012$. For $1 \leq k \leq n$, let b_k be the number of adjacent 0s to the right of a_k. Paul will carry out the following steps. In Step 1, Paul determines b_1 using one question. He chooses 1, and Peter's answer must be $2011 + 9b_1$. In step k, $2 \leq k \leq n$, Paul will determine a_{k-1} and b_k using a_{k-1} questions. Paul chooses a number which has a 2 in front of the known digits of Peter's number. If Peter's answer is 2010, Paul replaces the 2 by the 3. Continuing this way, when the first digit of Paul's choice is a_{k-1}, Peter will answer $2013 - a_{k-1}$. However, when Paul replaces a_{k-1} by $a_{k-1} + 1$, Peter's answer will be $2012 - a_{k-1} + 9b_k$. Now only a_n is left, but Paul does not know that unless $a_n = 1$. However, by the time he asks the $(a_n - 1)$st questions here, he will get 0 as a response and knows Peter's number. The total number of questions needed is $1 + a_1 + a_2 + \cdots + (a_n - 1) = 2012$. This justifies our claim.

We now prove that Peter can force Paul to use 2012 questions, by constructing his number one step at a time. Let b_1 be the number of digits in Paul's first choice. Peter chooses the last b_1 digits of his number to be 0 and put a 1 in front, referred to as a_1. His response will be $2011+9b_1$. Paul can only tell that Peter's number ends in exactly b_1 zeros. Paul's next choice must be a number with more digits. Peter adds b_2 zeros in front of a_1 to match the length of Paul's choice, and put another 1 in front, referred to as a_2. Paul can only determine the last b_2+1+b_1 digits of Peter's number, plus the fact that the digit in front is non-zero. Continuing this way, we have $a_1 = a_2 = \cdots = a_{2012} = 1$, and Paul needs one question to determine each of b_1, b_2, ..., b_{2012}.

Senior A-Level Paper

1. The sequence is not necessarily periodic, and here is a counter-example. For any positive integer k, let a_k be the highest power of 2 which divides k. Thus the sequence is 0,1,0,2,0,1,0,3,0,1,0,2,0,1,0,4,.... The choice $t(k) = 2^{a_k+1}$ satisfies the given condition and yet the sequence is not periodic as the maximum size of the terms increases without bound.

2. Chip puts 143 nuts into the first box, 286 nuts into the second one and 572 nuts into the third. By acting according to the chart below, he can keep the number nuts in the fourth box to at most 71.

Dale's Number	Chip's Action
$1 \le n \le 71$	Move $n \le 71$ from First
$72 \le n \le 143$	Move $143 - n \le 71$ from First
$144 \le n \le 214$	Move $n - 143 \le 71$ from Second
$215 \le n \le 286$	Move $286 - n \le 71$ from Second
$287 \le n \le 357$	Move $n - 286 \le 71$ from First
$358 \le n \le 429$	Move $429 - n \le 71$ from First
$430 \le n \le 500$	Move $n - 429 \le 71$ from Third
$501 \le n \le 572$	Move $572 - n \le 71$ from Third
$573 \le n \le 643$	Move $n - 572 \le 71$ from First
$644 \le n \le 715$	Move $715 - n \le 71$ from First
$716 \le n \le 786$	Move $n - 715 \le 71$ from Second
$787 \le n \le 858$	Move $858 - n \le 71$ from Second
$859 \le n \le 929$	Move $n - 858 \le 71$ from First
$930 \le n \le 1001$	Move $1001 - n \le 71$ from First

We claim that 71 is indeed the optimal value. For $1 \le i \le 3$, let Chip put m_i nuts in the ith box, such that $m_1 + m_2 + m_3 = 1001$ and $m_1 \le m_2 \le m_3$. Suppose $m_1 > 143$. Dale chooses 72 and Chip's best option is to move 72 or $m_1 - 72 \ge 72$ nuts from the first box. Henceforth, we assume that $m_1 \le 143$.

Suppose $m_3 < 572$. Then $m_2 > 286$. Dale chooses 215. Chip's best option is to move $215 - m_1 \geq 72$ or $m_2 - 215 \geq 72$ nuts from the second box. Henceforth, we assume that $m_3 \geq 572$. Suppose $m_2 < 286$. Dale chooses 500. Chip's best option is to move $500 - (m_1 + m_2) \geq 72$ or $m_3 - 500 \geq 72$ nuts from the third box. It follows that we must have $m_2 = 286$, $m_1 = 143$ and $m_3 = 572$.
(Compare with Problem 2 of the Junior A-Level Paper.)

3. (See Problem 5 of the Junior A-Level Paper.)

4. Let $ACDF$ be a convex quadrilateral. Let K and M be the respective midpoints of DF and AC. We claim that $KM < \frac{AF+CD}{2}$. Let L be the midpoint of AD, as shown in the diagram below on the left. By the Midpoint Theorem, $KL = \frac{AF}{2}$ and $LM = \frac{CD}{2}$. By the Triangle Inequality, $KM < KL + LM = \frac{AF+CD}{2}$, justifying our claim.

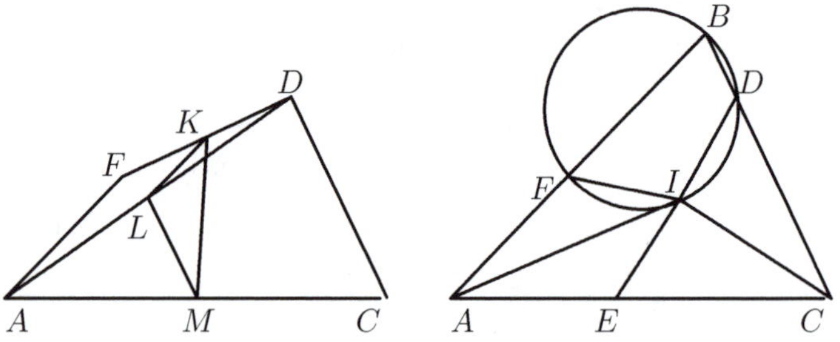

Let E be the point on AC such that $AE = AF$. as shown in the diagram above on the right. Let $\angle AEI = \alpha$ and $\angle CEI = \beta$, so that $\alpha + \beta = 180°$. Since AI bisects $\angle CAB$, triangles AEI and AFI are congruent. Hence $\angle AFI = \alpha$ and $\angle BFI = \beta$. Since $BDIF$ is cyclic, $\angle BDI = \alpha$ and $\angle CDI = \beta = \angle CEI$. Hence triangles CDI and CEI are also congruent, so that we have $CD = CE$. Let M be the midpoint of AC. By our earlier claim, $KM < \frac{AF+CD}{2} = \frac{AE+CE}{2} = \frac{AC}{2}$. Hence K lies within the semicircle with diameter AC, which implies that $\angle AKC$ is obtuse.

5. (See Problem 7 of the Junior A-Level Paper.)

6. (a) Let O be the centre of the given sphere. Let the three mutually perpendicular chords be D_1E_1, D_2E_2 and D_3E_3. Let their respective midpoints be B_1, B_2 and B_3. For $1 \leq k \leq 3$, the centre of mass of D_k and E_k is at B_k, with a weight of 2. Let O_1 be the centre of the cross-section of the sphere containing D_2E_2 and D_3E_3, O_2 be the centre of the cross-section of the sphere containing D_3E_3 and D_1E_1, and O_3 be the centre of the cross-section of the sphere containing D_1E_1 and D_2E_2. Then $AB_1O_3B_2 - B_3O_2OO_1$ forms a rectangular block, as shown in the diagram below on the left.

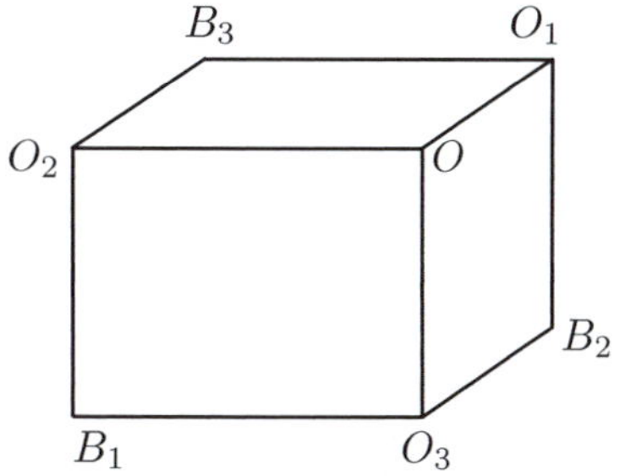

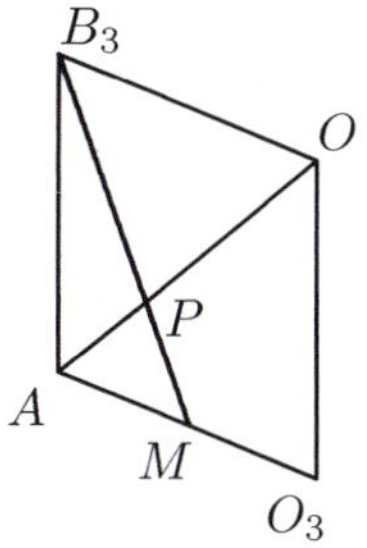

The overall centre of mass C is the centoid of triangle $B_1B_2B_3$. Let M be the midpoint of B_1B_2. Then C lies on MB_3. Since M is also the midpoint of AO_3, C lies on the plane AO_3OB_3, as shown in the diagram above on the right. Let P be the point of intersection of OA and MB_3. Then triangles APM and OPB_3 are similar. Note that $OB_3 = O_3A = 2AM$. Hence $B_3P = 2MP$ so that P coincides with C. Now we have $OC = 2AC$, so that C is a fixed point on OA.

(b) If we have a regular octahedron with centre A instead of a regular icosahedron with centre A, we get the result in (a). Let us call two opposite vertices of the regular octahedron T and B for top and bottom, and the other four N, E, W and S for north, east, west and south. A regular icosahedron may be obtained by splitting each vertex into two, in a direction parallel to a diagonal of the regular octahedron as follows. T and B become T_n, T_s, B_n and B_s. N and S become N_e, N_w, S_e and S_w. E and W become E_t, E_b, W_t and W_b. Now $\angle T_nAT_s = \angle N_2AN_w = \angle E_tAE_b$. Denote the common value by 2θ.

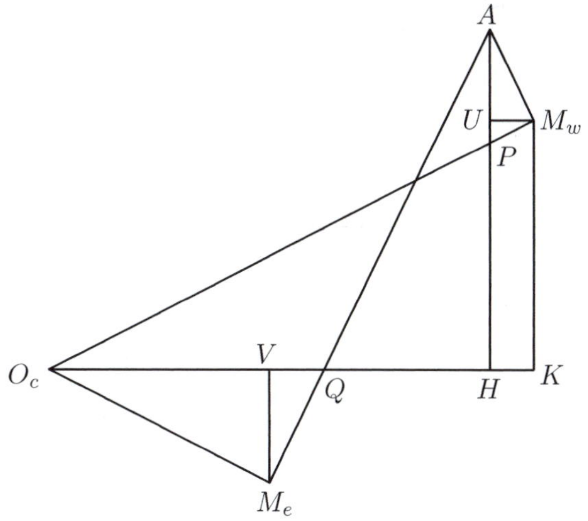

Set up a coordinate system with the origin at the centre O of the given sphere. Let the $x-$, $y-$ and $z-$axes be in the direction of AE, AN and AT. Let the coordinates of A be (a, b, c). Consider the plane $z = c$ which contains the points A, N_e, N_w, S_e and S_w. Let O_c be the centre of the circular cross-section, so that its coordinates are $(0, 0, c)$. Let M_e be the midpoint of N_eS_e and M_w be the midpoint of N_w and S_w. Let H be the point with coordinates $(a, 0, c)$. Then AH bisects $\angle M_eAM_w$ and intersects O_cM_w at P, while O_cH bisects $\angle M_eO_cM_w$ and intersects AM_e at Q. Finally, let U, V and K be the feet of perpendicular from M_w, M_e and M_w to AH, OH and OH respectively, as shown in the diagram above. Note that $\angle M_eO_cH = \angle M_wO_cH = \angle M_eAH = \angle M_wAH = \theta$. We have

$$\begin{aligned} PH &= a\tan\theta, \\ AP &= b - a\tan\theta, \\ AM_w &= b\cos\theta - a\sin\theta, \\ AU &= b\cos^2\theta - a\sin\theta\cos\theta. \\ M_wK &= b\sin^2\theta + a\sin\theta\cos\theta, \\ M_wU &= b\sin\theta\cos\theta - a\sin^2\theta, \\ O_cK &= a\cos^2\theta + b\sin\theta\cos\theta. \end{aligned}$$

Hence $(a\cos^2\theta + b\sin\theta\cos\theta, b\sin^2\theta + a\sin\theta\cos\theta, c)$ are the coordinates of M_w, and $(a\cos^2\theta - b\sin\theta\cos\theta, b\sin^2\theta - a\sin\theta\cos\theta, c)$ are the coordinates of M_e. It follows that the centre of mass of N_e, N_w, S_e and S_w is at the point whose coordinates are $(a\cos^2\theta,$ $b\sin^2\theta, c)$ with a weight of 4. Similarly, from the plane $y = b$, the centre of mass of T_n, T_s, B_n and B_s is at the point whose coordinates are $(a\sin^2\theta, b, c\cos^2\theta)$, and from the plane $x = a$, the centre of mass of E_t, E_b, W_t and W_b is at the point whose coordinates are $(a, b\cos^2\theta, c\sin^2\theta)$, also with weight 4. Hence the overall centre of mass is at the point whose coordinates are $(\frac{2a}{3}, \frac{2b}{3}, \frac{2c}{3})$. In fact, it is the same fixed point C as in (a). If we choose a different regular icosahedron with centre A, the coordinate system set up as above will be different, and the coordinates of A will become (a', b', c'). However, the coordinates of C will become $(\frac{2a'}{3}, \frac{2b'}{3}, \frac{2c'}{3})$, which yields the same point.

7. Note that our process is reversible because the positions of the soldiers prior to each iteration can be determined uniquely. Since there are finitely many permutations, our process must be periodic. We terminate it as soon as every solider in the first segment initially has already revisited the first segment, possibly without leaving.

We call a pair of soldiers *special* if they are neighbours in the first segment initially but revisit this segment for the first time after different numbers of iterations. Clearly, at some moment, they get separated. Mark the 99 borders between segments by flags, which remain in place during iterations. Consider the first moment when a special pair of soldiers gets separated. They are neighbours still, but now there is a flag between them. Let the pair of soldiers sign the flag. Assume that some flag F is signed by two special pairs A and B. Let A sign it after k iterations and B after $m > k$ iterations. Rewind the process by k iterations from the moment k when F is signed by A. This puts the soldiers from A back to their initial positions in the first segment. Now rewind the process by k iterations from the moment m when F is signed by B. Since the soldiers in B occupy the same places at the moment m as the soldiers in A at the moment k, they must be in the same positions at the moment $m - k$ as the soldiers in A initially, which are in the first segment. However, since $0 < m - k < m$, the soldiers in B have not yet separated but are yet back in the first segment. Thus they cannot be a special pair and cannot sign any flag. Therefore, the number of special pairs does not exceed 99. This means that going from one soldier to the next along the first segment in the initial arrangement, the numbers these soldiers have recorded can change no more than 99 times, so that there are at most 100 different numbers.

Spring 2013

Junior O-Level Paper

1. The answer is negative. Let D, E and F lie on BC, CA and AB respectively. We cannot form two disjoint triangles. In such a pair, the triangle containing A must also contain either F or E, say F. To them, we must add one of B, D and C, but not B. Now ADF intersects BCE at D and ACF intersects BDE at E.

2. Write the non-negative integers in nine columns as shown below, according to their remainders when divided by 9.

0	1	2	3	4	5	6	7	8
9	10	11	12	13	14	15	16	17
18	19	20	21	22	23	24	25	26
...	...	...	...	...	...	...	...	...
99	100	101	102	103	104	105	106	107
108	109	110	111	112	113	114	115	116
...	...	...	...	...	...	...	...	...

Observe that from any number, we can obtain all subsequent numbers in the same column by adding 9. Eventually, we get to some number with leading digit 1 followed by a number of 0s and at most one non-zero digit at the end. By deleting the leading digit 1, we can get to the first row of the chart. We now show that we can move among the columns, by the sequence of moves: 0,9,18,8,17,7,16,6,15,5,14,4,13,3,12,2,11,1,10,0. It follows that starting from any non-negative integer, we can obtain any other non-negative integer.

3. Let S be the total weight of the lightest 6 objects, and let T be the total weight of the other 5 objects. Let k be the weight of the heaviest object among the lightest 6. Then

$$\begin{aligned}
& (k-5)+(k-4)+(k-3)+(k-2)+(k-1)+k \\
\geq\ & S \\
>\ & T \\
\geq\ & (k+1)+(k+2)+(k+3)+(k+4)+(k+5).
\end{aligned}$$

Hence $k > 30$ and the weight of the heaviest object is at least $k+5 > 35$.

4. Each rook X attacks fourteen vacant squares, seven along its row and seven along its column. Each vacant square A attacked by X is attacked also by exactly one other rook, say Y. Suppose X attacks A along a row and Y attacks A along a column. There is a square B which X attacks

along a column and Y attacks along a row. The total distance from A to X and Y is equal to the total distance from B to X and Y. Hence exactly one square goes to X and one to Y. It follows that X claims the square on which it stands, plus one of every two vacant squares it attacks. Hence its total claim is eight squares, which is the same as that of any other rook.

5. Construct the parallelogram $ABCG$. CDG is an equilateral triangle since $CG = BA = CD$ and

$$\angle GCD = 90° - \angle GCB = 90° - (180° - 150°) = 60°.$$

Let E, F and H be the respective midpoints of AD, BC and DG. Then EH is parallel to AG and thus to BC. Note that we also have $EH = \frac{AG}{2} = \frac{BC}{2} = FC$. It follows that $EHFC$ is also a parallelogram. The acute angle between EF and BC is equal to

$$\angle BCH = \angle BCG + \angle GCH = (180° - 150°) + \frac{60°}{2} = 60°.$$

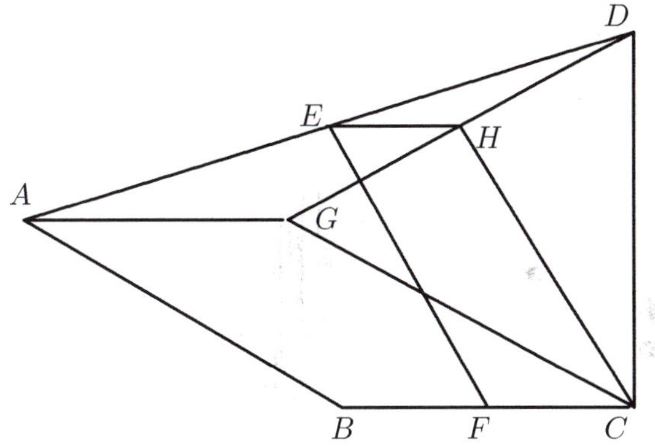

Senior O-Level Paper

1. (See Problem 2 of the Junior O-Level Paper.)

2. We have $\angle CAE = 90° - \angle ACE = \angle BCE$. Hence triangles CAE and BCE are similar, so that $\frac{AE}{CE} = \frac{AC}{BC} = \frac{AL}{CM}$. Since

$$\angle LAE = 90° + \angle CAE = 90° + \angle BCE = \angle MCE,$$

triangles LAE and MCE are also similar, so that $\angle LEA = \angle MEC$. Hence $\angle LEM = \angle MEA - \angle LEA = \angle MEA - \angle MEC = \angle CEA = 90°$.

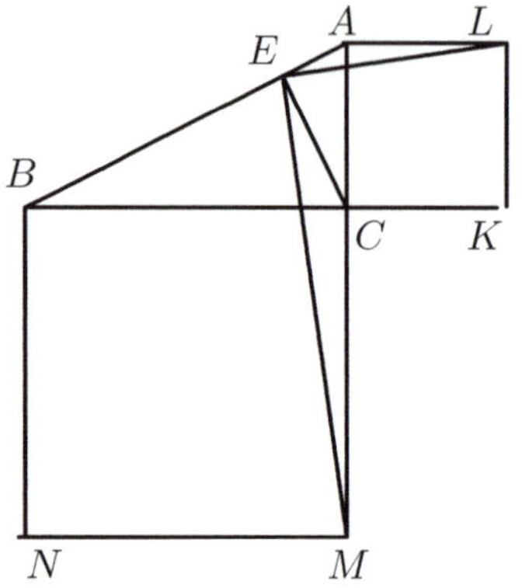

3. (See Problem 4 of the Junior O-Level Paper.)

4. Let the correct weight of the ith object be a_i with $a_1 < a_2 < \cdots < a_{100}$. Let the false label of the ith object indicate that its weight is b_i. A possible rearrangement of the labels is defined by $b_i = a_{i+1}$, $1 \leq i \leq 99$, and $b_{100} = a_1$. Any subset of objects not including the heaviest one will have total weight less than the sum of their labels. Any subset of objects including the heaviest one but not all 100 will have total weight greater than the sum of their labels.

5. Let $0 < |b| < 2013$. Consider a particular factorization of $|b|$ as a product of two integers m and n. Then the admissible polynomials associated with this factorization are $(x \pm m)(x \pm n)$ and their sum is $4x^2$ if $m \neq n$. If $m = n$, the admissible polynomials are $(x \pm n)^2$ and $(x - n)(x + n)$, and their sum is $3x^2 + n^2$. For $b = 2013$, the exception is when $\{m, n\} = \{1, 2013\}$. Here the admissible polynomials are only $(x-1)(x+2013)$ and $(x+1)(x-2013)$, with sum $2x^2 - 4016$. This is because $(x+1)(x+2013)$ and $(x - 1)(x - 2013)$ both result in a linear term whose coefficient is ± 2014. For $b = 0$, the admissible polynomials are $x^2 - kx$ where $-2013 \leq k \leq 2013$. Hence their sum is $4017x^2$. Summing all admissible polynomials, the linear term vanishes and the constant term is equal to $1^2 + 2^2 + \cdots + 44^2 - 4016 > 0$. Since the leading coefficient is also positive, there are no real roots.

Junior A-Level Paper

1. We may certainly have two numbers, say 1 and 3 with $1 + 3 = 4 = 2^2$. Suppose there are three numbers $a < b < c$. Then $a + b < a + c < b + c$ and they are different powers of 2. Now $b + c \geq 2(a + c)$ so that $b > c$, which is a contradiction. Hence there are at most two numbers.

2. The only countings arise from a girl sitting to the right of a boy, in which case they count each other. Hence the total number counted by the girls is equal to the total number counted by the boys. The number girls does not have to be equal to the number of boys.

3. A way to do so is shown in the diagram below.

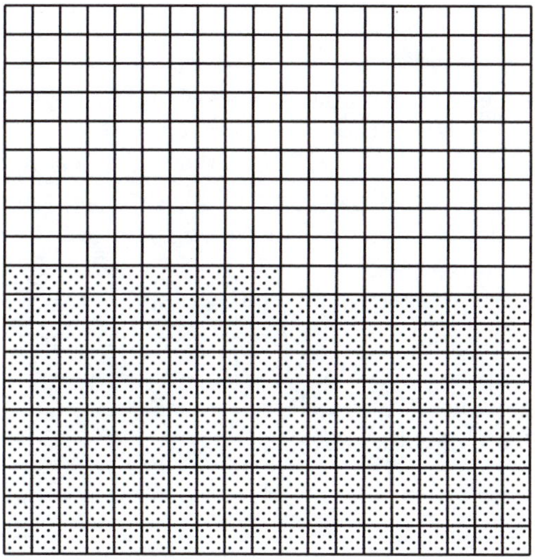

There are 100 10×10 boards in 10 rows of 10. The number of marked squares in the respective boards are given below.

10	9	8	7	6	5	4	3	2	1
20	19	18	17	16	15	14	13	12	11
30	29	28	27	26	25	24	23	22	21
40	39	38	37	36	35	34	33	32	31
50	49	48	47	46	45	44	43	42	41
60	59	58	57	56	55	54	53	52	51
70	69	68	67	66	65	64	63	62	61
80	79	78	77	76	75	74	73	72	71
90	89	88	87	86	85	84	83	82	81
100	99	98	97	96	95	94	93	92	91

4. Let a, b, c, d, e, f, g, h, i and j be ten numbers in succession along the circle, with a black. Then

$$\begin{array}{ll} c = \frac{b}{a} & d = c - b = \frac{b(1-a)}{a} \\ e = \frac{d}{c} = 1 - a & f = e - d = \frac{(1-a)(a-b)}{a} \\ g = \frac{f}{e} = \frac{a-b}{a} & h = g - f = a - b \\ i = \frac{h}{g} = a & j = i - h = b. \end{array}$$

It follows that the pattern will repeat in a cycle of length 8. The sum of the eight numbers in one cycle is

$$a+b+(1-a)+(a-b)+\frac{b+b(1-a)+(1-a)(a-b)+(a-b)}{a},$$

which simplifies to 3. Since $1000 = 8\times 125$, the sum of the 1000 numbers is $3 \times 125 = 375$.

5. Let ABC be a triangle with lattice points as vertices such that there are exactly two lattice points inside. Let D, E and F be the respective midpoints of AB, AC and BC. Suppose one of the lattice points P inside ABC is in fact inside one of AEF, BFD and CDE, say CDE. Then the point Q symmetric to C with respect to P is also a lattice point inside ABC, and PQ passes through C. Henceforth, we assume that both lattice points P and Q inside ABC are in fact inside DEF or on its boundary. If PQ is not parallel to a side of ABC, draw lines through A, B and C parallel to PQ. Exactly one of them, say the one through C, passes through the interior of ABC. The point R on this line inside ABC such that $PQ = CR$ is a lattice point inside CDE, which is contrary to our assumption.

6. Let $\angle A = 2\alpha$ and $\angle B = 2\beta$. Let CF be an altitude of triangle ABC. Let the line through D perpendicular to the bisector of $\angle A$ intersect this bisector at X and the extension of FC at Q. Since

$$\angle AFQ = 90^\circ = \angle AXQ,$$

$AFXQ$ is cyclic so that $\angle CQD = \angle FAX = \alpha$. Now $\angle FCB = 2\alpha$, so that $\angle CDQ = \alpha = \angle CQD$. It follows that $CD = CQ$.

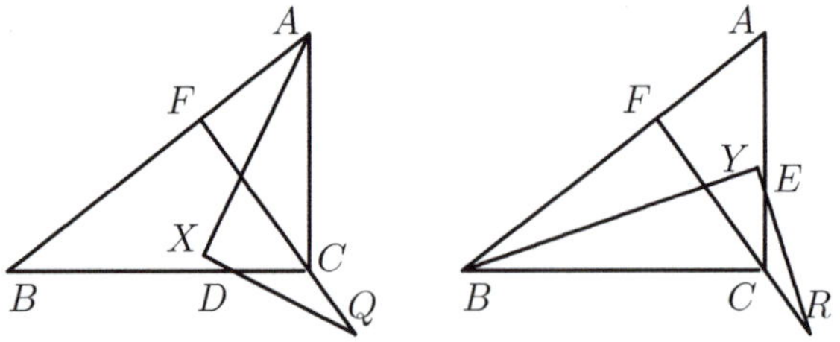

Now let the line through E perpendicular to the bisector of $\angle B$ intersect this bisector at Y and the extension of FC at R. As before, we have $CE = CR$. Note that $CD = CE$ as they are tangents from C to the incircle of ABC. Hence $CQ = CR$, so that Q and R are the same point P, which lies on CF.

7. Let there be m girls and n boys with $m > n$. Number the girls $G_1, G_2, \ldots, G_m$ and the boys $B_1, B_2, \ldots, B_n$ so that initially G_1 is playing B_1, G_i is ahead of G_{i+1} and B_j is ahead of B_{j+1} in the line.

Note that the players of the same gender, including the one who is playing, are always in the same cyclic order. Two players of opposite gender may trade places, but only just after they have played each other. Group the games into cycles each consisting of $m+n-2$ games. During the first cycle, $m+n-2$ players will go to the end of the line, all except G_m and B_n. Therefore these players start the second cycle. Similarly G_{m-1} and B_{n-1} start the third cycle, and so on. Thus each girl goes to the end of the line exactly $(m-1)$ times during m cycles and exactly $(m-1)n$ times during mn cycles. Similarly each boy goes to the end of the line exactly $m(n-1)$ times during mn cycles. Now $n(m-1)-m(n-1)=m-n\geq 1$. Assume that some girl G has never played against some boy B. During $2mn$ cycles G goes to the end of the line at least 2 more times than B. Therefore there are two consecutive exits of G such that between them there were no exits of B. Between these to exits of G, B remained in the line. However after the first exit of G, B is in front of G in the line, but when G enters the table again, B is behind her. This is a contradiction.

Senior A-Level Paper

1. (See Problem 1 of the Junior A-Level Paper.)

2. A space between two children of the same gender is called a brave spot. If a boy sits between two girls, he is brave. If he sits between a boy and a girl, he creates a brave spot. If he sits between two boys, he changes one brave spot into two. Hence each boy is either brave or creates a brave spot. By symmetry, the same can be said about each girl. At the beginning, there are no brave spots, which is also the case at the end. Hence of the twenty children who come later, half of them create a brave spot and the other half, namely ten of them, are brave.

3. (See Problem 5 of the Junior A-Level Paper.)

4. Let $A=\{1,2,\ldots,25,76,77,\ldots,100\}$ and $B=\{26,27,\ldots,75\}$. No two numbers from A could be adjacent. Since each set contains 50 elements, the numbers on the circle must alternate between elements of A and B. However, 26 could be adjacent only to 76 from A, making the circular arrangement impossible.

5. An equilateral triangle with one vertex in each of red, yellow and blue and in the correct orientation is said to be *good.* Let R_0, Y_0 and B_0 be the initial red, yellow and blue points initially. We may assume that they do not form a good triangle. Then $R_0Y_0B_1, Y_0B_0R_1$ and $B_0R_0Y_1$ are distinct good triangles.

We claim that R_0R_1, Y_0Y_1 and B_0B_1 make $60°$ angles with one another, and are concurrent at some point P. A $60°$-rotation about B_0 brings R_0 to Y_1 and R_1 to Y_0. Hence the angle between R_0R_1 and Y_0Y_1 is $60°$. Similarly, the angle between B_0B_1 and either R_0R_1 or Y_0Y_1 is also $60°$.

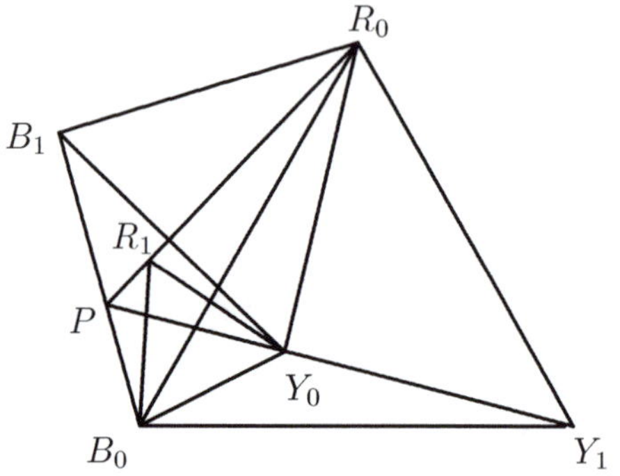

Let P be the point of R_0R_1 and Y_0Y_1. Then $\angle R_0PY_1 = 60° = \angle R_0B_0Y_1$. Hence $B_0PR_0Y_1$ is cyclic, so that $\angle B_0PY_1 = \angle B_0R_0Y_1 = 60°$. Since $\angle R_0PY_0 = 60° = \angle R_0B_1Y_0$, $R_0Y_0PB_1$ is also a cyclic quadrilateral. Hence $\angle B_1PR_0 = 60°$. Now $\angle B_1PR_0 + \angle R_0PY_1 + \angle Y_1PB_0 = 180°$. It follows that P also lies on B_0B_1.

We now prove that all subsequent red points lie on R_0R_1, all subsequent yellow points lie on Y_0Y_1, and all subsequent blue points lie on B_0B_1. Without loss of generality, we may take R as a point on R_0R_1 and Y as a point on Y_0Y_1. Let B be the point such that RYB is a good triangle. We claim that B must lie on B_0B_1. Note that $\angle RPY$ is either $60°$ or $120°$. In the former case, $RPBY$ is a cyclic quadrilateral. In the latter case, $RPYB$ is a cyclic quadrilateral. In either case, $\angle BPY = 60°$, and the claim is justified.

6. We shall just solve (b), which implies (a). Let $a < b < c < d < e$ be the numbers. Suppose $d < b+c$ and $e < b+d$. Then $a^2 < ab$, $b^2 < bc$, $c^2 < ce$, $d^2 < (b+c)d$ and $e^2 < (b+d)e$. It follows that the sum of the squares is less than the sum of all pairwise products. This is a contradiction. We have two cases.

   Case 1. $d \geq b + c$.

   Each of d and e is greater than or equal to each of $a+b$, $a+c$, $b+c$, and six desired triples are $\{a,b,d\}$, $\{a,c,d\}$, $\{b,c,d\}$, $\{a,b,e\}$, $\{a,c,e\}$ and $\{b,c,e\}$.

   Case 2. $e \geq b + d$.

   Five desired triples are $\{a,b,e\}$, $\{a,c,e\}$, $\{a,d,e\}$, $\{b,c,e\}$ and $\{b,d,e\}$. A sixth one may be $\{a,b,d\}$ or $\{c,d,e\}$. However, we may have $d < a+b$ and $e < c+d$. Then $a^2 < ab$, $b^2 < bc$, $c^2 < cd$, $d^2 < (a+b)d$ and $e^2 < (c+d)e$. It follows that the sum of the squares is less than the sum of all pairwise products. This is a contradiction.

7. Again we solve only (b). Let the wizards be $W_1, W_2, \ldots, W_{1000}$, and let n_i be the number on the hat of wizard W_i for $1 \leq i \leq 999$. Starting from the back, each wizard in turn will write down a permutation of $\{1, 2, \ldots, 1001\}$. W_{1000} can see $n_1, n_2, \ldots, n_{999}$, and write them down in order as the first 999 terms. He places the remaining two numbers in the last two places in the order which results in an even permutation. W_{999} can see $n_1, n_2, \ldots, n_{998}$ and can hear what W_{1000} calls out. So he puts what he hears down as the 1000th term, and places the remaining two numbers in the 999th and 1001st positions so that the resulting permutation is even. In fact, W_{999} has duplicated the permutation of W_{1000}, and thus he will correctly call out n_{999}. Similarly, each of the remaining wizards has two numbers to place, and he does it so that the resulting permutation is even. He will have duplicated the permutation of W_{1000} and will correctly call out his number. So every wizard except possibly W_{1000} can remain in the Council.

Tournament 35

Fall 2013

Junior O-Level Paper

1. In a wrestling tournament, there are 100 participants, all of different strengths. Each wrestler participates in two matches. In each match, the stronger wrestler always wins. A wrestler who wins both matches is given an award. What is the least possible number of wrestlers who win awards?

2. Can the ten digits 0, 1, 2, 3, 4, 5, 6, 7, 8 and 9 be arranged in a row so that no matter which six digits are removed, the remaining four digits, without changing their order, form a composite number?

3. Denote by $a \triangle b$ the greatest common divisor of a and b. Let n be a positive integer such that $n \triangle (n+1) < n \triangle (n+2) < \cdots < n \triangle (n+35)$. Prove that $n \triangle (n+35) < n \triangle (n+36)$.

4. In ABC, $AB = AC$. K and L are points on AB and AC respectively such that $AK = CL$ and $\angle ALK + \angle LKB = 60°$. Prove that $KL = BC$.

5. Eight rooks are placed on different squares of a board in such a way that no two rooks attack each other. All eight rooks are then moved simultaneously, each by a knight's move. Thus if two rooks are a knight's move apart, then it is possible for them to trade places. Prove that regardless of their initial placements, it is always possible to move the rooks so that afterwards, they are still on different squares and no two of them attack each other.

Note: The problems are worth 3, 4, 4, 5 and 6 points respectively.

Senior O-Level Paper

1. Can the ten digits 0, 1, 2, 3, 4, 5, 6, 7, 8 and 9 be arranged in a row so that no matter which six digits are removed, the remaining four digits, without changing their order, form a composite number?

2. ABC is an arbitrary triangle. X is a point on the same side of the line BC as A. Y is a point on the side of the line CA opposite to B. Z is a point on the side of the line AB opposite to C. The triangles XBC, YAC and ZBA are similar, with X, B and C corresponding to Y, A and C, and to Z, B and A respectively. Prove that $AYXZ$ is a parallelogram.

3. Denote by $a \bigtriangledown b$ the least common multiple of a and b. Let n be a positive integer such that $n \bigtriangledown (n+1) > n \bigtriangledown (n+2) > \cdots > n \bigtriangledown (n+35)$. Prove that $n \bigtriangledown (n+35) > n \bigtriangledown (n+36)$.

4. Eight rooks are placed on different squares of a board in such a way that no two rooks attack each other. All eight rooks are then moved simultaneously, each by a knight's move. Thus if two rooks are a knight's move apart, then it is possible for them to trade places. Prove that regardless of their initial placements, it is always possible to move the rooks so that afterwards, they are still on different squares and no two of them attack each other.

5. A spaceship lands on an asteroid, which is known to be either a sphere or a cube. The spaceship sends out an explorer which crawls on the surface of the asteroid. The explorer continuously transmits its current position in space back to the spaceship, until it reaches the point which is symmetric to the landing site relative to the centre of the asteroid. Thus the spaceship can trace the path along which the explorer is moving. Can it happen that these data are not sufficient for the spaceship to determine whether the asteroid is a sphere or a cube?

Note: The problems are worth 3, 4, 4, 5 and 6 points respectively.

Junior A-Level Paper

1. There are 100 red, 100 yellow and 100 green sticks. One can construct a triangle using any three sticks of different colours. Prove that we can choose one of the three colours and construct a triangle using any three sticks of that colour.

2. Ten consecutive positive integers are given. Each of Penny and Betty will divide them into five pairs, compute the product of each pair and add these products together. Prove that the girls can always form the pairs in different ways and yet come up with the same final sum.

3. In triangle ABC, $\angle C = 90°$ and its bisector intersects AB at M. N is the midpoint of the semicircular arc constructed with BC as diameter and lying outside ABC. Prove that AN passes through the midpoint of CM.

4. Penny chooses an interior point of one of the squares of an 8×8 board. Basil draws a subboard consisting of one or more squares such that its boundary is a single closed polygonal line which does not intersect itself. Penny will then tell Basil whether the chosen point is inside or outside this subboard. What is the minimum number of times Basil has to do this in order to determine whether the chosen point is black or white?

5. A 101-gon is inscribed in a circle. From each vertex, a perpendicular is dropped to the opposite side, intersecting it either on the side or on its extension. Prove that at least one such point is not on the extension.

6. Let n be a positive integer such that $3n+1$ is prime. When

$$1-\frac{1}{2}+\frac{1}{3}-\frac{1}{4}+\cdots+\frac{1}{2n-1}-\frac{1}{2n}$$

is expressed as an irreducible fraction, prove that the numerator is a multiple of $3n+1$.

7. Peter and Betty are playing a game with 10 stones in each of 11 piles. Peter moves first, and turns alternate thereafter. In his turn, Peter must take 1, 2 or 3 stones from any one pile. In her turn, Betty must take one stone from 1, 2 or 3 piles. Whoever takes the last stone overall is the winner. Which player has a winning strategy?

Note: The problems are worth 5, 5, 6, 7, 9, 10 and 12 points respectively.

Senior A-Level Paper

1. Penny chooses an interior point of one of the 64 squares of a standard board. Basil draws a subboard consisting of one or more squares such that its boundary is a single closed polygonal line which does not intersect itself. Penny will then tell Basil whether the chosen point is inside or outside this subboard. What is the minimum number of times Basil has to do this in order to determine whether the chosen point is black or white?

2. Determine all positive integers n with the following properties. For any two polynomials $P(x)$ and $Q(x)$ of degree n, there exist integers k and ℓ and real numbers a and b such that $0 \le k, \ell \le n$ and the graphs of $P(x)+ax^k$ and $Q(x)+bx^\ell$ have no common points.

3. ABC is an equilateral triangle with centre O. A line through C intersects the circumcircle of triangle BOA at points D and E. Prove that A, O and the midpoints of BD and BE are concyclic.

4. Can every integer be expressed as the sum of the cubes of distinct integers?

5. Do there exist two integer-valued functions f and g such that for every integer x, we have

   (a) $f(f(x))=x$, $g(g(x))=x$, $f(g(x))>x$ and $g(f(x))>x$;

   (b) $f(f(x))<x$, $g(g(x))<x$, $f(g(x))>x$ and $g(f(x))>x$?

6. Peter and Betty are playing a game with 10 stones in each of 11 piles. Peter moves first, and turns alternate thereafter. In his turn, Peter must take 1, 2 or 3 stones from any one pile. In her turn, Betty must take one stone from 1, 2 or 3 piles. Whoever takes the last stone overall is the winner. Which player has a winning strategy?

7. A closed polygonal line in the plane is such that exactly two segments intersect at a vertex and each segment has exactly one interior point in common with another segment. Is it possible that each segment is bisected by that interior point of intersection?

Note: The problems are worth 5, 6, 6, 7, 3+5, 9 and 14 points respectively.

Spring 2014

Junior O-Level Paper

1. When each of 100 numbers was increased by 1, the sum of their squares remained unchanged. Each of the new numbers is increased by 1 once more. How will the sum of their squares change this time?

2. Olga's mother baked 7 apple pies, 7 banana pies and 1 cherry pie. They are arranged in that exact order on a the rim of a round plate when they are put into the microwave oven. All the pies look alike, but Olga knows only their relative positions on the plate because it has rotated. She wants to eat the cherry pie. She is allowed to taste three of them, one at a time, before making up her mind which one she will take. Can she guarantee that she can take the cherry pie?

3. Each of the squares in a 5×7 table contains a number. Peter knows only that the sum of the numbers in the 6 squares of any 2×3 or 3×2 rectangle is 0. He is allowed to ask for the number in any position in the table. What is the minimum number of questions he needs to ask in order to be able to determine the sum of all 35 numbers in the table?

4. M is the midpoint of the side AB of triangle ABC, and L is a point on the side BC such that $\angle ALC = 45°$. If $AL = 2CM$, prove that AL is perpendicular to CM.

5. Forty Thieves are ranked from 1 to 40, and Ali Baba is also given the rank 1. They want to cross a river using a boat. Nobody may be in the boat alone, and no two people whose ranks differ by more than 1 may be in the boat at the same time. Is this task possible?

Note: The problems are worth 3, 4, 4, 5 and 6 points respectively.

Senior O-Level Paper

1. Inspector Gadget has 36 stones of masses 1 gram, 2 grams, ..., 36 grams. Doctor Claw has a superglue such that one drop of it can glue together two original stones, an original stone and a composite stone, or two composite stones. He wants to leave Inspector Gadget with a set of original and composite stones from which Inspector Gadget cannot get a subset with total mass 37 grams. What is the least number of drops of glue Doctor Claw needs to use?

2. $ABCD$ is a convex quadrilateral with perpendicular diagonals. M and N are points on the sides AD and CD, respectively, such that we have $\angle ABN = 90° = \angle CBM$. Prove that MN is parallel to AC.

3. Forty Thieves are ranked from 1 to 40, and Ali Baba is also given the rank 1. They want to cross a river using a boat. Nobody may be in the boat alone, and no two people whose ranks differ by more than 1 may be in the boat at the same time. Is this task possible?

4. The positive integers a, b, c and d are pairwise relatively prime. Moreover, $ab + cd = ac - 10bd$. Prove that one of these four numbers is the sum of two of the other three.

5. A park is in the shape of a convex quadrilateral $ABCD$. Alex, Ben and Chris are jogging there, each at his own constant speed. Alex and Ben start from A at the same time, Alex jogging along AB and Ben along AC. When Alex arrives at B, he immediately continues on along BC. At the time when Alex arrives at B, Chris starts from B, jogging along BD. Alex and Ben arrive at C at the same time, and Alex immediately continues on along CD. He and Chris arrive at D at the same time. Can it happen that Ben and Chris meet each other at the point of intersection of AC and BD?

Note: The problems are worth 4, 4, 5, 5 and 5 points respectively.

Junior A-Level Paper

1. During the Christmas Party, Santa hands out 47 chocolates and 74 marshmallows. Each girl gets 1 more chocolate than each boy, and each boy gets 1 more marshmallow than each girl. What are the possible values of the number of children at the party?

2. Anna paints several squares of a 5×5 board. The task of Boris is to cover up all of them by placing non-overlapping copies of the V-tromino so that each copy covers exactly three squares of the board. What is the minimum number of squares Anna must paint in order to prevent Boris from succeeding in his task?

3. A square tablecloth is placed flat on top of a square table which may be of a different size. None of the corners of the table is covered. A triangular part of the tablecloth hangs over each edge of the table. If the parts hanging over two adjacent edges of the table are congruent triangles, prove that the parts hanging over the other two edges are also congruent triangles.

4. On each of 100 cards, Anna writes down a positive integer. These numbers are not necessarily distinct. On $\binom{100}{2}$ cards, Anna writes down the sums of these numbers taken 2 at a time.

   On $\binom{100}{3}$ cards, she writes down the sums taken 3 at a time. She continues until she finally writes down the sum of all 100 numbers on 1 card. She is allowed to send some of these $2^{100}-1$ cards to Boris, no two of which may contain the same number. Boris knows the rules by which the cards are prepared. What is the minimum number of cards Anna must send to Boris in order for him to determine the original 100 numbers?

5. There are several white points and several black points. Each white point is joined to every black point by a segment. Each segment is labeled with a positive integer. Along any closed path, the product of the labels of the segments going from white points to black points is equal to the product of the labels of the segments going from black points to white points. Is it always possible to label the points with positive integers so that the label of each segment is equal to the product of the labels of its endpoints?

6. A $3 \times 3 \times 3$ block is made of 27 unit cubes, glued together along all common faces. Some of these unit cubes are removed so that the resulting structure does not fall apart into two or more pieces. Pieces hinged at a corner or along an edge will fall apart. What is the maximum number of unit cubes that may be removed so that the projection of the resulting structure onto each face of the block is a 3×3 square?

7. In cyclic order on a circle are marked the points A_1, A_2, …, A_{10} such that A_1A_6, A_2A_7, A_3A_8, A_4A_9 and A_5A_{10} are diameters. Initially, there is a grasshopper on each of these points. Every minute, one of them jumps over one of its neighbours and lands on a point on the circle, marked or otherwise, so that the distance between these two grasshoppers remains the same. A grasshopper may not jump over more than one grasshopper, and may not land on another grasshopper. After a while, there is a grasshopper on each of A_1, A_2, …, A_9, and the tenth grasshopper is on the arc A_1A_9 which contains A_{10}. Does it have to be at A_{10}?

Note: The problems are worth 3, 5, 6, 7, 7, 9 and 9 points respectively.

Senior A-Level Paper

1. Anna writes down several 1s, puts either a + sign or a $\times$ sign between every two adjacent 1s, adds several pairs of brackets and gets an expression equal to 2014. Boris takes Anna's expression and interchanges all the + signs with all the $\times$ signs. Is it possible that his expression is also equal to 2014?

2. Is it always possible to dissect a convex polygon by a straight cut into two polygons with equal perimeter such that the

   (a) longest;

   (b) shortest

   side of each has the same length?

3. On each of 100 cards, Anna writes down a positive integer. These numbers are not necessarily distinct. On $\binom{100}{2}$ cards, Anna writes down the sums of these numbers taken 2 at a time. On $\binom{100}{3}$ cards, she writes down the sums taken 3 at a time. She continues until she finally writes down the sum of all 100 numbers on 1 card. She is allowed to send some of these $2^{100} - 1$ cards to Boris, no two of which may contain the same number. Boris knows the rules by which the cards are prepared. What is the minimum number of cards Anna must send to Boris in order for him to determine the original 100 numbers?

4. We mark all the points (x, y) where x is an arbitrary integer and y is an integer such that $0 \leq y \leq 10$. At most how many marked points can lie on the graph of a polynomial of degree 20 and having integer coefficients?

5. A scalene triangle is given. In each move, Anna chooses a point on the plane, and Boris decides whether to paint it red or blue. Anna wins if she can get three points of the same colour forming a triangle similar to the given one. What is the minimum number moves Anna needs to force a win, regardless of the shape of the given triangle?

6. A computer directory lists all pairs of cities connected by direct flights. Anna hacks into the computer and permutes the names of the cities. It turns out that no matter which other city is renamed Moscow, she can rename the remaining cities so that the directory is perfectly correct. Later, Boris does the same thing. However, he insists on exchanging the names of Moscow with another city. Is it always possible for him to rename the remaining cities so that the directory is perfectly correct?

7. Consider a polynomial $P(x)$ such that $P(0) = 1$ and

   $$(P(x))^2 = 1 + x + x^{100}Q(x)$$

   for some polynomial $Q(x)$. Prove that the coefficient of x^{99} in the polynomial $(P(x) + 1)^{100}$ is 0.

Note: The problems are worth 3, 4+4, 6, 7, 8, 9 and 10 points respectively.

Solutions

Fall 2013

Junior O-Level Paper

1. Let each wrestler be matched with the wrestler immediately below in strength, except that the weakest wrestler is matched with the strongest one. Then each wrestler is in two matches. Everyone wins once and loses once except that the strongest one wins both matches while the weakest one loses both matches. Hence the smallest number of wrestlers who win awards is one, since the strongest wrestler always wins an award in any arrangement of matches.

2. Put 5 and the even digits in any order in the last six positions. If any of them remains, then the four-digit number is either divisible by 5 or by 2, and is hence composite. If all of them are removed, then the digits 1, 3, 7 and 9, which are in the first four positions in some order, must form a composite number. A simple way is to arrange them in the order 1, 3, 9 and 7, because the number 1397 is divisible by 11, and is hence composite.

3. Clearly, $n \triangle (n+1) = 1$. Hence $n \triangle (n+2) \geq 2$. On the other hand, since $n \triangle (n+2)$ must divide $(n+2) - n = 2$, we must have $n \triangle (n+2) = 2$. Similarly, $n \triangle (n+k) = k$ for $3 \leq k \leq 35$. It follows that n is divisible by the least common multiple of 2, 3, ..., 35. In particular, n is divisible by $4 \times 9 = 36$, which implies that $n + 36$ is also divisible by 36. Hence $n \triangle (n+36) = 36 > 35 = n \triangle (n+35)$.

4. Complete the equilateral triangle KLM, with M on the side of the line AC opposite to B. Note that $\angle ALM = 60^\circ - \angle ALK = \angle LKB$. Hence $\angle AKL = 180^\circ - \angle LKB = 180^\circ - \angle ALM = \angle CLM$. Since $AK = CL$ and $KL = LM$, triangles AKL and CLM are congruent, so that $AL = CM$ and $\angle LAK = \angle MCL$. It follows that KB is parallel to MC, and $KB = AB - AK = AC - CL = AL = MC$. Hence $BCMK$ is a parallelogram, so that $KL = KM = BC$.

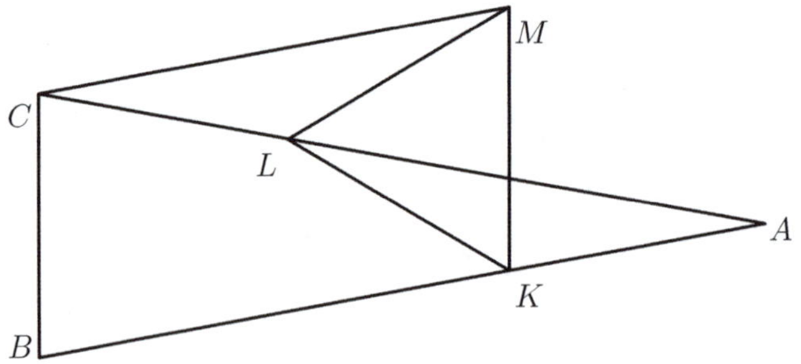

5. For $1 \le k \le 8$, let the rook in column k be in row y_k. In the simultaneous moves, the rook in column k moves to column $k+1$ if $k = 1, 3, 5, 7$ and to column $k-1$ if $k = 2, 4, 6, 8$. The rook in row y_k moves to row $y_k + 2$ if $y_k = 1, 2, 5, 6$ and to row $y_k - 2$ if $y_k = 3, 4, 7, 8$. Then each rook has made a knight's move, and in the new position, the rows they are in are distinct, and the columns they are in are also distinct.

Senior O-Level Paper

1. (See Problem 2 of the Junior O-Level Paper.)

2. Since triangle ZBA is similar to triangle XBC, we have $\frac{ZB}{XB} = \frac{AB}{CB}$ and $\angle ZBA = \angle XBC$. Hence

$$\angle ZBX = \angle ZBA + \angle ABX = \angle XBC + \angle ABX = \angle ABC,$$

so that triangle ZBX is similar to triangle ABC. Similarly, triangle YXC is also similar to triangle ABC. It follows that $\frac{XZ}{ZB} = \frac{AC}{AB} = \frac{CY}{XY}$. On the other hand, triangle ZBA is similar to triangle YAC. Hence $\frac{YA}{ZB} = \frac{AC}{AB} = \frac{CY}{ZA}$. It follows that $XZ = YA$ and $XY = ZA$, so that $AYXZ$ is a parallelogram.

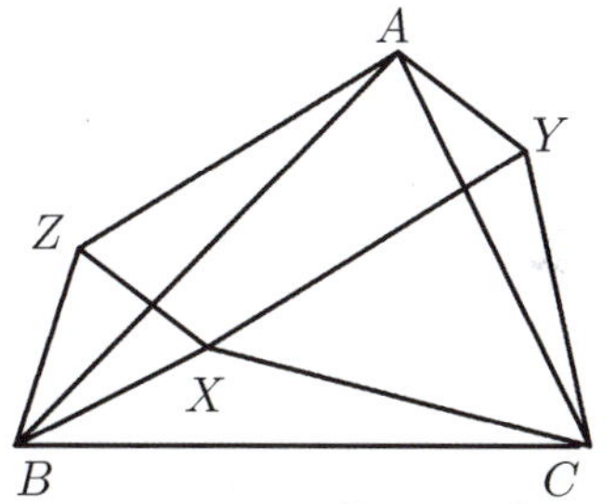

3. Denote by $a \bigtriangleup b$ the greatest common divisor of a and b. Then we have $(a \bigtriangleup b)(a \bigtriangledown b) = ab$. Since $n \bigtriangleup (n+1) = 1$, we have $n \bigtriangledown (n+1) = n(n+1)$. If $n \bigtriangleup (n+2) = 1$ also, then $n \bigtriangledown (n+2) = n(n+2) > n(n+1) = \bigtriangledown(n+1)$ is a contradiction. Hence $n \bigtriangleup (n+2) \ge 2$. On the other hand, since $n \bigtriangleup (n+2)$ must divide $(n+2) - n = 2$, we must have $n \bigtriangleup (n+2) = 2$ so that $n \bigtriangledown (n+2) = \frac{n(n+2)}{2}$. If $n \bigtriangleup (n+3) \le 2$, then

$$n \bigtriangledown (n+3) \ge \frac{n(n+3)}{2} > \frac{n(n+2)}{2} = n \bigtriangledown (n+2).$$

Hence $n \bigtriangleup (n+3) = 3$. Similarly, $n \bigtriangleup (n+k) = k$ for $4 \le k \le 35$. It follows that n is divisible by the least common multiple of 2, 3, ..., 35. In particular, n is divisible by $4 \times 9 = 36$, which implies that $n + 36$ is also divisible by 36. It follows that $n \bigtriangleup (n+36) = 36$ and $n \bigtriangledown (n+36) = \frac{n(n+36)}{36} < \frac{n(n+35)}{35} = n \bigtriangledown (n+35)$.
(Compare with Problem 3 of the Junior O-Level Paper.)

4. (See Problem 5 of the Junior O-Level Paper.)

5. Suppose a sphere intersects each of the six faces of a cube in its incircle. Then these six circles are hinged to one another at the midpoints of the edges of the cube. If the spaceship lands at some point on one of these circles, the destination is on the the circle on the opposite face. The explorer can get there by going from circle to circle. Since its path lies on the intersection of the surfaces of the cube and the sphere, the spaceship does not have enough information to determine the shape of the asteroid.

Junior A-Level Paper

1. Let the shortest stick overall have length s, and we may assume that it is red. Let the longest stick among the yellow and green ones have length ℓ, and we may assume that it is green. Then a triangle can be formed using the red stick of length s, the green stick of length ℓ and any yellow stick. Let the lengths of any three yellow sticks be $x \leq y \leq z$. Then $x + y \geq s + y > \ell \geq z$. Hence they can form a triangle.

2. Let the consecutive integers be $n,\ n+1,\ n+2,\ \ldots,\ n+9$ for some positive integer n. For now, allow $n = 0$. Penny may have

$$0 \times 7 + 1 \times 8 + 2 \times 3 + 4 \times 5 + 6 \times 9 = 88$$

while Betty may have

$$0 \times 9 + 1 \times 4 + 2 \times 5 + 3 \times 6 + 7 \times 8 = 88.$$

In general, Penny may have

$$n(n+7)+(n+1)(n+8)+(n+2)(n+3)+(n+4)(n+5)+(n+6)(n+9)$$

while Betty may have

$$n(n+9)+(n+1)(n+4)+(n+2)(n+5)+(n+3)(n+6)+(n+7)(n+8).$$

Since $0+1+2+\cdots+9 = 45$, both expressions are equal to $5n^2+45n+88$.

3. Extend AC to D so that $CD = CB$. Then DBC is a right isosceles triangle. Since CM bisects $\angle BCA$, DB is parallel to CM. Moreover, the midpoint of DB is also N since NBC is a right isosceles triangle. Since AN bisects DB, it must also bisect CM.

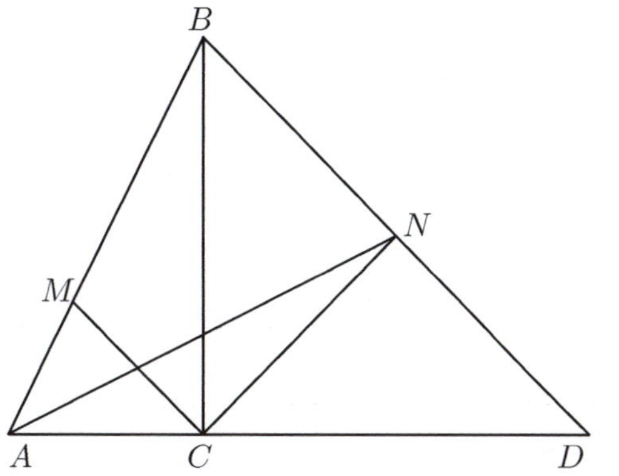

4. The diagram below shows two subboards which work. If the chosen point is inside a white square, it is either inside both subboards or outside both subboards. If the chosen point is inside a black square, it is inside one subboard and outside the other subboard. Black squares outside the subboards are marked with black dots.

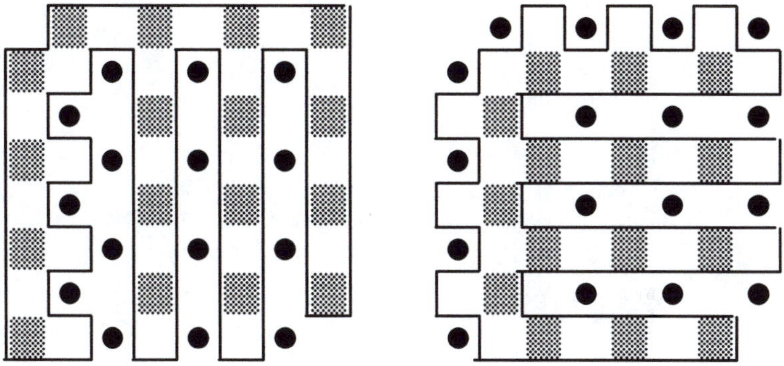

5. Let the polygon be $A_1A_2\ldots A_{101}$ with the vertices in clockwise order. Consider the 101 chords A_iA_{i+50}, $1 \le i \le 101$, with A_{101+k} interpreted as A_k. The chord is said to be *long* if the arc going clockwise from A_i to A_{i+50} is no less than a semicircle. Otherwise, it is said to be *short*. Suppose there are at least 51 long chords. Consider the 102 endpoints of these 51 chords. By the Pigeonhole Principle, one of them is an endpoint of two long chords. We may assume that they are A_1A_{51} and $A_{51}A_{101}$. Each of these two non-overlapping arcs is at least a semicircle, and yet their union is less than the whole circle. This is a contradiction. It follows that there are at least 51 short chords. As before, we may assume that A_1A_{51} and $A_{51}A_{101}$ are short. This means that both $\angle A_{51}A_1A_{101}$ and $\angle A_{51}A_{101}A_1$ are acute, so that the foot of the perpendicular from A_{51} to A_1A_{101} lies on the side and not its extension.

6. Note that n must be even, so that the number of integers from $n+1$ to $2n$ is even. We have

$$\begin{aligned}
& 1-\frac{1}{2}+\frac{1}{3}-\frac{1}{4}+\cdots+\frac{1}{2n-1}-\frac{1}{2n} \\
= \; & 1+\frac{1}{2}+\frac{1}{3}+\frac{1}{4}+\cdots+\frac{1}{2n-1}+\frac{1}{2n}-2\left(\frac{1}{2}+\frac{1}{4}+\cdots+\frac{1}{2n}\right) \\
= \; & \frac{1}{n+1}+\frac{1}{n+2}+\cdots+\frac{1}{2n-1}+\frac{1}{2n} \\
= \; & \left(\frac{1}{n+1}+\frac{1}{2n}\right)+\left(\frac{1}{n+2}+\frac{1}{2n-1}\right)+\cdots+\left(\frac{1}{\frac{3n}{2}}+\frac{1}{\frac{3n}{2}+1}\right) \\
= \; & (3n+1)\left(\frac{1}{2n(n+1)}+\frac{1}{(2n-1)(n+2)}+\cdots+\frac{1}{(\frac{3n}{2}+1)(\frac{3n}{2})}\right).
\end{aligned}$$

When combined as a single fraction, the denominator is the least common multiple of $2n(n+1)$, $(2n-1)(n+2)$, $\ldots$, $\frac{3n}{2}(\frac{3n}{2}+1)$. All of its factors are less than $3n+1$. During the reduction process, the factor $3n+1$ in the numerator, being prime, cannot be cancelled. When expressed as an irreducible fraction, the numerator is a multiple of $3n+1$.

7. Label the 110 stones as follows.

Pile A	00	01	02	03	04	05	06	07	08	09
Pile B	00	11	12	13	14	15	16	17	18	19
Pile C	01	11	22	23	24	25	26	27	28	29
Pile D	02	12	22	33	34	35	36	37	38	39
Pile E	03	13	23	33	44	45	46	47	48	49
Pile F	04	14	24	34	44	55	56	57	58	59
Pile G	05	15	25	35	45	55	66	67	68	69
Pile H	06	16	26	36	46	46	66	77	78	79
Pile I	07	17	27	37	47	57	67	77	88	89
Pile J	08	18	28	38	48	58	68	78	88	99
Pile K	09	19	29	39	49	59	69	79	89	99

Note that each label appears on two stones in different piles. Moreover, whenever two stones are in the same pile, the two stones which match their labels are not in the same pile. It follows that whichever stones Peter may take in a move, Betty can take the other stones which match their labels in her move. Hence Betty will get the last stone overall and win.

Senior A-Level Paper

1. (See Problem 4 of the Junior A-Level Paper.)

2. Suppose n is even. Let $R(x) = P(x) - Q(x)$. It is a polynomial with degree at most n. Hence it has at most $n+1$ coefficients. Let M be a positive real number which exceeds the absolute value of any of these coefficients. Now choose $k = n$, $\ell = 0$, $a = (n+1)M$ and $b = -(n+1)M$. We claim that the graphs of $P(x) + (n+1)Mx^n$ and $Q(x) - (n+1)M$ have no common points. Their difference is $R(x) + (n+1)M(x^n + 1)$. Our plan is to augment each term cx^h of $R(x)$ by $M(x^n + 1)$. Let x be any real number. Now $cx^h + M(x^n + 1) \geq cx^h + M|x|^h + M > M > 0$ if $|x| \geq 1$. If $|x| < 1$,

$$\begin{aligned} cx^h + M(x^n+1) &\geq cx^h + Mx^n + M \\ &> cx^h + Mx^n + M|x|^h \\ &> Mx^n \\ &> 0. \end{aligned}$$

This justifies the claim. Now suppose n is odd. For $n = 1$, choose $k = 1$ and $\ell = 0$. Choose a and b such that $P(x) + ax - Q(x) - b = 1$. Then the graphs of $P(x)+ax$ and $Q(x)+b$ have no common points. For odd $n > 1$, we give a counter-example. Let $P(x) = 2x^n$ and $Q(x) = x^n + x^{n-2}$. Consider $P'(x) = P(x) + ax^k$ and $Q'(x) = Q(x) + bx^\ell$. If neither k nor ℓ is 0, then $P'(0) = 0 = Q'(0)$. If neither k nor ℓ is n, then $P'(x) - Q'(x)$ is a polynomial of degree n, which is odd. Hence it has a real root r, and $P'(r) = Q'(r)$. Finally, suppose one of k and ℓ is 0 and the other is n. Now $P'(x) - Q'(x)$ is a polynomial of degree n or $n-2$, both of which are odd. Hence it has a real root r, and $P'(r) = Q'(r)$. In summary, the result holds for $n = 1$ and all even n.

3. Extend BA to P so that $BA = BP$. Extend BO to cut CP at Q. Extend CO to cut the circumcircle of BOA again at R.

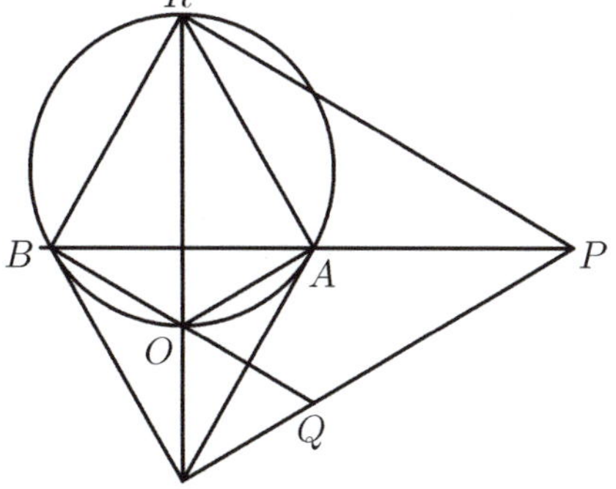

Since $\angle AOB = 120°$, $\angle ARB = 60°$. Since OR is a diameter perpendicular to AB, we have $RA = RB$, so that BAR is an equilateral triangle. From $AR = AB = AC = AP$, $\angle BCP = 90° = \angle CRP$. Hence

$\angle RCP = 60° = \angle CRP$, so that CPR is also an equilateral triangle. Hence $\angle OCQ = 60°$, and since $\angle COQ = \angle BOR = 60°$, COQ is another equilateral triangle. It follows that $OQ = OC = OB$ and $CQ \cdot CP = CO \cdot CR$. Moreover, $CO \cdot CR = CD \cdot CE$ since OR and DE are both chords of the circumcircle of BOA. Hence $CQ \cdot CP = CD \cdot CE$. This means that P, Q, D and E all lie on the same circle.

When this circle is contracted towards B by a factor of $\frac{1}{2}$, the smaller circle passes through A, O, the midpoint of BD and the midpoint of BE.

4. We have $0 = 0^3$ and $1 = 1^3$. We claim that every integer $n \geq 2$ is a sum of at most five distinct cubes. It will then follow that every negative integer is also a sum of at most five distinct cubes. For $n \geq 2$, consider the number $n^9 - n = n(n-1)(n+1)(n^2+1)(n^4+1)$. One of n and $n-1$ is even, and one of n, $n-1$ and $n+1$ is a multiple of 3. For $n \equiv 0$, 1, 2, 3 or 4 (mod 5), the respective factors n, $n-1$, n^2+1, n^2+1 and $n+1$ are divisible by 5. Hence $n^9 - n = 30k$ for some positive integer k. Since $n \geq 2$, we have $30k \geq 2^9 - 2 = 510$ so that $k \geq 17$. We also have $n^6 k \geq 64k > 2n + 60k$. Now

$$n = n^9 - 30k = (n^3)^3 + (k+2)^3 + (k-2)^3 + (-k+3)^3 + (-k-3)^3.$$

The first three terms are positive and the last two are negative. Clearly, $-k+3 > -k-3$ and $k+2 > k-2$. Finally, $n^3 = \frac{n+30k}{n^6} < \frac{k}{2} < k-2$. Thus the claim is justified.

5. (a) Such functions do not exist. Otherwise,we have

$$x = g(g(x)) = g(f(f(g(x)))) > f(g(x)) > x.$$

(b) Such functions do exist. Let

$$f(x) = \begin{cases} 2|x|+2 & x \text{ odd} \\ -(2|x|+2) & x \text{ even} \end{cases} \qquad g(x) = \begin{cases} -(2|x|+1) & x \text{ odd} \\ 2|x|+1 & x \text{ even} \end{cases}$$

Whether x is even or odd, $|f(x)|$ is the even number $2|x|+2$ and $|g(x)|$ is always the odd number $2|x|+1$. Hence

$$\begin{aligned} f(f(x)) &= -(2(2|x|+2)+2) &&= -4|x|-6 &&< x, \\ g(f(x)) &= 2(2|x|+2)+1 &&= 4|x|+5 &&> x, \\ g(g(x)) &= -(2(2|x|+1)+1) &&= -4|x|-3 &&< x, \\ f(g(x)) &= 2(2|x|+1)+2 &&= 4|x|+4 &&> x. \end{aligned}$$

6. (See Problem 7 of the Junior O-Level Paper.)

7. We first prove a preliminary result. A closed polygonal line with only simple self-intersection divides the plane into an infinite region and a number of polygonal regions, We claim that they can be painted black and white in checkerboard fashion, so that regions with a common border have opposite colours. We may assume that there are no vertical segments in the polygonal line.

For any region R, choose any point Q inside R and draw a vertical ray from P upwards. If this ray intersects the polygonal line an odd number of times, we paint R black. Otherwise we paint R white. This is independent of the choice of Q. If we move Q sideways inside R, then the number of intersections can only change when the ray passes through a vertex of the polygonal line, and by ± 2 if it does change. If we move Q up and down inside R, this number will not change. The colouring scheme is consistent overall since the number of intersections changes by ± 1 when Q moves vertically out of R into an adjacent region. (This number is not considered to change if the ray passes through a point of self-intersection of the polygonal line.) Thus our claim is justified. In particular, note that the infinite region is white, so that all black regions are polygons. The diagram below illustrates part of the overall configuration, showing segments AB and HI intersecting at X, segments BC and DE intersecting at Y and segments EF and GH intersecting at Z. The black polygonal regions are shaded. Each half segment separates a black region from a white region. Orient them so that the black region is on its left and the white region is on its right. Then the half segments are counterclockwise around the borders of all black regions.

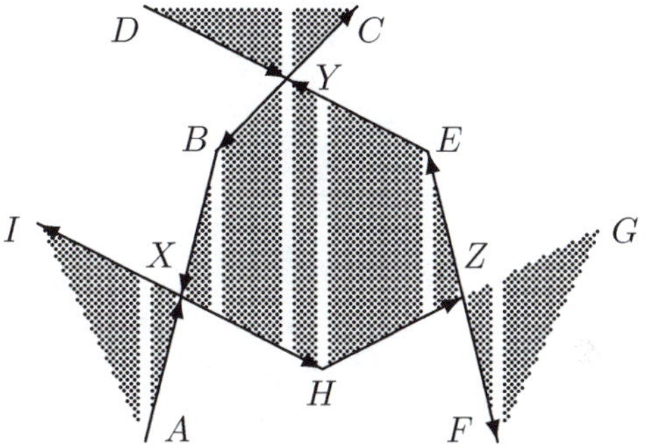

Now choose a point O in the infinite white region which does not lie on the extension of any segment. Let $[P]$ denote the area of the polygon P. For a half segment MN, compute the oriented area of triangle OMN. This is taken to be $[OMN]$ if MN is counterclockwise when viewed from O, and $-[OMN]$ otherwise. Consider the half segments on the border of one black region at a time, as illustrated in the diagram below. The total oriented area for the six half segments is

$$[OBX]+[OXH]+[OHZ]-[OZE]-[OEY]+[OYB]=[BXHZEY]>0.$$

Since the sum for each black region is positive, and each half segment is on the border of exactly one black region, the total oriented area summed over all half segments must be positive.

Consider now all the half segments at the same time. Each segment is divided into two half segments with opposite orientations. If each segment is bisected by the point of self-intersection, then the total oriented area summed over all half segments must be 0. This is a contradiction.

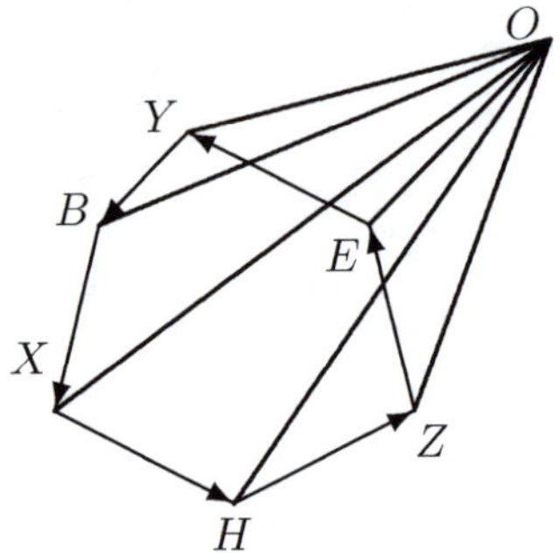

Spring 2014

Junior O-Level Paper

1. Let the numbers be $a_1,\ a_2,\ \ldots,\ a_{100}$. We are given that

$$\begin{aligned} 0 &= (a_1+1)^2-a_1^2+(a_2+1)^2-a_2^2+\cdots+(a_{100}+1)^2-a_{100}^2 \\ &= 2(a_1+a_2+\cdots+a_{100})+100. \end{aligned}$$

   Hence $a_1+a_2+\cdots+a_{100}=-50$. Now

$$\begin{aligned} &(a_1+2)^2-(a_1+1)^2+(a_2+2)^2-(a_2+1)^2 \\ &+\cdots+(a_{100}+2)^2-(a_{100}+1)^2 \\ =\ &2(a_1+a_2+\cdots+a_{100})+300. \end{aligned}$$

   Thus the sum increases by $2(-50)+300=200$.

2. Number the pies 1 to 15 in cyclic order, with the 7 banana pies following the 7 apple pies and preceding the cherry pie. Olga tries number 8. There are three cases.
   Case 1. Number 8 is the cherry pie.
   Then Olga gets what she wants.
   Case 2. Number 8 is an apple pie.
   Then number 15 must be a banana pie and the cherry pie is not between these two. Olga tries number 4. There are three subcases.
   Subcase 2(a). Number 4 is the cherry pie.
   Then Olga gets what she wants.
   Subcase 2(b). Number 4 is another apple pie.
   Then number 5 to number 7 are all apple pies. Olga tries number 2. If it is the cherry pie, she gets what she wants. If it is another apple pie, then number 1 is the cherry pie. If it is a banana pie, then number 3 is the cherry pie.
   Subcase 2(c). Number 4 is a banana pie.
   Then number 1 to number 3 are all banana pies. Olga tries number 6. The analysis is analogous to Subcase 2(b).
   Case 3. Number 8 is a banana pie.
   Then number 1 must be an apple pie and the cherry pie is not between these two. Olga tries number 12. The analysis is analogous to Case 2.

3. The diagram below shows a partition of the 5×7 table into six 2×3 or 3×2 rectangles, two of which overlap at the central square which is shaded. Hence the sum of 35 numbers is $6\times 10=60$ minus the number in the central square. Peter only needs to ask for the number in that square.

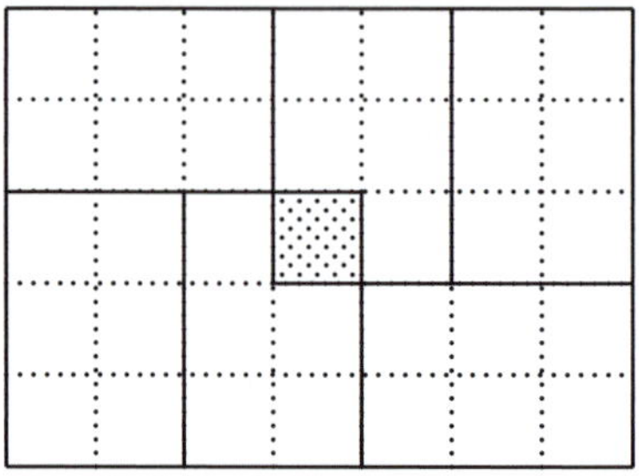

4. Let AL intersect CM at N, and let K be the midpoint of AL. Then $KL = CM$ and MK is parallel to LC. Hence $\angle LKM = \angle KLC = 45°$. Also, $\angle CMK = \angle MCL$, and let their common value be θ. If $\theta < 45°$, then $KL = KN + NL < MN + NV = CM$, which is a contradiction. If $\theta > 45°$, we also have a contradiction with all the inequalities reversed. Hence we must have $\theta = 45°$. Thus $\angle CNL = 180° - 45° - \theta = 90°$, and AL is indeed perpendicular to CM.

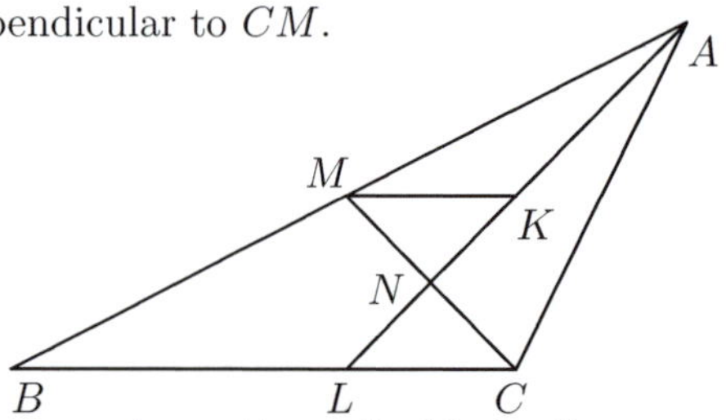

5. More generally, we prove by mathematical induction on n that the crossing is possible for any positive number n of thieves. For $n = 1$ or 2, the whole party can cross together. For $n = 3$, the crossing can be accomplished in the following five steps.

   (1) Ali Baba and the thieves ranked 1 and 2 go to the far shore.
   (2) The thieves ranked 1 and 2 come back to the near shore.
   (3) The thieves ranked 2 and 3 go to the far shore.
   (4) Ali Baba and the thief ranked 2 come back to the near shore.
   (5) Ali Baba and the thieves ranked 1 and 2 go to the far shore.

   Suppose the crossing is possible for $n = 1, 2, \ldots, k$ for some $k \geq 3$. Then the crossing for $n = k+1$ can be accomplished in the following five steps.

   (1) Ali Baba and the thieves ranked 1 to k go to the far shore.
   (2) The thieves ranked $k - 1$ and k come back to the near shore.
   (3) The thieves ranked k and $k + 1$ go to the far shore.
   (4) Ali Baba and the thieves ranked 1 to $k - 2$ come back to the near shore.
   (5) Ali Baba and the thieves ranked 1 to $k - 1$ go to the far shore.

Note that Step (1), Step (4) and Step (5) are possible by the induction hypothesis, with $n = k$, $k - 2$ and $k - 1$ respectively. This is why we need to include the case $k = 3$ as part of the basis.

Senior O-Level Paper

1. At least one original stone from each of 18 pairs with combined mass of 37 grams must be glued to some other stone. Hence at least 9 drops of glue are necessary. This is also sufficient since Doctor Claw can glue into 9 composite stones the 18 original stones whose masses are odd numbers of grams. This process leaves Inspector Gadget with 27 stones with masses being even numbers of grams. Hence he cannot get a subset with total mass 37 grams.

2. Draw two lines through N, one parallel to BC, cutting BD at H and the other parallel to AC, cutting AD at K. Then $\frac{DH}{DB} = \frac{DN}{DC} = \frac{DK}{AC}$. Hence HK is parallel to AB, and therefore perpendicular to AN. It follows that H is the orthocentre of triangle BNK. Hence BK is perpendicular to HN, and therefore to BC. It follows that K and M coincide, so that MN is parallel to AC.

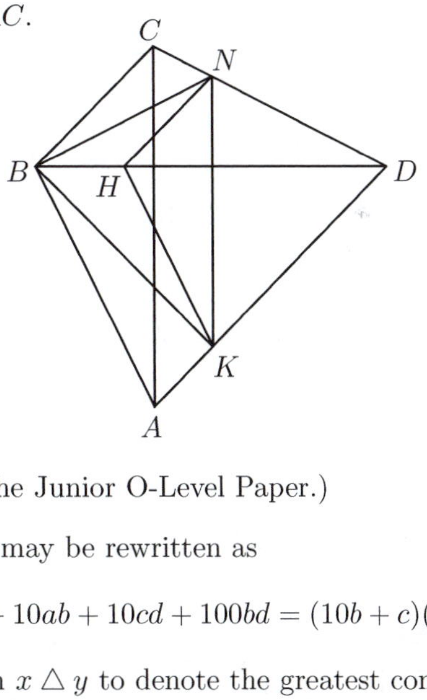

3. (See Problem 5 of the Junior O-Level Paper.)

4. The given equation may be rewritten as

$$11ac = ac + 10ab + 10cd + 100bd = (10b + c)(10d + a).$$

We use the notation $x \triangle y$ to denote the greatest common divisor of x and y. We consider three cases.
Case 1. $a \triangle 10 = 1 = c \triangle 10$.
Then $c \triangle (10b + c) = 1 = a \triangle (10d + a)$. Hence a must divide $10b + c$ and c must divide $10d + a$, while 11 divides exactly one of them. By symmetry, we may assume that $a = 10b + c$ and $11c = 10d + a$. It follows that $11c = 10d + 10b + c$ so that $c = d + b$.

Case 2. $a \triangle 10 > 1$ and $c \triangle 10 > 1$.
By symmetry, we may assume that $a \triangle 10 = 2$ and $c \triangle 10 = 5$. Then $a = 2h$ and $c = 5k$ for positive integers h and k with $h \triangle 5 = 1 = k \triangle 2$. The equation becomes $11hk = (2b + k)(5d + h)$. Hence h must divide $2b + k$, k must divide $5d + h$ while 11 divides exactly one of them. We may still assume by symmetry that $h = 2b + k$ and $11k = 5d + h$. Then $11k = 5d + 2b + k$ so that $10k = 5d + 2b$. This is impossible since b, being relatively prime to c, is not divisible by 5.
Case 3. $a \triangle 10 > 1$ and $c \triangle 10 = 1$, or $a \triangle 10 = 1$ and $c \triangle 10 > 1$.
The analysis is analogous to Case 2, again leading to a contradiction.

5. If Ben and Chris do meet each other at the point O of intersection of AC and BD, it must occur after Chris leaves B and before Ben arrives at C. Let the lengths of these two intervals be x and y respectively. Let the constant speeds of Alex, Ben and Chris be a, b and c respectively. Since $AB + BC > AC$, $a > b$. Since $BC + CD > BD$, $a > c$. Now $BC = a(x + y)$, $BO = cx$ and $OC = by$. We have

$$BO + OC = cx + by < a(x + y) = BC,$$

which is a contradiction. Hence Ben and Chris cannot have met each other at O.

Junior A-Level Paper

1. Let there be b boys and g girls. Let each boy get c chocolates and each girl get m marshmallows. Then $bc+g(c+1) = 47$ and $b(m+1)+gm = 74$. Adding the two equations, we have $(b+g)(c+m+1) = 121$. If $b+g = 1$, either $b = 1$, $g = 0$, $c = 47$ and $m = 73$ or $b = 0$, $g = 1$, $c = 46$ and $m = 74$. If $b + g = 11$, then $b = 8$, $g = 3$, $c = 4$ and $m = 6$. If $b + g = 121$, then $b = 74$, $g = 47$, $c = 0$ and $m = 0$.

2. Anna paints 9 squares as shown in the diagram below on the top left. Boris can put at most $\lfloor \frac{5\times 5}{3} \rfloor = 8$ copies of the V-tromino on the board. Each copy can cover at most 1 painted square. Hence Boris cannot succeed in his task.

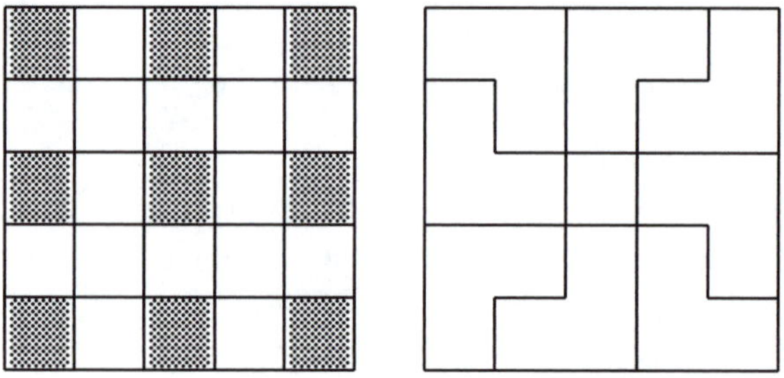

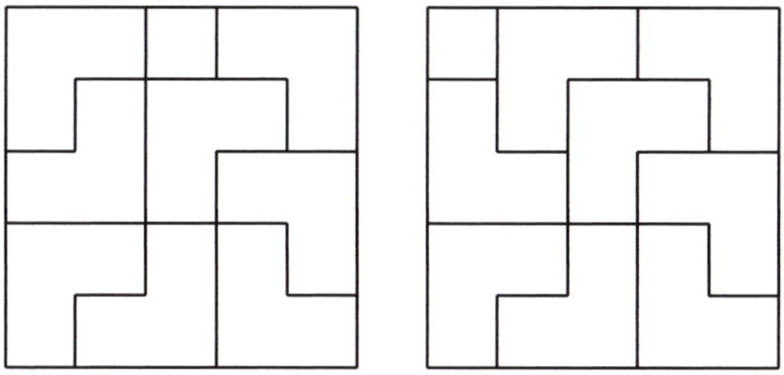

Suppose Anna paints 8 squares or fewer. Then at least one of the 9 painted squares in the diagram above on the top left is now unpainted. This unpainted square can be in the position shown in the diagram above on the bottom left, top right or bottom right. In each case, Boris can succeed in his task.

3. In the diagram below, $ABCD$ is the table and $KLMN$ is the tablecloth. The non-overlapping parts of these two squares consist of eight similar right triangles. Let the ratio of the lengths of the shorter leg, the longer leg and the hypotenuse be $x : y : z$. Let the actual lengths of the sides of the triangle with A as a vertex be ax, ay and az, those of the triangle with B as a vertex be bx, by and bz, and so on.

   Let the triangle with K as a vertex be congruent to the triangle with N as a vertex, so that $k = n$. From $AB = AD$, $ay+kz+bx = ax+nz+dy$, so that $(a-d)y = (a-b)x$. Similarly, from $KL = KN$, we have $(a-b)z = (\ell - k)y$, and from $KN = NM$, we have $(m-n)x = (a-d)z$. It follows that $\ell - k = m - n$. Since $k = n$, we have $\ell = m$, so that the triangle with L as a vertex is congruent to the triangle with M as a vertex.

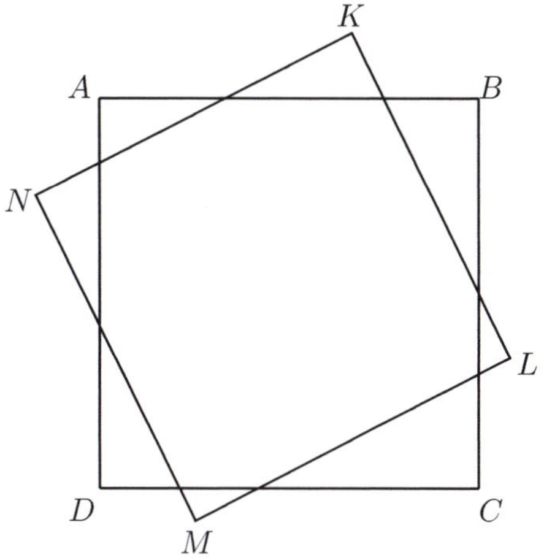

4. Anna chooses the numbers 2^k for $0 \le k \le 99$ and sends Boris these 100 cards along with the card which contains their sum, namely $2^{100} - 1$. Boris can tell that at least k of the original numbers are not exceeding 2^k, but their sum is not less than $2^{100} - 1$. It follows that the 100 numbers must be 2^k, $0 \le k \le 99$. Suppose Anna sends only 100 cards with numbers in increasing order. Each of the first 99 may contain one of the original numbers while the last may be the last original number, or the sum of the last number with another of the original numbers.

5. Let there be $m \ge 2$ white points and $n \ge 2$ black points. We may assume that the greatest common divisor of the labels of all the segments is 1. Let the label of each point be the greatest common divisor of all segments of which it is an endpoint. We prove by mathematical induction on $m+n$ that our labelling scheme works. For $m+n = 4$, we must have $m = n = 2$. Let W and X be the white points, and Y and Z be the black points. Let the labels of WY, WZ, XY abd XZ be a, b, c and d respectively. Let $w = a \triangle b$, where $\triangle$ denotes the greatest common divisor. Then $a = wy$ and $b = wz$ for some relatively prime positive integers y and z. Since $ad = bc$. we have $yd = zc$. Hence y divides c and z divides d. Let $c = xy$ for some positive integer x. Then $d = xz$ so that $c \triangle d = x$. Now w and x must be relatively prime as otherwise $a \triangle b \triangle c \triangle d \triangle = 1$. By our scheme, we label W, X, Y and Z with w, x, y and z respectively, and we have shown that this works. Suppose our scheme works for some $m + n \ge 4$. Consider the case with $m + n + 1$ points. We may assume that $m + 1$ of them are white and n of them are black. We apply our labelling scheme. We set aside a white point P and apply the induction hypothesis to conclude that for any segment not connected to P, its label is equal to the product of the labels of its endpoints. Let Q be any other white point and let R and S be any two black points. Let ' the labels of P, Q, R and S be p, q, r and s respectively. By the induction hypothesis, the label of QR is qr and the label of QS is qs. Now the label of PR is kr for some positive integer k. Hence the label of PS must be ks. Since their greatest common divisor is $k = p$. This completes the inductive argument.

6. Three cubes in a line form a row or a column if they are on the same level of the block. Otherwise, they form a stack. We can remove as many as 14 cubes, leaving behind 13 cubes represented by shaded squares in the diagram below. Suppose we leave behind only 12 cubes. Each level must contain at least 3 of them. Suppose one level contains exactly 3 cubes, which must be in different rows and different columns. In order for the structure to stay in one piece, each of these 3 cubes must be connected to a cube in an adjacent level in the same row and same column. Thus we have only 6 cubes to cover the remaining 6 stacks, and the structure will fall into pieces.

The only alternative is to have exactly 4 cubes on each level. As before, 3 of the 4 must be in different rows and different columns, so that these 4 cubes must be in at least two pieces. Hence at least two stacks must contain a cube on each of the top and middle levels, and at least two stacks must contain a cube on each of the middle and the bottom levels. If these four stacks are distinct, then we have only 4 cubes left to cover the remaining five stacks, which is impossible. Hence some stack must retain all 3 cubes. By symmetry, some row and some column must also do likewise, and the task is still impossible.

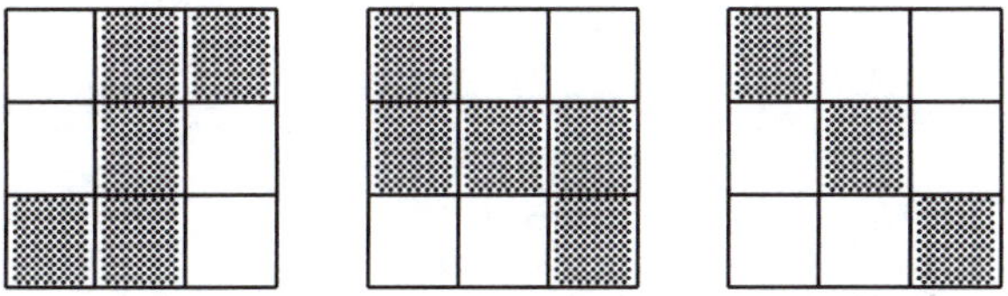

Bottom Level Middle Level Top Level

7. The 10 grasshoppers divide the circle into 10 arcs. Paint them alternatively black and white. Initially, the difference Δ between the total length of the white arcs and the total length of the black arcs is 0 since each of the five white arcs is symmetric to a black arc with respect to the centre of the circle. Let W, X, Y and Z be the positions of four grasshoppers, with the arcs WX and YZ painted white and the arc XY painted black. Suppose the grasshopper on X jumps over the one on Y, to land on a point T between Y and Z. After the jump, the arcs WY and TZ will be repainted black while the arc YT is repainted white. Since the arcs XY and YT have the same length, the value of Δ is unchanged. In the final configuration, the 9 grasshoppers on A_1 to A_9 determine 4 white arcs and 4 black arcs of known length, and the sum of the lengths of the 5th white arc and the 5th black arc is also known. Since Δ is an invariant, the position of the 10th grasshopper is uniquely determined. This position must be A_{10} because it yields the correct value $\Delta = 0$.

Senior A-Level Paper

1. Anna writes down 4027 1s and puts + signs and × signs alternately between them, adds a pair of brackets around the first two 1s and adds another pair of brackets around the last two 1s. In other words, her expression is $(1+1)\times 1+1\times 1+\cdots\times 1+(1\times 1)$. Its value is 2014 since there are 2013 + signs. Now $(1\times 1)+1\times 1+1\times\cdots+1\times(1+1)$ is the expression of Boris. Its value is also 2014 for the same reason.

2. (a) Let ℓ be a directed line which bisects the perimeter of a convex polygon. Let $f(\ell)$ be the length of the longest side of the piece to the left of ℓ, and $g(\ell)$ be the length of the longest side of the piece to the right of ℓ. If $f(\ell) = g(\ell)$, there is nothing further to prove. We may assume by symmetry that $f(\ell) > g(\ell)$. Rotate ℓ so that it continues to bisect the perimeter of the polygon. Now both $f(\ell)$ and $g(\ell)$ change continuously. When the new position of ℓ coincides with its original position but in the opposite direction, we have $f(\ell) < g(\ell)$. By continuity, equality must have occurred sometime during the rotation.

   (b) In triangle ABC, $BC = 4$, $CA = 3$ and $AB = 5$. The cut must divide the longest side so that it passes through a point P on AB. We consider two cases.
   Case 1. The cut passes through a point Q on BC.
   Note that PQ is shortest when $BP = BQ = 3$. Since $\cos B = \frac{4}{5}$,

$$PQ = \sqrt{BP^2 + BQ^2 - 2BP \cdot BQ \cos B} = \sqrt{\frac{18}{5}} > \frac{3}{2}.$$

   Now $AP + CQ = 3$. Hence PQ is not the shortest side of $ACQP$, nor is CA. We have $AP = 5 - BP = 5 - (6 - BQ) = BQ - 1$ while $CQ = 4 - BQ = 4 - (6 - BP) = BP - 2$. Hence the shortest side of $ACQP$ is shorter than the shortest side of BPQ.

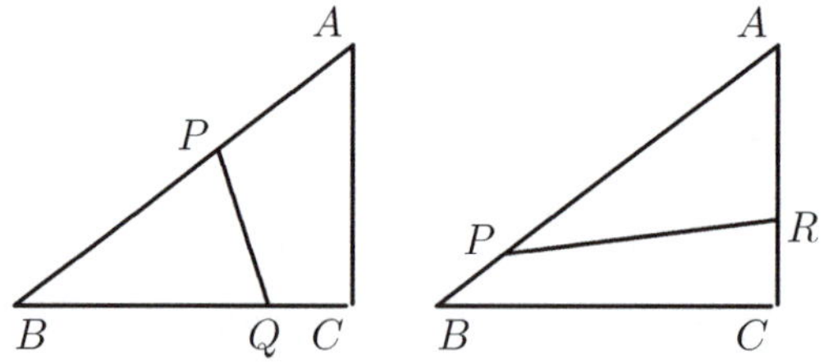

   Case 2. The cut passes through a point R on CA.
   Note that PR is shortest when $AP = AR = 3$. Since $\cos A = \frac{3}{5}$,

$$PR = \sqrt{AP^2 + AR^2 - 2AP \cdot AR \cos A} = \sqrt{\frac{36}{5}} > 1.$$

   Now $BP + CR = 2$. Hence PR is not the shortest side of $BCRP$, nor is BC. We have $BP = 5 - AP = 5 - (6 - AR) = AR - 1$ while $CR = 3 - AR = 3 - (6 - AP) = AP - 3$. Hence the shortest side of $BCRP$ is shorter than the shortest side of APR.

3. See Problem 4 of the Junior A-Level Paper.)

4. The polynomial $x(x-1)\cdots(x-19)$ of degree 20 has twenty roots $x = 0$, 1, ..., 19. Hence its graph passes through the 20 marked points (0,0), (1,0), ..., (19,0). We now prove that 20 is maximum. Suppose the graph of a polynomial $P(x)$ of degree 20 passes through 21 marked points (x_i, y_i), $1 \leq i \leq 21$, with $x_1 < x_2 < \cdots < x_{21}$ and $0 \leq y_i \leq 10$. The integer $x_{21} - x_1$ is at least 20 while the integer $|y_{21} - y_1|$ is at most 10. Since the former divides the latter, we must have $y_{21} = y_1$. Denote their common value by r, $0 \leq r \leq 10$. Then $P(x)$ is of the form $P(x) = (x - x_1)(x_{21} - x)Q(x) + r$. For $2 \leq i \leq 20$, the integer $(x_i - x_1)(x_{21} - x_i)$ is at least 19. Unless $Q(x_i) = 0$, we will either have $y_i \leq -19 + c < 0$ or $y_i \geq 19 + r > 10$. Hence $Q(x_i) = 0$ for $2 \leq i \leq 20$. However, this is impossible since $Q(x)$ is of degree 18.

5. Let the triangle be ABC. First Anna chooses all three vertices. Boris must paint two of them in one colour and the third one in the other colour. For definiteness, say Boris paints B and C red and A blue. Now Anna chooses D and E on the same side of BC as A, so that triangles ABC, BCD and BEC are similar. Then Boris must paint both of them blue. Now DCE is a triangle with three blue vertices. We claim that Anna has won after choosing these five points.

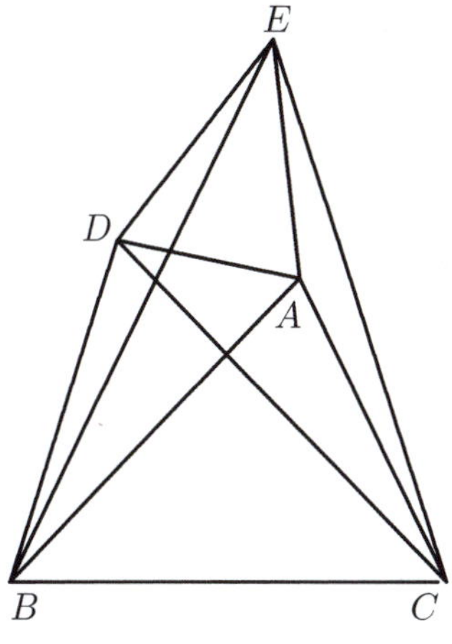

We have $\angle DBE = \angle DBC - \angle EBC = \angle BCE - \angle BCA = \angle ACE$ and $\frac{DB}{CA} = \frac{BC}{AB} = \frac{EB}{CE}$ by similar trianglesBy similar triangles, . Hence triangles DBE and ACE are similar, so that we have $\frac{DE}{AE} = \frac{BE}{CE}$ and $\angle BED = \angle CEA$. It follows that $\angle AED = \angle CEB$, so that triangles AED and CEB are similar. This justifies the claim. Finally, observe that if Anna chooses only four points, Boris can prevent Anna from winning by painting any two of them red and the other two blue.

6. The diagram below shows 10 cities linked by line segments representing direct flights. Since all cities are situated symmetrically, Anna can always rename the cities without affecting the directory.

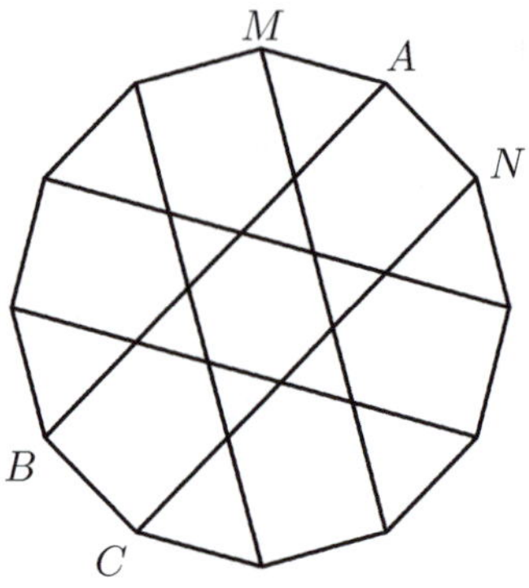

Let Moscow and Novgorod be marked M and N as shown in the diagram above. Then A, as the only city with direct flights to both M and N, must retain its own name after those of M and N have been exchanged. Now B, as the only other town with a direct flight to A, must also retain its own name. Before the exchange, A is the only city with direct flights to both M and B. After the exchange, both A and C have direct flights to both M and B. Thus the action of Boris must affect the directory.

7. Since $P(0) = 1$, the constant term of $P(x)$ must be 1. Hence $1 - P(x)$ is divisible by x, so that $(1 - P(x))^{100}$ is divisible by x^{100}. It follows that

$$(1 + P(x))^{100} + (1 - P(x))^{100} \equiv (1 + P(x))^{100} \pmod{x^{100}}.$$

Now the left side consists only of even powers of $P(x)$. Hence it is of the form $F((P(x))^2)$ where $F(x)$ is a polynomial of degree 50. Now $F((P(x))^2) = F(1 + x + x^{100}Q(x)) \equiv F(1 + x) \pmod{x^{100}}$. The highest power of x in $F(1+x))$ is x^{50}. Hence the coefficient of x^{99} in $(1+P(x))^{100}$ is 0.

Tournament 36

Fall 2014

Junior O-Level Paper

1. There are 99 sticks of lengths 1, 2, 3, ..., 99. Is it possible to use all of them to form the perimeter of a rectangle?

2. Do there exist ten distinct positive integers such that their average divided by their greatest common divisor is equal to

   (a) 6;

   (b) 5?

3. K and L are points on the sides AB and BC of a square $ABCD$ respectively, such that $KB = LC$. Let P be the point of intersection of AL and CK. Prove that DP and KL are perpendicular.

4. In the 40 tests Andrew had taken, he got 10 As, 10 Bs, 10 Cs and 10 Ds. A score is said to be *unexpected* if this particular score appeared for the kth time after the other three scores had appeared at least k times. Without knowing the order of these 40 scores, is it possible to determine the number of unexpected ones?

5. There are $n > 1$ right triangles. Adam chooses a leg from each of them and calculates the sum of those lengths. Then he calculates the sum of the lengths of the remaining legs. Finally, he calculates the sum of the lengths of the hypotenuses. If these three numbers are the side lengths of a right triangle, prove that the n triangles are similar to one another for

   (a) $n = 2$;

   (b) an arbitrary positive integer n.

Note: The problems are worth 3, 2+2, 5, 5 and 2+3 points respectively.

Senior O-Level Paper

1. Do there exist ten pairwise distinct positive integers such that their average divided by their greatest common divisor is equal to

   (a) 6;

   (b) 5?

2. The vertices of triangle ABC are in clockwise order. The triangle is rotated to $A_1B_1C_1$ about $A = A_1$ clockwise through an angle equal to $\angle A$. Then it is rotated to $A_2B_2C_2$ about $B_1 = B_2$ clockwise through an angle equal to $\angle B$. Then it is rotated to $A_3B_3C_3$ about $C_2 = C_3$ clockwise through an angle equal to $\angle C$. This is continued for another three rotations in the same manner, clockwise though angles equal respectively to $\angle A$, $\angle B$ and $\angle C$, until the triangle becomes $A_6B_6C_6$. Prove that $A_6B_6C_6$ coincides with ABC.

3. Peter writes down the sum of every subset of size 7 of a set of 15 distinct integers, and Betty writes down the sum of every subset of size 8 of the same set. If they arrange their numbers in non-decreasing order, can the two lists turn out to be identical?

4. There are $n > 1$ right triangles. Adam chooses a leg from each of them and calculates the sum of those lengths. Then he calculates the sum of the lengths of the remaining legs. Finally, he calculates the sum of the lengths of the hypotenuses. If these three numbers are the side lengths of a right triangle, prove that the n triangles are similar to one another for an arbitrary positive integer n.

5. At the beginning, there are some silver coins on a table. In each move, we can either add a gold coin and record the number of silver coins on a blackboard, or remove a silver coin and record the number of gold coins on a whiteboard. At the end, only gold coins remain on the table. Must the sum of the numbers on the blackboard be equal to the sum of the numbers on the whiteboard?

Note: The problems are worth 1+2, 4, 5, 5 and 5 points respectively.

Junior A-Level Paper

1. Half of all entries in a square table are plus signs, and the remaining half are minus signs. Prove that either two rows or two columns contain the same number of plus signs.

2. Prove that any polygon with an incircle has three sides that can form a triangle.

3. Is it possible to divide all positive divisors of 100!, including 1 and 100!, into two groups of equal size such that the product of the numbers in each group is the same?

4. On a circular road there are 25 equally spaced booths, each with a patrolman numbered from 1 to 25 in some order. The patrolmen switch booths by moving along the road, so that their numbers are from 1 to 25 in clockwise order. If the total distance travelled by the patrolmen is as low as possible, prove that one of them remains in the same booth.

5. In triangle ABC, $\angle A = 90°$. Two equal circles tangent to each other are such that one is tangent to BC at M and to AB, and the other is tangent to BC at N and to CA. Prove that the midpoint of MN lies on the bisector of $\angle A$.

6. A *uniform* number is a positive integer in which all digits are the same. Prove that any n-digit positive integer can be expressed as the sum of at most $n+1$ uniform numbers.

7. A spiderweb is a square grid with 100×100 nodes, at 100 of which flies are stuck. Starting from a corner node of the web, a spider crawls from a node to an adjacent node in each move. A fly stuck at the node where the spider is wi ll be eaten. Can the spider always eat all the flies in no more than

   (a) 2100 moves;

   (b) 2000 moves?

Note: The problems are worth 4, 5, 6, 7, 8, 8 and 5+5 points respectively.

Senior A-Level Paper

1. Prove that any polygon with an incircle has three sides that can form a triangle.

2. On a circular road there are 25 equally spaced booths, each with a patrolman numbered from 1 to 25 in some order. The patrolmen switch booths by moving along the road, so that their numbers are from 1 to 25 in clockwise order. If the total distance travelled by the patrolmen is as low as possible, prove that one of them remains in the same booth.

3. Gregory writes down 100 numbers on a blackboard and calculates their product. In each move, he increases each number by 1 and calculates their product. What is the maximum number of moves Gregory can make if the product after each move does not change?

4. The incircle of triangle ABC is tangent to BC, CA and AB at D, E and F respectively. It is given that AD, BE and CF are concurrent at a point G, and that the circumcircles of triangles GDE, GEF and GFD intersect the sides of ABC at six distinct points other than D, E and F. Prove that these six points are concyclic.

5. Peter prepares a list of all possible words consisting of m letters each of which is T, O, W or N, such that the numbers of Ts and Os is the same in each word. Betty prepares a list of words consisting of $2m$ letters each of which is T or O, such that the numbers of Ts and Os is the same in each word. Whose list contains more words?

6. Let PQR be a given triangle. $AFBDCE$ is a non-convex hexagon such that the interior angles at D, E and F all have measure $181°$. Moreover,

$$\begin{array}{ll} BD + DC = QR, & \angle EAF = \angle RPQ - 1°, \\ CE + EA = RP, & \angle FBD = \angle PQR - 1°, \\ AF + FB = PQ, & \angle DCE = \angle QRP - 1°. \end{array}$$

Prove that $\frac{BD}{DC} = \frac{CE}{EA} = \frac{AF}{FB}$.

7. Each day, positive integers m and n are chosen by the government. On that day, x grams of gold may be exchanged for y grams of platinum such that $mx = ny$. Initially, $m = n = 1001$. On each subsequent day, the government reduces exactly one of m and n by 1, and after 2000 days, both numbers are equal to 1. Someone has 1 kilogram of each of gold and platinum initially. Without knowing in advance which of m and n will be reduced the next day, can this person have a sure way of performing some clever exchanges, and end up with at least 2 kilograms of each of gold and platinum after these 2000 days?

Note: The problems are worth 4, 6, 6, 7, 7, 8 and 10 points respectively.

Spring 2015

Junior O-Level Paper

1. Is it possible to paint each of the six faces of a cube in one of three colours so that each colour is used, and any three faces sharing a common vertex do not all have different colours?

2. K and L are points on the side AB of triangle ABC such that we have $KL = BC$ and $AK = LB$. M is the midpoint of the side AC. Prove that $\angle KML = 90°$.

3. Basil computes the sum of several consecutive positive integers starting from 1. Patty computes the sum of 10 consecutive positive powers of 2, not necessarily starting from 1. Is it possible for the two sums to be equal?

4. The figure in the diagram below is to be dissected along the dotted lines into a number of squares which are not necessarily all of the same size or all of different sizes. What is the minimum number of such squares?

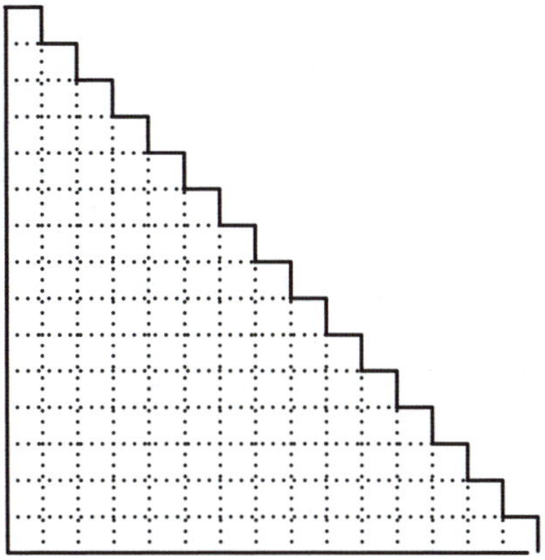

5. We have one copy of 0 and two copies of each of 1, 2, …, n. For which n can we arrange these $2n + 1$ numbers in a row, such that for each m between 1 and n inclusive, there are exactly m other numbers between the two copies of the number m?

Note: The problems are worth 3, 4, 4, 4 and 5 points respectively.

Senior O-Level Paper

1. Basil computes the sum of several consecutive positive integers starting from 1. Patty computed the sum of 100 consecutive positive powers of 2, not necessarily starting from 1. Is it possible for the two sums to be equal?

2. There are four points inside a 275×275 square. Is it possible to cut out a 100×100 square without containing any of the four points?

3. We have one copy of 0 and two copies of each of 1, 2, ..., n. For which n can we arrange these $2n + 1$ numbers in a row, such that for each m from 1 to n inclusive, there are exactly k other numbers between the two copies of the number m?

4. M is the midpoint of the side BC of triangle ABC. K and L are points on AM such that $AK = KL = LM$. P is a point on the same side of AM as C such that $\angle LKP = \angle CAB$ and $\angle KLP = \angle ACB$. Prove that P lies on line AC.

5. Along a circle are 2015 positive integers such that the difference between any two adjacent numbers is equal to their greatest common divisor. Determine the maximal value of a positive number which divides the product of these 2015 numbers.

Note: The problems are worth 3, 4, 4, 4 and 5 points respectively.

Junior A-Level Paper

1. E is a point is inside a parallelogram $ABCD$ such that $CD = CE$. Prove that DE is perpendicular to the segment joining the midpoints of AE and BC.

2. A prohibited area is in the shape of a non-convex polygon. A straight power line crossing this polygon does not pass through any of its vertices. There are 36 poles along the line. Some of them are inside the polygon, and the remaining ones are outside it. A spy moves once around the perimeter of the polygon so that the polygon is always on his right. Each time the spy crossed the power line, he counts the number of poles to the left of him. He can see all the poles all the time. Upon returning to his starting point, the spy has counted 2015 poles in total. Find the number of poles inside the fence.

3. (a) The integers x, x^2 and x^3 begin with the same digit. Does it imply that this digit is 1?

   (b) Answer (a) for the integers x, x^2, x^3, ..., x^{2015}.

4. For each side of some polygon, the line containing it contains at least one more vertex of this polygon. Is it possible that the number of vertices of this polygon is not greater than

   (a) 9;

   (b) 8?

5. (a) A $2 \times n$ table of numbers with $n > 2$ is such that the sums of all the numbers in the columns are different. Is it always possible to permute the numbers in the table so that the column sums are still different and now the row sums are also different?

   (b) Is this always possible for a 10×10 table?

6. A convex n-gon with equal sides is located inside a circle. Each side is extended in both directions up to the intersection with the circle so that it contains two new segments outside the polygon. Prove that one can paint some of these $2n$ new segments in red and the others in blue so that the sum of lengths of all the red segments would be the same as for the blue ones.

7. An Emperor invited 2015 wizards to a festival. Each wizard is either good or evil. A good wizard always tells the truth, while an evil wizard can say what he wants. Each wizard knows whether each of the other wizards is good or evil, but the Emperor does not. In each round, the Emperor may ask each wizard at most one yes/no question. Different questions may be used for different wizards, and the order in which the questions are asked is chosen by the Emperor. After listening to all the answers, the Emperor expels one of the wizards through a magic door. It will announce whether the expelled wizard is good or evil. In subsequent rounds, the Emperor may choose not to expel a wizard. In that case, the process is terminated. Prove that the Emperor can expel all the evil wizards while expelling at most one good wizard.

Note: The problems are worth 4, 6, 3+4, 4+5, 3+6, 9 and 10 points respectively.

Senior A-Level Paper

1. (a) The integers x, x^2 and x^3 begin with the same digit. Does it imply that this digit is 1?

   (b) Answer (a) for the integers $x,\ x^2,\ x^3,\ \ldots,\ x^{2015}$.

2. A point X is marked on the base BC of an isosceles triangle ABC, and points P and Q are marked on the sides AB and AC respectively so that $APXQ$ is a parallelogram. Prove that the point Y symmetrical to X with respect to the line PQ lies on the circumcircle of triangle ABC.

3. (a) A $2 \times n$ table of numbers with $n > 2$ is such that the sums of all the numbers in the columns are different. Prove that it is possible to permute the numbers in the table so that the column sums are still different and now the row sums are also different.

   (b) Is this always possible for a 10×10 table?

4. A convex n-gon with equal sides is located inside a circle. Each side is extended in both directions up to the intersection with the circle so that it contains two new segments outside the polygon. Prove that one can paint some of these $2n$ new segments in red and the others in blue so that the sum of lengths of all the red segments would be the same as for the blue ones.

5. Do there exist two polynomials with integer coefficients such that each polynomial has a coefficient with an absolute value exceeding 2015 but all coefficients of their product have absolute values not exceeding 1?

6. An Emperor invited 2015 wizards to a festival. Each wizard is either good or evil. A good wizard always tells the truth, while an evil wizard can say what he wants. Each wizard knows whether each of the other wizards is good or evil, but the Emperor does not. In each round, the Emperor may ask each wizard at most one yes/no question. Different questions may be used for different wizards, and the order in which the questions are asked is chosen by the Emperor. After listening to all the answers, the Emperor expels one of the wizards through a magic door. It will announce whether the expelled wizard is good or evil. In subsequent rounds, the Emperor may choose not to expel a wizard. In that case, the process is terminated. Prove that the Emperor can expel all the evil wizards while expelling at most one good wizard.

7. It is well-known that if a quadrilateral has a circumcircle and an incircle, and the centres of the two circles coincide, then the quadrilateral must be a square. Each face of a hexahedron is a convex quadrilateral. It has a circumsphere and an insphere, and the centres of the two spheres coincide. Is this hexahedron necessarily a cube?

Note: The problems are worth 2+3, 5, 2+6, 8, 10, 10 and 10 points respectively.

Solutions

Fall 2014

Junior O-Level Paper

1. We first use the sticks of lengths 1, 2, 3, 4, 5, 6 and 7 to form the perimeter of a square, with side length 7=6+1=5+2=4+3. Divide the remaining 92 sticks into groups of four: (8,9,10,11), (12,13,14,15), ..., (96,97,98,99). For each group, add the longest and the shortest sticks to the top edge of the rectangle and the other two sticks to the bottom edge.

2. (a) We may take the ten numbers to be 1, 2, 3, 4, 5, 6, 7, 8, 9 and 15. Their sum is 60 so that their average is 6. Their greatest common divisor is 1, and indeed $6 \div 1 = 6$.

   (b) This is impossible. The greatest common divisor appears everywhere and can be cancelled out. Hence we may take it to be 1. However, the smallest ten distinct positive integers are 1, 2, 3, 4, 5, 6, 7, 8, 9 and 10, with an average of 5.5. Thus the quotient can never be equal to 5.

3. Since triangles ABL and DAK are congruent, AL is perpendicular to DK. Since triangles CBK and DCL are congruent, CK is perpendicular to DL. Hence P is the orthocentre of triangle DKL, so that DP is perpendicular to KL.

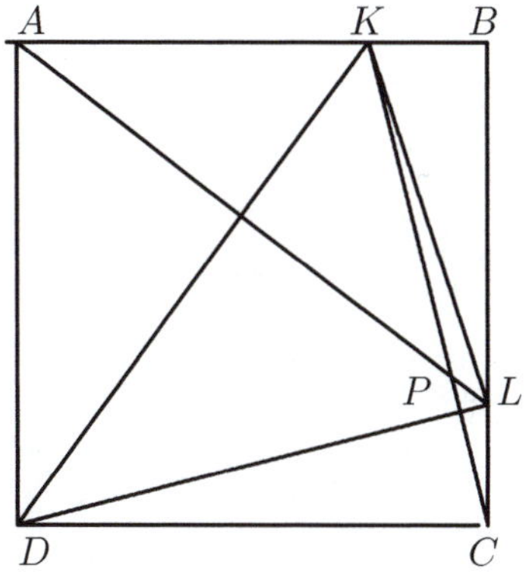

4. Consider the first A, the first B, the first C and the first D that Andrew gets. The last one to come along must be unexpected, and none of the other three can be unexpected. The same applies to the second A, the second B, the second C and the second D that he gets, and so on. It follows that exactly 10 of the scores are unexpected.

5. (a) For $i = 1$ or 2, let the side lengths of the ith triangle be a_i, b_i and c_i with $a_i^2 + b_i^2 = c_i^2$. From $(a_1 + a_2)^2 + (b_1 + b_2)^2 = (c_1 + c_2)^2$, we have $a_1a_2 + b_1b_2 = c_1c_2$. Hence

$$\begin{aligned} 0 &= a_1a - 2 + b_1b_2 - c_1c_2 \\ &= \frac{(a_1a_2 + b_1b_2)^2 - (a_1^2 + b_1^2)(a_2^2 + b_2^2)}{a_1a_2 + b_1b_2 + c_1c_2} \\ &= -\frac{(a_1b_2 - a_2b_1)^2}{a_1a_2 + b_1b_2 + c_1c_2}. \end{aligned}$$

It follows that $a_1b_2 - b_1a_2 = 0$, which is equivalent to $\frac{a_1}{b_1} = \frac{a_2}{b_2}$. Since both triangles are right triangles with proportional legs, they are similar to each other.

(b) For $1 \le i \le n$, let the side lengths of the ith triangle be a_i, b_i and c_i with $a_i^2 + b_i^2 = c_i^2$. From

$$(a_1 + a_2 + \cdots + a_n)^2 + (b_1 + b_2 + \cdots + b_n)^2 = (c_1 + c_2 + \cdots + c_n)^2,$$

we have

$$a_1a_2 + a_1a_3 + \cdots + a_{n-1}a_n + b_1b_2 + b_1b_3 + \cdots + b_{n-1}b_n = c_1c_2 + c_1c_3 + \cdots + c_{n-1}c_n.$$

Hence

$$\begin{aligned} 0 &= (a_1a_2 - b_1b_2 - c_1c_2) + (a_1a_3 - b_1b_3 - c_1c_3) \\ &\quad + \cdots + (a_{n-1}a_n - b_{n-1}b_n - c_{n-1}c_n) \\ &= \frac{(a_1a_2 + b_1b_2)^2 - (a_1^2 + a_2^2)(b_1^2 + b_2^2)}{a_1a_2 + b_1b_2 + c_1c_2} \\ &\quad + \frac{(a_1a_3 + b_1b_3)^2 - (a_1^2 + a_3^2)(b_1^2 + b_3^2)}{a_1a_3 + b_1b_3 + c_1c_3} \\ &\quad + \cdots + \frac{(a_{n-1}a_n + b_{n-1}b_n)^2 - (a_{n-1}^2 + a_n^2)(b_{n-1}^2 + b_n^2)}{a_{n-1}a_n + b_{n-1}b_n + c_{n-1}c_n} \\ &= -\frac{(a_1b_2 - a_2b_1)^2}{a_1a_2 + b_1b_2 + c_1c_2} - \frac{(a_1b_3 - a_3b_1)^2}{a_1a_3 + b_1b_3 + c_1c_3} \\ &\quad - \cdots - \frac{(a_{n-1}b_n - a_nb_{n-1})^2}{a_{n-1}a_n + b_{n-1}b_n + c_{n-1}c_n}. \end{aligned}$$

It follows that

$$0 = a_1b_2 - b_1a_2 = a_1b_3 - b_1a_3 = \ldots = a_{n-1}b_n - b_{n-1}a_n,$$

which is equivalent to $\frac{a_1}{b_1} = \frac{a_2}{b_2} = \cdots = \frac{a_n}{b_n}$. Since all triangles are right triangles with proportional legs, they are similar to one another.

Senior O-Level Paper

1. (See Problem 2 of the Junior O-Level Paper.)

2. Let ℓ be the line AC and let B be to its right. In the first rotation, B_1 is on ℓ and C_1 is to its left. In the second rotation, C_2 is on ℓ and A_2 is to its right. In the third rotation, A_3 is on ℓ and B_3 is to its left.

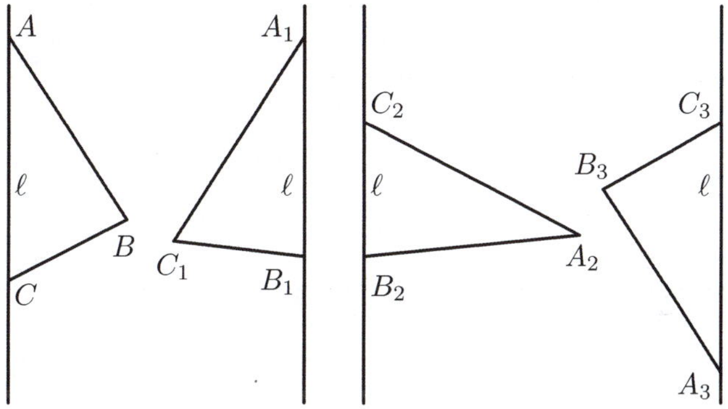

Since $AC = A_3C_3$, AA_3 and CC_3 have a common midpoint O, which must also be the midpoint of BB_3. Hence $A_3B_3C_3$ may be obtained directly from ABC by a half-turn about O. By symmetry, the next three rotations combine into another half-turn about O, which brings $A_3B_3C_3$ back to ABC.

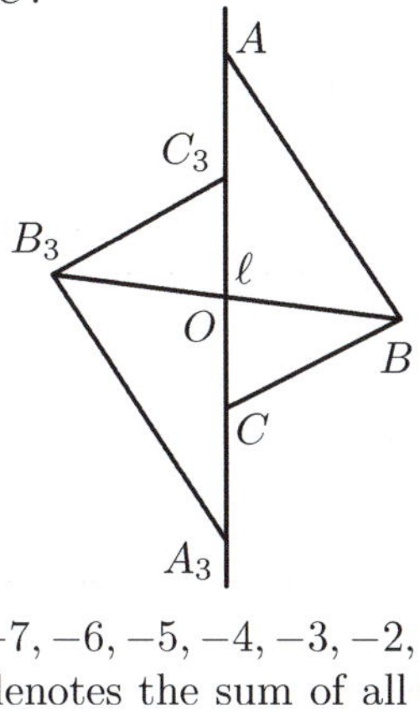

3. Take the set $S = \{-7, -6, -5, -4, -3, -2, -1, 0, 1, 2, 3, 4, 5, 6, 7\}$. Then $f(S) = 0$, where f denotes the sum of all the elements in the set. Let $A = \{a_1, a_2, \ldots, a_7\}$ be a subset of S. Let $B = \{b_1, b_2, \ldots, b_7\}$ be such that $b_k = -a_k$ for $1 \leq k \leq 7$. Then $f(B) = -f(A)$. Now $f(S - A) = f(S) - f(A) = -f(A) = f(B)$. Similarly, $f(S - B) = f(A)$. If $A \neq B$, then the two sums $f(A)$ and $f(B)$ on Peter's list correspond respectively to the two sums $f(S - B)$ and $f(S - A)$ on Betty's list. If $A = B$, then we must have $f(A) = 0$ so that $f(S - A) = 0$ also. In this case, the sum $f(A)$ on Peter's list corresponds to the sum $f(S - A)$ on Betty's list. It follows that the two lists are identical.

4. For $1 \le i \le n$, let the ith triangle be $A_iB_iC_i$, with $B_iC_i = a_i$, $C_iA_i = b_i$ and $A_iB_i = c_i$. Let ABC be a right triangle with $BC = a_1+a_2+\cdots+a_n$ and $CA = b_1+b_2+\cdots+b_n$. Overlay $A_1B_1C_1$ on ABC so that $B_1 = B$ and C_1 lies on BC. For $2 \le i \le n$, overlay $A_iB_iC_i$ on ABC so that $B_i = A_{i-1}$ and B_iC_i is parallel to BC. Then $A_n = A$ and C_n lies on CA, as shown in the diagram below. By the Triangle Inequality, $AB \le A_1B_1 + A_2B_2 + \cdots + A_nB_n = c_1 + c_2 + \cdots + c_n$. Since it is given that $AB = c_1 + c_2 + \ldots + c_n$, $A_{i-1} = B_i$ lie on AB for $2 \le i \le n$. It follows that $A_iB_iC_i$ is similar to ABC for $1 \le i \le n$, and hence to one another.

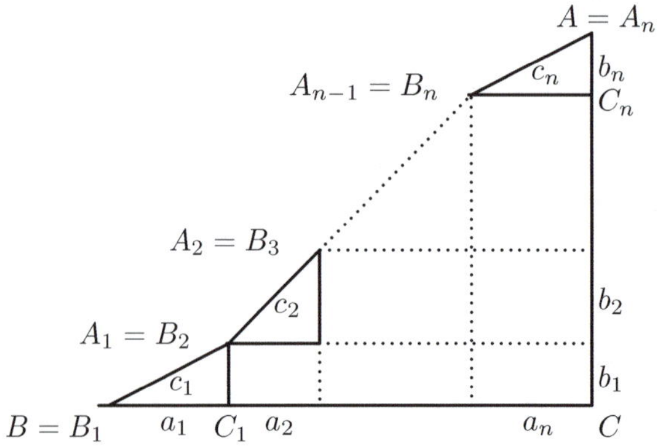

(Compare with Problem 5 of the Junior O-Level Paper.)

5. Suppose there are m silver coins at the beginning and n gold coins at the end. In each of the $m+n$ moves, we either remove a silver coin or add a gold coin. Place the $m+n$ coins in a row so that, for $1 \le k \le m+n$, the coin involved in the kth move is in the kth position from the left. Replace each silver coin by a girl facing left, and each gold coin by a boy facing right. Each girl counts the number of boys in front of her, and each boy counts the number of girls in front of him. For any girl and any boy, either both count the other or neither does. Hence the total count by the girls must be equal to the total count by the boys. The numbers counted by the girls are precisely those recorded on the blackboard and the numbers counted by the boys are precisely those recorded on the whiteboard. Hence the sum of the numbers on the blackboard must be equal to the sum of the numbers on the whiteboard.

Junior A-Level Paper

1. Since half of the entries are plus signs, the table must be $2n \times 2n$ for some positive integer n. Suppose the numbers of plus signs in the $2n$ rows are all different. Then they must be 0, 1, 2, ..., $2n$, with one omitted. Since the total number of plus signs is $2n^2$, the one omitted is n. Then there is a row with $2n$ plus signs and another row with 0 signs. This means

that we do not have a column with $2n$ plus signs, and we do not have a column with 0 plus signs. Hence the numbers of plus signs in the $2n$ columns cannot be all different.

2. Let BC be the longest side of the polygon and let AB and CD be the sides on either side of BC. Then we have $AB + BC > BC \geq CD$ and $BC + CD > BC \geq AB$. Let the incircle be tangent to AB, BC and CD at K, L and M respectively. Then

$$AB + CD > KB + CM = BL + CL + BC.$$

Hence these three sides of the polygon can form a triangle.

3. In the prime factorization of 100!, each of the primes 97 and 89 appears only once. Hence the number of positive divisors of 100! is divisible by $(1+1)(1+1) = 4$. They form an even number of pairs whose product is 100!. If we put half of the pairs in one group and the remaining pairs in the other group, then the two groups are of equal size and the product of the numbers in each group is the same.

4. Consider the motion plan which accomplishes the desired result in which the total distance covered by the patrolmen is minimum. Suppose all of them move. Let m be the number of those who move clockwise and n be the number of those who move counterclockwise. Then we have $m \neq n$ since $m + n = 25$. By symmetry, we may assume that $m < n$. Ask each patrolman who moves clockwise to go one booth farther and each patrolmen who moves counterclockwise to stop one booth earlier. Then the patrolmen's numbers will still be in order. However, the total distance they have covered will be reduced by $n - m$ times the distance between two booths. This contradicts the minimality assumption on the original motion plan.

5. Let P be the centre circle tangent to BC at M and Q be the centre of the circle tangent to BC at N. Let T be the point of tangency of these two circles and let D be the midpoint of MN. Drop perpendiculars from P to CA at E and from Q to AB at F, intersecting each other at R. Then $AERF$ is a square whose side length is equal to the common radii of the two circles. Hence $\angle RAF = \angle ARF = 45°$.

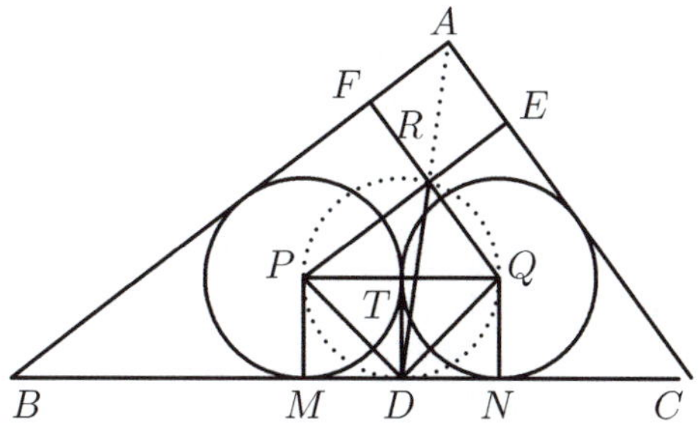

$DMPT$ and $DNQT$ are also squares. Hence $\angle PDQ = 90° = \angle QRP$. Hence $DPQR$ is a cyclic quadrilateral and $\angle DRQ = \frac{1}{2}\angle DTQ = 45°$. It follows that A, R and D are collinear, so that D lies on the bisector of $\angle A$.

6. Denote by d_n the n-digit number in which every digit is d. We prove by induction on n that every positive integer less than or equal to 1_n is the sum of at most n uniform numbers. The result is trivial for $n = 1$. Suppose it holds for some $n \geq 1$. Consider a number $m \leq 1_{n+1}$. Subtract from m the largest uniform number $u \leq m$. If $m = 1_{n+1}$, then $u = m$ and $m - u = 0$. If $9_n \leq m < 1_{n+1}$, then $u = 9_n$ and $m - u \leq 1_n$. If $d_n \leq m < (d+1)_n$ for some d, $0 \leq d \leq 8$, then $u = d_n$ and $m - u \leq 1_n$. In all cases, $m - u \leq 1_n$ and is a sum of at most n uniform numbers by the induction hypothesis. It follows that m is the sum of at most $m + 1$ uniform numbers. Since an n-digit number is less than 1_{n+1}, it is also the sum of at most $n + 1$ uniform numbers.

7. The answers to both parts are affirmative. We may assume that the spider starts from the top left corner. Partition the spiderweb into ten 100×10 vertical strips, which it will comb through one by one from left to right. All vertical moves are along either the first or the last column of a strip, downwards on odd-numbered strips and upwards on even-numbered strips. The total number of vertical moves is $10 \times 99 = 990$. All horizontal moves are within the same strip back and forth betwen the first and the last columns, at the horizontal levels which contains at least one fly. When the spider reaches the bottom row in an odd-numbered strip or the top row in an even-numbered strip, it moves over to the next strip. The number of horizontal moves used for gobbling up flies is at most $9 \times 100 = 900$ since there are 100 flies. The number of horizontal moves used for changing strips is 9. The number of horizontal moves used for getting to the correct column for chaning strips is at most $9 \times 10 = 90$. It follows that the total number of moves required is at most 990+900+9+90 =1989.

Senior A-Level Paper

1. (See Problem 2 of the Junior A-Level Paper.)

2. (See Problem 3 of the Junior A-Level Paper.)

3. Suppose $a_1, a_2, \ldots, a_{100}$ are real numbers for which there exists a real number k such that $(k+a_1)(k+a_2)\cdots(k+a_{100}) = a_1 a_2 \cdots a_{100}$. Treating this as an equation for k, there are at most 100 real roots, one of which is 0. It follows that Gregory can make at most 99 moves. This maximum can be attained if Gregory starts with the numbers from -99 to 0. The initial product is 0, and this value is maintained for the next 99 moves, until he hits 100! on the 100th move.

4. Let the circumcircle of triangle FGD intersect the line AB at U and the line BC at P. Let the circumcircle of triangle DGE intersect the line BC at Q and the line CA at R. Let the circumcircle of triangle EGF intersect the line CA at S and the line AB at T. Since $PDFU$ is cyclic, $BP \cdot BD = BU \cdot BF$. Since $BD = BF$, we have $BP = BU$. Since $QEDR$ is cyclic, $BQ \cdot BD = BG \cdot BE$. Since $TFGE$ is cyclic, $BG \cdot BE = BF \cdot BT$. From $BD = BF$, we have $BQ = BT$. It follows that $PQTU$ has a circumcircle. Similarly, we can show that $PRQU$ and $PSTU$ have circumcircles too. The three circles cannot be distinct as otherwise PQ, RS and TU are their pairwise radical axes and will be concurrent at the radical centre of the three circles. However, these three lines are the sides of ABC, and cannot be concurrent. It follows that two of these three circles must coincide. Then all three will coincide, and P, Q. R, S, T and U all lie on this common circle.

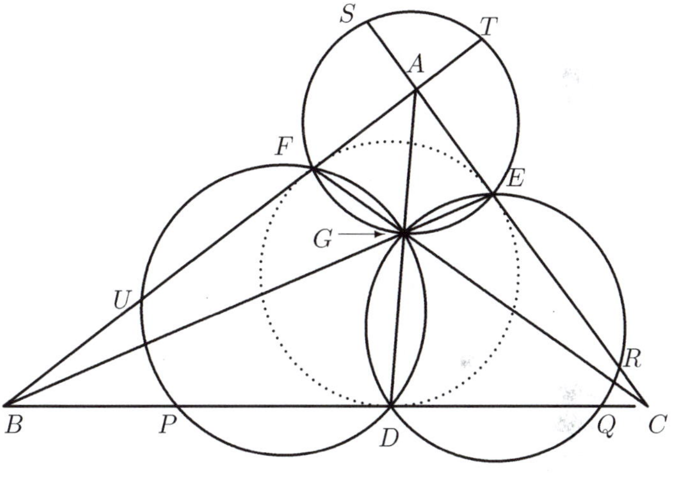

5. For each word Peter writes down, convert every T to TT, every O to OO, every W to OT and every N to TO. Since Peter's word has an equal number of Ts and Os, the new word has an equal number of TTs and OOs. Regardless of the numbers of OTs and TOs, the new word has an equal number of Ts and Os overall, and is therefore on Betty's list. By reversing this conversion process, every word Betty's writes down is on Peter's list. It follows that the two lists have equal length.

6. Let $BD + DC = QR$, $CE + EA = RP$ and $AF + FB = PQ$. We claim that $\frac{BD}{DC} = \frac{CE}{EA} = \frac{AF}{FB}$ if and only if $\angle EAF = \angle RPQ - 1°$, $\angle FBD = \angle PQR - 1°$ and $\angle DCE = \angle QRP - 1°$. We first assume that $\frac{BD}{DC} = \frac{CE}{EA} = \frac{AF}{FB}$. Then triangles BDC, CEA and AFB are similar to one another. Hence $\frac{BD+DC}{BC} = \frac{CE+EA}{CA} = \frac{AF+FB}{AB}$. It follows that $\frac{QR}{BC} = \frac{RP}{CA} = \frac{PQ}{AB}$, so that triangles ABC and PQR are similar.

Then

$$\begin{aligned}\angle EAF &= \angle CAB - \angle CAE - \angle ECA \\ &= \angle RPQ - \angle CAE - \angle ECA \\ &= \angle RPQ - 1^\circ.\end{aligned}$$

Similarly, $\angle FBD = \angle PQR - 1^\circ$ and $\angle DCE = \angle QRP - 1^\circ$. We now prove the converse, assuming that we have $\angle EAF = \angle RPQ - 1^\circ$, $\angle FBD = \angle PQR - 1^\circ$ and $\angle DCE = \angle QRP - 1^\circ$. Fix the points B, C and D. Let M be the fixed point on the same side of BC as D such that $MB = PQ$ and $\angle MBD = \angle PQR - 1^\circ$. Then F lies on the segment MB and we have $FM = BM - BF = PQ - BF = AF$. Moreover, $\angle AFM = 180^\circ - \angle AFB = 1^\circ$. Hence $\angle FMA = \frac{1}{2}(180 - 1)^\circ = 89\frac{1}{2}^\circ$. It follows that A lies on a fixed line through M.

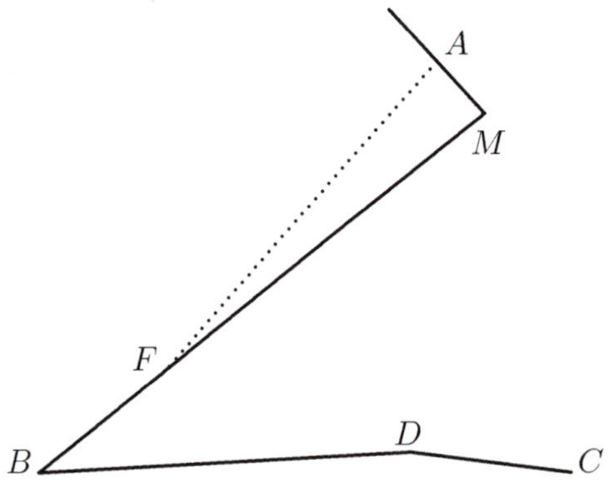

Let the fixed point N be be defined in an analogous way, using E instead of F and interchanging B and C. Then A also lies on a fixed line through N. Hence there is at most one possible position for A, and A can only exist if we indeed have $\frac{BD}{DC} = \frac{CE}{EA} = \frac{AF}{FB}$. This result holds if 181° amd 1° are replaced respectively by $180^\circ + \theta$ and θ, as long as $AFBDCE$ is still a non-convex hexagon. The diagram below illustrates the case $\theta = 10^\circ$, which is much easier to see than if $\theta = 1^\circ$.

7. On a day with given m and n, the virtual value of x grams of gold and y grams of platinum is taken to be $mx + ny$. This is well-defined since it remains constant regardless of any exchange between the precious metals within the day. Consider an amount of precious metal with virtual value 1 on this day. After $m+n-2$ days, both m and n are 1. Without knowing in advance the order in which m and n are reduced during this period, let $f(m, n)$ be the maximum virtual value which can be guaranteed by performing some clever exchanges.

We prove by induction on $m+n$ that $f(m,n) = \frac{m+n-1}{mn}$. Note that $f(1,1) = 1$ since no exchange matters, and indeed $\frac{1+1-1}{1^2} = 1$. For any $m > 1$, we have $f(m,1) = 1$. This is because only m can be reduced, and the virtual value is maximized by converting all platinum into gold. Indeed, $\frac{m+1-1}{m} = 1$. Similarly, $f(1,n) = 1$ for any $n > 1$. Suppose $m > 1$ and $n > 1$. Exchange the precious metals to end up with $\frac{\lambda}{m}$ grams of gold and $\frac{1-\lambda}{n}$ grams of platinum, where λ is some parameter to be determined. Note that we indeed have $m(\frac{\lambda}{m}) + n(\frac{1-\lambda}{n}) = 1$. On the following day, there are two possible scenarios. If m is reduced by 1, the new virtual value is $(m-1)(\frac{\lambda}{m}) + n(\frac{1-\lambda}{n}) = 1 - \frac{\lambda}{m}$. Then we have $f(m,n) = f(m-1,n)(1-\frac{\lambda}{m})$. However, if n is reduced by 1, the new virtual value is $m(\frac{\lambda}{m}) + (n-1)(\frac{1-\lambda}{n}) = 1 - \frac{1-\lambda}{n}$. Then we have $f(m,n) = f(m,n-1)(1-\frac{1-\lambda}{n})$. Since we do not know which scenario will take place, we choose the parameter λ so that

$$f(m-1,n)(1-\frac{\lambda}{m}) = f(m,n-1)(1-\frac{1-\lambda}{n}).$$

By the induction hypothesis,

$$\frac{m+n-2}{(m-1)n}\left(\frac{m-\lambda}{m}\right) = \frac{m+n-2}{m(n-1)}\left(\frac{n-1+\lambda}{n}\right).$$

This simplifies to $\frac{m-\lambda}{m-1} = \frac{n-1+\lambda}{n-1}$, which yields $\lambda = \frac{n-1}{m+n-2}$. It follows that

$$\begin{aligned} f(m,n) &= \frac{m+n-2}{(m-1)n}\left(\frac{m-\frac{n-1}{m+n-2}}{m}\right) \\ &= \frac{m^2+mn-2m-n+1}{mn(m-1} \\ &= \frac{(m-1)^2-n(m-1)}{mn(m-1)} \\ &= \frac{m+n-1}{mn}. \end{aligned}$$

This completes the inductive argument. Initially, we have $m = n = 1001$ and $x = y = 1000$, so that the virtual value is 2002000. Hence the final virtual value which can be guaranteed is

$$2002000f(1001,1001) = 2002000 \times \frac{2001}{1001^2} = \frac{4002000}{1001} < 4000.$$

Since we have $m = n = 1$ now, $x+y < 4000$ so that it is impossible to obtain 4 kilograms of precious metals in any combination.

Spring 2015

Junior O-Level Paper

1. Paint one face red, the opposite face blue and the remaining faces yellow.

2. Let N be the midpoint of AB. Then it is also the midpoint of KL. Now MN is equal half the length of BC, so that $KM = LM = NM$. Hence M is the circumcentre of triangle KLN so that $\angle KML = 90°$.

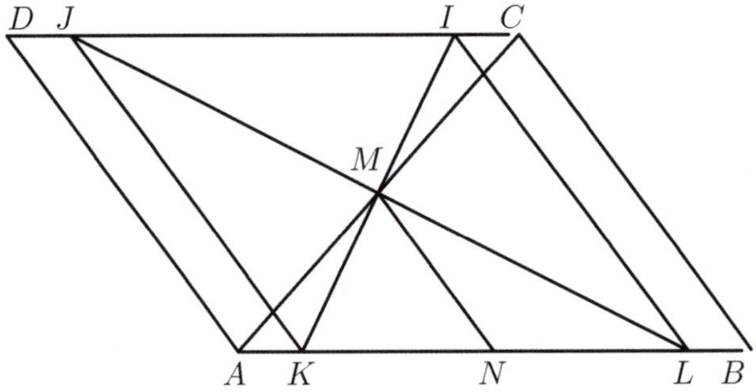

3. Basil's sum is $\frac{n(n+1)}{2}$ for some positive integer n. Suppose Patty starts from 2^k. Then her sum is $2^{k+10} - 2^k = 2^k(2^{10} - 1)$. If $k = 9$ and $n = 1023$, then

$$1 + 2 + \cdots + 1023 = 523776 = 512 + 1024 + \cdots + 262144.$$

4. No two of the 15 unit squares along the diagonal can belong to the same square. The diagram below shows a dissection of the figure into exactly 15 squares.

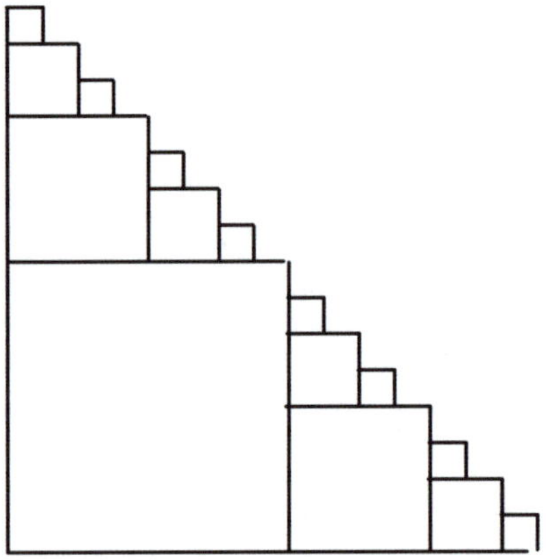

5. The construction is possible for all positive integer n. If $n = 2k$, we take

$$2k-1, 2k-3, \ldots, 1, 2k, 1, 3, \ldots, 2k-1,$$

$$2k-2, 2k-4, \ldots, 2, 0, 2k, 2, 4, \ldots, 2k-2.$$

If $n = 2k+1$, we take

$$2k-1, 2k-3, \ldots, 1, 2k+1, 1, 3, \ldots, 2k-1,$$

$$2k, 2k-2, \ldots, 2, 0, 2k+1, 2, 4, \ldots, 2k.$$

The first few cases are 101, 12102, 1312032, 314132042, 31513420524 and 5316135420624.

Senior O-Level Paper

1. Basil's sum is $\frac{n(n+1)}{2}$ for some positive integer n. Suppose Patty starts from 2^k. Then her sum is $2^{k+100} - 2^k = 2^k(2^{100} - 1)$. If $k = 99$ and $n = 2^{99} - 1$, then $1 + 2 + \cdots + (2^{100} - 1) = 2^{99} + 2^{100} + \cdots + 2^{198}$.

2. The diagram below shows that five disjoint 100×100 squares can be cut out from a 275×275 square. By the Pigeonhole Principle, at least one of them does not contain any of the four points.

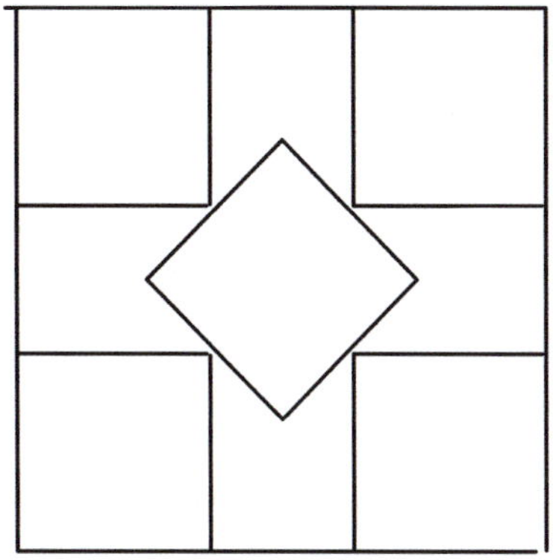

3. (See Problem 5 of the Junior O-Level Paper.)

4. Let Q be the point of intersection of AC with the line LP. Then $\angle ALQ = \angle KLP = \angle BCA$. Hence triangles ALQ and ACM are similar, so that $\frac{LQ}{CM} = \frac{AL}{AC}$. From the similarity between triangles KPL and ABC, we have $\frac{LP}{CB} = \frac{KL}{AC}$. It follows from $AL = 2KL$ and $BC = 2MC$ that $AC(LQ - LP) = AL \cdot CM - KL \cdot BC = 0$. Hence P and Q coincides, so that P indeed lies on AC.

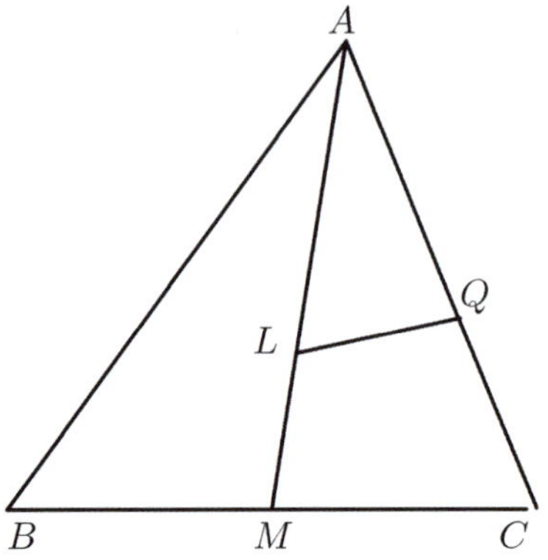

5. Let the first 2012 numbers be alternately 2s and 1s. Let the last three numbers be 2, 4 an 3. Then all conditions are satisfied, and the product of the 2015 numbers is 3×2^{1009}. We claim that this number divides the product of any set of 2015 numbers satisfying all conditions. No two adjacent numbers can both be odd. Hence among the first 2012 numbers, there must be at least 1006 even numbers. Among the last 3 numbers, there must be at least 2 even numbers, at least one of which must be a multiple of 4. Hence the product of the 2015 numbers is divisible by 2^{1009}. We claim that it is also divisible by 3. Suppose each of the 2015 numbers is congruent modulo 3 to 1 or 2. Then these two types of numbers must alternate as otherwise the difference between some adjacent pairs will be divisible by 3. However, alternation is impossible since 2015 is odd. It follows that the maximum value of the number which divides the product of the 2015 numbers is indeed 3×2^{1009}.

Junior A-Level Paper

1. Let K, M and N be the respective midpoints of DE, BC and AE. KN is parallel to DA and hence to CM. Also, $KN = \frac{1}{2}DA = \frac{1}{2}CB = CM$. It follows that $CKNM$ is a parallelogram. Since $CD = CE$, DE is perpendicular to CK, and hence to MN.

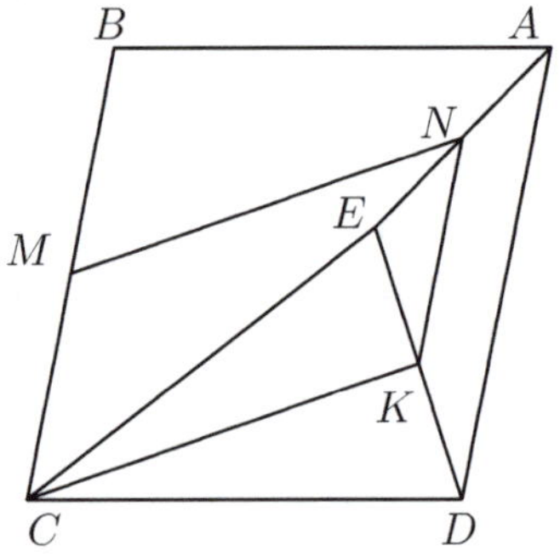

2. The powerline intersects the perimeter of the polygon in an even number of points. Label them from left to right A_1, B_1, A_2, B_2, $\ldots$, A_n, B_n for some positive integer n. By symmetry, we may assume that the segments A_iB_i are inside the polygon for $1 \leq i \leq n$. The diagram below illustrates the case $n = 3$.

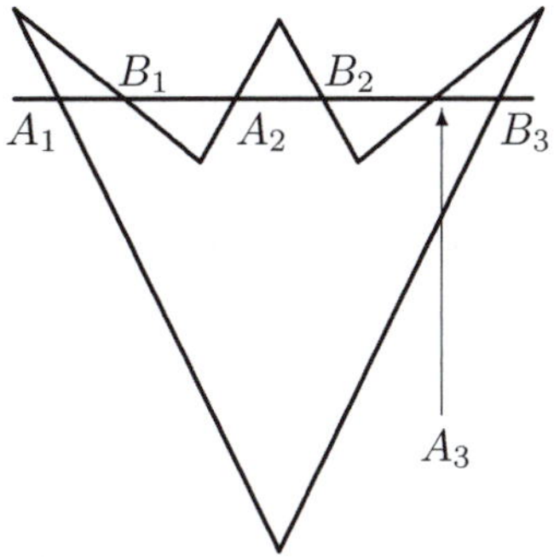

Let the number of poles on the segment A_iB_i be x_i. Let y_0 be the number of poles to the left of A_1, y_i be the number on the segment B_iA_{i+1} be y_i and y_n be the number to the right of B_n. Crossing A_1, the spy counts y_0 poles. Crossing B_1, the spy counts $36 - y_0 - x_1$ poles. The subtotal is $36 - x_1$. In general, the subtotal for A_i and B_i is $36 - x_i$. Let x be the total number of poles inside the polygon. Then $2015 = 36m - x$ for some positive integer m. Now $2015 = 36 \times 56 - 1$. It follows that $x = 1$.

3. (a) Let $x = 99$. Then $x^2 = 9801$ and $x^3 = 970299$. Hence the conclusion is false.

   (b) Let $x = 10k - 1$ where k is a positive integer n. Then

$$\begin{aligned} x^{2015} &= 10^{2015k} - \binom{2015}{1} x^{2014k} + \binom{2015}{2} 10^{2013k} - \cdots \\ &= 10^{2015k} - f(k) \end{aligned}$$

   for some positive integer $f(k)$ dependent on k. If k is sufficient large, $f(k) < 10^{2015k-1}$. Then all of x, x^2, x^3, $\ldots$, x^{2015} begin with the digit 9. Hence the conclusion is also false.

4. (a) It is possible. Such a polygon is shown in the diagram below on the left.

   (b) It is still possible. Such a polygon is shown in the diagram below on the right.

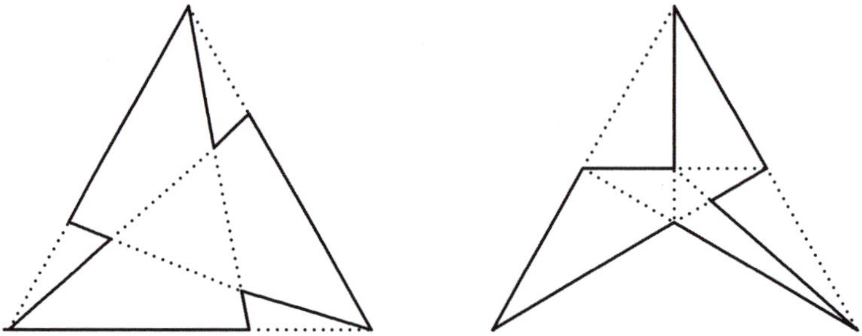

5. (a) If the two row sums are also different initially, there is nothing to be done. Suppose they are not. We permute the columns so that the column sum increases from left to right. We then permute the numbers within the columns so that the larger number is at the bottom. If the two rows sums are different, the task is accomplished. If this is not the case, then every column contains two equal numbers. Since $n > 2$, let the numbers in the first three columns be a, b and c respectively. If we permute them so that a and b are in the first column, a and c are in the second column while b and c are in the third column, the row sums will be different.

1	1	1	1	1	1	1	1	1	0
1	1	1	1	1	1	1	1	0	0
1	1	1	1	1	1	1	0	0	0
1	1	1	1	1	1	0	0	0	0
1	1	1	1	1	0	0	0	0	0
1	1	1	1	0	0	0	0	0	0
1	1	1	0	0	0	0	0	0	0
1	1	0	0	0	0	0	0	0	0
1	0	0	0	0	0	0	0	0	0
0	0	0	0	0	0	0	0	0	0

(b) This is not always possible. Consider the 10×10 table above with 50 copies of 0 and 50 copies of 1. The column sums are distinct, being 0, 1, 2, 3, 4, 6, 7, 8, 9 and 10 from left to right. Suppose the desired permutation exists. Since the sum of all the numbers is 50 and the maximum sum of the numbers in any row is at most 10, the distinct row sums must also be 0, 1, 2, 3, 4, 6, 7, 8, 9 and 10. However, a column sum of 10 cannot coexist with a row sum of 0. This is a contradiction.

6. We may assume that the common side length is 1. Label the sides 1 to n in clockwise order. For the ith side, let the new segment on the left have length a_i and the new segment on the right have length b_n. By the Power of a Point Theorem, $b_i(1+a_i) = a_{i+1}(1+b_{i+1})$ for $1 \leq i \leq n-1$, and $b_n(1+b_n) = a_1(1+b_1)$. Summing these n equations and cancelling common terms, we have $b_1 + b_2 + \cdots + b_n = a_1 + a_2 + \cdots + a_n$. If we paint all the segments on the left blue and all the segments on the right red, we have the desired result.

7. In the first round, the Emperor chooses a wizard W and asks every other wizard whether W is good. Suppose everyone says no. The Emperor expels W. If W is evil, the Emperor proceeds to the next round with one less wizard to worry about. If W is good, then one good wizard has been expelled. However, this means that every other wizard is evil. In the subsequent rounds, the Emperor expels all of them one at a time. Suppose wizard X says yes in the first round. The Emperor expels X. If X is evil, the Emperor proceeds to the next round with one less wizard to worry about. If X is good, then one good wizard has been expelled. However, this means that W is good. In the next round, the Emperor asks W whether wizard A is good. If the answer is no, then A is indeed evil. The Emperor expels A and proceeds to the next round with one less wizard to worry about, and still knowing that W is good. If the answer is yes, then A is also good. The Emperor then asks A whether wizard B is good, and so on. If at least one answer is no, there will be an expulsion. If all the answers are yes, then no further expulsion is needed.

Senior A-Level Paper

1. (See Problem 3 of the Junior A-Level Paper.)

2. Since triangles PBX and ABC are similar, $PB = PX$. Since triangles QPX and QPY are congruent, $PX = PY$. Hence $PB = PY$ so that $\angle PBX = \angle PBY$. Now $\angle QYA = \angle BAY$ and $\angle PYQ = \angle PBQ = \angle CAB$.

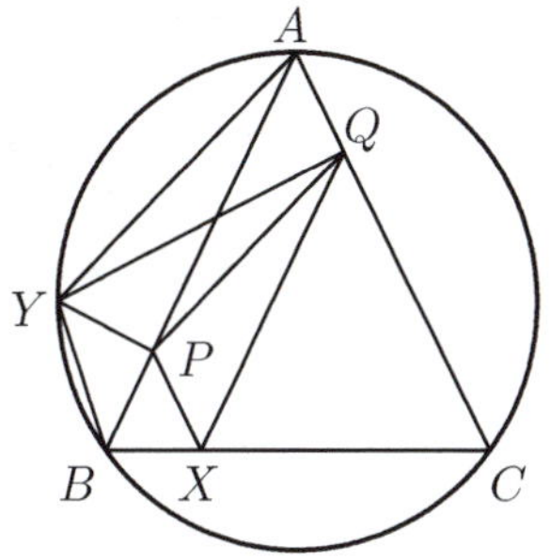

Hence $2(\angle ABY + \angle BAY) + \angle PYQ = 180° = \angle CAB + 2\angle BCA$, so that $\angle ABY + \angle BAY = \angle BCA$. Thus

$$\angle CAY + \angle CBY = \angle CAB + \angle ABC + \angle BCA = 180°.$$

It follows that $ACBY$ is a cyclic quadrilateral.

3. (See Problem 5 of the Junior A-Level Paper.)

4. (See Problem 6 of the Junior A-Level Paper.)

5. We call a polynomial *nice* if each of its coefficients is either 0 or 1. Observe that if $P(x)$ is a nice polynomial of degree n then $(x^m+1)P(x)$ with $m > n$ is also a nice polynomial. If we start from the polynomial $x+1$ and multiply it recursively by 2019 polynomials (x^m+1) with increasing odd values of m, we arrive at a nice polynomial $f(x)$ divisible by $g(x) = (x+1)^{2020}$ since (x^m+1) with odd m is divisible by $(x+1)$. Observe that $g(x) = x^{2020} + 2020x^{2019} + \cdots$. Let $f(x) = g(x)h(x)$ with $h(x) = x^k + ax^{k-1} + \cdots$. Since the second coefficient of $f(x)$, which is either 0 or 1, is equal to $2020 + a$, we conclude that $a \leq -2019$.

6. (See Problem 7 of the Junior A-Level Paper.)

7. We construct a counterexample. Let $O(0,0,0)$ be the origin of the three dimensional space. Take the following eight points as the vertices of our hexahedron.

$$\begin{array}{cccc} A(4,1,2) & B(4,-1,2) & C(-4,-1,2) & D(-4,1,2) \\ A'(1,4,-2) & B'(1,-4,-2) & C'(-1,-4,-2) & D'(-1,4,-2) \end{array}$$

The distance from O of each vertex is $\sqrt{4^2+1^2+4} = \sqrt{21}$. It follows that the hexahedron $ABCDA'B'C'D'$ has a circumsphere with centre O and radius $\sqrt{21}$. Both bases of the hexahedron are 8×2 rectangles. All four lateral faces are congruent isosceles trapezoids. Consider one of them, say $ABA'B'$. Let M and M' be the respective midpoints of AB and $A'B'$. M has coordinates $(4,0,2)$ while M' has coordinates $(1,0,-2)$. In the plane $y=0$, the equation of the line MM' is $\frac{z-2}{x-4} = \frac{4}{3}$. The equation of the line through O perpendicular to MM' is $\frac{z}{x} = -\frac{3}{4}$. The two lines intersect each other at the point N whose coordinates are $(1.6, 0, -1.2)$. Since $ON = \sqrt{1.6^2 + 0^2 + 1.2^2} = 2$, the hexahedron has an insphere with centre O and radius 2.

Tournament 37

Fall 2015

Junior O-Level Paper

1. Is it true that every positive integer can be multiplied by one of 1, 2, 3, 4 or 5 so that the resulting number starts with 1?

2. A rectangle is dissected into congruent scalene right triangles. Must the dissection contain a rectangle consisting of two such triangles?

3. Three players play the game "rock-paper-scissors". In every round, each player simultaneously makes one of these signs. Rock beats scissors, scissors beat paper, while paper beats rock. In a round in which exactly two distinct signs are made, meaning that one of them is made twice, one point is awarded to the one player or each of the two players who makes the winning sign. In all other rounds, no points are awarded. After several rounds, the total number each sign has been made is the same. Prove that the total number of points awarded up to this moment is a multiple of 3.

4. In ABC, $\angle C = 90°$. K, L and M are points on the sides CA, BC and AB respectively such that $AK = BL$, $KM = LM$ and $\angle KML = 90°$. Prove that $AK = KM$.

5. In a country there are 100 cities. Every two cities are connected by direct flight in both directions, both costing the same amount. We wish to visit all the other 99 cities and then return to our home city, such that the average cost per flight on our trip is not greater than the average cost of all flights.

   (a) Is it always possible to do so?

   (b) Is it always possible to do so if we leave out one of the other 99 cities?

Note: The problems are worth 4, 4, 5, 5 and 3+3 points respectively.

Senior O-Level Paper

1. Let p be a prime number. Determine the number of positive integers n such that pn is a multiple of $p + n$.

2. C is the midpoint of a semicircle with diameter AB and D is another point on the semicircle. The bisector of $\angle ADB$ intersects AB at K. Prove that the circumcentre of triangle ACK lies on the line AD.

3. Three players play the game "rock-paper-scissors". In every round, each player simultaneously makes one of these signs. Rock beats scissors, scissors beat paper, while paper beats rock. In a round in which exactly two distinct signs are made, meaning that one of them is made twice, one point is awarded to the one player or each of the two players who makes the winning sign. In all other rounds, no points are awarded. After several rounds, the total number each sign has been made is the same. Prove that the total number of points awarded up to this moment is a multiple of 3.

4. In a country there are 100 cities. Every two cities are connected by direct flight in both directions, both costing the same amount. We wish to visit all the other 99 cities and then return to our home city, such that the average cost per flight on our trip is not greater than the average cost of all flights.

   (a) Is it always possible to do so?

   (b) Is it always possible to do so if we leave out one of the other 99 cities?

5. An infinite increasing arithmetic progression is partitioned into blocks of consecutive terms. Can the sums of the blocks, taken in order, form an infinite geometric progression?

Note: The problems are worth 3, 4, 4, 2+2 and 5 points respectively.

Junior A-Level Paper

1. A polyomino is said to be *amazing* if it is not a rectangle and several copies of it can be assembled into a larger copy of it. The diagram below shows that the V-tromino is amazing.

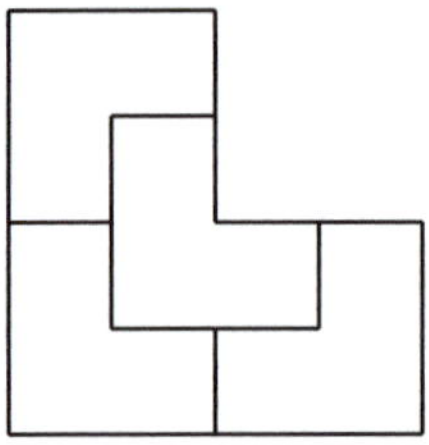

   (a) Does there exist an amazing tetromino?

   (b) Determine all $n > 4$ such that there exists an amazing n-omino.

2. From $\{1,2,3,\ldots,100\}$, k integers are removed. Among the numbers remaining, do there always exist k distinct integers with sum 100 if

   (a) $k = 9$;

(b) $k = 8$?

3. Prove that the sum of the lengths of any two medians in an arbitrary triangle is
   (a) not greater than $\frac{3}{4}$ of the perimeter of the triangle;
   (b) not less than $\frac{3}{8}$ of the perimeter of the triangle.

4. A 9×9 grid square is made of 180 matches. Pete and Basil alternately remove one match at a time, Peter going first. A player wins if after his move, no 1×1 square remains. Which of the two players has a winning strategy?

5. In triangle ABC, medians AD, BE and CF intersect at point G. Let P, Q, R and S be the circumcentres of triangles GDE, GCE, GDF and GBF respectively. Prove P, Q, R, S and G are concyclic.

6. Several distinct real numbers are written on a blackboard. Peter wants to make a single expression with values that are exactly these numbers. To make such an expression, he may use any real numbers, brackets, and usual signs $+$, $-$ and $\times$. He may also use a special sign $\pm$. When computing the values of the resulting expression, Peter uses every combination of $+$ or $-$ in place of all $\pm$. For instance, the expression 5 ± 1 results in $\{4, 6\}$, and $(2 \pm 0.5) \pm 0.5$ results in $\{1, 2, 3\}$. Can Peter construct such an expression

   (a) if the numbers on the blackboard are 1, 2, 4;
   (b) for any collection of 100 distinct real numbers on a blackboard?

7. Santa Claus has n sorts of candies, k candies of each sort. He distributes them at random among k gift bags, n candies in each, and gives a bag to each of k children. The children know what they have in the bags and decide to trade. Two children can trade one candy for one candy if each of them gets a candy of the sort that he or she lacks. Is it true that a sequence of trades can always be arranged so that in the end every child has candies of each sort?

Note: The problems are worth 2+3, 2+4, 3+5, 8, 8, 3+7 and 10 points respectively.

Senior A-Level Paper

1. A geometrical progression consists of 37 positive integers. The first and the last terms are relatively prime numbers. Prove that the 19th term of the progression is the 18th power of an integer.

2. A 10×10 board is partitioned into 20 pentominoes by 80 unit segments lying on common sides of adjacent squares. What is the maximum number of different kinds of pentominoes among these 20?

3. Each coefficient of a non-constant polynomial is an integer of absolute value not exceeding 2015. Prove that every positive root of this polynomial is greater than $\frac{1}{2016}$.

4. The extensions of opposite sides of a cyclic quadrilateral intersect at points P and Q. M and N are the midpoints of the diagonals. Prove that $\angle PMQ + \angle PNQ = 180°$.

5. Several distinct real numbers are written on a blackboard. Peter wants to make a single expression with values that are exactly these numbers. To make such an expression, he may use any real numbers, brackets, and usual signs $+$, $-$ and $\times$. He may also use a special sign $\pm$. When computing the values of the resulting expression, Peter uses every combination of $+$ or $-$ in place of all $\pm$. For instance, the expression 5 ± 1 results in $\{4, 6\}$, and $(2 \pm 0.5) \pm 0.5$ results in $\{1, 2, 3\}$. Can Peter construct such an expression

   (a) if the numbers on the blackboard are 1, 2, 4;

   (b) for any collection of 100 distinct real numbers on a blackboard?

6. Basil makes three pairwise perpendicular cuts into a sphere with a 20 cm. diameter. Each cut is of depth h cm, producing a circular segment with height h in the plane of the cut. Does it necessarily follow that the sphere is divided into two or more pieces if

   (a) $h = 17$;

   (b) $h = 18$?

7. In a line are n children of different heights. In each move, the line is first split into the least possible number of groups so that the children within each group are in ascending order of height. It is possible that a group may consist of a single child. Then the children within each group are rearranged so that they are now in descending order of height. Prove that after at most $n - 1$ moves, all children in the line are in descending order of height.

Note: The problems are worth 3, 6, 6, 7, 2+6, 6+6 and 12 points respectively.

Spring 2016

Junior O-Level Paper

1. There is at least one boy and at least one girl among twenty children in a circle. None of them is wearing more than one T-shirt. For each boy, the next child in the clockwise direction is wearing a blue T-shirt. For each girl, the next child in the counterclockwise direction is wearing a red T-shirt. Is it possible to determine the exact number of boys in the circle?

2. H is the orthocentre of triangle ABC with $\angle BCA = 60°$. The circle with centre H and passing through C cuts the lines CA and CB at M and N respectively. Prove that AN is parallel to BM or they coincide.

3. Do there exist 2016 integers whose sum and product are both 2016?

4. In a 10×10 board, the 25 squares in the upper left 5×5 subboard are black while all remaining squares are white. The board is dissected into a number of connected pieces of various shapes and sizes such that the number of white squares in each piece is three times the number of black squares in that piece. What is the maximum number of pieces?

5. A median, an angle bisector and an altitude are drawn from some combination of the vertices of some triangle. Is it possible for these three lines to enclose an equilateral triangle?

Note: The problems are worth 3, 4, 5, 5 and 5 points respectively.

Senior O-Level Paper

1. A point inside a convex quadrilateral is joined to all four vertices and to one point on each side, dividing the quadrilateral into eight triangles. If the circumradii of all eight triangles are equal, prove that the original quadrilateral is cyclic.

2. Do there exist 2016 integers whose sum and product are both 2016?

3. In a 10×10 board, the 25 squares in the upper left 5×5 subboard are black while all remaining squares are white. The board is divided into a number of connected pieces of various shapes and sizes such that the number of white squares in each piece is three times the number of black squares in that piece. What is the maximum number of pieces?

4. There are 100 positive real numbers each with at most two digits after the decimal point. In each move, two of the numbers are replaced by their sum, and any digit after the decimal point in the sum is erased. After 99 moves, only one number remains. How many different values are possible for this final number?

5. Must a sphere pass through the midpoints of all 12 edges of a cube if it passes through at least

   (a) 6 of them;

   (b) 7 of them?

Note: The problems are worth 4, 4, 4, 6 and 3+3 points respectively.

Junior A-Level Paper

1. The integers from 1 to 1000000 inclusive are written in an arbitrary order on a tape without spacing. The tape is then cut into pieces, each consisting of two digits. Prove that all two-digit numbers appear on these pieces regardless of the initial order of the integers.

2. Do there exist integers a and b such that the equation $x^2 + kax + b = 0$ has no real roots, and the equation $\lfloor x^2 \rfloor + kax + b = 0$ has at least one real root, where

   (a) $k = 1$;

   (b) $k = 2$?

3. Dissect a 10×10 square into 100 congruent quadrilaterals which have circumcircles of diameter $\sqrt{3}$.

4. Each face of a cube is divided into 25 squares and each square is painted in one of red, white or black. Two squares sharing an edge, whether they lie on the same face or not, must have different colours. What is the minimum number of black squares?

5. Let k be a positive integer and $p > 10^k$ be a prime number. Pete inserts a k-digit number between two adjacent digits of a multiple of p, and obtains a new multiple of p. Then he inserts another k-digit number between two adjacent digits of the first insert and obtains again another multiple of p. Prove that the digits of the two inserts are permutations of each other.

6. A circle is rolling without slipping inside another circle twice its radius. The two circles are tangent to each other at all times. Prove that every point on the small circle moves in a straight line.

7. There are m good batteries and $n > 2$ bad batteries. They are not distinguishable until used to light an electrical torch, the proper functioning of which requires two good batteries. What is the minimum number of attempts in order for the torch to function properly, if

   (a) $m = n + 1$;

   (b) $m = n$?

Note: The problems are worth 4, 2+3, 6, 8, 8, 9 and 5+5 points respectively.

Senior A-Level Paper

1. The integers from 1 to 1000000 inclusive are written in an arbitrary order on a tape without spacing. The tape is then cut into pieces, each consisting two digits. Prove that all two-digit numbers appear on these pieces regardless of the initial order of the integers.

2. Dissect a 10×10 square into 100 congruent quadrilaterals which have circumcircles of diameter $\sqrt{3}$.

3. In triangle ABC, $AB = AC$. D is the midpoint of BC. E and F are points on CA and AB respectively, such that $\angle CDE = \angle DFE = \theta$ and $CE \neq BF$. Determine $\angle BFD$ in terms of θ.

4. Some pairs of 64 towns are linked by telegraph service, and only a commissioner knows these connections. In each question, we name two towns and the commissioner will tell us if they are so connected. We wish to know if it is possible to pass a message by telegraph from any town to any other town. The message may be relayed through other towns. Prove that we may not know if we ask the commissioner less than 2016 questions.

5. On the blackboard are several monic polynomials of degree 37, with non-negative coefficients. In each move, we may replace two of them by two other monic polynomials of degree 37, such that either the sum or the product of new pair is equal to the sum or the product, respectively, of the old pair. Coefficients are not required to be non-negative. Prove that after any finite number of moves, at least one polynomial does not have 37 distinct positive roots.

6. We have an unlimited supply of red cards with *abc*, *bca* and *cab* printed on them. We also have an unlimited supply of blue cards with *acb*, *bac* and *cba* printed on them. We want to build a "word" which reads the same in either direction, starting with a card *abc*. In each move, we may add a card to either end of the existing word, or insert it between two letters of the existing word.

(a) Is it possible to accomplish the task using only red cards?

(b) If it is possible to accomplish the task using red and blue cards, must the number of red cards used, including the initial one, be equal to the number of blue cards used?

7. On a spherical planet are n great circles, each of length 1, which serve as railways. On each railway, several trains run continuously at the same positive constant speed. The trains are great arcs of the sphere but without their endpoints. If the trains never stop and never collide, what is the maximum total length of the trains, where

   (a) $n = 3$;

   (b) $n = 4$?

Note: The problems are worth 4, 5, 6, 8, 8, 4+6 and 4+6 points respectively.

Solutions

Fall 2015

Junior O-Level Paper

1. If the first digit of the given number is 1, multiplying by 1 will do. If it is 2 or 3, we can multiply by 5. If it is 4, we can multiply by 4. If it is 5, 6, 7, 8 or 9, we can multiply by 2. It is easy to check that in each case, the first digit of the product cannot reach 2.

2. The diagram below shows such a dissection of a $3 \times \sqrt{3}$ rectangle which does not contain a forbidden rectangle.

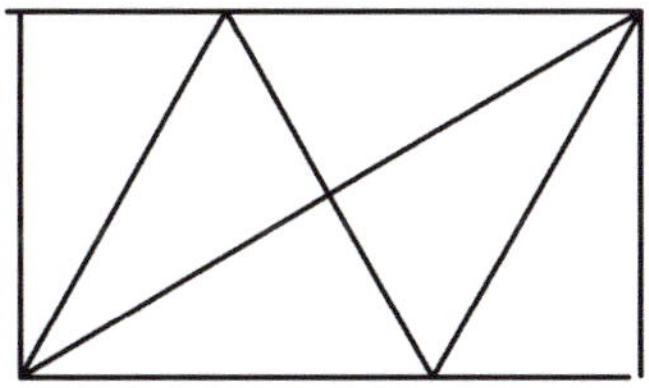

3. Let R, S and P denote rock, scissor and paper respectively. We use mathematical induction on the number n of rounds. For $n = 1$, since we have an equal number of Rs, Ss and Ps, this must be PRS. So 0 points are awarded, and 0 is a multiple of 3. Suppose the result holds for some $n \geq 1$. Consider now $n+1$ rounds. If 0 points are awarded every round, there is nothing to prove. If one of the rounds is PRS, we can discard it and apply the inductive hypothesis. Hence we may assume that there are no PRS rounds, and there is at least one round in which 1 or 2 points are awarded. Suppose this round is RRS. Since we have an equal number of Rs, Ss and Ps overall, there must be another round with at least one P. This may be PPP, PPR, PPS, PRR or PSS. We switch a P in this round with one of the Rs in RRS. The chart below summarizes the changes.

Original Round	Points Awarded	New Round	Points Awarded	Net Change
PPP	0	PPR	2	2
PPR	2	PRR	1	-1
PRR	1	RRR	0	-1
PPS	1	PRS	0	-1
PSS	2	RSS	1	-1

Whether the net change is 2 or -1, the total numbers of points awarded remains a multiple of 3. This is because 2 points are awarded in the chosen round RRS, and it has been changed to PRS and 0 points are now awarded. Moreover, we can now discard it and apply the induction hypothesis. The situation is similar if the chosen round is PPR, PPS, PRR, PSS or RSS. This completes the inductive argument.

4. Since $\angle BCA = 90°$, $\angle KAM + \angle LBM = 90°$. Since $\angle KML = 90°$, $\angle KMA + \angle LMB = 90°$. If $AK > KM$, then $BL > LM$, so that $\angle KAM < \angle KMA$ and $\angle LBM < \angle LMB$. This is a contradiction. If $AK < KM$, we have an analogous contradiction. It follows that we must have $AK = KM$.

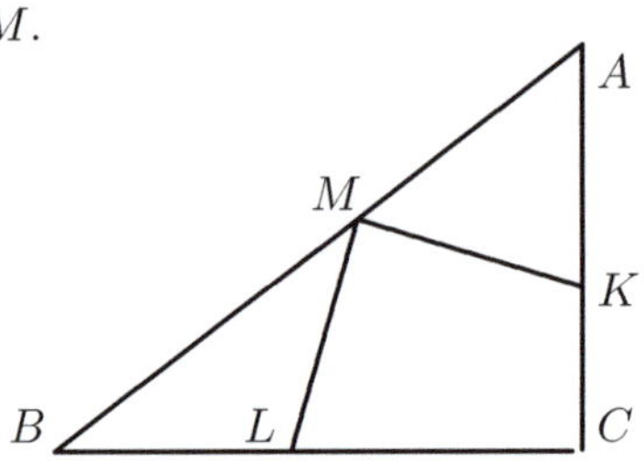

5. (a) It is always possible to do so. Consider all possible round trips of the 100 cities. If we take the one with the lowest total cost, the average cost per flight on our trip cannot exceed the average cost of all flights.

   (b) It is not always possible to do so. Let the cost of each flight from our home city be 5.9 units and let the cost of any other flight be 0.9 units. Then the average cost of all the flights is $\dfrac{5.9(99) + 0.9\binom{99}{2}}{\binom{100}{2}} = 1$ unit while the average cost per flight on our trip is $\frac{5.9\times2+0.9\times97}{99} > 1$ unit.

Senior O-Level Paper

1. Suppose we have $pn = k(p+n)$ for some positive integer k. Then $p^2 = p^2 + pn - kp - kn = (p+n)(p-k)$. Since $p+n > p$ and p is prime, we must have $p+n = p^2$ so that $n = p(p-1)$ is the unique positive integer with the desired properties.

2. We first consider the case where D lies on the minor arc BC, as shown in the diagram below on the left. Let the lines AD and BC intersect L. Let the lines AC and BD intersect at M. Let the lines LM and AB intersect at H. Now AD is perpendicular to BM and BC is perpendicular to AM. Hence L is the orthocentre of triangle AMB, so that MH is perpendicular to AB. The circle with diameter AL certainly passes through C and H.

Note that $BDLH$ is cyclic since $\angle BDL = 90° = \angle LHB$. Hence $\angle DBL = \angle DHL$ so that

$$\begin{aligned}\angle BDH &= 180° - \angle LBH - (\angle DBL + \angle DHB) \\ &= 180° - 45° - 90° \\ &= 45°.\end{aligned}$$

It follows that DH bisects $\angle ADB$, so that H coincides with K. The desired result follows.

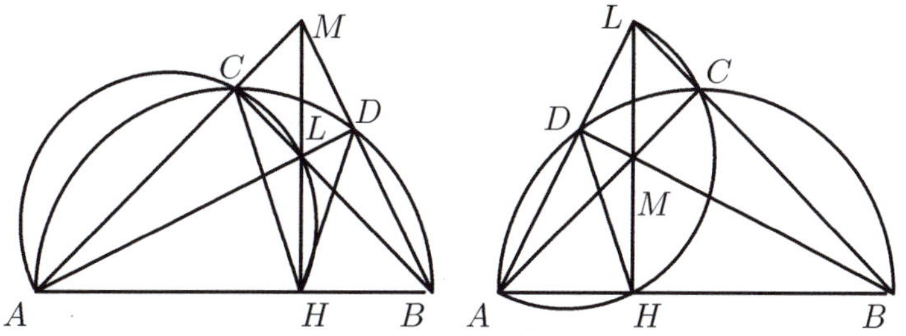

The argument is essentially the same for the case where D lies on the minor arc AC, as shown in the diagram above on the right.

3. (See Problem 3 of the Junior O-Level Paper.)
4. (a) It is always possible to do so. Put our home city at the centre of a regular 99-gon and the other cities at the vertices. The flights from our home city to the others are represented by radial lines, while the other flights are represented by the sides and diagonals of the 99-gon. Each side or diagonal is perpendicular to exactly one radial line. Divide the flights into 99 groups each consisting of a radial line, the side perpendicular to it and all diagonals perpendicular to it. Note that the flights represented by each group visits each city exactly once. Hence our trip can be made by using alternately the flights in two groups. If we arrange the groups in ascending order of total cost and take the first two, the average cost per flight on our trip cannot exceed the average cost of all flights.

   (b) It is not always possible to do so. Let the cost of each flight from our home city be x and let the cost of any other flight be y. Then the total cost of all flights is $99x + \binom{99}{2}y$ and the total cost of our trip is $2x + 97y$. We want $\dfrac{2x+97y}{99} \le \dfrac{99x+\binom{99}{2}y}{\binom{100}{2}}$, which simplifies to $x \le y$. Hence if we choose $x > y$, then the desired trip is impossible.
5. (see Problem 5 of the Junior O-Level Paper.)

Junior A-Level Paper

1. (a) The diagram below shows that the L-tetromino is amazing.

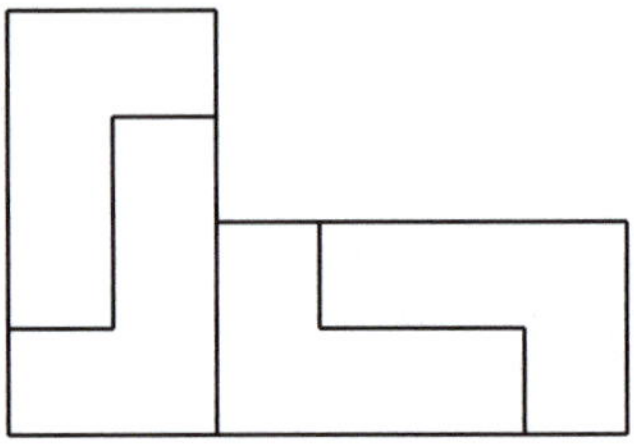

   (b) Amazing n-ominoes exist for all $n > 4$. Two copies of the L-shaped n-omino can be assembled into a $2 \times n$ rectangle, and $2n$ copies of this rectangle can be assembled into a $2n \times 2n$ square. Finally, n copies of this square can be assembled into a larger copy of the n-ominno.

2. (a) The answer is no. If we remove the numbers 1 to 9 inclusive, then the sum of any nine of the remaining numbers is at least $10 + 11 + \cdots + 18 = 126$.

   (b) The answer is yes. The sum of any four of the twelve pairs (1,24), (2,23), ..., (12,13) is 100. Removing 8 numbers affects at most eight of the pairs.

3. (a) Let ABC be a triangle in which BE and CF are the longest two medians. Extend BE to H and CF to K so that $EH = BE$ and $FK = CF$. Applying the Triangle Inequality to BAH and CAK, we have

$$\begin{aligned} AB + BC &= AB + AH > BH = 2BE, \\ AB + BC &= AC + AK > CK = 2CF. \end{aligned}$$

Addition yields

$$\begin{aligned} BE + CF &< \frac{1}{2}(AB + AC + 2BC) \\ &= \frac{1}{2}(AB + AC + BC) + \frac{1}{4}BC + \frac{1}{4}BC \\ &< \frac{1}{2}(AB + AC + BC) + \frac{1}{4}(AB + AC) + \frac{1}{4}BC \\ &= \frac{3}{4}(AB + AC + BC). \end{aligned}$$

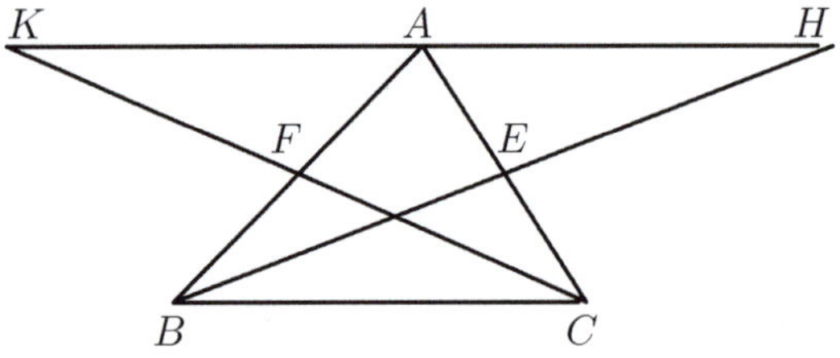

(b) Let ABC be a triangle in which BE and CF are the shortest two medians, intersecting at the centroid G. Extend BE to H and CF to K so that $EH = GE$ and $FK = GF$. Applying the Triangle Inequality to GBC, BAH and CAK, we have $BG + CG > BC$, $2BG+CG = BH+AH > AB$ and $2CG+BG = CK+AK > AC$. Addition yields $AB+AC+BC < 4(BG+CG) = 4\times\frac{2}{3}(BE+CF)$. It follows that we have $BE + CF > \frac{3}{8}(AB + AC + BC)$.

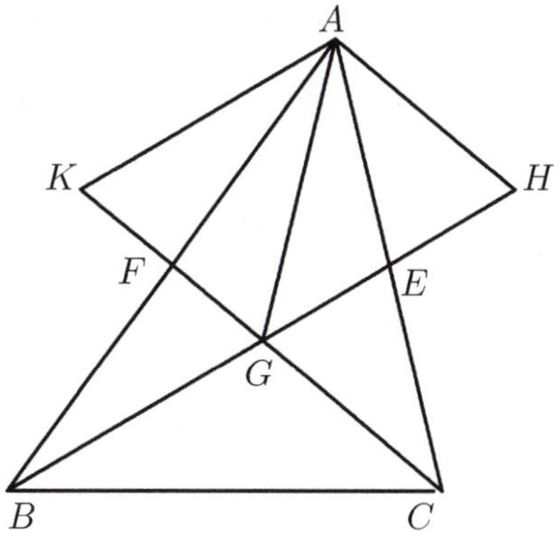

4. Basil has a sure win. He plays randomly until at the point when it is his turn, there are either 2 or 3 intact squares left. This must happen since the removal of any match destroys at most two intact squares. Since the game starts with an even number of matches, the number of remaining matches must be odd. The position must be equivalent to one of the following five cases.

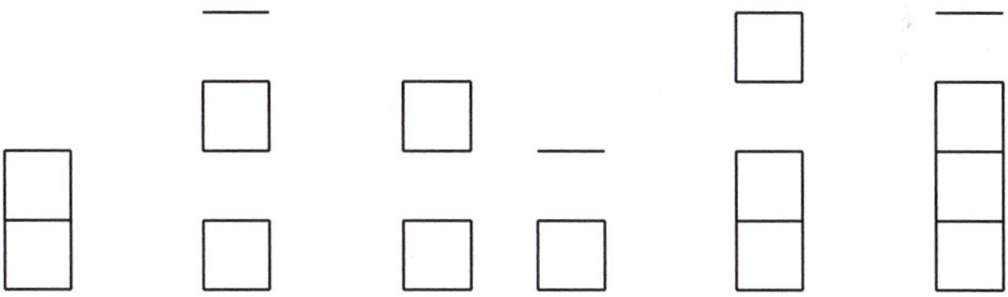

We define a junk match as a match which is no longer a side of an intact square. The actual number of junk matches is irrelevant. What is important is the parity. This is because whenever Peter takes a junk match, Basil will do likewise, maintaining the parity, so that at most one junk match needs to be considered. In the first case, Basil destroys both squares simultaneously. In the second case, Basil takes the junk match. Peter must destroy one of the squares, and Basil destroys the other. In the third case, Basil takes the junk match. Peter must destroy a square, creating three junk matches which can be considered as one, so that we are back to the second case.

In the fourth case, Basil destroys one of the two adjacent squares. Disregarding the two junk matches, we have the same situation as in the second case. In the fifth case, Basil takes the junk match. If Peter destroys two squares simultaneously, Basil destroys the remaining square. If Peter destroys either end square, Basil destroys the remaining two squares simultaneously. If Peter destroys only the middle square, he will create a junk match for Basil to take and we have the same situation as in the second case.

5. PR is the perpendicular bisector of GD, intersecting it at its midpoint X. PQ is the perpendicular bisector of GE, intersecting it at its midpoint Y. It follows that $GXPY$ is a cyclic quadrilateral, so that $\angle QPR + \angle EGD = 180°$. Since FD is parallel to EC, $\angle GFD = \angle GCE$. Since R and Q are the respective circumcentres of GFD and GCE, $\angle GRD = \angle GQE$. Hence $\angle DGR = \angle EGQ$ so that $QGR = \angle EGD$. It follows that $\angle QPR + \angle QGR = 180°$, so that Q lies on the circumcircle of GRP. Similarly, we can prove that S also lies on the circumcircle of GRP, so that G, P, Q, R and S are conncyclic.

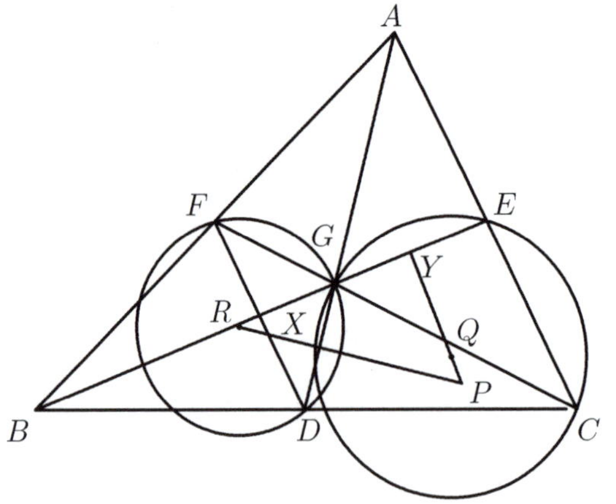

6. (a) We take $(1.5 \pm 0.5) \times (1.5 \pm 0.5)$. If we take two + signs, the value is 4. If we take two − signs, the value is 1. If we take one of each, the value is 2.

   (b) We prove by induction on n that this works for any n real number. The base $n = 1$ is trivial. Suppose the result holds for some $n \geq 1$. Consider $n + 1$ real numbers $a_1, a_2, \ldots, a_{n+1}$. By the induction hypothsis, there exists an exprression $E(a_1, a_2, \ldots, a_n)$ which yields all values and only values among $a_1, a_2, \ldots, a_n$. Consider the expression $a_{n+1} + (0.5 \pm 0.5)E(a_1 - a_{n+1}, a_2 - a_{n+1}, \ldots, a_n - a_{n+1})$. If we take − for ± in front of E, the value is a_{n+1}. If we take + instead, the expression yields all values and only values among $a_1, a_2, \ldots, a_n$. This completes the inductive argument.

7. Construct a $k \times n$ matrix where the rows represent the children and the columns represent the types of candies. The entry in the ith row and the jth column is the current number of candies of the jth kind that the ith child has. The sum of the numbers in each row is n and the sum of the numbers in each column is k. We desire a matrix consisting of kn 1s. The measure of a matrix is the number of 1s in it. The index of a row is the number of 0s in it. A row with index 0 consists of n 1s. Permute the rows so that their indices are in increasing order. If there are no rows with positive indices, then the measure of the matrix is at its maximum value kn and the desired matrix has been obtained. Suppose the measure of the matrix is less than kn, so that there is at least one row with a positive index. Let the row with the minimum positive index correspond to, say, the child Peter. We may assume that he does not have any candies of the 1st kind. Then some other child, say Betty, must have at least 2 candies of the 1st kind. Since the row corresponding to her has an index no smaller than that of Peter's row, there must be a 0 in her row such that the number in the same column in Peter's row is not 0. We may assume that this is the 2nd column. Now a trade can be arranged, with Betty taking a candy of the 2nd kind from Peter while giving him a candy of the 1st kind. We have created a new 1 in each of these two rows. Even if the original number in the 2nd column in Peter's row is also 1, the measure of the matrix has still increased. Since the measure cannot increase without bound, the desired matrix will be obtained eventually.

Senior A-Level Paper

1. Let the terms of the geometric progression be a_k for $1 \leq k \leq 37$. Then the common ratio is a positive rational number $\frac{p}{q}$ where p and q are relatively prime positive integers. Now $a_{37} = a_1(\frac{p}{q})^{36}$ is an integer. It follows that $a_1 q^{-36}$ is some positive integer b. Hence $a_1 = bq^{36}$ while $a_{37} = bp^{36}$. Since a_1 and a_{37} are relatively prime, we must have $b = 1$. Indeed, we have $a_{19} = a_1(\frac{p}{q})^{18} = (pq)^{18}$.

2. Each segment on the perimeter of the board is on the perimeter of exactly one of the pentominoes. Each partitioning segment is on the perimeter of exactly two of them. Hence the perimeter of each pentomino is given by $(40 + 2 \times 80) \div 20 = 10$. Of the twelve pentominoes, the P-pentomino is the only one with perimeter 10. Hence all 20 pentominoes are copies of the P-pentomino. Such a partition exists because two copies of the P-pentomino can form a 2×5 rectangle.

3. Let the polynomial be $P(x) = a_0x^n + a_1x^{n-1} + \cdots + a_{n-}x + a_n$ with $n \geq 1$. We may assume that $a_n \neq 0$ since we can ignore roots equal to 0. Then for $0 < x \leq \frac{1}{2016}$, we have

$$\begin{aligned} |P(x)| &\geq |a_n| - |P(x) - a_n| \\ &\geq 1 - 2015(x^n + x^{n-1} + \cdots + x) \\ &= 1 - \frac{2015(x - x^{n+1})}{1 - x} \\ &> 1 - \frac{2015x}{1 - x}. \end{aligned}$$

This is a decrreasing function of x, and at $x = \frac{1}{2016}$, it is equal to 0. Since $|P(x)| > 0$ for $0 < x \leq \frac{1}{2016}$, all positive roots of $P(x)$ are greater than $\frac{1}{2016}$.

4. Let $ABCD$ be the cyclic quadrilateral, with M and N the respective midpoints of AC and BD. Let the extensions of BA and CD meet at P, and the extensions of BC and AD meet at Q.

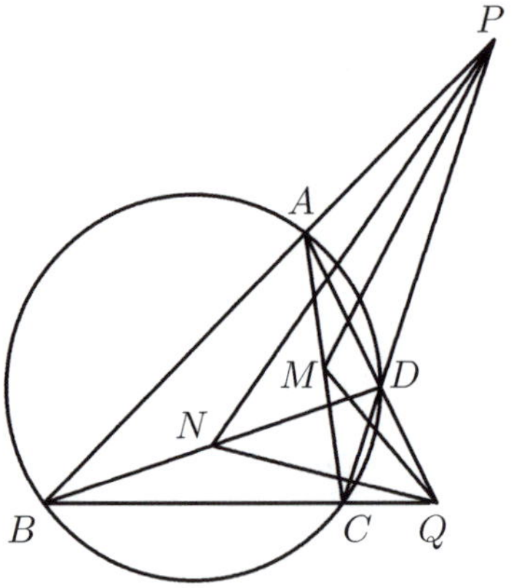

Then $\angle PCA = \angle PBD$, so that triangles PCA and PBD are similar. Hence $\frac{PA}{PD} = \frac{CA}{BD} = \frac{MA}{ND}$. Since $\angle PAM = \angle PDN$, triangles PAM and PDN are similar, so that $\angle PMA = \angle PND$. Similarly, we have $\angle QMC = \angle QND$. It follows that

$$\begin{aligned} \angle PMQ + \angle PNQ &= \angle PMQ + \angle PND + \angle QND \\ &= \angle PMQ + \angle PMA + \angle QMC \\ &= 180°. \end{aligned}$$

5. (See Problem 6 of the Junior A-Level Paper.)

6. We solve (b) in the affirmative, which implies that the answer to (a) is also affirmative. Let the sphere be $x^2 + y^2 + z^2 = 100$. The first cut is in the plane $z = 0$, consisting of a circle of radius 10 and ending in the chord joining $(-7\sqrt{2}, -\sqrt{2}, 0)$ and $(-\sqrt{2}, -7\sqrt{2}, 0)$. The distance from the point $(5\sqrt{2}, 5\sqrt{2})$ to this chord is exactly 18. The second cut is along the plane $y = -4$, consisting of a circle of radious $2\sqrt{21}$ and ending in the chord joining $(\sqrt{72\sqrt{21} - 324}, -4, 18 - 2\sqrt{21})$ and $(-\sqrt{72\sqrt{21} - 324}, -4, 18 - 2\sqrt{21})$. The distance from the point $(-4, -4, -2\sqrt{21})$ to this chord is exactly 18. The third cut is along the plane $x = -4$, consisting of a circle of radious $2\sqrt{21}$ and ending in the chord joining the points $(-4, \sqrt{72\sqrt{21} - 324}, 2\sqrt{21} - 18)$ and $(-4, -\sqrt{72\sqrt{21} - 324}, 2\sqrt{21} - 18)$. The distance from the point $(-4, -4, 2\sqrt{21})$ to this chord is exactly 18. When the last two cuts intersect, there is a bit of each which extends beyond the other, so that the corresponding circular arcs do not intersect. This means that the surface of the sphere remains connected, so that the sphere itself also remains connected.

7. First, it should be emphasized that we are not an active participant in the rearrangement, which is completely determined by the initial permutation. Our role is to observe that the whole process takes at most $n - 1$ moves. Let the children be numbered 1 to n is descending order of height. The final permutation is then $\{1, 2, \ldots, n\}$. For any k, $1 \leq k \leq n$, we claim that in at most $n - 1$ moves, child k will occupy one of the first k positions in the final permutation. Since this is true for all k, the final permutation is reached in at most $n - 1$ moves. We now justify our claim for each k. Label S for short all children shorter than child k. Label T for tall all children taller than child k. Child k is also labeled T. Thus the initial permutation corresponds to some sequence of k Ts and $n - k$ Ss. We need show that in at most $n - 1$ moves, the Ts will occupy the first k positions in the sequence.

   Let us first given an illustrative example, using $\{621534\}$ as the initial permutation. The brackets around groups of size 1 are omitted.

Permutation	$k = 1$	$k = 2$	$k = 3$
(621)(53)4 Rank	S**ST**SSS 2,0,0	**STT**SSS 2,0,0	**STTSTS** 3,2,0
12(63)(54) Rank	TSSSSS 0,0,0,0	TTSSSS 0,0,0,0	TT**ST**SS 0,0,1,0
123(64)5 Rank			TTTSSS 0,0,0,0,0
1234(65)			
123456			

Permutation	$k = 4$	$k = 5$	$k = 6$
(621)(53)4	**ST**T**ST**T	**ST**TTTT	TTTTTT
Rank	4,3,0	5,0,0	0,0,0
12(63)(54)	TT**STST**	TT**ST**TT	
Rank	0,0,3,2	0,0,3,0	
123(64)5	TTT**ST**S	TTT**ST**T	
Rank	0,0,0,1,0	0,0,0,2,0	
1234(65)	TTTTSS	TTTT**ST**	
Rank	0,0,0,0,0	0,0,0,0,1	
123456		TTTTTS	
Rank		0,0,0,0,0,0	

An ST pair is an S followed immediately by a T. Such a pair must be within a group. For each group, we define its rank as follows. If it has no ST pairs, its rank is 0. Otherwise there will be exactly one ST pair in it. The rank of this group is equal to 1 plus the sum of the number of Ss before the ST pair and the number of Ts after it. The rank of the permutation is defined as the maximum value of the ranks of its groups. During a move, the two children corresponding to an ST pair reverse their relative positions. If this S forms an ST pair in a new group, we have the same number of Ss before it but at least 1 less T after it. Hence its rank must decrease by at least 1. It follows that the rank of the permutation must also decrease by at least 1 in each move. When the rank of the permutation reaches 0, the final permutation is achieved. Since there are k Ts and $n-k$ Ss, the rank of a group is maximum when all the other $n-k-1$ Ss are before it, and all the other $k-1$ Ts are after it. Its rank is then $1+(n-k-1)+(k-1)=n-1$. It follows that the task can be accomplished in at most $n-1$ moves.

Spring 2016

Junior O-Level Paper

1. We claim that the boys and girls must alternate along the circle. Suppose to the contrary that two girls are next to each other. Then for some boy, the two children clockwise from him are both girls. The first girl must be wearing a blue T-shirt because of the boy, and a red T-shirt because of the other girl. This is a contradiction. Similarly, we cannot have two boys next to each other, and our claim is justified. It follows that the number of boys must be 10. In fact, all the boys are wearing red T-shirts and all the girls are wearing blue T-shirts.

2. Suppose $CA = CB$. Then ABC is an equilateral triangle, M coincides with A and N with B. Henceforth, we assume that $CA < CB$. Then M is on the extension of CA while N is on CB. Let AD and BE be altitudes. Then each of CAD and CBE is half an equilateral triangle, so that both CAN and CBM are equilateral triangles. It follows that AN is parallel to BM.

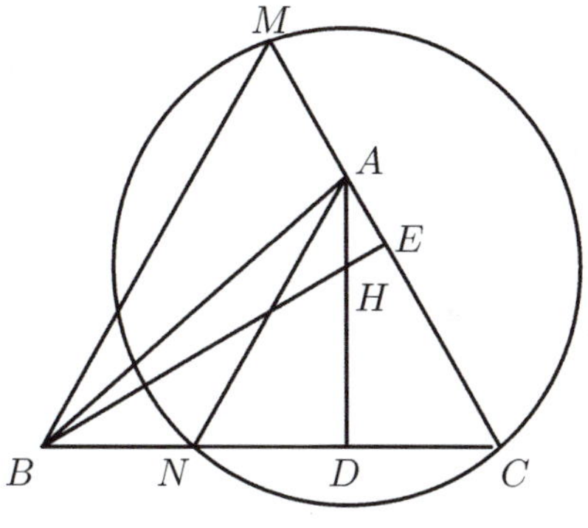

3. More generally, we show that for any positive integer k, there exist $8k$ integers whose sum and product are both $8k$. Take one copy of $4k$, one copy of 2, m copies of 1 and n copies of -1. Since the number of integers is $8k$, we have $m + n = 8k - 2$. Since the sum of the integers is $8k$, we have $m - n = 8k - 4k - 2 = 4k - 2$. Hence $m = 6k - 2$ and $n = 2k$. Since n is even, the product of the integers is $2(4k) = 8k$.

4. Since there are only 9 black squares which share common sides with white squares, the number of pieces is at most 9. The diagram below shows that we can have 9 pieces.

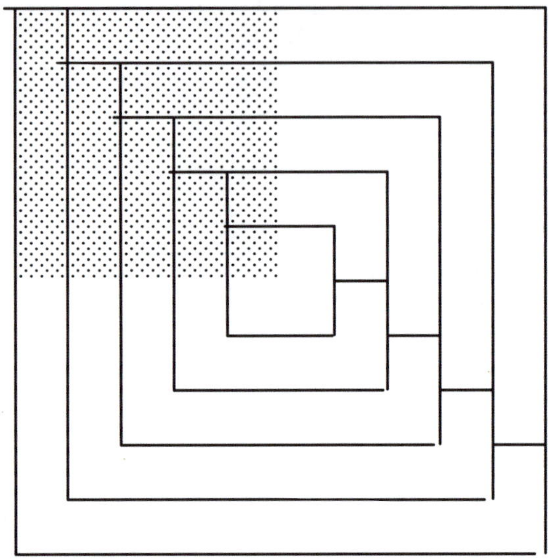

5. This is possible. First draw a segment AB. From B, draw a ray forming a $60°$ angle with AB. From A, draw a line perpendicular to this ray, intersecting it at D. The BAD is half an equilateral triangle. Let P be the point on AD such that $AP = 2PD$. Then BP bisects $\angle ABD$. From A, draw a line perpendicular to AB, cutting the ray at M and the extension of BP at Q. Take the point C on the ray such that $BM = MC$. In triangle ABC, AM is a median, AD is an altitude and BQ is an angle bisector, and APQ is an equilateral triangle.

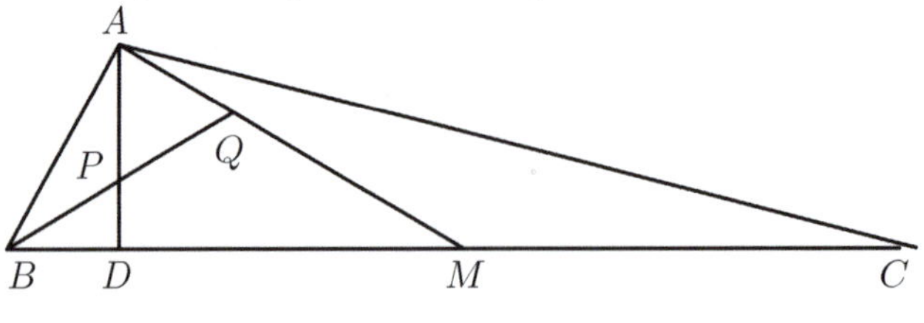

Senior O-Level Paper

1. Let O be the point inside the quadrilateral $ABCD$ and let E be the point on AB such that the circumradii of triangles OAE and OBE are equal. Then $\angle OAB = \angle OBA$ since they are subtended by OE in equal circles. Hence $OA = OB$. Similarly, $OB = OCB = OD$. It follows that O is the circumcentre of $ABCD$.

2. (See Problem 3 of the Junior O-Level Paper.)

3. (See Problem 4 of the Junior O-Level Paper.)

4. The final number is equal to the sum of the integer parts of the 100 numbers along with any carry-over from their decimal parts during addition. Thus the number of possible values of the final number depends only on the carry-over when the original 100 numbers are added two at a time. Since there are at most 50 additions with carry-over, the number may range from 0 to 50, so that there are at most 51 different values for the final number. We now construct an example to show that we can get as many as 51. Take one copy each of 0.1 and 0.9 and 98 copies of 0.5. We can certainly get 50 carry-over by adding 0.1 to 0.9 and the other numbers to one another in pairs. To get 0 carry-over, add 0.1 to 0.5 and add the resulting 0 to the other numbers in turn. To get 1 to 48 carry-over, start by adding 1 to 49 pairs of 0.5 and then use the 0.1 to kill the rest. To get 49 carry-over, start by adding 48 pairs of 0.5 and then add 0.1 to 0.5 and 0.9 to the last 0.5.

5. There is a sphere Ω which passes through the midpoints of all 12 edges of the cube. Stand the cube on a vertex so that the space diagonal joining it to the opposite vertex is vertical. Then the 12 points lie on three horizontal planes in a 3-6-3 distribution.

   (a) The answer is "No". The 6 points in the horizontal plane in between the other two are vertices of a regular hexagon. There are infinitely many spheres other than Ω which pass through all of them.

   (b) The answer is "Yes". Suppose a sphere passes through 7 of these 12 points. By the Pigeonhole Principle, one of the three horizontal planes must contain 3 of them. Moreover, another of the 7 points must lie on a different plane. Hence the sphere has 4 non-coplanar points in common with Ω, and must therefore coincide with Ω.

Junior A-Level Paper

1. Take the two-digit number 54. In the six-digit number 154154, one of the two copies of 54 will survive the cutting. This is true of all other two-digit numbers.

2. (a) We can take $a = -3$ and $b = 3$. The discriminant of $x^2 - 3x + 3 = 0$ is $(-3)^2 - 4 \times 3 = -3$. Hence it has no real roots. However, $\lfloor (\frac{4}{3})^2 \rfloor - 3(\frac{4}{3}) + 3 = 1 - 4 + 3 = 0$.

   (b) Suppose that a and b are integers such that $x^2 + 2ax + b = 0$ has no real roots but $\lfloor x^2 \rfloor + 2ax + b = 0$ has at least one real root. Then $(2a)^2 - 4b < 0$ so that $b > a^2$.

Note that $2ax$ is an integer. We have a contradiction since

$$\begin{aligned} 0 &= \lfloor x^2 \rfloor + 2ax + b \\ &> \lfloor x^2 \rfloor + 2ax + a^2 \\ &= \lfloor x^2 + 2ax + a^2 \rfloor \\ &= \lfloor (x+a)^2 \rfloor \\ &\geq 0. \end{aligned}$$

3. First divide the 10×10 square into twenty-five 2×2 squares. Draw a circle with radius $\sqrt{\frac{3}{2}}$ centred at the centre of the square, cutting its perimeter in eight points. Join four alternate points in opposite pairs by two perpendicular segments, dividing the square into four congruent quadrilaterals having two right angles. This quadrilateral is cyclic, and its diameter joins the other two vertices. Its length is $\sqrt{\frac{3}{2}+\frac{3}{2}}=\sqrt{3}$.

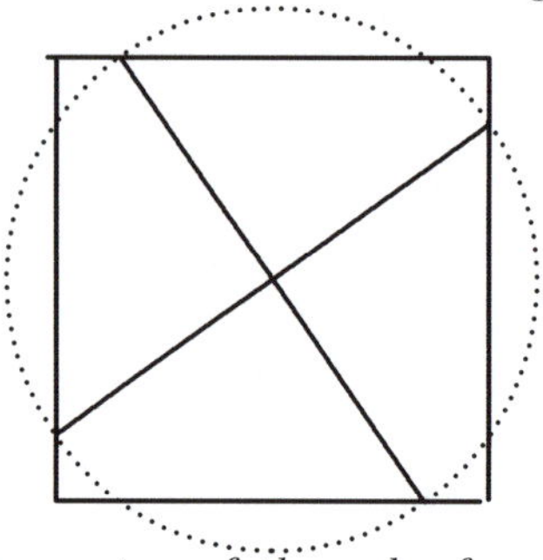

4. The squares around a vertex of the cube form a three-cycle so that exactly one of them must be black. The squares in the next layer form a nine-cycle, so that at least one of them must be black. Since there are 8 three-cycles and 8 nine-cycles all independent of one another, we need at least 16 black squares. The remaining squares contain two independent fifteen-cycles, requiring another 2 black squares. The colouring scheme in the diagram below shows that 18 black squares are sufficient.

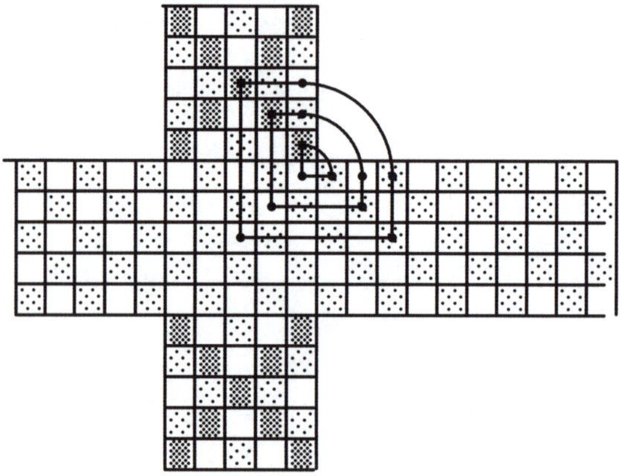

5. Let the initial multiple of p be $q_0 = 10^j \times a + b$, where b is a j-digit number and the first insertion A is placed between $10^j \times a$ and b. The new multiple is therefore $q_1 = 10^{k+j} \times a + 10^j \times A + b$. Let $A = 10^i \times c + d$, where d is an i-digit number, $i < k$, and the second insertion B is placed between $10^{j+i} \times c$ and $10^j \times d$. Hence
$$q_1 = 10^{k+j} \times a + 10^{j+i} \times c + 10^j \times d + b$$
and the final multiple of p is
$$q_2 = 10^k(10^{k+j} \times a + 10^{j+i} \times c) + 10^{j+i} \times B + 10^j \times d + b.$$
It follows that in modulo p, $q_1 - q_0 \equiv q_2 - q_1 \equiv 0$, so that
$$10^j \times a(10^k - 1) + 10^{j+i} \times c + 10^j \times d \equiv 0, \quad (1)$$
$$(10^{k+j} \times a + 10^{j+i} \times c)(10^k - 1) + 10^{j+i} \times B \equiv 0. \quad (2)$$
Multiplying (1) by 10^k, we have, still in modulo p,
$$10^{k+j} \times a(10^k - 1) + 10^{k+j+i} \times c + 10^{k+j} \times d \equiv 0. \quad (3)$$
Subtracting (2) from (3) and cancelling 10^{j+i}, we have $c + 10^{k-i} \times d \equiv B$ (mod p). Since both sides are less than p and greater than 0, we have equality. It follows that B is obtained from A by moving the block of digits forming d from the end to the front. In other words, the digits of B are a permutation of the digits of A.

6. Let O be the centre of the large circle. Note that it always lies on the small circle. Let XY be a diameter of the large circle and let A be the point on the small circle which coincides with X initially. Let P be the centre and C be some point of the small circle. Suppose the small circle rotates around P counterclockwise so that it rolls along the inside of the large circle clockwise. Consider the moment when the new position of C is a point D on the large circle. Then the new position Q of P lies on OD. Let B be the point of intersection of the small circle with XY. Since there is no slipping, the arcs AC and XD have the same length. Note that $\angle BQD = 2\angle BOD$. The length of the arc XD is equal to $OX\angle XOD$ while the length of the arc BD is equal to $QB\angle BQD$. Since $OX = 2QB$, the two arcs have the same length. It follows that the arcs AC and BD also have the same length, so that B must be the new position of A. Thus A always lies on XY.

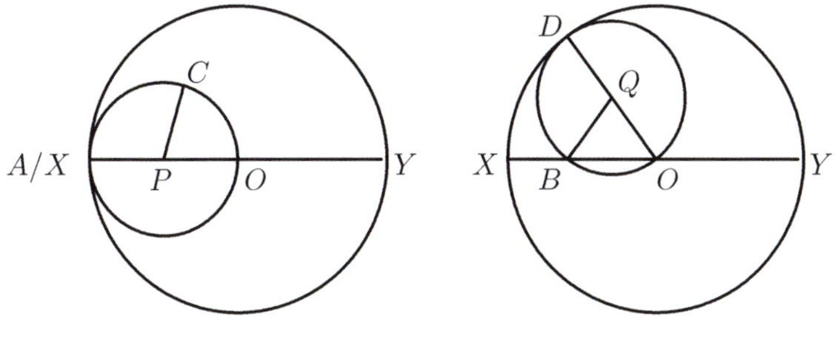

7. (a) The lower bound is $n+2$. Test the $2n+1$ batteries in pairs, leaving one off. We may as well assume that nothing works. Then each pair consists of a good battery and a bad one. Moreover, the one left off is good. Test it with both batteries in any pair, and we will have a working combination. Consider any plan with $n+1$ attempts. Construct a graph with $2n+1$ vertices representing the batteries, and $n+1$ edges representing the attempts. Since the total degree is $2n+2$, at least one vertex V has degree 2. Choose V and either vertex in any edge not incident with V, and we have chosen at most n vertices. If they represent the bad batteries, none of the $n+1$ attempts will be successful.

   (b) The lower bound is $n+3$. Test all three pairs of three batteries. We may assume that nothing works. This means that either all three are bad batteries, or two are bad and one is good. Now test the remaining $2n-3$ batteries in pairs, leaving one off. If the initial three batteries are all bad, one pair now will work. If none of the pairs work, then the battery left off is good, and we can test it with both batteries in any pair. Thus we will find a working combination in at most $3+(n-2)+2=n+3$ attempts. Consider any plan with $n+2$ attempts. Construct a graph with $2n$ vertices representing the batteries, and $n+2$ edges representing the attempts. Remove the vertex V with maximum degree, along with all edges incident with it. Since V has degree at least 2, the resulting subgraph has at most n edges. If it has at most $n-1$ edges, we can choose either vertex of each. If it has n edges, then one of its vertices U has degree at least 2. We can then choose U along with either vertex of any edge not incident with U. In either case, we have chosen at most $n-1$ vertices of the subgraph such that every edge contains at least one of them. Along with V, we can choose n vertices to represent the bad batteries, so that none of the $n+2$ attempts will be successful.

Senior A-Level Paper

1. (See Problem 1 of the Junior A-Level Paper.)

2. (See Problem 3 of the Junior A-Level Paper.)

3. Construct the circumcircle of triangle DEF. From $\angle CDE = \angle DFE$, the circle is tangent to BC at D. Since D is the midpoint of BC, the circle intersects AB at another point G with $BG = CE$, so that GE is parallel to BC. Hence $\angle GED = \angle CDE = \theta$. Now $\angle BFD = \angle GED = \theta$ since $\angle BFD$ is an exterior angle of the cyclic quadrilateral $DEGF$.

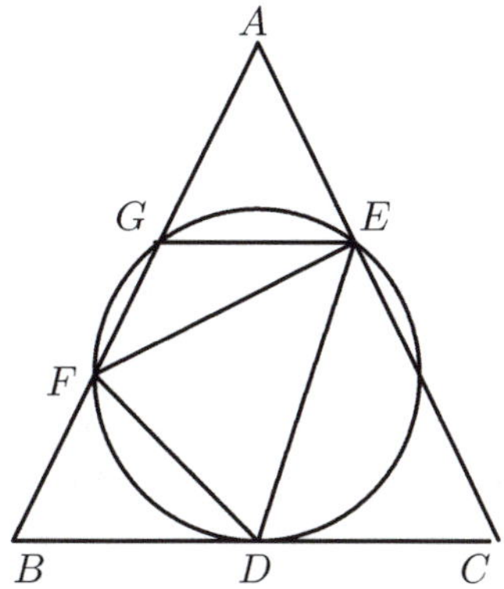

4. Note that $\binom{64}{2} = 2016$. If we ask less than 2016 questions, we will not know if two of towns A and B are linked by telegraph service or not. It may happen that A is not linked to any other town except possibly B while B is linked to every town except possibly A. Then we will not know whether messages can be sent from or received by A.

5. We seek an invariant amidst all the changes, and this is the sum of the coefficients of the x^{36} terms of all the polynomials, which is also the sum of all 37 roots of each of them. When we replace two polynomials by two others with the same product, the combined set of roots of the old polynomials is the same as the combined set of roots of the new polynomials. When we replace two polynomials by two others with the same sum, the sum of the coefficients of their x^{36} terms is unchanged. Initially, all coefficients are non-negative. Hence this invariant is also non-negative. However, if after a finite number of moves, every polynomial has 37 distinct real roots, then this invariant must be negative. We have a contradiction.

6. In a word, we score 1 every time an a precedes a b, a b precedes a c, or a c precedes an a, not necessarily immediately. We score -1 every time a b precedes an a, an a preceds a c, or a c precedes a b, also not necessarily immediately. By itself, each red card generates a net score of 1 while each blue card generates a net score of -1. Suppose a card is added to an existing word. Note that the card contains each of a, b and c once. Any a in the existing word preceding the card generates a score of 1 with the added b and a score of -1 with the added c, for a net score of 0. The same can be said about any other letter in the existing word, whether it precedes or follows the added card.

   (a) Suppose a palindrome can be obtained using k red cards. Then the net score is k. However, by the symmetry of the palindrome, the net score must be 0. We have a contradiction.

   (b) By the symmetry of the palindrome, the net score must be 0. Hence we must have an equal number of red and blue cards.

7. Let P be a point of intersection of two of the railways. The trains on the two railways cannot occupy P simultaneously. Since all trains have the same uniform speed, the total length of the trains in any two railways is at most 1. If the sum of the lengths of the trains on each railway is at most $\frac{1}{2}$, then the total length is at most $\frac{1}{2}$ times the number of railways. Suppose the sum of the lengths of the trains on at least one railway exceeds $\frac{1}{2}$. Then the sum of the lengths of the trains on it and on another railway is at most 1, while the sum of the lengths of the trains on each remaining railway must be less than $\frac{1}{2}$. Hence the total length is still at most $\frac{1}{2}$ times the number of railways.

   (a) We now construct an example with 3 railways on which the total length of the trains is $\frac{3}{2}$. Inscribe in the sphere a regular octahedron. Its Schlegel diagram, a planar representation of its skeleton, is shown below. Each edge represents an arc of length $\frac{1}{4}$.

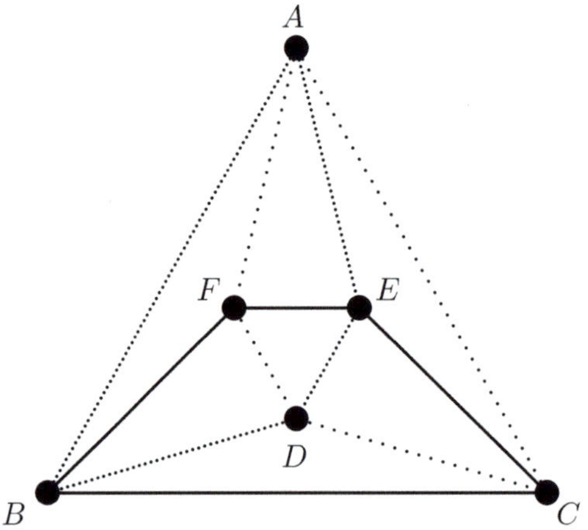

   Note that there are three interlocking 4-cycles, $ABDE$, $BCEF$ and $CAFD$ which constitute the railways. We have six trains each of length $\frac{1}{4}$, all running counterclockwise and completing the lap in one hour. The trains on the first railway start on AB and DE, the trains on the second railway start on BC and EF, while the trains on the third railway start on CA and FD. During the first fifteen minutes, they pass through vertices B, E, C, F, A and D respectively. Hence there is no collision. The situation is symmetric in each of the remaining three blocks of fifteen minutes in the hour.

   (b) We now construct an example with 4 railways on which the total length of the trains is 2. Inscribe in the sphere an cuboctahedron, an Archimedean solid in which every vertex is surrounded by an opposite pair of equilateral triangles and an opposite pair of squares. Its Schlegel diagram is shown below, where each edge represents an arc of length $\frac{1}{6}$.

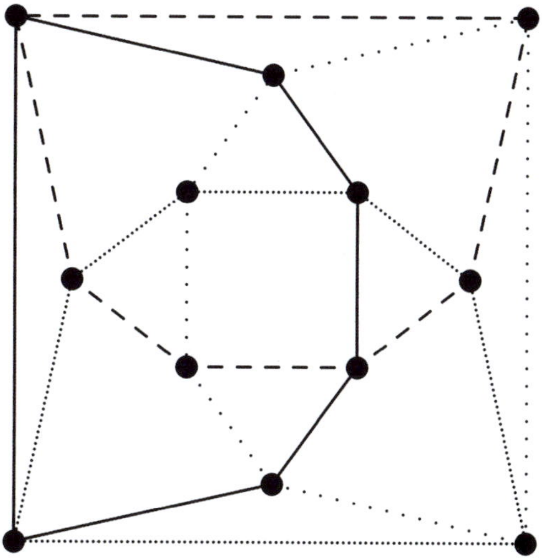

Note that there are four interlocking 6-cycles which constitute the railways. We have twelve trains each of length $\frac{1}{6}$, all running counterclockwise and completing the lap in one hour. The trains on the railways start on alternate edges so that each passes through a different vertex in each block of ten minutes. Hence there is no collision.

Tournament 38

Fall 2016

Junior O-Level Paper

1. Do there exist five positive integers such that their ten pairwise sums end in different digits?

2. Four points on a line plus a fifth point not on the line determine six triangles. At most how many of them can be isosceles?

3. On a circle are 100 points labeled with the positive integers 1 to 100 in some order.

   (a) Prove that these points can be joined in pairs by 50 non-intersecting chords such that the sum of the labels of the two endpoints of each chord is odd.

   (b) Can one always join these points in pairs by 50 non-intersecting chords such that the sum of the labels of the two endpoints of each chord is even?

4. $ABCD$ is a parallelogram. K is a point such that $AK = BD$ and M is the midpoint of CK. Prove that $\angle BMD = 90°$.

5. One hundred bear-cubs have 1, 2, ..., 2^{99} berries, respectively. A fox chooses two bear-cubs and divides their berries equally between them. If a berry is left over, the fox eats it. What is the least number of berries the fox can leave for the bear-cubs?

Note: The problems are worth 4, 4, 2+2, 5 and 5 points respectively.

Senior O-Level Paper

1. Two parabolas are the graphs of quadratic trinomials with leading coefficients p and q respectively. They have different vertices which lie on each other. What are the possible values of $p + q$?

2. Of the triangles determined by 100 points on a line plus an extra point not on the line, at most how many of them can be isosceles?

3. One hundred bear-cubs have 1, 2, ..., 2^{99} berries respectively. A fox chooses two bear-cubs and divides their berries equally between them. If a berry is left over, the fox eats it. What is the least number of berries the fox can leave for the bear-cubs?

4. A 100-omino can be dissected into two congruent 50-ominoes as well as 25 congruent tetrominoes. Is it always possible to dissect it into 50 dominoes?

5. Prove that in a right triangle, the altitude to the hypotenuse passes through the orthocentre of the triangle determined by the points of tangency of the incircle with the right triangle.

Note: The problems are worth 4, 5, 5, 5 and 5 points respectively.

Junior A-Level Paper

1. Each of ten boys has 100 cards. In each move, one of the boys gives one card to each of the other boys. What is the minimal number of moves before each boy has a different number of cards?

2. There are 64 positive integers in the squares of an 8×8 board. Whenever the board is covered by 32 dominoes, the sum of the two integers covered by each domino is unique. Is it possible that the largest integer on the board does not exceed 32?

3. The diagram shows an arbitrary triangle dissected into congruent triangles by lines parallel to its sides. Prove that the orthocentres of the six shaded triangles are concyclic.

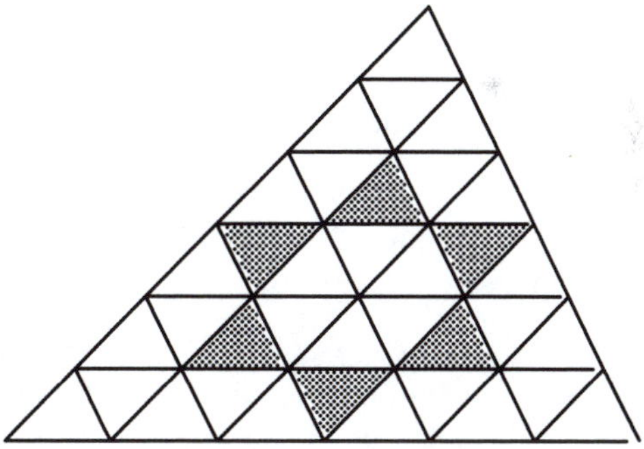

4. In a 7×7 box, each of the 49 pieces is either dark chocolate or white chocolate. In each move, Alex may eat two adjacent pieces along a row, a column or a diagonal, provided that they are of the same kind.

   What is the maximum number of pieces Alex can guarantee to be able to eat, regardless of the initial arrangement the pieces?

5. The three pairwise sums of three numbers are recorded on a piece of paper. They are distinct and positive. The three pairwise products of the same three numbers are recorded on another piece of paper. They are also distinct and positive. Later, it is forgotten which piece is which. Is it still possible to determine which piece is which?

6. Let $A_1A_2\ldots A_{2n}$, $n \geq 5$, be a regular $2n$-gon with the centre O. Diagonals A_2A_{n-1} and A_3A_n intersect at F while diagonals A_1A_3 and A_2A_{2n-2} intersect at E. Prove that $EF = EO$.

7. An examination consists of 20 questions with k multiple choices. For any 10 questions and any of the k^{10} combination of answers to them, some student has given precisely these answers to these questions. Must there exist two students who have given different answers to all 20 questions, where

   (a) $k = 2$;

   (b) $k = 12$?

Note: The problems are worth 5, 5, 6, 8, 8, 9 and 5+6 points respectively.

Senior A-Level Paper

1. Each of ten boys has 100 pokemon cards. In each move, one of the boys gives one card to each of the other boys. What is the minimal number of moves before each boy has a different number of cards?

2. There are 64 positive integers in the squares of an 8×8 board. Whenever the board is covered by 32 dominoes, the sum of the two integers covered by each domino is unique. Is it possible that the largest integer on the board does not exceed 32?

3. The circumcentre O of a quadrilateral $ABCD$ does not lie on either diagonal. The circumcircle of AOC passes through the midpoint of BD. Prove that the circumcircle of BOD passes through the midpoint of AC.

4. The 2016 pairwise sums of 64 numbers are recorded on one piece of paper. They are distinct and positive. The 2016 pairwise products of the same 64 numbers are recorded on another piece of paper. They are also distinct and positive. Later, it is forgotten which piece is which. Is it still possible to determine which piece is which?

5. Is it possible to cut a 1×1 square into two pieces which can cover a disk of diameter slightly greater than 1?

6. Alice chooses a polynomial $P(x)$ with integer coefficients. In each move, Bob gives Alice an integer a, and Alice tells him the number of different integer solutions of the equation $P(x) = a$. Bob may not give Alice the same number twice. Determine the minimal number of moves for Bob to make Alice tell him a number that she has told him before, regardless of the polynomial chosen by Alice.

7. A finite number of frogs are placed on distinct integer points on the real line. At each move, a single frog jumps by 1 to the right provided that the new location is unoccupied. Altogether, the frogs make n moves, and this can be done in m ways. Prove that if they jump by 1 to the left instead of to right, they can still make n moves in m ways.

Note: The problems are worth 5, 5, 7, 8, 9, 9 and 12 points respectively.

Spring 2017

Junior O-Level Paper

1. Find the smallest positive multiple of 2017 such that its first four digits are 2016.

2. Prove that the graph of any monic quadratic polynomial with a repeated root passes through a point (p, q) such that x^2+px+q also has a repeated root.

3. O is the circumcentre of an acute-angled triangle ABC with $\angle A = 60^\circ$. The bisector of $\angle A$ intersects BC at D. Prove that $\angle ADB = \angle ODC$.

4. One hundred children of distinct heights stand in a line. At each step, a group of 50 consecutive children is chosen and they are rearranged in an arbitrary way. Is it always possible, in 6 such steps, to arrange the children so that their heights decrease from left to right?

5. A circle is drawn with each side of an n-gon as diameter. Is it possible that all these circles pass through a point which is not a vertex of the n-gon, where

   (a) $n = 10$;

   (b) $n = 11$?

Note: The problems are worth 3, 4, 5, 5 and 2+3 points respectively.

Senior O-Level Paper

1. $A_1A_2 \ldots A_{12}$ is a regular 12-gon. Is it possible to choose seven of the vectors A_1A_2, A_2A_3, $\ldots$, $A_{11}A_{12}$ and $A_{12}A_1$ with zero sum?

2. O is the common centre of two circles, and P is a point inside the smaller circle. A and B are points on the larger circle and on the same side of OP, such that $\angle AOB = \angle APB$. Let PA and PB intersect the smaller circle at C and D respectively. Prove that $\angle COD = \angle CPD$.

3. There is a number in each square of a 1000×1000 table. The sum of the numbers in each rectangle consisting of n of the squares is the same. Find all values of n for which all the numbers in the table must be equal.

4. Ten children of distinct height stand in a circle. In each step, one of them moves to a new place in the circle between two children. What is the minimum number of steps in which the children can always be arranged so that their heights increase in clockwise order?

5. The graphs of two quadratic polynomials intersect in two points. In each of these points, the tangents to the graphs are perpendicular. Is it necessarily true that the graphs have a common axis of symmetry?

Note: The problems are worth 3, 4, 5, 5 and 6 points respectively.

Junior A-Level Paper

1. A tournament has 10 participants. Five games are played simultaneously in each round. In nine rounds, each participant has played every other participant. At least half of the games are played between two participants from the same town. Prove that in each round, there is a game played between two participants from the same town.

2. A polygon consisting of a number of squares of a piece of graph paper is divided into two congruent pieces by a cut which runs along the grid lines. For each figure in the diagram below, find a polygon for which the cut has precisely the shape shown by the solid lines.

(a) (b) (c)

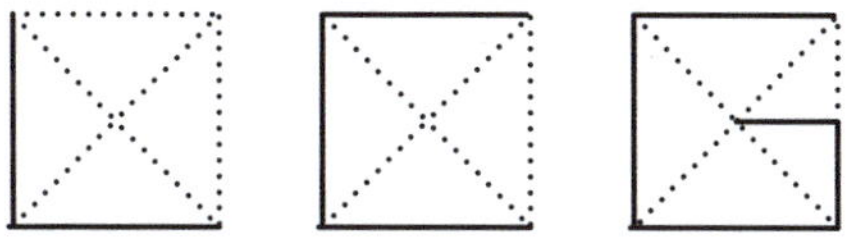

3. A set of positive numbers has sum a_1. For any integer $k \geq 2$, a_k is the sum of the kth powers of the numbers in the set. Is it possible that

   (a) $a_1 > a_2 > a_3 > a_4 > a_5$ and $a_5 < a_6 < a_7 < \cdots$;

   (b) $a_1 < a_2 < a_3 < a_4 < a_5$ and $a_5 > a_6 > a_7 > \cdots$?

4. In a convex equilateral hexagon $ABCDEF$, $AD = BE = CF$. Prove that the hexagon has an incircle.

5. Each token weighs a non-integral number of grams. Any object of integer weight from 1 gram to 40 grams can be balanced by placing some tokens on the same pan. What is the smallest possible number of tokens?

6. On a $1 \times n$ board, a grasshopper can jump to the 8th, the 9th or the 10th square in either direction. Find an integer $n \geq 50$ such that starting from some square of the board, the grasshopper cannot visit every square exactly once.

7. An 8×8 squares is marked on a piece of paper. Place n non-overlapping dominoes on the piece of paper such that centre of each domine is strictly inside the square, where

(a) $n = 40$;

(b) $n = 41$;

(c) $n > 41$.

Note: The problems are worth 5, 1+2+4, 4+4, 8, 8, 10 and 6+3+3 points respectively.

Senior A-Level Paper

1. On the plane are ten lines and one triangle. Every line is equidistant from two of the triangle's vertices. Prove that either at least two of these lines are parallel or at least three of them are concurrent.

2. A set of positive numbers has sum a_1. For any integer $k \geq 2$, a_k is the sum of the kth powers of the numbers in the set. Is it possible that

   (a) $a_1 > a_2 > a_3 > a_4 > a_5$ and $a_5 < a_6 < a_7 < \cdots$;

   (b) $a_1 < a_2 < a_3 < a_4 < a_5$ and $a_5 > a_6 > a_7 > \cdots$?

3. Is it possible to dissect a cube into two pieces which can be reassembled into a convex polyhedron with only triangular and hexagonal faces?

4. Pete painted each square of a 1000×1000 board in one of ten colours. He also constructed a frame enclosing 10 squares such that no matter how it is placed on the board, the ten squares inside the frame are of distinct colours. Is it necessary for this frame to be a rectangle?

5. In triangle ABC, $\angle A = 45°$. When the median from A is reflected across the altitudes from B and from C, the two images intersect at a point X. Prove that $AX = BC$.

6. Find all positive integers n which have a multiple with digit sum k for any integer $k \geq n$.

7. Each of 36 gangsters belongs to several gangs. There are no two gangs with the same membership. Two gangsters are enemies if they do not both belong to the same gang. Each gangster has at least one enemy in every gang to which he does not belong. What is the largest possible number of gangs?

Note: The problems are worth 4, 3+3, 7, 8, 9, 10 and 12 points respectively.

Solutions

Fall 2016

Junior O-Level Paper

1. Consider the sum of these ten sums. It must be even since each of the five original numbers appears four times. However, the sum of ten numbers ending in different digits must be odd. This is a contradiction. Hence no such five positive integers exist.

2. We have $\angle ABE = \angle BAC = \angle CAD = \angle DAE = \angle AEB = 36°$ in the diagram below. It follows that $BD = AB = AE = EC$ as well as $BC = CA = AD = DE$. Hence all of ABC, ACD, ADE, EAC, BAD and ABE are isosceles. Thus six triangles are possible.

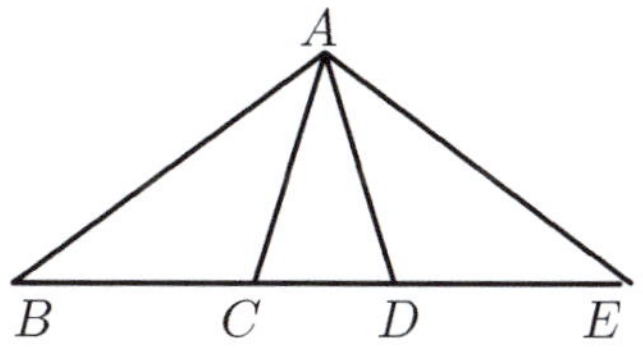

3. Since only parity matters, we may replace all even labels by 0s and all odd labels by 1s. We consider the general case where there are $2n$ points for some positive integer n, half of them labeled 0 and the other half labeled 1.

   (a) We use induction on n. The basis $n = 1$ is trivial. Suppose the result is true for some $n \geq 1$. Consider $2(n+1)$ points. A point labeled 0 must be adjacent to a point labeled 1. Join them by a chord which cannot intersect any other chord. We can apply the induction hypothesis to the remaining $2n$ points.

   (b) Let the 0s and 1s be arranged alternately along the circle. Then we can only join two points with the same label. However, this will leave an odd number of other points on each side of this chord, and some other chord must intersect it.

4. Let O be the centre of $ABCD$. Now K lies on a circle with centre A and radius $AK = BD$. Then M lies on a circle with centre O and radius $\frac{1}{2}BD$. Hence BD is a diameter of this circle, and $\angle BMD = 90°$.

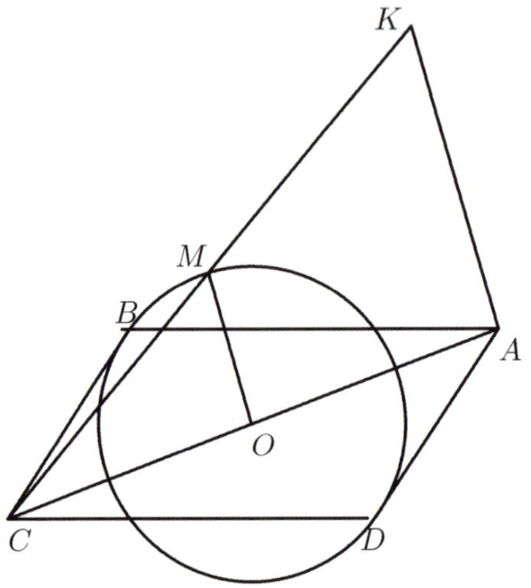

5. Since each bear-cub starts with at least 1 berry, it will end up with at least 1 berry. Thus the least number of berries the fox can leave for the bear-cubs is 1 each. We show more generally that if there are n bear-cubs with 1, 2, ..., 2^{n-1} berries initially, then the fox can accomplish this task. Arrange the bear-cubs in ascending order of the number of berries, so that we start with $(1,2,\dots,2^{n-1})$. All sharings are between adjacent bear-cubs, so that their order does not change. We claim that for all $n \geq 1$, (1) $(1,2,\dots,2^{n-1})$ can be converted to $(1,1,\dots,1,2^{n-1})$, and (2) $(1,1,\dots,1,2^{n-1})$ can be converted to $(1,1,\dots,1)$. The basis $n=2$ is trivial. Assume that (1) and (2) holds for some $n \geq 2$. Consider $(1,2,\dots,2^n)$. Applying the induction hypothesis to the first n terms, we can convert $(1,2,\dots,2^n)$ to $(1,1,\dots,1,2^n)$ by (1) and then (2). A sharing between the last two bear-cubs produces $(1,1,\dots,1,2^{n-1},2^{n-1})$. Applying the induction hypothesis to the first n terms, we have $(1,1,\dots,1,2^{n-1})$ by (2). Applying the induction hypothesis to the last n terms, we have $(1,1,\dots,1)$ by (2). This completes the inductive argument.

Senior O-Level Paper

1. Translation of a parabola does not affect the coefficient of its quadratic term. If we translate the two parabolas as a single entity, their vertices will still lie on each other. We translate them so that one of them has vertex (0,0). Its equation will be $y=px^2$. Let (a,b) be the vertex of the other parabola after translation. Its equation will be $y=q(x-a)^2+b$. Since the two vertices do not coincide, $a \neq 0$. Putting (0,0) into the second equation and (a,b) into the first, we have $0=qa^2+b$ and $b=pa^2$. Addition yields $0=a^2(p+q)$. Since $a \neq 0$, we must have $p+q=0$.

2. We have $\angle AB_0E_0 = \angle B_0AC = \angle CAD = \angle DAE_0 = \angle AE_0B_0 = 36°$ in the diagram below. Then we have $B_0D = AB_0 = AE_0 = E_0C$ and $B_0C = CA = AD = DE_0$. Hence all of AB_0C, ACD, ADE_0, E_0AC, B_0AD and AB_0E_0 are isosceles triangles. We now add B_1 and E_1 so that $B_1B_0 = B_0A = AE_0 = E_0E_1$. Then we have three more isosceles triangles, namely, AB_1B_0, AE_0E_1 and AB_1E_1. Continuing this way, we add points B_n and E_n up to $n = 48$. This gives us $4 + 2 \times 48 = 100$ points on the line and $6 + 3 \times 48 = 150$ isosceles triangles.

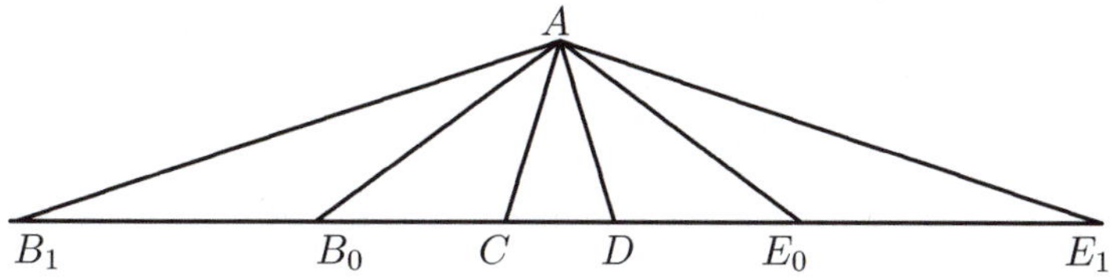

We now prove that 150 is indeed the maximum. Call the vertex the pivot of an isosceles triangle if it is the common vertex of the two equal sides. There are two kinds of isosceles triangles, those whose pivot is the point A outside the line, and those whose pivot is on the line. Since only 100 segments have A as an endpoint, the number of isosceles triangles of the first kind is at most 50. Each point P on the line is a non-pivot of at most one isosceles triangle of the second kind, since the pivot must lie on the perpendicular bisector of AP, which intersects the line at a unique point. It follows that the number of such triangles is at most 100.

3. (See Problem 5 of the Junior O-Level Paper.)

4. Of the five tetrominoes, four can be dissected into two dominoes. Assume to the contrary that the desired task is not possible. Then the 100-omino must be dissected into 25 copies of the T-tetromino. When placed on a board with the standard black and white colouring, it covers three squares of one colour and one of the other colour. Hence 25 copies of it will cover an odd number of white squares and an odd number of black squares. Now the 100-omino can also be dissected into two congruent 50-ominoes. If black squares in one copy correspond to white squares in the other copy, then the figure must cover 50 white squares and 50 black squares. If black squares in one copy correspond to black squares in the other copy, the figure must still cover an even number of white squares and an even number of black squares. Either way, we have a contradiction. Thus the desired task is always possible.

5. Let the triangle be ABC with a right angle at C. Let the incircle touch BC, CA and AB at D, E and F respectively. Let the altitudes DP and EQ of DEF intersects at the orthocentre H of DEF.

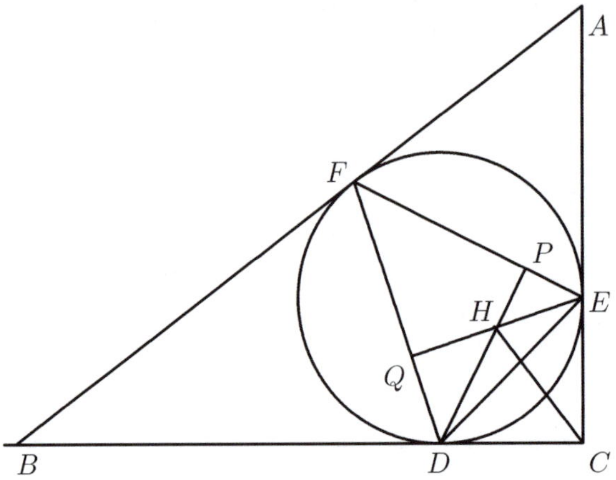

Note that $\angle BFD + \angle AFE = \frac{1}{2}(2 \times 180° - \angle ABC - \angle BAC) = 135°$. Hence $\angle DFE = 45°$ so that $\angle PDF = 45° = \angle FEQ$. Since $FPHQ$ is cyclic, $\angle DHE = \angle PHQ = 135°$, so that $\angle DHE + \frac{1}{2}\angle DCE = 180°$. It follows that H lies on the circle with centre C and passing through D and E. Now $\angle EHC = \angle HEC = 135° - \angle PEA = 135° - \angle PFA = \angle BFD$. Hence $\angle HCA = \angle FBD$, so that CH is perpendicular to AB.

Junior A-Level Paper

1. Suppose after n moves, every boy has a different number of Pokemon cards. For $1 \le k \le 10$, let the kth boy give away cards in m_k moves and receives cards in $n - m_k$ moves. Then he will end up with a total of $100 + (n - m_k) - 9m_k = 100 + n - 10m_k$ cards. Each boy will end up with a different number of cards if and only if m_k are distinct. Now $n = m_1 + m_2 + \cdots + m_{10}$. Since each m_k is a non-negative integer, the minimum value of n is $0 + 1 + 2 + \cdots + 9 = 45$.

2. It is possible. Take the board to be the standard board. Put the numbers from 1 to 32 in any order into the white squares, each number appearing exactly once. Put the same number, arbitrarily chosen from 1 to 32, in each of the black squares. For any two dominoes, the numbers in the white squares which they cover are different while the numbers in the black squares which they cover are the same. Hence the sums are different.

3. Let A, B, C, D, E and F be the orthocentres in cyclic order. Let B be the orthocentre of triangle PQR as shown in the diagram. Then QR is parallel to CE and RP is parallel to EA. B is the point of intersection of the altitudes PX and QY, which pass through C and A respectively. Since $\angle BCE = 90° = \angle BAE$, the circumcircle of triangle ACE passes through B. Similarly, it passes through D and F.

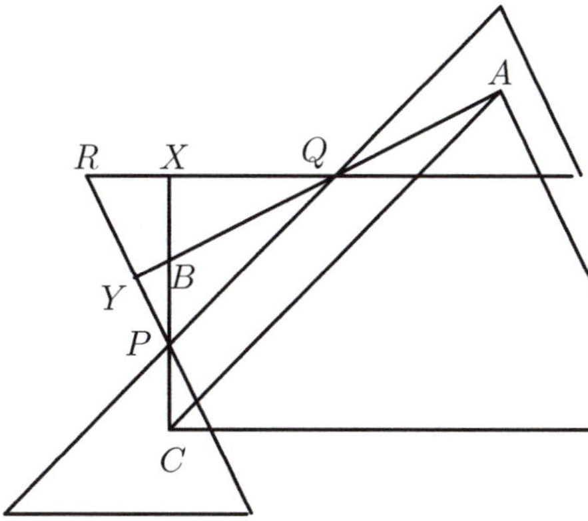

4. If the initial arrangement is as shown in the diagram below on the left, Alex cannot eat any of the 16 dark chocolates. Since there are 33 white chocolates and he can eat 2 pieces at a time, the most he can eat is 32 pieces. He can guarantee to eat 32 pieces because the box contains 16 disjoint V-trominoes, as shown in the diagram below on the right. The 3 pieces in each V-tromino are adjacent to one another, and by the Pigeonhole Principle, 2 of the pieces must be of the same kind. Hence Alex can eat 2 pieces from each of the 16 V-trominoes.

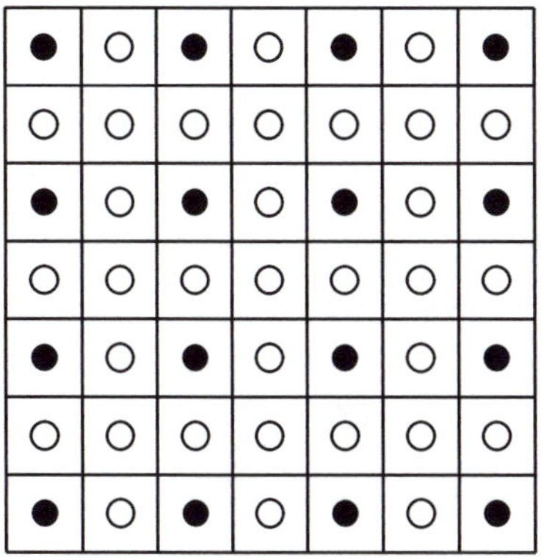

5. Let the three numbers be x, y and z. Since their pairwise products are all positive, they are either all positive or all negative. Since their pairwise sums are also positive, they are all positive. We may assume that $x > y > z > 0$. Let one piece of paper contain the numbers $a > b > c > 0$ and the other piece contain the numbers $d > e > f > 0$. We first assume that a, b and c are the pairwise sums. Then $x + y = a$, $x + z = b$ and $y + z = c$. Hence $x = a + b - c$, $y = c + a - b$ and $z = b + c - a$. If xy, xz and yz do not agree with d, e and f, then the piece containing a, b and c records the pairwise products.

We must not jump to conclusion that if they agree, then the piece containing a, b and c records the pairwise sums. We must still rule out the possibility that d, e and f are the pairwise sums of three other numbers $u > v > w > 0$. We will then have

$$\begin{array}{lll} x+y=a=uv & x+z=b=uw & y+z=c=vw \\ xy=d=u+v & xz=e=u+w & yz=f=v+w \end{array}$$

Then $y-z=a-b=u(v-w)=u(d-e)=ux(y-z)$, so that $ux=1$. Similarly, $x-y=b-c=w(u-v)=w(e-f)=wz(x-y)$, so that $wz=1$ also. However, $\frac{1}{x}=u>w=\frac{1}{z}$ implies that $z>x$, which is a contradiction. It follows that ambiguity cannot arise, and we can determine which piece of paper records the pairwise sums.

6. We solve this problem in three steps.
   Step 1.
   Note that triangles OA_2A_{n-1} and OA_3A_n are congruent. Hence we have $\angle OA_2F=\angle OA_3F$, so that OA_2A_3F is cyclic. If C is the circumcentre, then $\angle FCA_3=2\angle A_{n-1}A_2A_3=\angle A_{n-1}OA_3=(n-4)\angle A_2OA_3$. Also, $2\angle OA_2A_{n-1}=180^\circ-\angle A_2OA_{n-1}=3\angle A_2OA_3$.

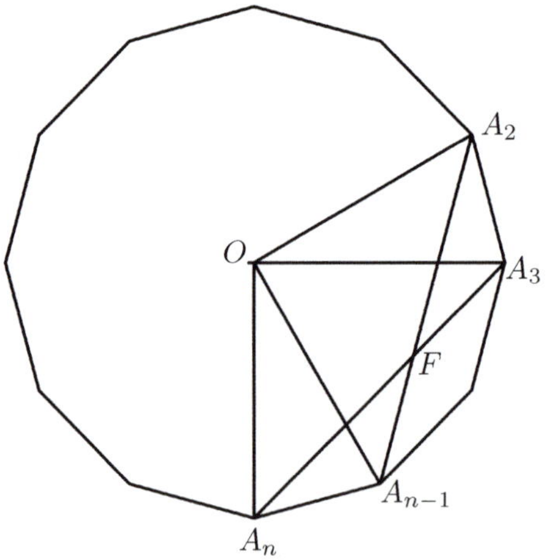

   Step 2.
   Note that EA_2A_3C is cyclic since

$$\begin{aligned} \angle A_2EA_3 &= \angle A_{2n-2}A_2A_1+\angle A_3A_1A_2 \\ &= \frac{1}{2}(\angle A_{2n-2}OA_1+\angle A_3OA_2) \\ &= \frac{1}{2}(3\angle A_3OA_2+\angle A_3OA_2) \\ &= 2\angle A_3OA_2 \\ &= \angle A_3CA_2. \end{aligned}$$

Hence

$$\begin{aligned}\angle A_3CE &= 180^\circ - \angle A_{n-2}A_2A_3\\ &= \angle A_{2n-2}A_nA_3\\ &= \frac{1}{2}\angle A_{2n-2}OA_3\\ &= \frac{5}{2}\angle A_2OA_3.\end{aligned}$$

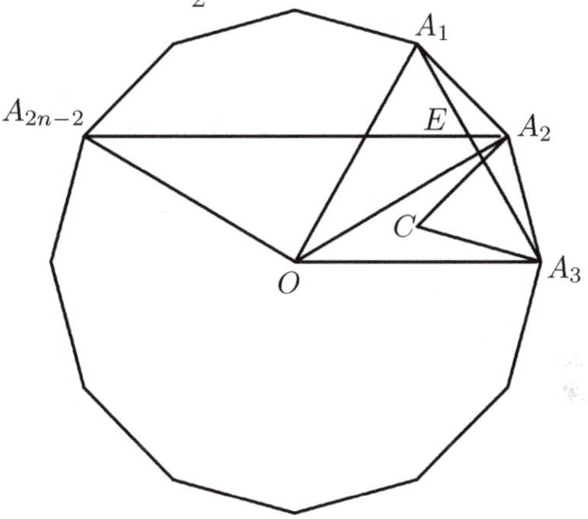

Step 3.

We have $\angle OCF = 2\angle OA_2A_{n-1} = 3\angle A_2OA_3$ by Step 1. By Steps 1 and 2,

$$\begin{aligned}\angle FCE &= \angle FCA_3 + \angle A_3CE\\ &= (n-4)\angle A_2OA_3 + \frac{5}{2}\angle A_2OA_3\\ &= \frac{1}{2}(2n-3)\angle A_2OA_3.\end{aligned}$$

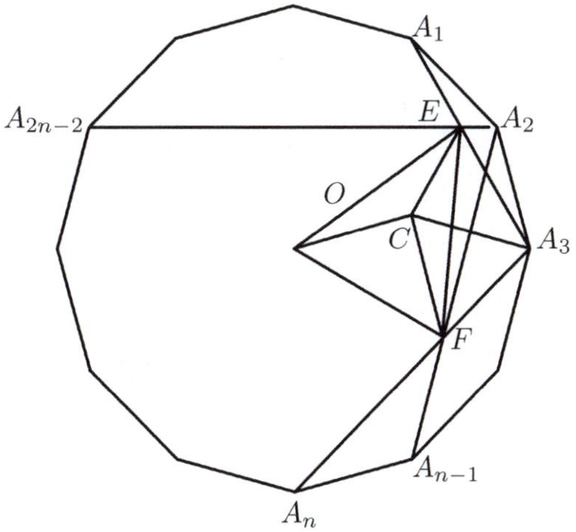

Hence

$$\begin{aligned} \angle OCE &= 360^\circ - \angle OCF - \angle FCE \\ &= (2n - 3 - \frac{1}{2}(2n - 3))\angle A_2OA_3 \\ &= \angle FCE. \end{aligned}$$

It follows that $OE = FE$.

7. (a) Let the choices be 0 and 1. A student is said to be odd if the sum of her choices to the first eleven questions is odd. For any ten questions and any of the $2^{10} = 1024$ possible choices, there is a student who makes those choices. Every student can make choices for the remaining 10 questions so that she becomes an odd student. Suppose there exist two students who make different choices for each question. In particular, they will make different choices for the first eleven questions. Then the sum of their choices for these questions will be 11, which means that one of them is not odd. This is a contradiction.

   (b) Let the choices be 0, 1, 2, ..., 11. For $1 \leq k \leq 11$, there is a student whose choice for each of the last ten questions is k. These eleven students will be called the Committee. For each of the first ten questions, there is a choice which is not made by any member of the Committee. On the other hand, there is a student, Charlie, who makes those choices for the first ten questions. For the last ten questions, Charlie makes at most ten different choices. Hence there is a member of the Committee whose choices differ from Charlie's for every question.

Senior A-Level Paper

1. (See Problem 1 of the Junior A-Level Paper.)

2. (See Problem 2 of the Junior A-Level Paper.)

3. Let the extensions of OM and DB meet at P. Then

$$\begin{aligned} \angle OCN &= 180^\circ - \angle ONC - \angle CON \\ &= \angle OAC - \angle CON \\ &= 90^\circ - \angle MOC - \angle CON \\ &= 90^\circ - \angle MON \\ &= \angle OPN. \end{aligned}$$

Hence P lies on the circle passing through A, O, N and C. Moreover, OP is a diameter of this circle. Since $\angle OCP = 90°$, CP is a tangent to the circumcircle of $ABCD$. It follows that $CP^2 = PB \cdot PD$. Also, triangle POC is similar to triangle PCM, so that $CP^2 = PO \cdot PM$. From $PB \cdot PD = PO \cdot PM$, O, M, B and D are concyclic.

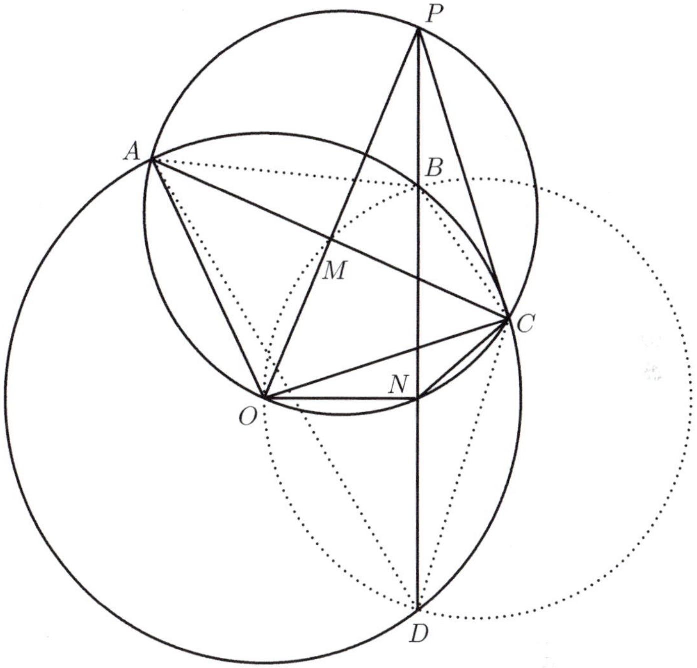

4. Let one piece of paper contain the numbers $a_1 > a_2 > \cdots > a_{2016} > 0$ and the other piece contain the numbers $b_1 > b_2 > \cdots > b_{2016} > 0$. Since the pairwise products of the 64 numbers are all positive, they are either all positive or all negative. Since their pairwise sums are also positive, they are all positive. We first show that ambiguity cannot arise. Let $x_1 > x_2 > \cdots > x_{64} > 0$ be such that their pairwise sums are the as and their pairwise products are the bs. Let $y_1 > y_2 > \cdots > y_{64} > 0$ be such that their pairwise sums are the bs and their pairwise products are the as. Then $x_1 + x_2 = a_1 = y_1y_2$, $x_1 + x_3 = a_2 = y_1y_3$, $y_1 + y_2 = b_1 = x_1x_2$ and $y_1 + y_3 = b_2 = x_1x_3$. Hence

$$\begin{aligned} a_1 - a_2 &= x_2 - x_3 \\ &= y_1(y_2 - y_3) \\ &= y_1(b_1 - b_2) \\ &= x_1y_1(x_2 - x_3) \\ &= x_1y_1(a_1 - a_2). \end{aligned}$$

It follows that $x_1y_1 = 1$. Similarly, we can show that $x_{64}y_{64} = 1$. This is a contradiction since $x_1y_1 > x_{64}y_{64}$. Let $x_1 > x_2 > \cdots > x_{64} > 0$ be the numbers. We assume that the as are the pairwise sums. Then $x_1 + x_2 = a_1$, $x_1 + x_3 = a_2$, $x_1x_2 = b_1$ and $x_1x_3 = b_2$. Then we have $a_1 - a_2 = x_2 - x_3$ while $b_1 - b_2 = x_1(x_2 - x_3)$. Hence $x_1 = \frac{b_1-b_2}{a_1-a_2}$. Now we check whether $a_1 - x_1 = \frac{b_1}{x_1}$ and $a_2 - x_1 = \frac{b_2}{x_1}$. If so, our assumption is correct. Otherwise, the as are the pairwise products.

5. In the diagram below, $ABCD$ is a unit square. A circle ω of diameter slightly greater than 1 is tangent to BC at a point E and intersecting AD at two points, F being the one closer to A. There is a slight separation of ω from AB. $A'B'C'D'$ is obtained from $ABCD$ first by a 45° rotation about its centre, then a reflection about $A'C'$ and finally a translation so that the separation from ω of $A'B'$ is the same as that of AB, and ω is tangent to $B'C'$ at a point E'. F' is the point of intersection of ω and $A'D'$ closer to A'. The curvilinear quadrilateral $ABEF$ is cut off from $ABCD$, reflected over and placed on $A'B'E'F'$. The two pieces can cover ω.

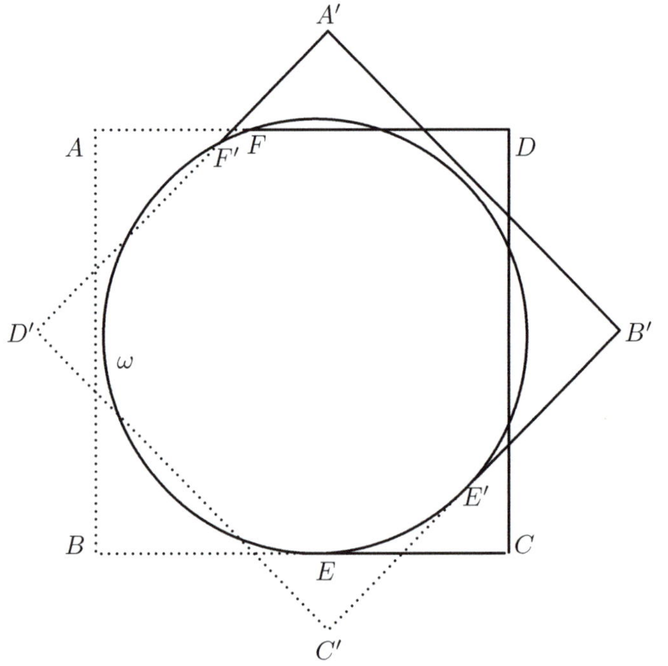

6. Bob can guarantee a win in 4 moves but not in 3. In 3 moves, whatever numbers Bob gives, Alice can always answer 0, 1 and 2. Let Bob's numbers a, b and c in that order. Alice can show Bob that her polynomial is $P(x) = (c-b)x^{2n} + b$ for some suitably chosen positive integer n. When $P(x) = c$, $(c-b)(x^{2n}-1) = 0$. Since $c \neq b$, $x^{2n} = 1$ and there are indeed two distinct integer roots, namely, ± 1. When $P(x) = b$, $(c-b)x^{2n} = 0$. Since $c \neq b$, $x^{2n} = 0$ and $x = 0$ is the only root. Finally, when $P(x) = a$, the only roots are $\sqrt[2n]{\frac{a-b}{c-b}}$. If $\frac{a-b}{c-b} < 1$, the roots are either non-real or non-integral. Since $a \neq c$, $\frac{a-b}{c-b} \neq 1$. Suppose $\frac{a-b}{c-b} > 1$. Since the positive integer n is arbitrary, it can always be chosen so that none of the roots are integral. We now prove that 4 moves are sufficient. Note that for integers x and y, $|P(x) - P(y)| = 1$ implies $|x - y| = 1$ since $P(x) - P(y)$ is divisible by $x - y$. Bob starts by calling 0. We consider four cases.
Case 1. Alice's answer is at least 3.
This means that $P(x) = 0$ has at least 3 distinct integral roots. Now every root of $P(x) = \pm 1$ must differ by exactly 1 from every root of $P(x) = 0$. Clearly, $P(x) = \pm 1$ cannot have integral roots. By calling 1 and -1, Bob wins in 3 moves.
Case 2. Alice's answer is 2.
Bob can win as in Case 1 unless the two roots of $P(x) = 0$ are integers differing by 2. Suppose Bob does not win by calling 1 and -1. We may assume by symmetry that $P(x) = 1$ has no integral roots and $P(x) = -1$ has one. Now $P(x) = -2$ can have at most 2 distinct integral roots, and Bob wins by calling -2.
Case 3. Alice's answer is 1.
Then each of $P(x) = 1$ and $P(x) = -1$ has at most 2 distinct integral roots. If Bob does not win after calling 1 and -1, we may assume by symmetry that $P(x) = 1$ has no integral roots while $P(x) = -1$ has two. Then $P(x) = -2$ has at most one integral root, and Bob wins by calling -2.
Case 4. Alice's answer is 0.
Bob calls -1. We consider four subcases.
Subcase 4(a). Alice's answer is 0.
Then Bob has won already.
Subcase 4(b). Alice's answer is 1.
Bob calls -2. Since $P(x) = -2$ has at most two distinct integral roots, Bob wins by calling -2 unless it has exactly two such roots. Now Bob wins by calling -3.
Subcase 4(c). Alice's answer is 2.
Then $P(x) = -2$ has at most one integral root, and Bob wins by calling -2 unless it has exactly one integral root. Now Bob wins by calling -3.
Subcase 4(d). Alice's answer is at least 3.
Bob wins by calling -2 since $P(x) = -2$ cannot have any integral roots.

7. Let $S = \{a_1, a_2, \ldots, a_n\}$, where each a_k, $1 \leq k \leq n$, is either ℓ or r. It denotes a sequence of jumps where the kth jump is to the left if $a_n = \ell$ and to the right if $a_k = r$. Let $f(S)$ be the number of possible ways of carrying out S from the initial configuration of frogs. We claim that $f(\ell\ell\ldots\ell) = f(rr\ldots r)$. Note that $f(S\ell) = f(Sr)$ for any S. Clearly, we have equality up to the last jump, and the number of possible last jumps depends only on the number of groups of adjacent frogs at that point. We also have $f(S\ell rS') = f(Sr\ell S')$ for any S and S'. Again, we have equality up to the completion of S. Consider $S\ell rS'$. If the two switched jumps are made by different frogs, we can just switch their order and continues. If they are made by the same frog but not the leftmost one, we can replace them with jumps r and ℓ by the frog immediately to its left, with at least one space in between. Finally, if both jumps are made by the leftmost frog, we can replace them with jumps r and ℓ by the rightmost frog. These transformations allow us to conclude that $f(S)$ depends only on the length of S.

Spring 2017

Junior O-Level Paper

1. Out of n consecutive numbers exactly one must be a multiple of n. Since 20170, 201700, and 2017000 are all divisible by 2017, then $\overline{2016a}$, $\overline{2016ab}$ and $\overline{2016abc}$ cannot be divisible by 2017 respectively for any digits a, b and c. Therefore the smallest multiple must take the form $2016 \cdot 10^4 + k$ where K is an integer, $0 \leq k \leq 2016$. Since $2016 \cdot 10^4 = (2017-1)(4 \cdot 2017 + 1932)$, picking $k = 1932$ will give us the multiple. In conclusion 20161932 is the smallest multiple of 2017 whose digits start with 2016.

2. Let the polynomial be $y = (x-r)^2$. Since it passes through (p, q), $q = (p-r)^2$. Since $x^2 + px + q$ also has repeated roots, $p^2 - 4q = 0$. Hence $p^2 = 4q = 4(p-r)^2$. If $p = 2p - 2r$, then $p = 2r$ and $q = r^2$. This is the original polynomial. If $p = 2r - 2p$, then $p = \frac{2}{3}r$ and $q = \frac{1}{9}r^2$, so that $x^2 + px + q = (x + \frac{1}{3}r)^2$.

3. The extension of AD cuts the circumcircle at the midpoint K of the arc BC. Hence OBK and OCK are congruent isosceles triangles. Since $\angle BOC = 2\angle BAC$, both triangles are in fact equilateral. Hence triangles OBD and KBD are also congruent, so that $\angle ODB = \angle KDB = \angle ADC$.

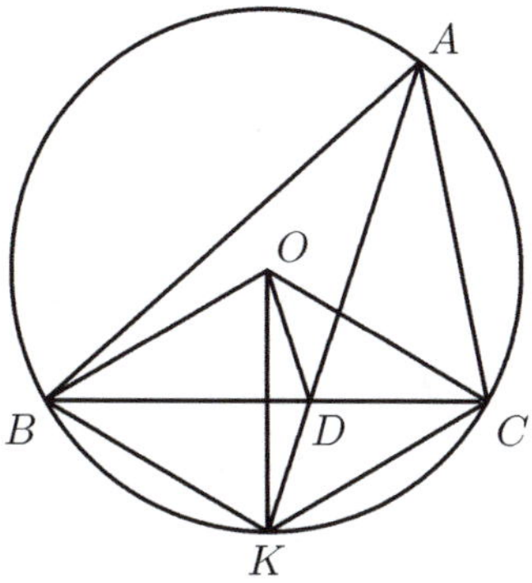

4. Label the children's heights 1 to 100, and the seats 1 to 100 going from right to left. On the first move, order the children in seats 51 to 100. On the second move, order the children in seats 1 to 50. On the third move, order the children in seats 26 to 75. Repeat this process for the fourth, fifth and sixth moves. If the child is of height 1 to 25, then after the second move, it must be sitting in one of the seats labeled 1 to 50. Therefore after the fifth move, they will be sitting in the correct spot. Similarly after the fourth move any child of heigh 76 to 100 will be sitting in the correct spot. Therefore the children of heights 26 to 75 are sitting in the seats labeled 26 to 75, but perhaps not in the right order. The sixth move guarantees everyone moves into their correct seat.

5. The vertices of the polygon must lie alternately on two perpendicular lines passing through the common point of the circles.

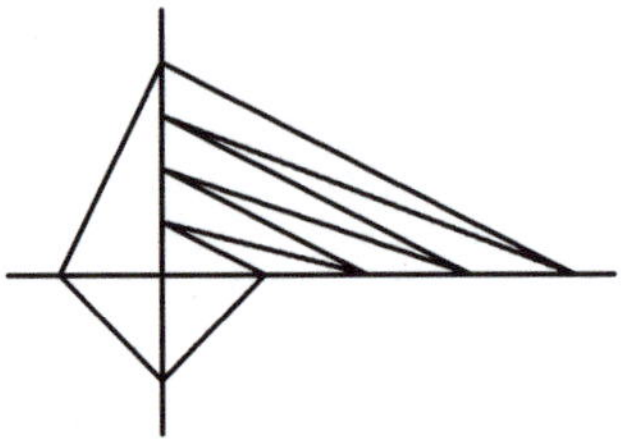

   (a) The diagram above shows an example for $n = 10$.
   (b) We cannot have $n = 11$ or any other odd values as the polygonal line will not close.

Senior O-Level Paper

1. The vectors A_1A_2, A_5A_6 and A_9A_{10} have zero sum. The vectors A_2A_3 and A_8A_9 have zero sum. The vectors A_4A_5 and $A_{10}A_{11}$ have zero sum. Hence these seven vectors have zero sum.

2. Note that $\angle APB = \angle AOB$. Hence $ABOP$ is a cyclic quadrilateral so that $\angle PAO = \angle PBO$. We also have $OA = OB$ and $OC = OD$. Either triangles AOC and BOD are congruent, or $\angle AOC + \angle BOD = 180°$. The latter clearly cannot hold. It follows that $\angle AOC = \angle BOD$, so that $\angle CPD = \angle APB = \angle AOB + \angle AOC - \angle BOD = \angle COD$.

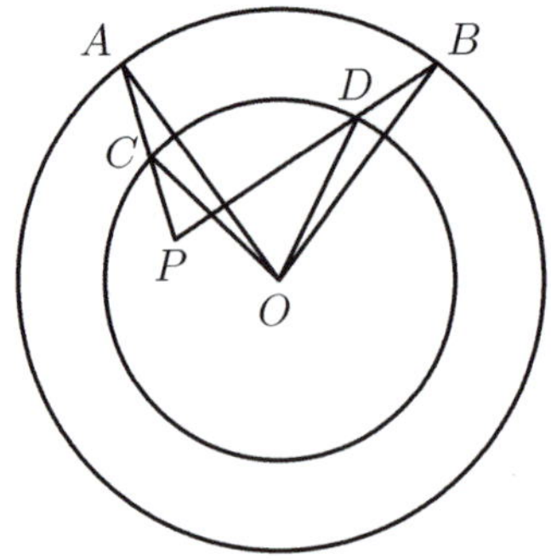

3. Obviously, if $n = 1$, all the numbers in the table must be equal. This is the only case. If $n > 1$, let p be the smallest prime divisor of n. All numbers on any diagonal from the northwest to the southeast are equal. Start with 1 in the southwest corner square, 2s on the next diagonal, 3s on the diagonal above that, and so on. After a diagonal with ps, the nesxt diagonal will have 1s again. Not all the numbers in the table are equal. However, in any rectangle consisting of n squares, the numbers consist of $\frac{n}{p}$ sets of 1, 2, ..., p, and they have constant sum.

4. Take any eight children out of the circle, and put them back in order around the two that did not move. This shows that eight moves are sufficient. Arrange the children so their heights increase in counterclockwise order. Now no matter which seven children are moved, the three that do not move will always be in the wrong order.
(Compare with Problem 4 of the Junior O-Level Paper.)

5. Let one of the polynomials be $y = -x^2$ with axis $x = 0$. Let one of the points of intersection be $(1, -1)$. The slope of the tangent at this point is -2. Let the other polynomial be $y = x^2 + bx + c$ with axis $x = -\frac{b}{2}$. Since it passes through $(1, -1)$, $b + c = -2$. The slope of the tangent at this point is $b + 2$. Since the two tangents are perpendicular, $-2(b+2) = -1$. Hence $b = -\frac{3}{2}$ and $c = -\frac{1}{2}$. Solving $-x^2 = x^2 - \frac{3}{2}x - \frac{1}{2}$, we have $0 = 4x^2 - 3x - 1 = (x - 1)(4x + 1)$. Hence the other point of intersection is $(-\frac{1}{4}, \frac{1}{16}$. At this point, the slopes of the tangents are $-2(-\frac{1}{4}) = \frac{1}{2}$ and $2(-\frac{1}{4}) - \frac{3}{2} = -2$. Their product is indeed -1. However, since $b \neq 0$, the two graphs do not have a common axis.

Junior A-Level Paper

1. Suppose that at most five of them are from the same town. Even if we have two groups of five, the number of games played among them is only $2 \times \binom{5}{2} = 20$. This is less than half of the total number of games, which is $\binom{10}{2} = 45$. Hence there must be six participants from the same town. Thus in each round, some two of them will play each other by the Pigeonhole Principle.

2. (a) (b)

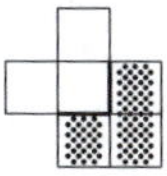

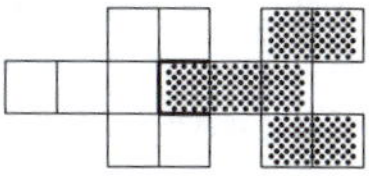

(c)

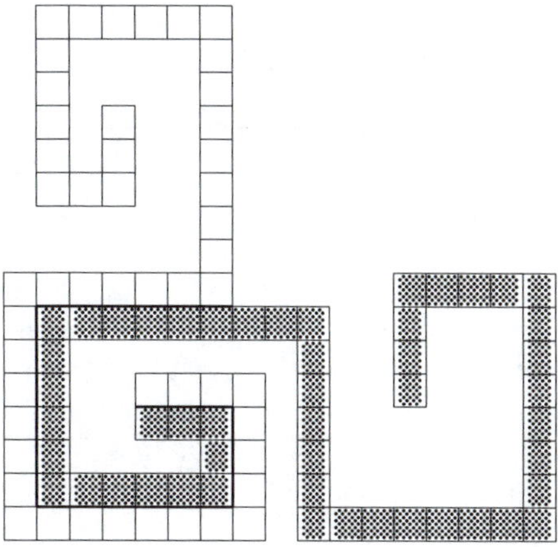

3. (a) It can happen. Take the given positive integers to be $\frac{1}{2}$ and $1+\frac{1}{64}$. Since there are only two numbers the sequence can only change between increasing and decreasing at most once, so it suffices to check $a_4 > a_5$ and $a_5 < a_6$. For the first inequality, notice that $(1+\frac{1}{2^6})^4 < 2$. Then

$$\begin{array}{rcl} a_4 - a_5 & = & \left(\frac{1}{2}\right)^5 + \left(1+\frac{1}{2^6}\right)^5 - \left(\frac{1}{2}\right)^4 - \left(1+\frac{1}{2^6}\right)^4 \\ & = & \frac{1}{2^4}\left(\frac{1}{2}-1\right) + \frac{1}{2^6}\left(1+\frac{1}{2^6}\right)^4 \\ & = & \frac{1}{2^6}\left(1+\frac{1}{2^6}\right)^4 - \frac{1}{2^5} \\ & < & 0. \end{array}$$

For the second inequality, notice that $(1+\frac{1}{2^6})^5 > 1$. Then, by a calculation similar to the above, $a_6 - a_5 = \frac{1}{2^6}(1+\frac{1}{2^6})^5 - \frac{1}{2^6} > 0$.

(b) It cannot happen. Out of the given positive integers, at least one must be larger than 1 as otherwise the sequence will be decreasing for all terms. The nth power of such a number gets larger without bound as n increases. Hence for large enough n, a_n must always grow without bound, so it is impossible to get an eventually decreasing sequence.

4. Let AD and BE intersect at I. We consider two cases.
Case 1. I is the midpoint of AD.
If $BI \neq EI$, then $AB \neq DE$. Hence I must also be the midpoint of BE. It follows that AB is parallel to DE. Since $AF = EF$ and $BC = DC$, both C and F lie on the line halfway between AB and DE, so that CF also passes through I. The same reasoning shows that I is also the midpoint of CF. Hence triangles IAB, IBC, ICD, IDE, IEF and IFA are congruent to one another, so that I is equidistant to the six sides of $ABCDEF$. Thus the hexagon has an incircle centred at I.
Case 2. I is not the midpoint of AD.
Then it is not the midpoint of BE either. We may assume that $AI < DI$. Since $AB = DE$, we must have $BI > EI$. In fact, we must hve $AI = EI$ and $BI = DI$. Once again, symmetry forces CF to pass through I, and we must have $CI = AI$ and $FI = DI$. As in Case 1, we have six congruent triangles and I is the incentre of $ABCDEF$.

5. Seven tokens are sufficient. Let their weights be 0.5, $0.5 \times 2 - 0.5 = 0.5$, $0.5\times2+0.5 = 1.5$, $1.5\times2-0.5 = 2.5$, $2.5\times2+0.5 = 5.5$, $5.5\times2-0.5 = 10.5$ and $10.5 \times 2 + 0.5 = 21.5$ grams respectively. It is routine to verify that every object with integral weight up to 41 grams can be balanced. Suppose we only have six tokens. Put one of them aside. Then there are $2^5 = 32$ sets formed from the other five tokens. We associate each set with a companion set formed by adding the token set aside. Since this token has a non-integral weight, at most one set from each associated pair can balance an object with integral weight. Since $32 < 40$, six tokens are not sufficient.

6. If there are 63 squares, it is impossible for the grasshopper to complete its task. Number the squares 1 to 63 and divide them into seven blocks of 9 squares. colour the blocks alternately red and blue. The grasshopper can jump from a red square to a red square at most 7 times, namely between the pairs (1,9), (9,19), (19,27), (27,37), (37,45), (45,55) and (55,62). There are 27 blue squares. Even if the grasshopper never jumps from a blue square to another blue square, it can only jump from a blue square to a red square 27 times. The starting square may be red, but this accounts for only 27+7+1=35 red squares. We have a contradiction since there are 36 red squares.

7. We answer all three parts together by showing that 42 dominoes can be placed on the board according to the rules. We divide the board into two halves by a diagonal. We claim that the configuration of 21 dominoes shown in the diagram below can fit on each half.

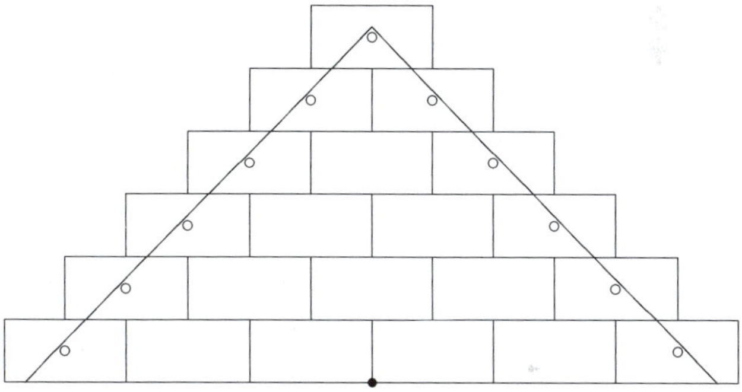

Place this configuration so that the black dot coincides with the centre of the board and the long edge runs along the dividing diagonal. The slanting lines shown are two edges of the board. The centres of the dominoes which are in somewhat insecure positions are marked with white dots. The distance between the black dot and the top white dot is 5.5 while the distance between the centre of the board and a corner is $4\sqrt{2}$. Since $11^2 = 121 < 128 = 2(8^2)$, $11 < 8\sqrt{2}$ so that $5.5 < 4\sqrt{2}$. Since the white dots form two lines of slopes ± 1 from the top one, all of them stay on the board as claimed.

Senior A-Level Paper

1. A line equidistant from two vertices of a triangle is either parallel to the side they determine, or passing through the midpoint of that side. If there are at least four lines of the first type, two of them must be parallel since each of them is parallel to one of three sides. If there are at most three lines of the first type, then there are at least seven lines of the second type. Three of them must be concurrent since each of them passes through one of three midpoints.

2. (See Problem 3 of the Junior A-Level Paper.)

3. A cube can be dissected into two congruent pieces as shown in the diagram below on the left. Each consists of one hexagonal face, three pentagonal faces (hidden) and three triangular faces.

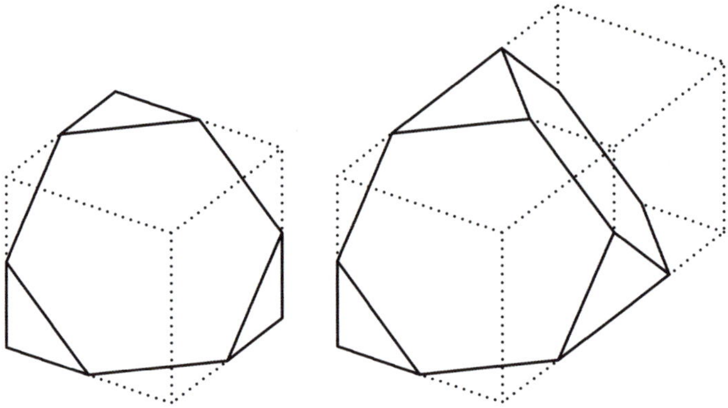

When the two pieces are reassembled by abutting one pentagonal face from each piece, two pairs of pentagonal faces merge into two hexagonal faces (hidden) and two pairs of triangular faces merge into two triangular faces. The resulting convex polyhedron, as shown in the diagram above on the right, has four hexagonal faces and four triangular faces.

4. The frame may enclosed the ten shaded squares in the diagram below, which also shows the corresponding painting of the board in colours 0, 1, 2, 3, 4, 5, 6, 7, 8 and 9.

0	2	4	6	8	**1**	3	5	7	9
1	3	5	7	9	**0**	**2**	4	6	8
0	2	4	6	8	1	**3**	**5**	7	9
1	3	5	7	9	0	2	**4**	**6**	8
0	2	4	6	8	1	3	5	**7**	**9**
1	3	5	7	**9**	**0**	2	4	6	**8**
0	2	4	**6**	**8**	1	3	5	7	9
1	3	**5**	**7**	9	0	2	4	6	8
0	**2**	**4**	6	8	1	3	5	7	9
1	**3**	5	7	9	0	2	4	6	8

5. Let PQR be the triangle such that A, B and C are the respective midpoints of QR, RP and PQ. Then the median from A passes through P. The altitude from B is perpendicular to CA and therefore to RP. Hence the reflection of the median AP at the point Y on this altitude passes through R. Similarly, the reflection of the median AP at the point Z on the altitude from C passes through Q. Let the extension of QZ intersect RY at X. Now $\angle XZY = \angle QZA = 2(90° - \angle QZC) = 2\angle CPY$. Similarly, $\angle XYZ = 2\angle BPY$. It follows that

$$\begin{aligned} \angle ZXY &= 180° - \angle XYZ - \angle XZY \\ &= 180° - 2\angle BPC \\ &= 180° - 2\angle CAB \\ &= 90°. \end{aligned}$$

Hence A is the circumcentre of the right triangle XQR, so that we have $AX = \frac{1}{2}QR = BC$.

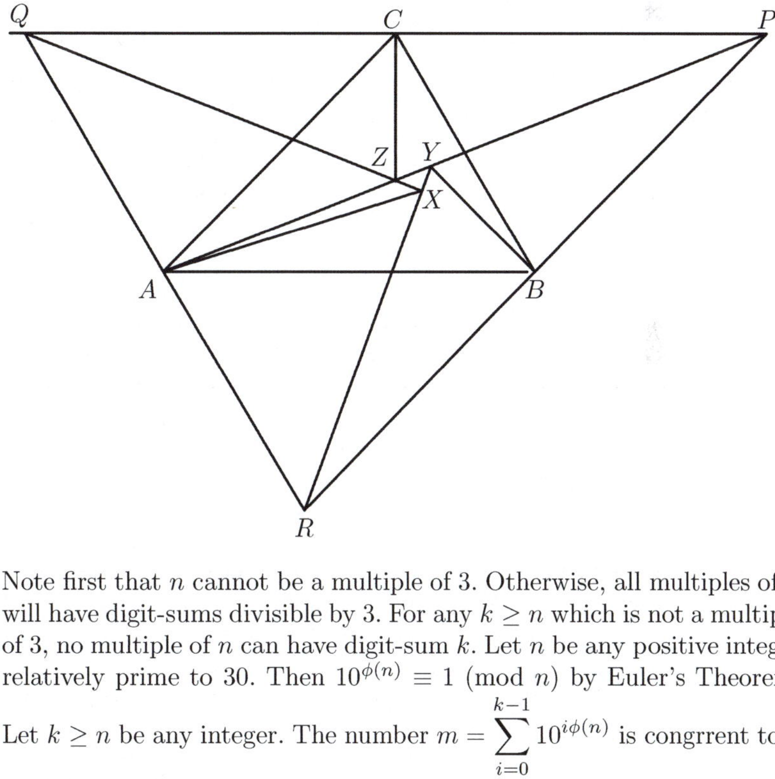

6. Note first that n cannot be a multiple of 3. Otherwise, all multiples of n will have digit-sums divisible by 3. For any $k \geq n$ which is not a multiple of 3, no multiple of n can have digit-sum k. Let n be any positive integer relatively prime to 30. Then $10^{\phi(n)} \equiv 1 \pmod{n}$ by Euler's Theorem. Let $k \geq n$ be any integer. The number $m = \sum_{i=0}^{k-1} 10^{i\phi(n)}$ is congrrent to k modulo n. It has digit-sum k since it consists of k copies of 1 separated by copies of 0. If k is a multiple of n, then m is the desired multiple. Otherwise, there exists a positive integer $d < n$ such that $k + 9d$ is a multiple of n.

Let each of the last d copies of 1 in m trade places with the copy of 0 to the left. Equivalently, we replace $10^{i\phi(n)} \equiv 1 \pmod{n}$ by $10^{i\phi(n)+1} \equiv 10 \pmod{n}$ for $0 \leq i \leq d-1$. The new number still has digit-sum k, but is now congruent to $k + 9d \equiv 0 \pmod{n}$. Finally, let n be any positive integer relatively prime to 3. Let ℓ be the number obtained from n by removing the 2s and the 5s from its prime factorization, and let m be the desired multiple for ℓ as constructed above. By adding a sufficient number of 0s at the end, this will be a multiple of n which still has digit-sum k. In the special case $\ell = 1$, take m to be the number consisting of k copies of 1.

7. More generally, let there be $n \geq 2$ gangsters and let $g(n)$ be the largest possible number of gangs. Define

$$f(n) = \begin{cases} 3^k & n = 3k, \\ 4(3^{k-1}) & n = 3k+1, \\ 2(3^k) & n = 3k+2. \end{cases}$$

Note that if n is partitioned into any number of positive integers, the largest product of these integers is $f(n)$. If we take $f(0) = f(1) = 1$, then $f(k)f(n-k) \leq f(n)$ for $0 \leq k \leq n$. We claim that $g(n) = f(n)$ for all $n \geq 2$. In particular, $g(36) = 3^{12}$. We first prove that $g(n) \geq f(n)$ for all $n \geq 2$. Partition the n gangsters into groups of 3. If there is one left over, add him to an existing group. If there are two left over, start a new group. Now form all possible gangs with one member from each group. The number of gangs is $f(n)$, and these gangs satisfy all conditions of the problem. Hence $g(n) \geq f(n)$. We now use mathematical induction on n to prove that $g(n) \leq f(n)$ for all $n \geq 2$. We may take $g(0) = 1 = f(0)$ and $g(1) = 1 = f(1)$. It is routine to verify that $g(n) = n = f(n)$ for $2 \leq n \leq 4$. For $n \geq 5$, assume that $g(k) = f(k)$ for $0 \leq k \leq n-1$. We shall prove that $g(n) \leq f(n)$. Construct a graph with n vertices representing the n gangsters. Two vertices are joined by an edge if and only if the gangsters they represent are enemies. If the graph is not connected, let there be k vertices in one component, $1 \leq k \leq n-1$. Then each gang is a union of a subgang consisting of some gangsters represented by vertices in this component and a subgang consisting of some of the remaining gangsters. There are at most $g(k)$ subgangs of the first type, and at most $g(n-k)$ subgangs of the second type. Hence $g(n) \leq g(k)g(n-k)$ since no two gangs can have identical membership. By the induction hypothesis, $g(k)g(n-k) \leq f(k)f(n-k) \leq f(n)$. Henceforth, assume that the graph is connected.

Suppose it is in fact a cycle or a path. Then there are three consecutive vertices each of degree 2. Let them represent the gangsters B, C and D, with B also hostile to A and D to E. If a gang does not contain any of B, C and D, then C has no enemies in it, which is a contradiction. Consider the gangs which contain B. Then it can contain neither A nor C. Thus there are at most $g(n-3)$ of them. Similarly, there are at most $g(n-3)$ gangs which contain C, and $g(n-3)$ gangs which contain D. It is possible that a gang may contain both B and D, but this will only reduce the total number of gangs. Then $g(n) \leq 3g(n-3) \leq 3f(n-3) \leq f(n)$ by the induction hypothesis. Finally, suppose there is a vertex of degree at least 3. Let it represent a gangster W who has at least three enemies X, Y and Z. The number of gangs not containing W is at most $g(n-1)$. If a gang contains W, it cannot contain any of X, Y and Z. Hence the number of gangs containing W is at most $g(n-4)$. By the induction hypothesis, $g(n) \leq g(n-1) + g(n-4) \leq f(n-1) + f(n-4) \leq f(n)$.

Tournament 39

Fall 2017

Junior O-Level Paper

1. Five nonzero numbers are added in pairs. Five of the sums are positive and the other five are negative. If they are multiplied in pairs, find the numbers of positive and negative products.

2. Do there exist 99 consecutive positive integers such that the smallest one is divisible by 100, the next by 99, the third by 98, an so on, so that the last one is divisible by 2?

3. Among 100 coins in a row are 26 fake ones which form a consecutive block. The other 74 coins are real, and they have the same weight. All fake coins are lighter than real ones, but their weights are not necessarily equal. What is the minimum number of weighings on a balance to guarantee finding at least one fake coin?

4. A treasure is buried in one of the squares of an 8×8 board. A metal detector is on one of the corner squares. In each move, it can only go from one square to another with a common side. The metal detector beeps if the treasure is in the same square or one of the four squares sharing a common side. Is it possible to identify the square with the treasure in at most 26 moves?

5. A circle of radius 1 is drawn on a checkerboard so that it contains an entire white square of side 1. Prove that the parts of the circumference passing through white squares have total length at most one third of the total length of the circumference.

Note: The problems are worth 3, 4, 4, 5 and 5 points respectively.

Senior O-Level Paper

1. Do there exist non-integer numbers x and y such that $\{x\}\cdot\{y\} = \{x+y\}$, where $\{x\} = x - \lfloor x \rfloor$?

2. Let M be the midpoint of the side AC of triangle ABC. Let L be the point on AB such that CL bisects $\angle BCA$. The line through M perpendicular to AC intersects CL at K. Prove that the circumcircles of triangles ABC and AKL are tangent.

3. Twenty-one non-zero numbers are added in pairs. Half of the sums are are positive and the other half are negative. If the numbers are multiplied in pairs, what is the largest possible number of positive products?

4. A sphere intersects each face of a regular tetrahedron in a circle. Is it possible for the radii of these circles to be 1, 2, 3 and 4 respectively, if the radius of the sphere

   (a) can been chosen freely;

   (b) must be 5?

5. On a 100×100 board, an ant starts from the bottom left corner, visits the top left corner and ends on the top right corner. It goes between squares sharing a common side. The moves are alternately vertical and horizontal, with the first move horizontal. Must there exist two adjacent squares such that the ant has gone from one to the other at least twice?

Note: The problems are worth 4, 4, 4, 2+3 and 5 points respectively.

Junior A-Level Paper

1. We have a faulty balance with which equilibrium may only be obtained if the ratio of the total weights in the left pan and in the right pan is 3:4. We have a token of weight 6 kg, a sufficient supply of sugar and bags of negligible weight to hold the sugar. In each weighing, you may put the token or any bags of sugar of known weight on the balance, and add sugar to a bag so that equilibrium is obtained. Is it possible to obtain a bag of sugar of weight 1 kg?

2. There are two circular coins with radius 1 cm, two with radius 2 cm and two with radius 3 cm. They are to be placed on a flat surface, without overlapping, one at a time in any order. The first two must be tangent to each other. Coins placed subsequently must be tangent to at least two circles already placed. Is it possible to place coins so that the centres of three of them are collinear?

3. A financial advisor predicts the exchange rate for the Euro. His initial prediction is correct. In each of three months, his prediction is correct in the percentage, which is between 0% and 100%, but wrong in the direction. However, his final prediction is correct. Has the Euro risen or fallen over the three months?

4. One hundred doors and one hundred keys are numbered 1 to 100 respectively. Each door is opened by a unique key whose number differs from the number of the door by at most one. Is it possible to match the keys with the doors in n attempts, where

(a) $n = 99$;
(b) $n = 75$;
(c) $n = 74$?

5. The digits of two integers greater than 1 are in reverse order of each other. Is it possible that every digit of their product is 1?

6. The incircle of triangle ABC is tangent to BC, CA and AB at K, M and N respectively. The extensions of MN and MK intersect the exterior bisector of $\angle ABC$ at R and S respectively. Prove that RK and S intersect on the incircle of ABC.

7. A Lego structure consists of stacks on an $m \times n$ board. Initially, there are mn stacks of height 1. Two stacks whose bases share a common side may be merged into a single stack of their combined height. The shorter stack goes on top of the taller one. If they are of the same height, either can be put on top of the other. This continues until there are no more adjacent stacks. What is the smallest possible number of stacks in the final structure if

(a) $m = n = 20$;
(b) $m = 50$ and $n = 90$?

Note: The problems are worth 4, 4, 6, 1+3+4, 9, 9 and 5+5 points respectively.

Senior A-Level Paper

1. One hundred doors and one hundred keys are numbered 1 to 100 respectively. Each door is opened by a unique key whose number differs from the number of the door by at most one. Is it possible to match the keys with the doors in n attempts, where

(a) $n = 99$;
(b) $n = 75$;
(c) $n = 74$?

2. Six circles of radius 1 with centres at the vertices of a regular hexagon are drawn, so that the centre O of the hexagon lies inside all six circles. An angle with measure θ and vertex O cuts out six arcs in these circles. Prove that the sum of the lengths of these arcs is equal to 6θ.

3. A financial advisor predicts the exchange rate for the Euro. His initial prediction is correct. In each of twelve months, his prediction is correct in the percentage, which is between 0% and 100%, but wrong in the direction. However, his final prediction is correct. Has the Euro risen or fallen over the twelve months?

4. Prove that in any infinite sequence $\{a_n\}$ in which each term is either 1 or -1, we can choose n and k such that

$$|a_0a_1\cdots a_k + a_1a_2\cdots a_{k+1} + \cdots + a_na_{n+1}\cdots a_{n+k}| = 2017.$$

5. Let a positive number be given. A piece of cheese cut so that a new part is created with each cut. Moreover, after every cut, the ratio of the weight of any piece to the weight to any other one must be at least r.

   (a) Prove that if $r = 0.5$, we can cut the cheese so that the process will never stop.

   (b) Prove that if $r > 0.5$, then at some point we will have to stop cutting.

   (c) What is the greatest number of parts we can obtain if $r = 0.6$?

6. The excircles of triangle ABC opposite A and B are tangent to BC and CA at D and E, respectively. K is the point of intersection of AD and BE. Prove that the circumcircle of triangle AKE passes through the midpoint of JC, where J is the excentre of ABC opposite C.

7. The streets of an $n \times n$ city run from north to south and from west to east, dividing the city into 1×1 blocks. A man walks every day from the south-west corner to the north-east corner, moving only north or east, and then returns, moving only south or west. Whether going forth or going back, he always chooses his path so as to maximize the total length of its parts not previously covered. Prove that in n days he will cover the entire length of every street.

Note: The problems are worth 1+2+3, 5, 6, 8, 3+4+4, 10 and 10 points respectively.

Spring 2018

Junior O-Level Paper

1. Six rooks are placed on the squares of a 6×6 board so that no two of them attack each other. Is it possible for every empty squares to be attacked by two rooks

   (a) at the same distance away;

   (b) at different distances away?

2. In triangle ABC, $\angle BCA = 90°$. K is a point on AB where $AK = AC$. L is a point on AC where $CL = BK$. Prove that triangle CLM is isosceles.

3. In a 4×4 square, the sum of the numbers in each row and in each column is the same. Nine of the numbers are erased, leaving behind the seven shown below. Is it possible to recover uniquely

   (a) one;

   (b) two

   of the erased numbers?

1			2
	4	5	
	6	7	
3			

4. Each of three positive integers is a multiple of the greatest common divisor of the other two, and a divisor of the least common multiple of the other two. Must these three numbers be the same?

5. On the plane are 30 green points and 7 red lines such that none of the lines passes through any of the points. Is it possible that every segment joining two green points intersects at least one red line?

Note: The problems are worth 1+2, 4, 2+2, 4 and 5 points respectively.

Senior O-Level Paper

1. The angle bisector and the altitude from one vertex of a triangle divide the opposite side into three segments. Is it possible for these three segments to form the sides of a triangle?

2. Each of four positive integers is a multiple of the greatest common divisor of the other three, and a divisor of the least common multiple of the other three. Prove that the product of the four integers is the square of an integer.

3. Two circles are tangent to each other at T and to a line at A and B respectively. The common tangent at T intersects AB at M. AC is a diameter of one circle and O is the centre of the other circle. Prove that CM is perpendicular to AO.

4. A corner square of an 8×8 board is painted, and a counter is put on it. Peter and Basil take turns moving the counter, Peter going first. In his turn, Peter moves the counter once as a queen to an unpainted square, and Basil moves the counter twice as a king to unpainted squares. The square visited by Peter and both squares visited by Basil are then painted. The player without a move loses the game. Which player has a winning strategy?

5. In a polyhedron, exactly three faces meet at each vertex. Each face is painted red, yellow or blue. Prove that the number of vertices surrounded by three faces of different colours is even.

Note: The problems are worth 3, 4, 4, 5 and 5 points respectively.

Junior A-Level Paper

1. Thirty-nine nonzero numbers are written in a row. The sum of any two adjacent numbers is positive, while the sum of all the numbers is negative. Is the product of all these numbers negative or positive?

2. Aladdin starts with a positive number of dollars. In each move, the Genie gives him 1000 dollars and then takes away half the money Aladdin then has. It turns out that in every move, the amount of money taken away by the Genie is an integral number of dollars. Is it possible for Aladdin to have more money than what he has initially after ten moves?

3. Do there exist 2018 positive irreducible fractions, each with a different denominator, so that the denominator of the difference of any two, reduced to the lowest terms, is less than the denominator of any of the 2018 fractions?

4. O is the circumcentre of triangle ABC. H is the foot of the altitude from A to BC, and P is the foot of the perpendicular from A to CO. Prove that the line HP bisects AB.

5. On a street are 50 pairs of houses. In each pair, the houses are directly across the street from each other. On the east side of the street, the house numbers are even numbers, while on the west side, they are odd. On each side of the street, the house numbers increase from south to north, but those of adjacent houses may differ by more than 2. In each pair, the house number on the west side is subtracted from the house number on the east side. All 50 differences are distinct. What is the smallest possible value of the largest house number on this street?

6. Each of ten people standing at the vertices of a regular decagon is either a knight who always tells the truth, or a knave who always lies. At least one of them is a knave. In each move, an outsider marks a point outside the decagon and ask the people, "What is the distance from my point to the nearest vertex on which a knave stands?" All ten people give their answers. What is the minimal number of moves the outsider needs to separate the knights from the knaves?

7. One of the expressions $a \otimes b$ and $a \odot b$ stands for $a + b$. The other stands for either $a - b$ or $b - a$, but consistently. Write an expression equal to $20a - 18b$, using only a, b and these two operations along with brackets.

Note: The problems are worth 4, 5, 6, 6, 8, 10 and 12 points respectively.

Senior A-Level Paper

1. Aladdin starts with a positive number of dollars. In each move, the Genie gives him 1000 dollars and then takes away half the money Aladdin then has. It turns out that in every move, the amount of money taken away by the Genie is an integral number of dollars. After ten moves, can Aladdin have more money than what he started with?

2. Do there exist 2018 positive irreducible fractions, each with a different denominator, so that the denominator of the difference of any two, reduced to the lowest terms, is less than the denominator of any of the 2018 fractions?

3. One hundred different numbers are written in the squares of a 10×10 table. In each move, you can take out a rectangle formed by the squares, perform a half-turn and then put it back. Is it always possible to arrange the numbers in the square so that they increase in every row from left to right, and in every column from top to bottom, in no more than 99 moves?

4. An equilateral triangle lying in the plane α is orthogonally projected onto a plane β, which is not parallel to α. The resulting triangle is again orthogonally projected onto a plane γ, and it turns out that the image is again an equilateral triangle.

(a) Prove that the dihedral angle between the planes α and β is equal to the dihedral angle between the planes β and γ;

(b) Prove that the plane β intersects the planes α and γ along perpendicular lines.

5. One of the expressions $a \otimes b$ and $a \odot b$ stands for $a + b$. The other stands for either $a - b$ or $b - a$, but consistently. Write an expression equal to $20a - 18b$, using only a, b and these two operations along with brackets.

6. $ABCD$ is a cyclic quadrilateral. The tangents α, β and δ to the circumcircle ω at A, B and D are drawn. P is the point of intersection the extensions of BA and CD. The line through P parallel to δ cut α and β respectively at U and V. Prove that the circumcircle of triangle CUV is tangent to ω.

7. Each of n wizards in a column wears a white hat or a black hat chosen with equal probability. Each can see the hats of the wizards in front of him, but not his own. Starting from the last wizard, each in turn guesses the colour of his own hat. Also, each wizard except the first one announces a positive integer which is heard by everyone. The wizrds can agree in advance on a collective strategy on what number each should announce, in order to maximize the number of correct guesses. Unfortunately, some of the wizards do not care for this, and may do as they like. It is not known who they are, but that there are exactly k of them. What is the maximal number of correct guesses which can be guaranteed by some collective strategy, despite the possible actions of the uncaring wizards?

Note: The problems are worth 4, 5, 6, 4+4, 10, 10 and 12 points respectively.

Solutions

Fall 2017

Junior O-Level Paper

1. Suppose there are more positive numbers than negative ones among the five. If four or more are positive, then at least six sums must be positive. Hence there are exactly three positive numbers. It follows that we have four positive and six negative products. If there are more negative numbers than positive ones among the five, there must be exactly three negative numbers. Once again, we have four positive and six negative products.

2. We can take the 99 consecutive numbers $100! - 100$, $100! - 99$, $\ldots$, $100! - 2$.

3. Weigh #1 to #25 against #76 to #100. If there is equilibrium, all these 50 coins are real. It follows that #50 and #51 are both fake. If there is no equilibrium, we may assume by symmetry that #76 to #100 are heavier. Then these 25 coins are real, whereas at least one of #1 to #25 is fake. It follows that #25 and #26 are both fake.

4. Starting from the square at the top left corner marked with +, the path follows the boldfaced letters in alphabetical order in the diagram below. For each square, the letters denote the locations where the machine will blip if the treasure is in that square. If the machine never blips, the treasure is in the blank square at the bottom left corner.

+ A	+,B	C	D	E	F	G	H
A +,B	**B** A,C	**C** B,D	**D** C,E	**E** D,F	**F** E,G	**G** F,H	**H** G,I
A	B,V	C,W	D,X	E,Y	F	G,I	**I** H,J
V	**V** U,W	**W** V,X	**X** W,Y	**Y** X,Z	Y	J	**J** I,K
U	**U** T,V	U,W	X,Z	**Z** Y	Z	K	**K** J,L
T	**T** S,U	R,T	Q	P,Z	O	L,N	**L** K,M
S	**S** R,T	**R** Q,S	**Q** P,R	**P** O,Q	**O** N,P	**N** M,O	**M** L,N
	S	R	Q	P	O	N	M

Except for four pairs, each square has a unique set of letters which can be used to identify the location of the treasure. If the machine blips for the first time on the square F, O, Q or S, the treasure is in one of the two squares in the corresponding ambiguous pair. We can then abandon the planned tour for the machine, and just check one of the two squares.

5. The circumference of a circle of radius 1 and enclosing the white square must intersect each of the four black squares adjacent to the white square along arcs linking two opposite sides of those squares. Hence the length of the chord joining the endpoints of each arc is at least 1 since the black square is of side 1. Since the radius of the circle is 1, the central angle subtended by each arc is at least 60°. It follows that the sum of the central angles subtended by the arcs passing through white squares is at most 120°, so that the total length of these arcs is at most one third that of the whole circumference.

Senior O-Level Paper

1. Since $0 < \{y\} < 1$, $\{x\} \cdot \{y\} < \{x\} < \{x\} + \{y\}$. Thus there are no solutions if $\{x\} + \{y\} = \{x + y\}$. If this is not the case, then we have $\{x\} \cdot \{y\} = \{x\} + \{y\} - 1$. Hence $(\{x\} - 1)(\{y\} - 1) = 0$, and we still cannot have solutions.

2. Construct the tangent to the circumcircle of ABC at A, as shown in the diagram below. Then $\angle TAB = \angle ACB$.

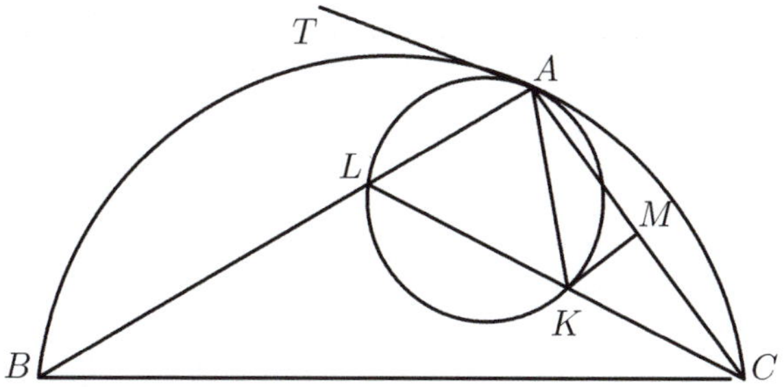

Since K lies on the perpendicular bisector of AC, $\angle KCA = \angle KAC$. Since CL is the bisector of $\angle BCA$, $\angle KCA = \angle KCB$. It follows that $\angle KAC = \angle KCB$. Now

$$\angle AKL = \angle KAC + \angle KCA = \angle KCB + \angle KCA = \angle ACB = \angle TAL.$$

Hence TA is also tangent to the circumcircle of AKL at A, so that the two circumcircles are tangent to each other.

3. By symmetry, we may assume that there are more positive numbers than negative ones among the twenty-one. If sixteen or more are positive, then at least 120 sums must be positive. Hence there are at most fifteen positive numbers. The number of negative products is the product of the numbers of positive and negative numbers respectively. The sum of these two numbers is twenty-one, and the number of negative products is minimum when these two numbers are as far apart as possible. It follows that in the optimal case, we have fifteen positive and six negative numbers. It follows that the maximum number of positive products is $210 - 15 \times 6 = 120$.

4. We shall answer (b) in the affirmative, so that (a) is also answered in the affirmative. Let $VABC$ be a regular tetrahedron. Construct a plane parallel to VAB, at a distance 3 away and on the same side of VAB as C. Construct a second plane parallel to VBC, at a distance 4 away and on the same side of VBC as A. Construct a third plane parallel to VCA, at a distance $\sqrt{21}$ away and on the same side of VCA as B. These three planes will intersect at a point O. Construct a sphere with radius 5, and lengthen VA, VB and VC at the same rate, so that the sphere is tangent to ABC. The sphere will intersect VAB in a circle of radius $\sqrt{5^2 - 3^2} = 4$, VBC in a circle of radius $\sqrt{5^2 - 4^2} = 3$ and VCA in a circle of radius $\sqrt{5^2 - 21} = 2$. At this point, the sphere intersects ABC along a circle of radius 0. If we parallel shift ABC towards V until it passes through the centre of the sphere, it will intersect the sphere along a circle of radius 5. At some point along this continuous motion, it will intersect the sphere along a circle of radius 1.

5. Paint the squares of the board black and white in the checkerboard pattern, with a black square at the bottom left corner. Then all horizontal moves are from black squares to white squares, while all vertical moves are from white to black squares. Since the square at the top left corner is white, the ant moves to it from the right. The existing path separates the board into a right part containing the square at the top right corner, and a left part which is not empty because the ant must move down from the square at the top left corner. In order to reach the square at the top right corner, the ant must recross the earlier path.

Suppose the first square on the path it revisits is black. Then during its earlier visit, the move is also horizontal but in the opposite direction. In both visits, a right angle turn is made in the next move. If they are in opposite directions, then in the second visit, the ant returns to the left part and must try again. To reach the right part, it must make a right angle turn in the same direction at some point. This is the desired conclusion.

Junior A-Level Paper

1. Put the token in the left pan and a bag of sugar in the right pan to obtain equilibrium. The weight of the bag is 8 kg. Replace the token by a bag of sugar to obtain equilibrium. The weight of the bag is 6 kg. Put the token and the 6-kg bag in the right pan. Put the 8-kg bag in the left pan and add a bag of sugar to the left pan to obtain equilibrium. The weight of this bag is 1 kg.

2. Place a coin of radius 1 cm and a coin of radius 2 cm so that they are tangent to each other. On each side of them, place a coin of radius 3 cm so that it is tangent to both of them. The centres form two triangles with side lengths 3 cm, 4 cm and 5 cm. Hence both are right triangles, so that the centres of the two circles of radius 3 are collinear with the centre of the circle of radius 1 cm.

3. Let the initial value of the Euro be 1. Let the percentage change for the ith month be p_i, where $0 < p_i < 100$ for $1 \le i \le 3$. Then the square of the value of the Euro at the end of the 3 months is

$$\frac{(100+p_1)(100-p_1)(100+p_2)(100-p_2)(100+p_3)(100-p_3)}{100^6}$$

$$= \frac{(100^2-p_1^2)(100^2-p_2^2)(100^2-p_3^2)}{100^6} < 1.$$

4. (a) For $1 \le k \le 99$, try door k with key k. If all attempt are successful, we know everything. Suppose the kth attempt fails for some k. Then key k must open door $k+1$ while key $k+1$ must open door k. This actually saves us one attempt.

(b) We claim that 3 attempts can settle 4 doors with 4 keys. Try door 3 with key 3. If the attempt is successful, one more attempt will settle doors and keys 1 and 2. Even if the doors and keys go beyond 4, one more attempt will settle door 4 and key 4. Suppose the attempt fails. Try door 2 with key 3. If the attempt is successful, then key 1 must open door 1 and key 2 must open door 3. One more attempt will settle door 4 and key 4. Suppose the attempt fails. Then key 3 must open door 4 and key 4 must open door 3. One more attempt will settle doors and keys 1 and 2. Iterating this process, we can accomplish the task in 75 attempts.

(c) Suppose that key k opens door k for $1 \le k \le 100$ but we do not know that. We claim that we need at least 75 attempts. For doors and keys 1 to 4, there are at least five possible scenarios.
(1) Key k opens door k for $1 \le k \le 4$.
(2) There is a switch between 1 and 2.
(3) There is a switch between 2 and 3.
(4) There is a switch between 3 and 4.
(5) There is a double switch between 1 and 2 as well as between 3 and 4.
Since two attempts can distinguish among at most four scenarios, we need at least three attempts to settle doors and keys 1 to 4. The same argument may be applied to each successive block of four doors and keys, justifying our claim.

5. Suppose such a pair of positive integers exist. Let the digits of one of them be a_1, a_2, $\dots$, a_{k-1}, a_k in that order. Since the product of the two numbers ends in a 1, we must have $(a_1, a_k) = (1, 1)$, (9,9), (3,7) or (7,3). Since the product also starts with a 1, only the first case is possible. Now, the product is less than $(2 \times 10^k - 1)(2 \times 10^k - 1)$, which simplifies to $4 \times 10^{2k} - 2$, so that it is a $2k - 1$ digits. In the second column from either side when the multiplication is performed, we must have $a_k a_2 + a_{k-1} a_1 + a_2 + a_{k-1} = 1$ since any carrying over will make the leftmost digit of the product greater than 1. Using an analogous argument on successive pairs of columns towards the middle, we arrive at the conclusion that the middle digit must be $a_1^2 + a_2^2 + \cdots + a_k^2 \ge 2 \ne 1$. This is a contradiction.

6. Let RK cut the incirle ω at P. Let $\angle NKM = \alpha$, $\angle KMN = \beta$ and $\angle MNK = \gamma$. Then $\alpha + \beta + \gamma = 180°$. Now $\angle RPN = \beta$ since it is an exterior angle of $KMNP$. We also have $\angle BNK = \beta$ since BN is tangent to ω. Finally, since NK and RS are both perpendicular to the bisector of $\angle ABC$, $\angle RBN = \beta = \angle RPN$, so that $BPNR$ is cyclic. Note that $\angle MSR = \alpha$ and $\angle MRS = \gamma$. Hence $\angle NPK = 180° - \alpha$ and $\angle NPB = 180° - \gamma$. It follows that $\angle BPK = 180° - \alpha$, so that $BPKS$ is also cyclic. We have $\angle KPS = \angle KBS = \beta = \angle RPN$. Since P lies on RK, it also lies on NS.

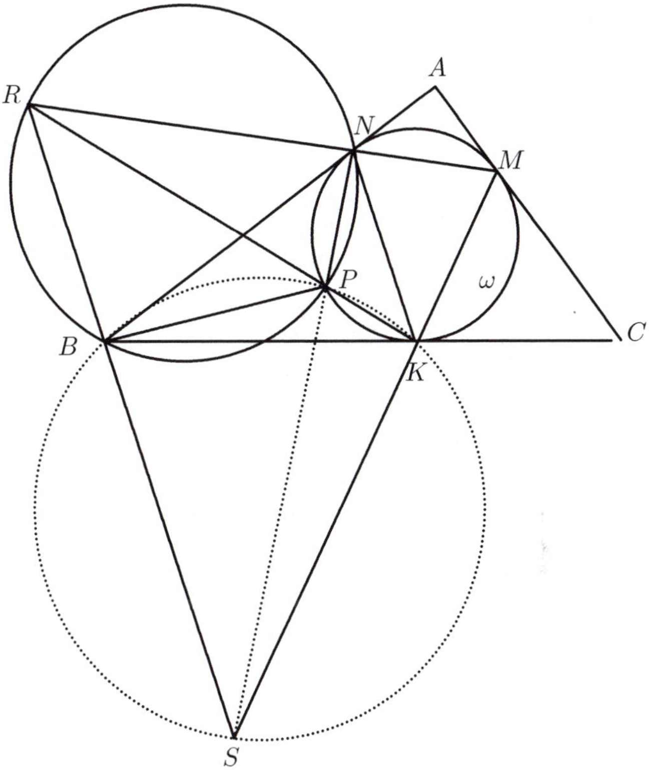

7. We first carry out some basic constructions. The 4 stacks of height 1 in a 2×2 board can be combined into a single stack of height 4.

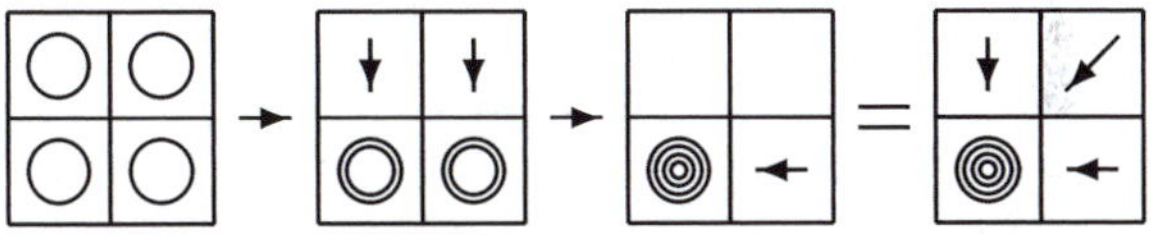

Similarly, the 16 stacks of height 1 in a 4×4 board can be combined into a single stack by applying the above process in two stages. The same result may be obtained for the following two configurations which are incomplete 4×4 boards.

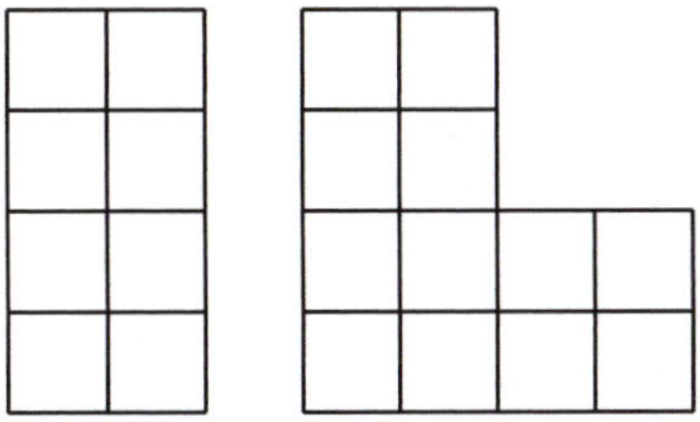

Finally, the 16 stacks of height in the configuration in the diagram below can be combined into a single stack of height 16.

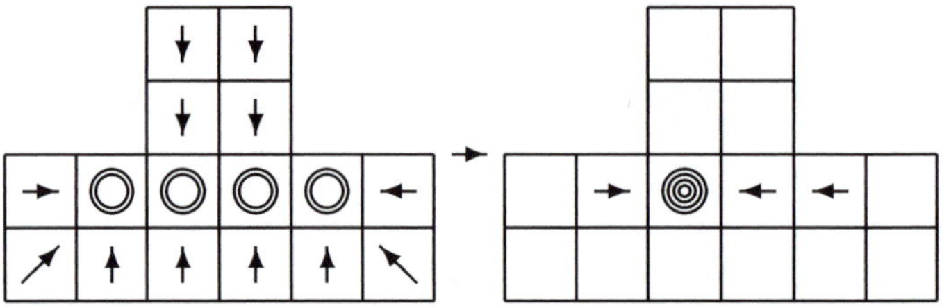

(a) The height of a stack can increase at most four times, and each time at most double. Thus the maximum height is 16. In a 20×20 board, the total height is 400. Hence the minimum number of stacks is $400 \div 16 = 25$. This can be attained by partitioning the 20×20 board into 25 copies of the 2×2 board.

(b) The minimum number of stacks in a 50×90 board is $\lceil \frac{50\times 90}{16} \rceil = 282$. We first note that the 160 stacks of height 1 in a 10×16 board may be combined into 10 stacks of height 16 by applying our basic constructions as shown in the diagram below.

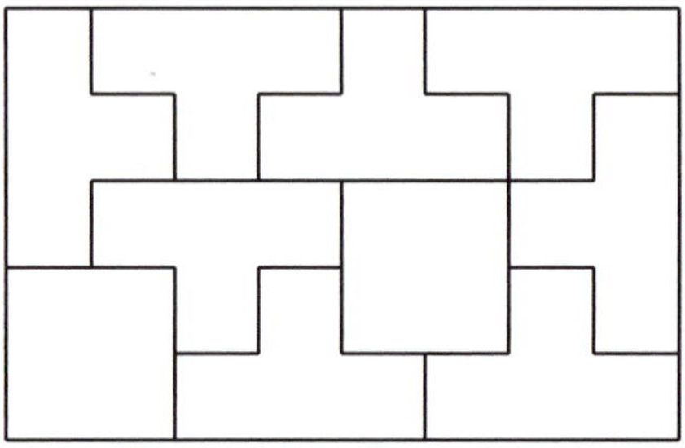

Now 25 copies of this board can cover a 50×80 subboard while the remaining 10×50 subboard can be covered by 2 more copies along with 1 copy of a 10×18 board. The 180 stacks of height 1 in the latter can be combined into 12 stacks as shown in the diagram below. The total number of stacks is $10 \times 27 + 12 = 282$.

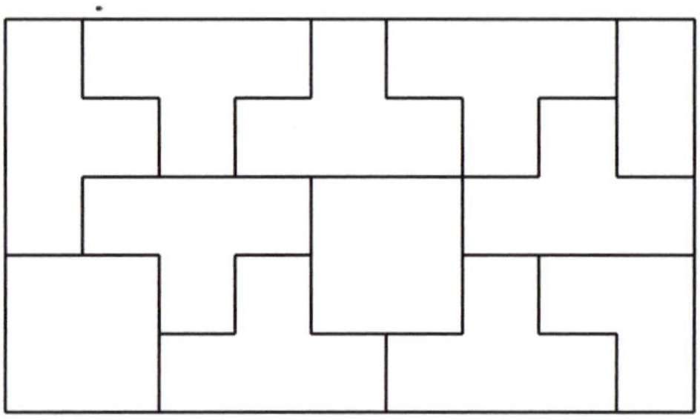

Senior A-Level Paper

1. See Problem 4 of the Junior A-Level Paper.)

2. Consider the six circles in three pairs, with O as the midpoint of the segment joining the centres of the two circles.

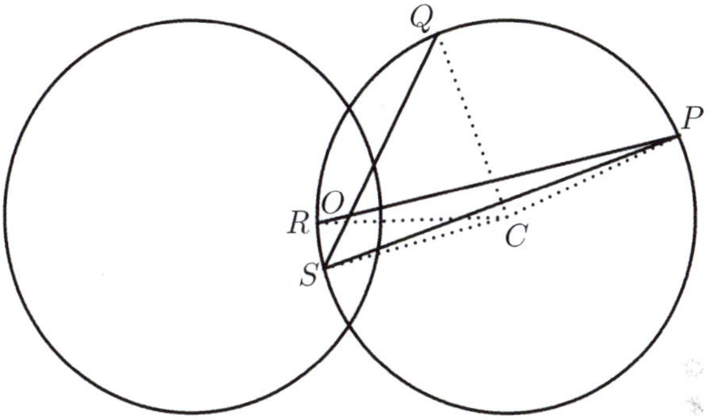

Let the arms of the given angle cut one of the circles, with centre C, at P and Q. Extend PO and QO to cut this circle again at R and S respectively. By symmetry, the arc RS is equal in length to the arc on the other circle intercepted by the arms of the given angle. Now $\angle PCQ + \angle RCS = 2(\angle PSQ + \angle RPS) = 2\angle POQ = 2\theta$, from which the desired conclusion follows.

3. Let the initial value of the Euro be 1. Let the percentage change for the ith month be p_i, where $0 < p_i < 100$ for $1 \le i \le 12$. Then the square of the value of the Euro at the end of the 12 months is

$$\frac{(100+p_1)(100-p_1)\cdots(100+p_{12})(100-p_{12})}{100^{24}}$$

$$= \frac{(100^2-p_1^2)\cdots(100^2-p_{12}^2)}{100^{24}} < 1.$$

4. Let $A_n = \{a_n, a_{n+1}, \ldots, a_{n+4033}\}$ for any n. A_n can only take 2^{4034} different forms, so that $A_i = A_j$ for some $i < j$. Choose $k = j - i - 1$ and let $p_n = a_n a_{n+1} \cdots a_{n+k}$. Since $a_n = a_{n+k+1}$ for $i \le n \le i + 4033$, $p_i = p_{i+1} = \cdots = p_{i+4033}$. Let $s_n = p_1 + p_2 + \cdots + p_n$ and consider the infinite sequence $s_1,\ s_2,\ s_3,\ \ldots$. Each term differs from the preceding one by 1. If $|s_{i-1}| \ge 2017$, then there exists n, $0 \le n \le i-1$, such that $|s_n| = 2017$. Suppose that $|s_{i-1}| < 2017$. Then $|s_{i+4033}| > 2017$ since in going from s_{i-1} to s_{i+4033} in 4034 steps, we have either increased it by 1 in every step or decreased it by 1 in every step. Now there exists n, $i \le n \le i + 4033$, such that $|s_n| = 2017$.

5. (a) Let the initial piece be of weight 1. We can cut it into two pieces of weight $\frac{1}{2}$. We then cut each into two pieces of weight $\frac{1}{4}$. Continuing this way, we will always have pieces of only two different weights, one being half of the other.

   (b) Let $r > 0.5$. Suppose to the contrary that cutting can continue forever. Consider any two pieces A and B with respective weights a and b where $a \geq b$. Then A must be cut before B, as otherwise one of the pieces resulting from cutting B has weight at most $\frac{b}{2}$ and hence less than ra. Suppose the pieces resulting from cutting A have weights c and d with $c \geq d$ and $c + d = a$. Then $d \leq \frac{a}{2}$, so that $r < \frac{d}{b} < \frac{a}{2b}$. It follows that $\frac{a}{b} > 2r > 1$. Suppose at some stage we have n pieces with respective weights $a_1 > a_2 > \cdots > a_n$. Then $a_1 > 2ra_2 > (2r)^2 a_3 > \cdots > (2r)^{n-1} a_n$. Hence $(\frac{1}{2r})^{n-1} > \frac{a_n}{a_1} > r$. This cannot hold forever as $\frac{1}{2r} < 1$.

   (c) From (b), n pieces may be obtained for $r = 0.6$ if $(\frac{1}{2\times 0.6})^{n-1} > 0.6$. This holds for $n-1 = 3$ but not for $n-1 = 4$. It follows that at most 4 pieces may be obtained. We now show that 4 pieces may in fact be obtained. We may take the weight of the initial piece of cheese to be 32. We first cut it into two pieces with respective weights 18 and 14, observing that $\frac{14}{18} > 0.6$. Next, we cut the heavier piece into two, each with weight 9, and observe that $\frac{9}{14} > 0.6$. Finally, we cut the piece with weight 14 into two, each with weight 7.

6. Let the excircle with centre J touch the extension of CA at P, the extension of CB at Q, and the side AB at F.

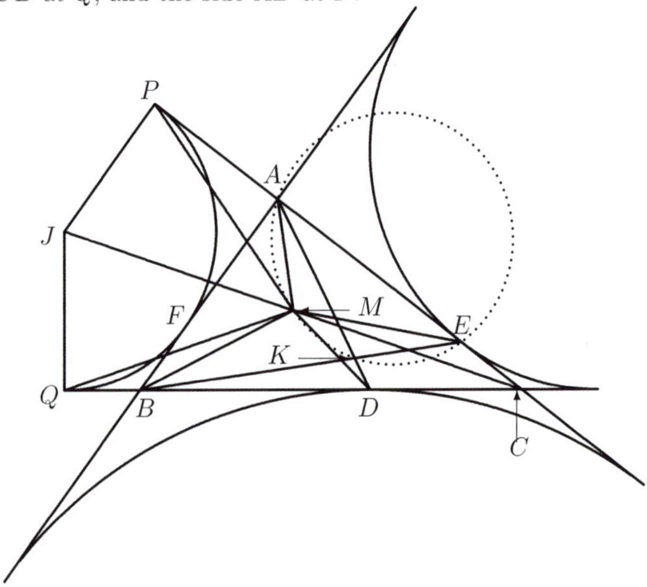

Then $CA + AF = CA + AP = CP = CQ = CB + BQ = CB + BF$. Hence CF bisects the perimeter of ABC. Thus AD and BE also bisect the perimeter of ABC, so that $AE = BD$. Similarly, we can prove that $BQ = CE$ and $DC = AP$. Let M be the midpoint of CJ. Then CM bisects $\angle BCA$. Since M is the circumcentre of the quadrilateral $CPJQ$, $MP = MC = MQ$ so that $\angle MPC = \angle MCP = \angle MCQ = \angle MQC$. Perform a rotation about M equal to $\angle CMP = \angle CMQ$. Then the line BC lands on the line CA, with Q landing on C, B on E, D on A and C on P. It follows that $MA = MD$, $MB = ME$, and both $\angle AMD$ and $\angle BME$ are equal to the angle of rotation. Hence AMD and BME are similar isosceles triangles. Thus $\angle DAM = \angle BEM$, so that $AEKM$ is a cyclic quadrilateral.

7. Convert the map of the city into a directed graph. The $(n+1)^2$ intersections of the streets are the vertices. Each segment of a street is an edge, directed towards east or north. Each of the vertices $(x, 0)$ and $(0, y)$, $1 \leq x, y \leq n-1$, has outdegree 2 and indegree 1. These vertices are called starters. Each of the vertices (x, n) and (n, y), $1 \leq x, y \leq n-1$, has outdegree 1 and indegree 2. These vertices are called finishers. Apart from (0,0) and (n, n), the remaining vertices all have equal outdegrees and indegrees. Initially, all the edges are green. We shall paint each edge yellow when we cover it for the first time, and red the second time.

Lemma.

On day k, $1 \leq k \leq n$, we divide the graph into three parts by drawing the diagonal connecting all points (x, y) with $x+y = k$ and the diagonal connecting all points (x, y) where $x+y = 2n-k$. Then exactly $4(n-k+1)$ edges have been painted yellow on that day, and at the end of the day, no edge in the middle part is red and no edge in the other two parts is green.

Proof:

We use mathematical induction on k. On day $k = 1$, the dividing diagonals are $x+y = 1$ and $x+y = n-$. We take any two edge-disjoint paths from (0,0) to (n, n). Following one going and the other coming back, we will have painted $4n$ edges yellow. They include both edges from (0,0) to (1,0) and (0,1), as well as both edges to (n, n) from $(n-1, n)$ and $(n, n-1)$. There are no red edges as no edge has yet been covered more than once. Thus the result holds for $k = 1$. Suppose it holds up to day $k-1$, $2 \leq k \leq n$. On day k, consider any path from (0,0) to (n, n). Its first $k-1$ and the last $k-1$ edges are not green by the induction hypothesis. Hence this path can contain at most $2(n-k+1)$ green edges.

Claim:
There exist two edge-dispjoint paths each with exactly $2(n-k+1)$ green edges on day k.

Justification:
Consider the starter $(n-1,0)$ and the finisher $(n,1)$. At least one edge from $(n-1,0)$ is green and at least one edge to $(n,1)$ is green. Either both edges to and from $(n,0)$ are green, or neither is. Thus we have a green path from $(n-1,0)$ to $(n,1)$. We paint this path blue. Consider now the starter $(n-2,0)$ and the finisher $(n,2)$. Again, at least one edge from $(n-2,0)$ is green and at least one edge to $(n,2)$ is green. Starting from $(n-2,0)$, we follow green edges until this path P can go no further. If we arrive at $(n,2)$, we paint the edges in P blue. If not, P must terminate in some other finisher. Start from $(n,2)$ and follow green edges, going against their directions. This path cannot terminate in $(n-1,0)$ as its extra outgoing edge has been painted blue. Hence it must terminate somewhere along P. Now start from $(n-2,0)$ and go along P until we can switch to the other path. We will arrive at $(n,2)$ after all. We paint this path from $(n-2,2)$ to $(n,2)$ blue. Continuing this way, we can construct a path from the starter $(k-1,0)$ to the finisher $(n,k-1)$, consisting of $2(n-k+1)$ green edges. Similarly, we can construct a path from the starter $(0,k-1)$ to the finisher $(k-1,n)$, consisting of $2(n-k+1)$ green edges. Thus our claim is justified.
We now restore the green colour to all edges painted blue. Consider the three regions partitioned by the dividing diagonals on day k. By the induction hypothesis, all edges in the outer parts are no longer green, but none of the edges in the middle part is red. On the dividing diagonals are 4 vertices of degree 3 and $2(k-1)$ vertices of degree 4. Thus there are $4k+4$ edges in the middle part that are incident with these vertices. The $2k$ paths on the first k days cover $4k$ of them, and these are different edges as none of them is red. Thus the 4 extra edges are green, and each is incident with one of the vertices of degree 3. These 4 vertices are just the starters $(k,0)$ and $(0,k)$, as well as the finishers $(n-k,n)$ and $(n,n-k)$. Thus we can form two edge-disjoint paths from (0,0) to (n,n) through these 4 vertices, which are linked in pairs by green edges in the middle part. It is easy to verify that the induction hypothesis holds at the end of day k. This completes the proof of the Lemma.
By the Lemma, $4(n+(n-1)+\cdots+1)=2n(n+1)$ edges have been covered during the n days. This coincides with the total number of the edges of the graph. The desired conclusion follows.

Spring 2018

Junior O-Level Paper

1. (a) If we put the six rooks on one long diagonal, every vacant square will be attacked by two rooks from the same distance away.

   (b) The task is equivalent to placing six non-attacking queens on a 6×6 board. This can be done as shown in the diagram below. Every vacant square will be attacked by two rooks from difference distances away.

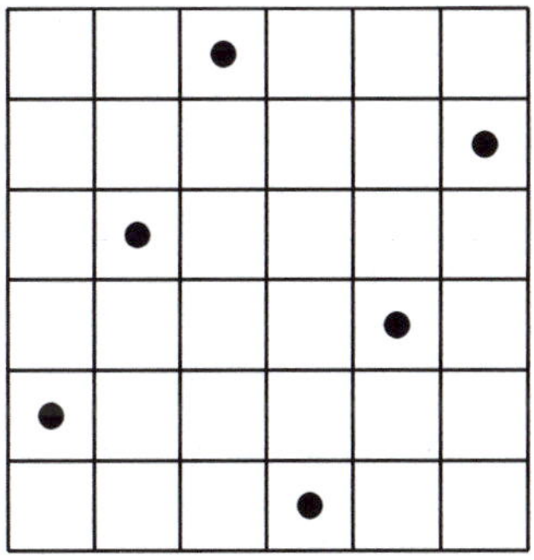

2. Extend CK to N so that BN is parallel to AC. Then $\angle BAC = \angle ABN$. We have $\angle ACK = \angle AKC = \angle BKN$. Hence $\angle BNK$ also has the same measure, so that $BN = BK = CL$. It follows that $BCLN$ is a rectangle and M is the point of intersection of its diagonals. Hence $MC = ML$.

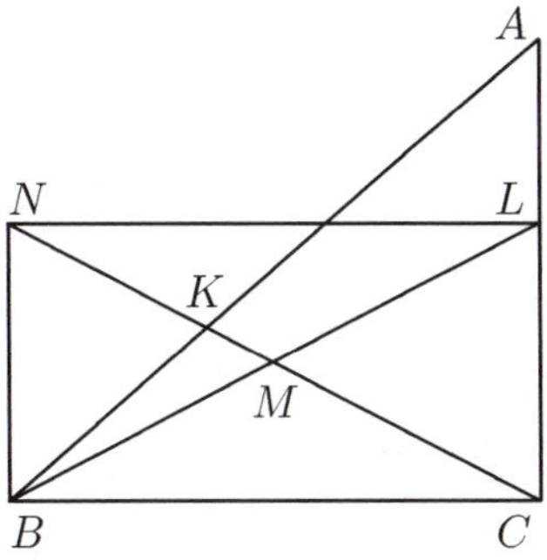

3. Denote by n the constant sum of the rows and columns.

   (a) We can determine uniquely the number in the square marked with a black dot in the diagram below. The sum of the numbers in the second row, the third row, the second column and the third column is $4n$. This is the sum of twice 4+5+6+22 plus the sum of the numbers in the eight shaded squares.

The sum of the numbers in the first row, the fourth row, the first column and the fourth column is also $4n$. This is the sum of twice 1+2+3=6 plus twice the number we wish to determine plus the sum of the numbers in the eight shaded squares. It follows that the number we wish to determine is equal to $22 - 6 = 16$.

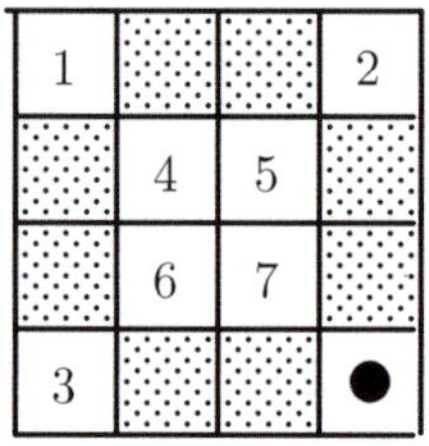

(b) It is not possible to uniquely determine any of the numbers in the eight shaded squares. Suppose we find a set of values for them. By adding 1 to each number, we have another set of values which will work. The constant sum of the rows and columns in the modified table is 2 greater than before.

4. We may assume that no prime number divides all three integers. The greatest common divisor of two of them cannot be divisible by a prime number as otherwise that prime number must divide the third integer. Hence the pairwise greatest common divisors must all be equal to 1. Now each integer is relatively prime to the other two, and therefore to their least common multiple. Since it divides the least common multiple, the integer must be equal to 1. Hence all three integers must be the same.

5. We claim that seven red lines divide the plane into at most 29 convex regions, some of which are infinite. Since we have 30 marked points, the Pigeonhole Principle guarantees that at least two of them will be in the same region. The line segment joining two marked points in the same convex region will not cross any red line. Thus the desired scenario is impossible. We now justify the claim. Let the maximum number of regions obtained from n lines be denoted by a_n. We have $a_1 = 2$. In order to maximize the number of regions, every two lines should intersect at a distinct point. After $n - 1$ such lines have been drawn, we have a_{n-1} regions. The nth line will intersect the other lines at $n-1$ distinct points, dividing itself into n segments. Each segment divides an existing region into two. Hence $a_n = a_{n-1} + n$. Iteration yields $a_2 = 4$, $a_3 = 7$, $a_4 = 11$, $a_5 = 16$, $a_6 = 22$ and $a_7 = 29$.

Senior O-Level Paper

1. Since the bisector AK and the altitude AD divide the side BC of triangle ABC into three segment, $AB \neq AC$. We may assume by symmetry that $AB > AC$. This means that $\angle ABC < \angle BCA$, so that we have $\angle AKB > \angle AKC$.

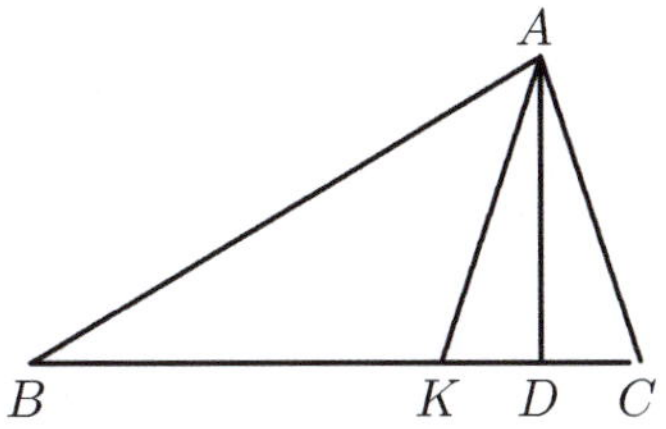

Hence $\angle AKB$ is obtuse and D lies on KC. From the Angle Bisector Theorem, $\frac{KB}{KC} = \frac{AB}{AC} > 1$. Hence $KD + DC = KC < KB$, and the Triangle Inequality does not hold. Thus the desired task is impossible.

2. We may assume that no prime number divides all four integers. No prime number can divide three of the integers, as their greatest common divisor will not divide the fourth number. If no prime number divide two of the integers, then all four integers must be equal to 1, and the desired conclusion follows trivially. Let the integers be a, b, c and d, and suppose a prime number p divides a and b but not c or d. Suppose the power of p in a is higher than the power of p in b. Then a cannot divide the least common multiple of b, c and d. By symmetry, p must appear in equal powers in a and b, and an even number of times in $abcd$. Since this argument applies to every prime number, the desired conclusion follows. (Compare with Problem 4 of the Junior O-Level Paper.)

3. Let Q be the centre of the first circle, and let P be the foot of the perpendicular from O to QA. Let $QA = a$ and $OB = b$. Then we have $QO = a + b$ and $PQ = a - b$. Hence

$$AB = PO = \sqrt{(a+b)^2 - (a-b)^2} = 2\sqrt{ab}.$$

Since $MA = MT = MB$, $MA = \sqrt{ab}$. Now $\frac{MA}{CA} = \frac{\sqrt{ab}}{2a} = \frac{\sqrt{b}}{2\sqrt{a}}$. On the other hand, $\frac{OB}{AB} = \frac{b}{2\sqrt{ab}} = \frac{\sqrt{b}}{2\sqrt{a}}$ also. Hence the right triangles MCA and OAB are similar, so that $\angle MCA = \angle OAB$. It follows that MC is perpendicular to OA.

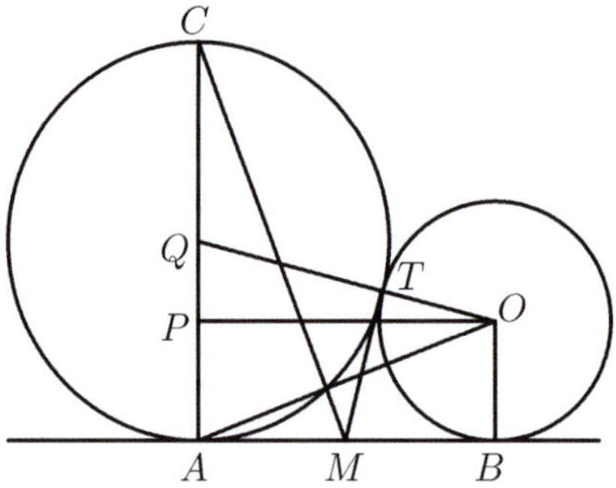

4. Basil has a winning strategy. He divides the board except the initial painted corner square into 21 V-trominoes as shown in the diagram below. Peter must move the counter to a V-tromino with three unpainted squares. Basil then moves the counter to the other two squares of the V-tromino one after the other. Then all three squares are painted, and Peter has to move to a new V-tromino. Eventually, the whole board will be painted, and it will still be Peter's move. Hence he must lose.

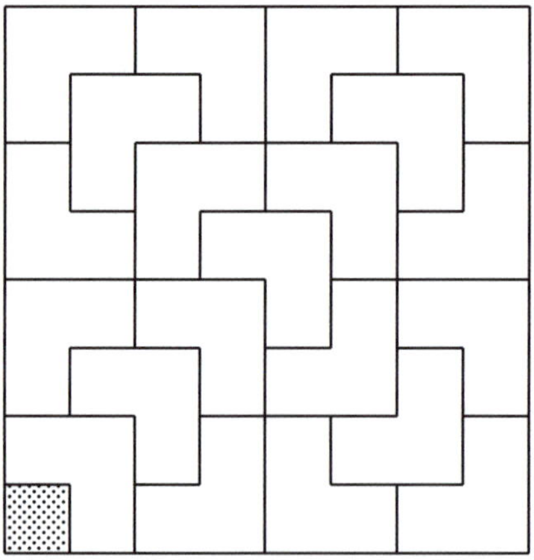

5. Let the colours be red, yellow and green. A vertex is said to be tri-coloured if it is surrounded by one face of each colour. Paint an edge red if it separates a red face from a green face, and yellow if it separates a yellow face from a green face. Consider the three edges at any vertex. They may all be unpainted. Suppose one is painted. We may assume by symmetry that it is red, meaning that it separates a red face from a green face. If the third face at that vertex is red or green, one of the other two edges is red and the third one is unpainted. If the third face is yellow, one of the other two edges is yellow and the third one is unpainted. It follows that the painted edges form disjoint cycles.

Clearly, vertices not on these cycles cannot be tri-coloured. A vertex on such a cycle is tri-coloured if and only if the colour of the edges changes there. Going once around a cycle, the colour must change an even number of times. Hence the number of tri-coloured vertices on the cycle is even. Since this is true of every cycle, the total number of tri-coloured vertices is even.

Junior A-Level Paper

1. Since each pair of adjacent numbers has positive sum and yet the overall sum is negative, the last number must be negative, and the second last one must be positive. Removing these two numbers, we have a shorter row with the same two properties as the original row. It follows that in the original row, the negative and positive numbers alternate. Hence there are twenty negative numbers and nineteen positive numbers, making the overall product positive. For a complete solution, we should exhibit such a row of numbers. Let it be of the form $(-b, a, -b, a, \ldots, -b, a, -b)$ where a and b are positive integers. Then $b < a$ while $19a < 20b$. These yield $b \leq a+1$ and $19a \leq 20b+1$, so that $19a - 1 \leq 20b \leq 20a + 20$. Hence $a \geq 21$, and the row of numbers may be $(-20, 21, -20, 21, \ldots, -20, 21, -20)$.

2. If Aladdin starts with 1000 dollars, he will stay at that sum. If he starts with more than 1000 dollars, his total wealth will actually decrease. Suppose he has more money at the end, then he must start with less than 1000 dollars, say $1000-x$ dollars for some $x < 1000$. After ten operations, he will have $1000 - \frac{x}{2^{10}}$ dollars. Since the Genie never takes a half-dollar, Aladdin always has a whole number of dollars. Hence $x \geq 1024$, but this contradicts $x < 1000$.

3. Let $p > 2^{2017}$ be a prime. For $1 \leq k \leq 2018$, the fractions $\frac{p-2^k}{2^k p}$ is irreducible since the numerator is odd and less than p. These denominators are distinct, and the smallest one among them is $2p$. For $1 \leq i < j \leq 2018$,
$$\frac{p-2^i}{2^i p} - \frac{p-2^j k}{2^j p} = \frac{2^{j-i}(p-2^i) - (p-2^j)}{2^j p} = \frac{2^{j-i}-1}{2^j}.$$
The largest denominator among them is $2^{2018} < 2p$.

4. Let PH intersect AB at F. We have
$$\angle ACO = \frac{1}{2}(180^\circ - \angle AOC) = 90^\circ - \angle ABC = \angle BAH.$$
Since $\angle AHC = 90^\circ - \angle APC$, $ACHP$ is a cyclic quadrilateral, so that $\angle AHP = \angle ACP = \angle BAH$. Hence $\angle PHB = 90^\circ - \angle AHP = \angle ABC$. It follows that $FA = FH = FB$.

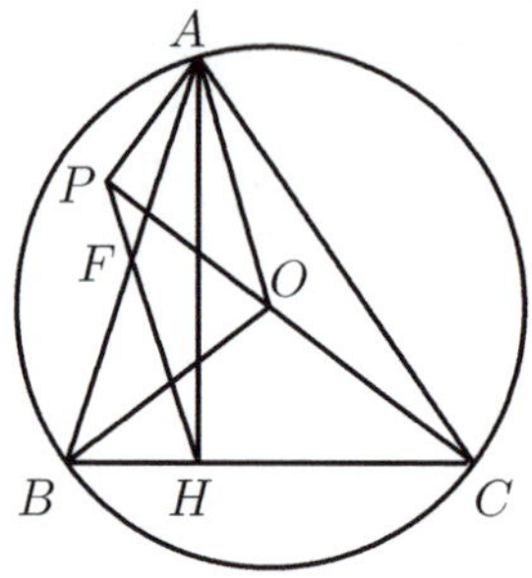

5. We first show that the largest house number can be as small as 197. Let the odd house numbers be those in the arithmetic progression 1, 5, 9, ..., 193, 197. Let the even house numbers be those in the arithmetic progression 50, 52, 54, ..., 146, 148. The differences form the arithmetic progression -49, -47, -45, ..., 47, 49. Note that all differences are odd numbers. Suppose the smallest is $-(2k+1)$ for some k. Then the largest is at least $98-(2k+1)=97-2k$. The smallest even house number is at least $1+(2k+1)=2k+2$ and the largest is at least $98+2k+2=100+2k$. It follows that the largest odd house number is at least $97-2k+100+2k=197$.

6. Choose any point outside the decagon. If all ten give the same answer, they are all knaves. Otherwise, two adjacent ones, say A and B, give different answers. Then at least one of them is a knave. Choose a point P outside the decagon arbitrarily close to the midpoint of AB, such that $PA=PB$. Then the correct answer to the question is unique determined. Anyone who gives the correct answer is a knight and everyone else is a knave. Thus two moves are sufficient. They are also necessary. With no prior information, the only question with a unique correct answer requires placing the point at the centre of the decagon, but this is not permitted.

7. The expression $(a\otimes a)\odot(a\otimes a)$ either means $(a+a)-(a+a)=0$ or $(a-a)+(a-a)=0$. Hence it is 0 regardless of the interpretation of the operators. This allows us to use 0 in addition to a and b. Now $(a\otimes 0)\otimes(0\otimes b)$ either means $(a-0)-(0-b)=a+b$, $(b-0)-(0-a)=a+b$ or $(0+a)+(0+b)=a+b$. This allows us to use $+$ in addition to $\otimes$ and $\odot$. Finally, consider $0\otimes((0\odot(a\odot 0))\otimes 0)$. It means $0+((0-(a-0))+0)=-a$, $0-((0+(a+0))-0)=-a$ or $(0-(0+(a+0))-0=-a$. This allows us to use the unary operation of negation. Now we can add eighteen copies of $a+(-b)$ to two copies of a to obtain $20a-18b$.

Senior A-Level Paper

1. (See Problem 2 of the Junior A-Level Paper)

2. (See Problem 3 of the Junior A-Level Paper)

3. A Ferrers domain in a rectangular table is a subset of the squares such that all squares in the same row are adjacent and start from the left edge, and each row has at least as many squares as the next. The Ferrers domain is said to be sound if the numbers increase in every row from left to right, and in every column from top to bottom. We shall build an expanding Ferrers domain which is sound, until it encompasses the entire table. We may assume that the one hundred numbers are 1 to 100. In the first move, we put 1 at the top left corner as a Ferrers domain with one square. It is clearly sound. Suppose after k moves, the squares containing the numbers 1 to k form a sound Ferrers domain. Consider the square containing the number $k+1$. Shift its left edge to the left until it is stopped by the boundary of the Ferrers domain. Then shift its top edge upwards until it is stopped again. Along with its right edge and bottom edge, a rectangle is defined on which we perform the half-turn. Then the square containing $k+1$ can be admitted into the Ferrers domain, which remains sound since $k+1$ is larger than any number already inside. After 99 moves, only the square containing the number 100 remains outside the Ferrers domain, and it must be at the bottom right corner. It can be admitted free of charge.

4. Let ABC be an equilateral triangle in the plane α. Let G be the centroid of ABC. Let D, E and F be the respective reflections of A, B and C across G. Then DEF is also an equilateral triangle inscribed in the circumcircle ω of ABC. In the first projection, ω turns into an ellipse such that its major axis is parellel to the line of intersection of α and β. Its length is unchanged while the length of the minor axis is scaled by a factor of $\cos\theta$, where θ is the dihedral angle between α and β. After the second projection, let $A'B'C'$ be the equilateral triangle on γ which is the image of ABC. Since projections preserve midpoints, they preserve medians, so that the image G' of G is also the centroid of $A'B'C'$. Let D', E' and F' be the respective reflections of A', B' and C' across G'. Then $D'E'F'$ is the image of DEF, and it is also an equilateral triangle inscribed in the circumcircle ω' of $A'B'C'$. Since there is a unique conic passing through five points, ω' must be the image of the ellipse in the second projection. Since ω' is a circle, the length of the minor axis of the ellipse must be parallel to the line of intersection of β and γ, so that its length is unchanged while the length of the major axis is scaled by a factor of $\cos\phi$, where ϕ is the dihedral angle between β and γ.

It follows that $\theta = \phi$ and β intersects α and γ along lines parallel to the major and minor axis of the ellipse, which means that they are perpendicular lines.

5. (See Problem 7 of the Junior A-Level Paper)

6. **Lemma.**
P is a fixed point on a fixed line ℓ. A variable line through P intersects a fixed circle ω at A and B. The tangents to ω at A and B intersect ℓ at U and V respectively. Prove that $\frac{1}{PU} - \frac{1}{PV}$ is constant.

Proof:

Let O be the centre and r be the radius of ω. Let H be the foot of the perpendicular from O to ℓ. Let T be the point of intersection of the two tangents. Note that $\angle PAU = \angle ABT = \angle PBV$. We also have $\angle APU + \angle BPV = 180°$. Hence the ratio of the areas of triangles PAU and PBV is equal to $\frac{PA \cdot AU}{PB \cdot BV}$ as well as $\frac{PA \cdot PU}{PB \cdot PV}$. It follows that $\frac{PU}{PV} = \frac{AU}{BV}$. Now

$$\begin{aligned}\frac{PU^2}{PV^2} &= \frac{OU^2 - r^2}{OV^2 - r^2} \\ &= \frac{OH^2 + (PU + PH)^2 - r^2}{OH^2 + (PV - PH) - r^2} \\ &= \frac{OP^2 - r^2 + 2PU \cdot PH + PU^2}{OP^2 - r^2 - 2PV \cdot PH + PV^2} \\ &= \frac{OP^2 - r^2 + 2PU \cdot PH}{OP^2 - r^2 - 2PV \cdot PH}.\end{aligned}$$

We have $PU^2(OP^2 - r^2 - 2PV \cdot PH) = PV^2(OP^2 - r^2 + 2PU \cdot PH)$. This may be rewritten as

$$(PU + PV)(PU - PV)(OP^2 - r^2) = 2PU \cdot PV \cdot PH(PU + PV).$$

Hence $\frac{1}{PV} - \frac{1}{PU} = \frac{PU - PV}{PU \cdot PV} = \frac{2PH}{OP^2 - r^2}$, which is constant.

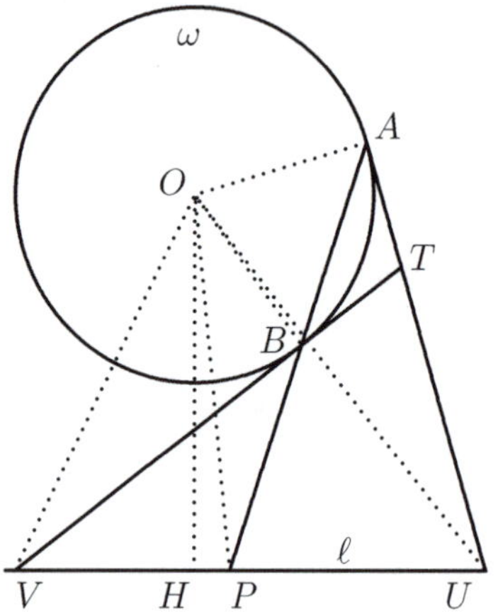

We now return to our problem. Let D be the point such that the tangent to ω at D is parallel to ℓ. Let C be the point of intersection of PD with ω. Let the tangent to ω at C intersect ℓ at W. Let X be the point of intersection of the tangents at C and D. Note that the tangent at D intersects ℓ at a point in infinity. Applying the Lemma to the line CD, we have $-\frac{1}{PW} = \frac{2PH}{OP^2-r^2} = \frac{1}{PV} - \frac{1}{PU}$. This simplifies to $PU \cdot PV - PV \cdot PW + PW \cdot PU = 0$. It follows that we have $WU \cdot WV = (PW + PU)(PW - PV) = PW^2$. We have $WP = WC$ since $\angle WPC = \angle XDC = \angle XCD = \angle WCP$. Hence $WC^2 = WU \cdot WV$. It follows that the circumcircle of triangle CUV is tangent to WX, and therefore to ω.

7. We first show that no matter what the collective strategy is, the number of incorrect guess may be as high as $k + 1$. This can happen if the last k wizards in the column are the uncaring ones, and they make k wrong guesses. The caring wizard immediately before them has no useful information to guide him, and may guess wrong too. We now show a collective strategy which limits the number of incorrect answers to at most $k + 1$. Let the wizards be numbered 1 to n from the front of the column to the back. We may assume that the last wizard always guess incorrectly. For $2 \leq j \leq n$, the jth wizard should announce a $(j - 1)$-digit number. For $1 \leq i \leq j - 1$, the ith digit is 0 if the ith wizard is wearing a white hat, and 1 otherwise. These announcements are heard by everyone. At his turn, a caring wizard always makes a correct announcement. He then checks to see if all announcements so far about hat colours of the wizards in front of him are correct. Those making incorrect announcements are uncaring and can be dismissed. Also, anyone not guessing according to the collective strategy, based on information available to him, is also uncaring. Suppose the announcements about the colour of his hat are consistent. The caring wizard will then guess according to the collective strategy. If not, he will accept the announcements of the majority. Let us see what damage an uncaring wizard can do. If he does not guess according to the collective strategy, his true nature is revealed. If he does, and make the correct announcement, he is not doing any damage. If he guesses according to the collective strategy but makes an incorrect announcement, he will cause at most a caring wizard before him to guess wrong. Since there are k uncaring wizards, they cause collectively at most k incorrect guesses. Along with the guess of the wizard last in the line, the number of incorrect guesses is at most $k + 1$. Thus the number of correct guesses is at least $n - k - 1$.

Tournament 40

Fall 2018

Junior O-Level Paper

1. In triangle ABC, $\angle B = 90°$. A circle passing through B and the mid-point of AC intersects BC at D and AB at F. If $AC = 2DF$, prove that $BD = CD$ and $AF = BF$.

2. Determine all positive integers n for which the numbers $1, 2, \ldots, 2n$ can be arranged in pairs so that if the sum of each pair is computed, the product of the sums is the square of an integer.

3. A 7×14 board is constructed from copies of the O-tetromino and the V-tromino. Is it possible that

   (a) the same number of copies of each piece is used;

   (b) more copies of the O-tetromino are used than copies of the V-tromino?

4. Kate has three real coins of the same weight and two fake coins whose total weight is the same as that of two real coins. However, one of them is heavier than a real coin while the other is lighter. Can Kate identify the heavier coin as well as the lighter coin in three weighings on a balance? She must decide in advance which coins are to be weighed without waiting for the result of any weighing.

5. Prove that there exist at least 1000 multiples of 37 each consisting of nine different digits.

Note: The problems are worth 4, 4, 1+3, 5 and 5 points respectively.

Senior O-Level Paper

1. Is it possible to place a line segment XY inside a regular pentagon $ABCDE$ so that

$$\angle XAY = \angle XBY = \angle XCY = \angle XDY = \angle XEY?$$

2. Determine all positive integers n for which the set of numbers $1, 2, \ldots, 2n$ can be arranged in pairs so that if the sum of each pair is computed, the product of the sums is the square of an integer.

3. In the parallelogram $ABCD$, $\angle A$ is acute. N is the point on AB such that $CN = CD$. If the circumcircle of the triangle CBN is tangent to the line AD, prove that D is the point of tangency.

4. Prove that there exist at least 2018 multiples of 37 each consisting of nine different digits.

5. Pete is placing 500 kings on a 100×50 board so that no two attack each other. Basil is placing 500 kings on the white squares of a 100×100 board so that no two attack each other. Who has more ways to place the kings?

Note: The problems are worth 3, 4, 5, 5 and 5 points respectively.

Junior A-Level Paper

1. D is the midpoint of the side BC of triangle ABC. E is a point on the side CA, strictly between A and C. If $BE \geq 2AD$, prove that ABC is an obtuse triangle.

2. On an island with 2018 inhabitants each person is either a knight, a knave or a kneejerk. Everyone knows who everyone else is. Each inhabitant is asked in trun whether there are more knights than knaves on the island. A knight always tells the truth, a knave always lies, and a kneejerk agrees with the majority of those who speak before him. In case of a tie, such as when a kneejerk speaks first, he answers "Yes" and "No" at random. If exactly 1009 answers are "Yes", at most how many kneejerks are on the island?

3. We can form expressions using any number of the digit 7, as well as brackets and the operations of addition, subtraction, multiplication, division, exponentiation (such as 7^7) and concatenation (such as 77). Does there exist a positive integer n such that the number consisting of n copies of 7 can be the value of an expression using less than n copies of the digit 7?

4. An invisible alien spaceship in the shape of an O-tetromino may land on our 7×7 airfield, occupying four of the 49 squares. Sensors are to be placed in certain squares of the airfield. A sensor will send a signal if it is in a square on which the spaceship lands. From the locations of all the sensors which send signals, we must be able to determine exactly on which four squares the spaceship has landed. What is the smallest number of sensors we have to place?

5. In a cyclic quadrilateral $ABCD$ with circumcentre O, AD is parallel to BC and $AB = CD$. The line BO intersects the segment AD at the point E. Prove that a circle passes through O, C and the circumcentres of triangles ABE and BDE.

6. (a) Prove that any integer of the form $3k + 1$, where k is an integer, is the sum of a square and two cubes of integers.

   (b) Prove that any integer is the sum of a square and three cubes of integers.

7. Some pairs of $n \geq 2$ towns are connected by single roads which only meet in towns. A town is said to be a dead-end if it is impossible to start from it and return without using the same road twice. Anna and Boris play the following game. Anna begins by making each road one way, and put a tourist in one of the towns. Turns alternate thereafter. In his turn, Boris reverses the permitted direction on one of the roads at the town where the tourist is. In her turn, Anna moves the tourist to another town along a road in the permitted direction. Boris wins if in Anna's turn, the tourist cannot move.

   (a) Prove that Boris does not have a winning strategy if every town is a dead-end.

   (b) Prove that Boris has a winning strategy if at least one town is not a dead-end.

Note: The problems are worth 5, 6, 8, 8, 8, 7+3 and 5+7 points respectively.

Senior A-Level Paper

1. On an island with 2018 inhabitants each person is either a knight, a knave or a kneejerk. Everyone knows who everyone else is. Each inhabitant is asked in trun whether there are more knights than knaves on the island. A knight always tells the truth, a knaver always lies, and a kneejerk agrees with the majority of those who speak before him. In case of a tie, such as when a kneejerk speaks first, he answers "Yes" and "No" at random. If exactly 1009 answers are "Yes", at most how many kneejerks are on the island?

2. ABC is an acute triangle without equal sides. BE and CF are altitudes. The points X and Y are symmetric to the points E and F with respect to the midpoints of the sides CA and AB, respectively. Prove that XY is bisected by the line AO, where O is the circumcentre of ABC.

3. (a) Prove that any integer of the form $3k+1$, where k is an integer, is the sum of a square and two cubes of integers.

   (b) Prove that any integer is the sum of a square and three cubes of integers.

4. A finite number of squares of a white infinite board are painted black. A connected subboard, not necessarily rectangular and containing at least two squares, may not be rotated, but may be translated in any direction for any distance, so that it still covers the same number of squares. If exactly one of the covered squares is white, it may be painted black. Prove that the exists a white square which will never be painted black.

5. The three medians of a triangle divide its angles into six angles. At most how many of them can be greater than $30°$?

6. On the number line, infinitely many positive integers are marked. When a wheel rolls along the line, each marked point leaves an imprint on the wheel. Prove that one can choose the radius of the wheel so that if it starts rolling from 0 along the line, every arc of size $1°$ on the wheel will have an imprint.

7. There are $n \geq 2$ cities, each with the same number of citizens. Initially, every citizen has exactly 1 dollar. In a game between Anna and Boris, turns alternate. Anna chooses one citizen from every city, and Boris redistributes their wealth so that the at least one citizen has a different number of dollars from before. Anna wins if there is at least one citizen in every city with no money. Which player has a winning strategy if the number of citizens in each city is

   (a) $2n$;

   (b) $2n-1$?

Note: The problems are worth 5, 7, 6+2, 8, 8, 9 and 10+4 points respectively.

Spring 2019

Junior O-Level Paper

1. The sum of all terms of a sequence of positive integers is 20. No term is equal to 3, and the sum of any number of consecutive terms is not equal to 3. Can such a sequence have more than 10 terms?

2. Initially, all $2n+1$ coins on a circle are showing heads. Moving clockwise, $2n+1$ flips are performed as follows. Flip 1 coin, skip over 1 coin, flip 1 coin, skip over 2 coins, flip 1 coin, skip over 3 coins, and so on, until $2n$ coins are skipped over before the final flip. Prove that exactly one coin shows tails.

3. The product of two positive integers $m \geq n$ is divisible by their sum. Prove that $m + n \leq n^2$.

4. The points P, Q and R are on the sides BC, AB and CD, respectively, of a rectangle $ABCD$. If $PQ = PR$ and $\angle RPQ$ has a fixed value, prove that the midpoint of QR is a fixed point.

5. As the assistant watches, the audience puts a coin in each of two of 12 boxes in a row. The assistant opens one box that does not contain a coin and exits. The magician enters and opens four boxes simultaneously. Does there exist a method that will guarantee that both coins are in the four boxes opened by the magician?

Note: The problems are worth 3, 4, 4, 5 and 5 points respectively.

Senior O-Level Paper

1. The distances from a point inside a regular hexagon to three consecutive vertices are 1, 1 and 2, respectively. Determine the side length of the hexagon.

2. The positive integers a and b are such that $a^{n+1} + b^{n+1}$ is divisible by $a^n + b^n$ for infinitely many positive integers n. Is it necessarily true that $a = b$?

3. Prove that any triangle can be cut into 2019 quadrilaterals with both incircles and circumcircles.

4. As the assistant watches, the audience puts a coin in each of two of 13 boxes in a row. The assistant opens one box that does not contain a coin and exits. The magician enters and opens four boxes simultaneously. Does there exist a method that will guarantee that both coins are in the four boxes opened by the magician?

5. Several positive integers are arranged in a row. Their sum is 2019. None of them is equal to 40 and the sum of any block of adjacent numbers is not equal to 40. What is the maximum length of this row?

Note: The problems are worth 4, 4, 4, 5 and 5 points respectively.

Junior A-Level Paper

1. Do there exist seven distinct positive integers with sum 100 such that they are determined uniquely by the fourth largest among them?

2. There are 2019 grasshoppers on a line. In each move, one of them jumps over another and lands at the point the same distance away from it. Jumping only to the right, the grasshoppers are able to position themselves so that some two of them are exactly 1 mm apart. Prove that the grasshoppers can achieve the same, starting from the initial positions and jumping only to the left.

3. Two fixed circles are disjoint and of equal size. A variable triangle ABC not containing either circle is such that AB is tangent to the first circle and AC is tangent to the second circle. Prove that the bisector of $\angle CAB$ passes through a fixed point.

4. Each of 100 points on a circle is labeled with a real number such that the sum of the squares of these numbers is equal to 1. Each chord joining two of them is labeled with the product of their labels. A chord is painted red if the number of other points between its endpoints is even, and blue otherwise. The sum of the labels of the blue chords is subtracted from the sum of the labels of the red chords. What is the largest possible value of the difference?

5. For which integer $n > 1$ is it possible to fill the squares of an $n \times n$ table with the first n^2 positive integers such that two consecutive integers must be adjacent in the same row or the same column, and two integers congruent modulo n must be in different rows and different columns?

6. In triangle ABC, $AB = BC$. K is a point inside such that $KC = BC$ and $\angle KAC = 30°$. Determine $\angle AKB$.

7. Initially, Pete has 0 points and 100 piles each consisting of 400 counters. At each move, Pete chooses two piles, adds the nonnegative difference between the numbers of counters in them to his score, and then removes 1 counter from each of these two piles. What is the largest possible score Pete can get when no more counters can be removed?

Note: The problems are worth 5, 7, 7, 8, 9, 9 and 12 points respectively.

Senior A-Level Paper

1. Prove that for any two adjacent digits of a positive multiple of 7, there exists a digit such that no matter how many times it is inserted between these two digits, the resulting number is still a multiple of 7.

2. There are 2019 grasshoppers on a line. In each move, one of them jumps over another and lands at the point the same distance away from it. Jumping only to the right, the grasshoppers are able to position themselves so that some two of them are exactly 1 mm apart. Prove that the grasshoppers can achieve the same, starting from the initial positions and jumping only to the left.

3. Two fixed circles are disjoint. A variable angle with a fixed ray from the vertex between the two arms is such that the circles are outside the angle, one arm of the angle is tangent to the first circle and the other arm is tangent to the second circle. Prove that the fixed ray can be chosen so that it always passes through a fixed point.

4. For which integer $n > 1$ is it possible to fill the squares of an $n \times n$ table with the first n^2 positive integers such that two consecutive integers must be adjacent in the same row or the same column, and two integers congruent modulo n must be in distinct rows and distinct columns?

5. The orthogonal projection of a tetrahedron onto the plane of its base is a quadrilateral with exactly one pair of parallel sides and of area 1. Is it possible that the orthogonal projection of this tetrahedron onto the plane of a lateral face is a square of area
   (a) 1;
   (b) $\frac{1}{2019}$?

6. There are $\binom{10}{5}$ cards each containing a different subset of size 5 of the variables $x_1,\ ,x_2,\ \ldots,\ x_{10}$. Anna takes a card, Boris takes a card, and then turns alternate until all the cards have been taken. Boris then chooses the values for the variables, provided that $0 \le x_1 \le \cdots \le x_{10}$. Can he ensure that the sum of the products of the numbers on his cards is greater than the sum of the products of the numbers on Anna's cards?

7. Starting at (0,0), each segment of a polygonal line either goes up or to the right, and can change directions only at lattice points. Associated with each polygonal line is a board consisting of all unit squares that share at least one point with the polygon line. Prove that for any integer $n > 2$, the number of polygonal lines whose associated boards can be dissected into dominoes in exactly n different ways is equal to $\phi(n)$, the number of positive integers that are less than n and relatively prime to n.

Note: The problems are worth 5, 6, 7, 8, 4+4, 8 and 12 points respectively.

Solutions

Fall 2018

Junior O-Level Paper

1. Let E be the midpoint of AC. Then E is the circumcentre of triangle ABC, so that $AC = AE + CE = 2BE$. Suppose BE is a diameter of the circle passing through B and E. Then D and F are indeed the respective midpoints of BC and AB, as shown in the diagram below. Suppose BE is not a diameter of this circle. Since $\angle FBD = 90°$, FD is a diameter, so that $2DF > 2BE = AC$. We have a contradiction.

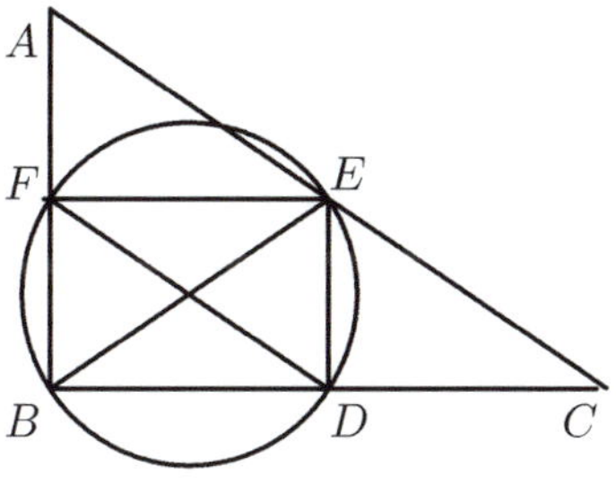

2. If $n = 2k$, we have $(1+4k)(2+(4k-1))\cdots(2k+(2k+1)) = ((4k+1)^k)^2$. Clearly, the task is impossible for $n = 1$. It is possible for $n = 3$ since $(1+5)(2+4)(3+6) = 18^2$. Suppse the task is possible for $n = 2k-1$, resulting in the product m^2 for some integer m. Then the task is also possible for $n = 2k+1$, the prouct being

$$((4k-1)+(4k+2))(4k+(4k+1))m^2 = ((8k+1)m)^2.$$

In summary, the task is possible except for $n = 1$.

3. (a) This is possible, as shown in the diagram below.

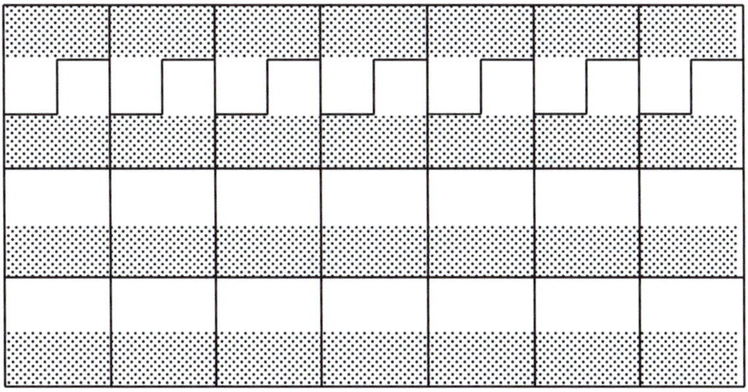

(b) This is impossible. Paint the seven rows alternately black and white, as shown in the diagram above. Then there are 14 extra black squares. A copy of the O-tetromino always covers two black squares and two white squares. A copy of the V-tromino covers two squares of one colour and one square of the other colour. Hence we need at least 14 copies of the V-tromino.

4. Label the coins A, B, C, D and E. First weigh A and B against C and D. Then weigh A and C against B and D. Finally, weigh A and D against B and C. The chart below shows the conclusions drawn based on the weighing results.

AB vs CD	AC vs BD	AD vs BC	Heavy	Light
$>$	$>$	$>$	A	E
$>$	$>$	$<$	E	D
$>$	$<$	$>$	E	C
$>$	$<$	$<$	B	E
$<$	$>$	$>$	E	B
$<$	$>$	$<$	C	E
$<$	$<$	$>$	D	E
$<$	$<$	$<$	E	A

AB vs CD	AC vs BD	AD vs BC	Heavy	Light
$=$	$>$	$>$	A	B
$=$	$>$	$<$	C	D
$=$	$<$	$>$	D	C
$=$	$<$	$<$	B	A
$>$	$=$	$>$	A	C
$>$	$=$	$<$	B	D
$<$	$=$	$>$	D	B
$<$	$=$	$<$	C	A
$>$	$>$	$=$	A	D
$>$	$<$	$=$	B	C
$<$	$>$	$=$	C	B
$<$	$<$	$=$	D	A

5. Since 111 is a multiple of 37, so is $1111111110 \div 9 = 123456790$. It consists of nine different digits. Five other such multiples are

$$2 \times 123456790 = 246913580 \qquad 4 \times 123456790 = 493827160$$
$$5 \times 123456790 = 617283950 \qquad 7 \times 123456790 = 864197530$$
$$8 \times 123456790 = 987654320.$$

Let $\overline{abcdefghi}$ be a multiple of 37 where a, b, $\ldots$, i are different digits. We can permute the digits a, e and i, and the resulting number is still a multiple of 37. This is because the difference of the two numbers is the sum of $\pm 99900000(a-d)$, $\pm 99900(d-g)$ and $\pm 99999900(a-g)$. Similarly, we can permute b, e and h, as well as c, f and i. Thus multiples of 37 consisting of nine different digits come in groups of $(3!)^3 = 216$. The total count is $6 \times 216 = 1296$. After removing those with a leading 0, we still have well over 1000.

Senior O-Level Paper

1. Clearly, X and Y cannot be collinear with any of the vertices. Hence three of them, say A, B and C, are on one side of XY. Then X and Y both lie on the circumcircle of triangle ABC. This is impossible since there are inside the circumcircle of $ABCDE$.

2. If $n = 2k$, we have $(1+4k)(2+(4k-1))\cdots(2k+(2k+1)) = ((4k+1)^k)^2$. Clearly, the task is impossoble for $n = 1$. It is possible for $n = 3$ since $(1+5)(2+4)(3+6) = 18^2$. Suppse the task is possible for $n = 2k-1$, resulting in the product m^2 for some integer m. Then the task is also possible for $n = 2k+1$, the product being

$$((4k-1)+(4k+2))(4k+(4k+1))m^2 = ((8k+1)m)^2.$$

In summary, the task is possible except for $n = 1$.

3. Suppose that the point T of tangency is distinct from D, as shown in the diagram below. Then $\angle ATN = \angle TCN$. Also,

$$\angle TAN = 180^\circ - \angle NBC = \angle NTC.$$

Hence $\angle ANT = \angle TNC$. Now

$$\angle CDN = \angle DNA > \angle ANT = \angle TNC > \angle DNC.$$

This contradicts $CN = CD$.

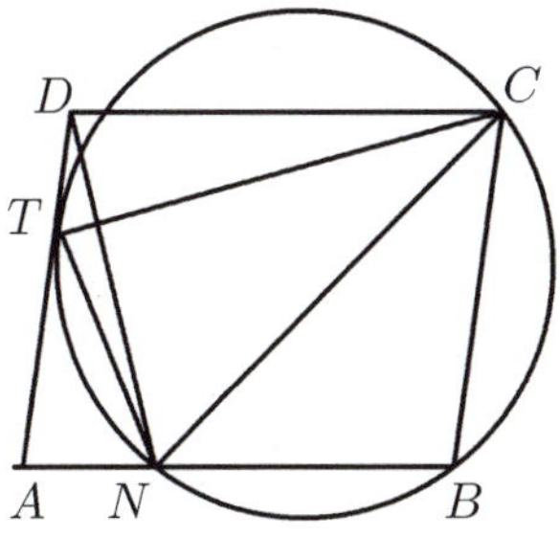

4. Let $\overline{abcdefghi}$ be a multiple of 37 where $a,\ b,\ \ldots,\ i$ are different digits. We can permute the digits a, e and i, and the resulting number is still a multiple of 37. This is because the difference of the two numbers is the sum of $\pm 99900000(a-d)$, $\pm 99900(d-g)$ and $\pm 99999900(a-g)$. Similarly, we can permute b, e and h, as well as c, f and i. Thus multiples of 37 consisting of nine different digits come in groups of $(3!)^3 = 216$. Such a multiple is $1111111110 \div 9 = 123456790$. Five other such multiples are

$$\begin{array}{ll} 2 \times 123456790 = 246913580 & 4 \times 123456790 = 493827160 \\ 5 \times 123456790 = 617283950 & 7 \times 123456790 = 864197530 \\ 8 \times 123456790 = 987654320 & \end{array}$$

To take the total count above 2018, we need four other multiples of 37 consisting of nine different digits. They are

$$\begin{array}{cc} 123456790 + 11100 = 123467890 & 123467890 \times 2 = 246935780 \\ 123467890 \times 4 = 493871560 & 987654320 - 1110 = 987653210 \end{array}$$

(Compare with Problem 5 of the Junior O-Level Paper.)

5. Consider any valid placement of 500 kings on the 100×50 board by Pete. For each column, Basil will transfer all the kings to the 100×100 board, so that the king in the kth square of the old column will be in the kth white square of the new column. This will be a valid placement by Basil. Hence he has at least as many valid placements as Pete. He actually has more. If he places a king on each white square of the odd-numbered columns, he will have a valid placement not derived from the transfer process.

Junior A-Level Paper

1. We take BC to be the longest side. Suppose that $\angle CAB \leq 90°$. Then the circumcentre O of ABC lies on BC or on the same of BC as A. It follows that $2AD \geq 2AO = OB + OC \geq BC > BE$, which is a contradiction.

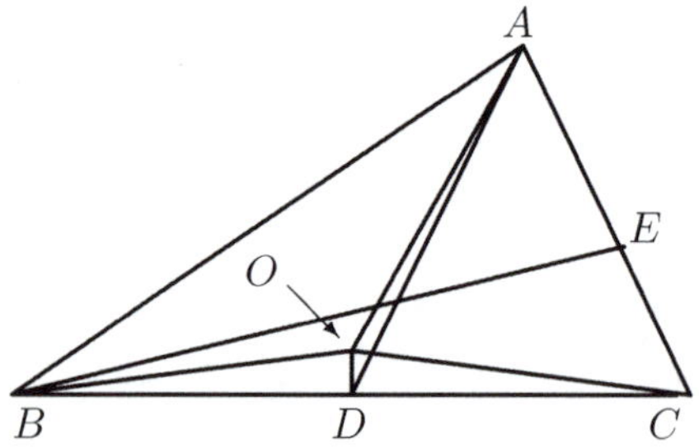

2. We may have as many as 1009 kneejerks. They answer first and all say "No". The other 1009, all knights, then say "Yes". We now show that there cannot be more than 1009 kneejerks. Let the questioner take a break when the running total of "Yes" answers and "No" answers are the same. Consider what happens between two breaks. In order to get to the next break, the number of "Yes" answers must be the same as the number of "No" answers. Moreover, all the answers by the kneejerks must be the same because the lead does cannot change. It follows that at most half of the answers are by the kneejerks. Since this is true between any two breaks, the total number of kneejerks is at most half the total populaation, namely 1009.

3. The expression $(((77-7)\div 7)^{77}-7\div 7)\div(77-7-7)\times 7\times 7$, which uses 7+7 copies of 7, is equal to the number consisting of 77 copies of 7.

4. The diagram below shows a placement of 16 sensors which works.

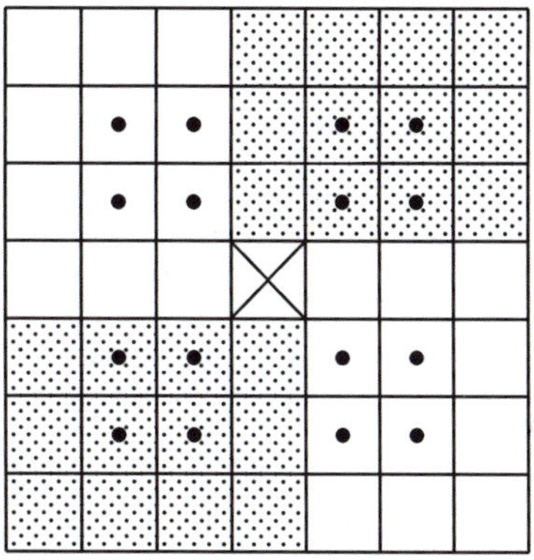

Suppose we place only 15 sensors. Divide the airfield into the central square and four 3×4 regions. One of the regions will contain at most 3 sensors, so that there is at most 1 sensor in a 3×2 half region. We cannot determine the exact location of an alien spaceship which lands within that half region.

5. Let AC intersect the circumcircle of triangle OAE at P. Then we have $\angle PEO = \angle PAO = \angle PCO$, so that

$$\angle PEA = \angle BEA - \angle OEP = \angle OCB - \angle OCP = \angle ACB = \angle PAE.$$

Hence $PA = PE$. On the other hand,

$$\angle BOP = \angle PAE = \angle PEA = \angle POA.$$

Hence triangles POB and POA are congruent, so that $PB = PA$. It follows that P is the circumcentre of triangle ABE. Let Q be the circumcentre of triangle BED. Then OQ and PQ are perpendicular to BD and BE respectively. Hence $\angle OQP = \angle DBE = \angle OCP$, so that $CPOQ$ is cyclic.

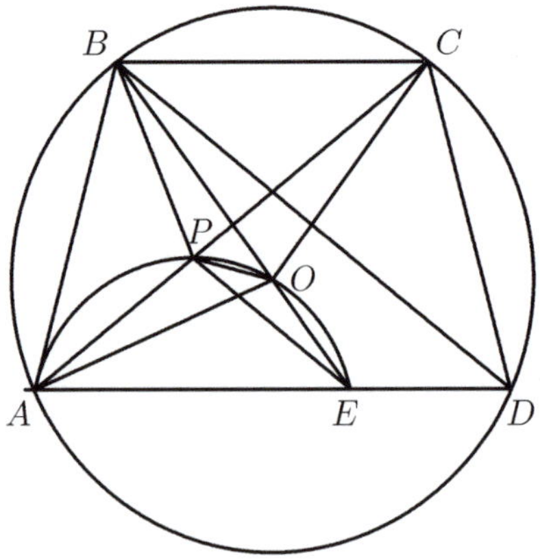

6. (a) We have $(3k+8)^2 + (k+1)^3 + (-k-4)^3 = 3k+1$.
   (b) By (a), $3k+1 = a^2+b^3+c^3$ for some integers a, b and c. It follows that $3k = a^2+b^3+c^3+(-1)^3$ and $3k+2 = a^2+b^3+c^3+1^3$.

7. We use the language of graph theory. Let the towns be represented by vertices and the roads by edges.

   (a) If every vertex is a dead-end, the graph is a tree. Anna puts the tourist at any vertex A and direct all the edges towards A. Boris must reverse the direction of an edge directed to A, say AB. Anna moves the tourist to B, and the situation is the same as before. Hence Boris does not have a winning strategy.

   (b) Suppose some vertex is not a dead-end. Then it lies on some cycle. The first step of a winning strategy for Boris was to force Anna to move the tourist on to this cycle by redirecting all edges towards it. The second step is to redirect the edges on the cycle so that they all go the same way. Once this is accomplished, Boris wins by reversing the direction of the edge on the cycle ahead of the tourist. Of course, Boris may have won earlier. The strategy just guarantees ultimate victory.

Senior A-Level Paper

1. (See Problem 2 of the Junior A-Level Paper.)

2. Let M and N be the respective midpoints of CA and AB. Then $AMON$ is cyclic. Since MN is parallel to CB and MO is parallel to EB, $\angle NAO = \angle NMO = \angle CBE$. Hence triangles ANO and BEC are similar, so that $\frac{AO}{NO} = \frac{BC}{EC} = \frac{BC}{AX}$. Triangles AMO and CFB are also similar, so that $\frac{AO}{MO} = \frac{CB}{FB} = \frac{CB}{AY}$. It follows that $NO \cdot AY = MO \cdot AX$, so that triangles OAX and OAY have the same area. The midpoint of XY therefore lies on the common side OA of the two triangles.

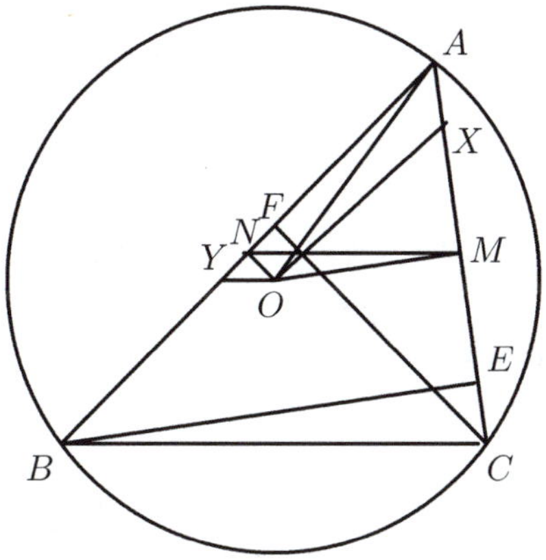

3. (a) We have $(3k+8)^2 + (k+1)^3 + (-k-4)^3 = 3k+1$.

   (b) By (a), $3k+1 = a^2+b^3+c^3$ for some integers a, b and c. It follows that $3k = a^2+b^3+c^3+(-1)^3$ and $3k+2 = a^2+b^3+c^3+1^3$.

4. We consider three cases.
Case 1. The subboard is a $1\times n$ or an $n\times 1$ rectangle where $n \geq 2$. With a $1\times n$ subboard, the best we can do is to paint black all rows which contain at least one black square initially. Since there are only finitely many black squares initially, rows without any remain white. The same goes for an $n\times 1$ subboard, with columns replacing rows in the argument.
Case 2. The convex hull of the subboard is an $m\times n$ rectangle with $m, n \geq 2$.
Since there are only finitely many black squares initially, there is a horizontal line above all of them. Since it is impossible to place the subboard so that exactly one of its squares is above this line, all squares above it will remain white.

Case 3. The convex hull of the subboard is non-rectangular.
At least one of its support lines is neither horizontal nor vertical. Let this line ℓ be to the southeast, directed as illustrated in the diagram below. It must contain the southeast vertices of at least two squares of the subboard. This means that the subboard must cross ℓ at least two squares at a time. If all initially black squares are on the left side of ℓ, no squares on the right side of ℓ can ever be painted black.

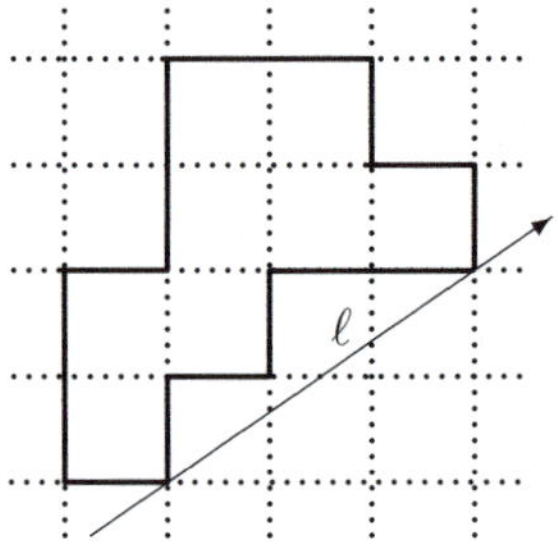

5. Let AD, BE and CF be the medians of triangle ABC. Let AL, BM and CN be the altitudes. Then $AL \leq AD$, $BM \leq BE$ and $CN \leq CF$. We may assume that $AL \leq CN \leq BM$.

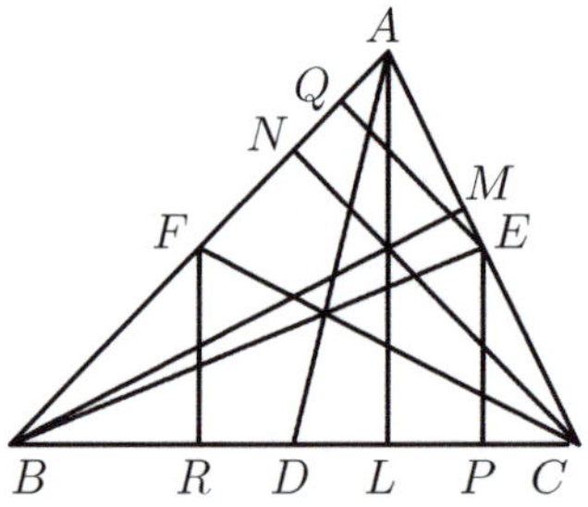

Drop the perpendiculars EP and EQ from E to BC and AB respectively. Then $2EP = AL \leq BM \leq BE$ and $2EQ = CN \leq BM \leq BE$. Hence $\angle CBE \leq 30°$ and $\angle ABE \leq 30°$. Similarly, $\angle BCF \leq 30°$ since $2FR = AD \leq CN \leq CF$, where R is the foot of the perpendicular from F to BC. Hence at most three of the six angles can exceed $30°$.

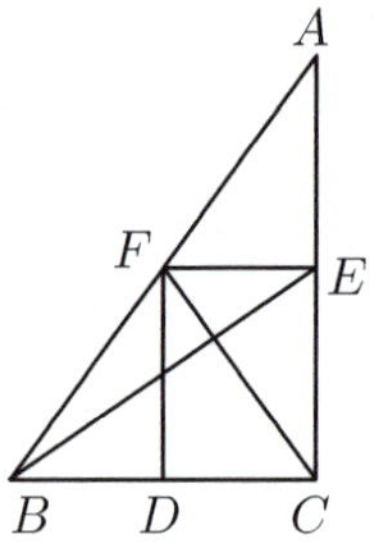

We now construct a triangle ABC in which three of the relevant angles exceed 30°. Let $AB = 10$, $BC = 6$ and $CA = 8$, so that $\angle BCA = 90^\circ$. Let AD, BE and CF be medians. Then we have $\angle CBE > 30^\circ$ since $\frac{CE}{BC} = \frac{4}{6} > \frac{1}{\sqrt{3}}$. Since $\frac{FD}{CF} = \frac{4}{5} > \frac{1}{2}$, $\angle BCF > 30^\circ$. Since $\frac{FE}{CD} = \frac{3}{5} > \frac{1}{2}$, $\angle ACF > 30^\circ$.

6. Mark 720 arcs of size $\frac{1}{2}^\circ$ on the wheel, indexed from 1 to 720. Place the wheel on the number line so that 0 is at the common endpoint of the 1st and the 720th arcs. From the infinitely many marked points, choose an infinite sequence $\{x_n\}$ such that $x_{n+1} > 10000x_n$. Define A_n as the set of positive integers m such that x_n will leave an imprint on the nth arc of a wheel of radius $\frac{1}{m}$. We claim that $A_1 \cap A_2 \cap \cdots \cap A_{720}$ is nonempty. Then we can choose m in the intersection, and a wheel of radius $\frac{1}{m}$ will have the desired property. We now prove by induction on n that $A_1 \cap A_2 \cap \cdots \cap A_n$ is nonempty for all $n \geq 1$. The basis $n = 1$ is trivial. Suppose the result holds for some $n \geq 1$. Then $A_1 \cap A_2 \cap \cdots \cap A_n$ contains an interval of length at least $x_1(10000)^{n-1}$. Since the period of A_{n+1} is less than $x_1(10000)^{n-1}$, $A_1 \cap A_2 \cap \cdots \cap A_{n+1}$ is nonempty.

7. We solve only (b), which implies (a). Anna constructs a $(2n-1) \times n$ table where the numbers in the jth column, $1 \leq j \leq n$, are the wealth of the citizens in the jth city in non-descending order. We use as a critical measure the $(2n-1)$-tuple $(x_1, x_2, \ldots, x_{2n-1})$, where x_i is the sum of the numbers in the ith row. We have $x_1 \leq x_2 \leq \cdots \leq x_{2n-1}$. We say that $(a_1, a_2, \ldots, a_{2n-1}) < (b_1, b_2, \ldots, b_{2n-1})$ if there exists an index k such that $a_i = b_i$ for all $i < k$ and $a_k < b_k$. We claim that Anna can decrease $(x_1, x_2, \ldots, x_{2n-1})$ until $x_1 = 0$ and wins. We consider two cases.
Case 1. $x_i \neq i$ for at least one i.
Suppose the x_i are distinct. We may assume that $x_1 \geq 1$. Since $x_2 > x_1$, $x_2 \geq 2$. Similarly, $x_i \geq i$ for all i. Inequality holds at least once. Hence $x_1 + x_2 + \cdots + x_{2n-1} > 1 + 2 + \cdots + (2n-1) = n(2n-1)$, which is the total wealth of all the citizens. It follows that there exists an index k such that $x_k = x_{k+1}$. Now the kth row and the $(k+1)$st row must be identical, since in each column, the number in the kth row is less than or equal to the number in the $(k+1)$st row. Anna will ask Boris to redistribute the wealth in the kth row. Since Boris cannot leave everything alone, at least one number has increased and at least one number has decreased. Anna will rearrange the numbers in each column in non-descending order.Those numbers which have decreased will either remain in kth row or move higher. Those numbers which have increased will either remain in the kth row or move lower. The $(2n-1)$-tuple will decrease.

Case 2. $x_i = i$ for all i.
We claim that there exists an index k such that the kth row contains k people with 1 dollar while the $(k+1)$st row contains at most k people with positive wealth. Note that since $x_1 = 1$, there is 1 person with 1 dollar in the 1st row. If both dollars in $x_2 = 2$ are in the hand of one person in the 2nd row, then $k = 1$. Otherwise, we move onto the 3rd row. By the time we reach the $(n+1)$st row, there are only n people altogether, so that at most n of them can have positive wealth. Anna will ask Boris to redistribute the wealth in the kth row. If in some column, Boris creates a positive number in the kth row above a 0 in the $(k+1)$st row, or a 0 below a 1 in the $(k-1)$st row, the $(2n-1)$-tuple will decrease. Suppose neither situation occurs. Still, the $(2n-1)$-tuple cannot increase. If it does not change, then the $(k-1)$st row contains $k-1$ people with 1 dollar while the kth row contains at most $k-1$ people with positive wealth. The above argument can be repeated with k replaced by $k-1$. Eventually, we back all the way up to the 1st row. Then Boris must create a positive number there above a 0 in the 2nd row.

Spring 2019

Junior O-Level Paper

1. Such a sequence with 11 terms is 1, 1, 4, 1, 1, 4, 1, 1, 4, 1, 1.

2. Number the coins consecutively from 1, whether they are flipped or skipped over, and continue the numbering round the circle. The kth coin flipped over is numbered $c_k = \frac{k(k+1)}{2}$. The last two coins flipped over are numbered

$$c_{2n} = \frac{2n(2n+1)}{2} = n(2n+1)$$

and

$$c_{2n+1} = \frac{(2n+1)(2n+2)}{2} = (n+1)(2n+1).$$

They are the same coin since $c_{2n} \equiv c_{2n+1} \pmod{2n+1}$. Now

$$\begin{aligned} c_{n+1} - c_{n-1} &= n + (n+1) = 2n+1, \\ c_{n+2} - c_{n-2} &= 2(2n+1), \\ \cdots &= \cdots, \\ c_{2n-1} - c_1 &= (n-1)(2n+1). \end{aligned}$$

Hence every coin is flipped over an even number of times except c_n. Sicne the total number $2n+1$ of flips is odd, this will be the only coin which shows tails.

3. We have $\frac{mn}{m+n} = \frac{mn+n^2-n^2}{m+n} = n - \frac{n^2}{m+n}$. Since $\frac{n^2}{m+n}$ is an integer, $n^2 \geq m+n$.

4. Let M be the midpoint of QR. Choose P to be the midpoint of BC, so that $\angle MPC = 90° = \angle MRD$. Let X be another point on BC, and let the line through M perpendicular to MX cut AB at Y and CD at Z. Then $XY = XZ$ and M is also the midpoint of YZ. Since $\angle PMX = 90° - \angle XMR = \angle RMZ$, PMX and RMZ are similar right triangles. Hence $\frac{PM}{MR} = \frac{XM}{MZ}$, so that PMR and XMZ are also similar right triangles. It follows that $\angle ZXY = 2\angle ZXM = 2\angle RPM = \angle RPQ$.

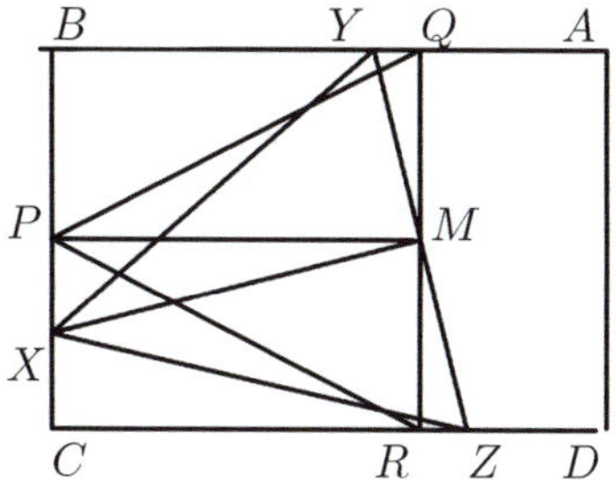

5. The assistant mentally arranges the boxes in clockwise order, dividing the circle into twelve unit arcs. The length of a chord is measured by the number of unit arcs on the minor arc it cuts off. If the chord is a diameter, its length is 6. Consider the quadrilateral in the diagram below. The lengths of its four sides and two diagonals are 1, 2, 3, 4, 5 and 6 in some order, covering all possible chord lengths. When the audience chooses two of the boxes, the assistant determines the length of the chord joining them. Then he rotates the quadrilateral about the centre of the circle until this chord coincides with a side or a diagonal of the quadrilateral. There is a unique position for the quadrilateral except when the chord is a diameter. Then there are two possible positions. In either case, the assistant will open the box immediately preceding the two adjacent boxes on the quadrilateral. For example, if the boxes chosen by the audience are numbered 1 and 6, then the quadrilateral is already in its correct position, and the assistant will open the box numbered 11. When the magician comes in, she looks at the number n of the open box. Then she opens the boxes numbered $n+1$, $n+2$, $n+5$ and $n+7$, reduced modulo 12. Continuing the example, she will open the boxes numbered 11+1=12, $11+3 \equiv 1$, $11+5 \equiv 4$ and $11+7 \equiv 6 \pmod{12}$.

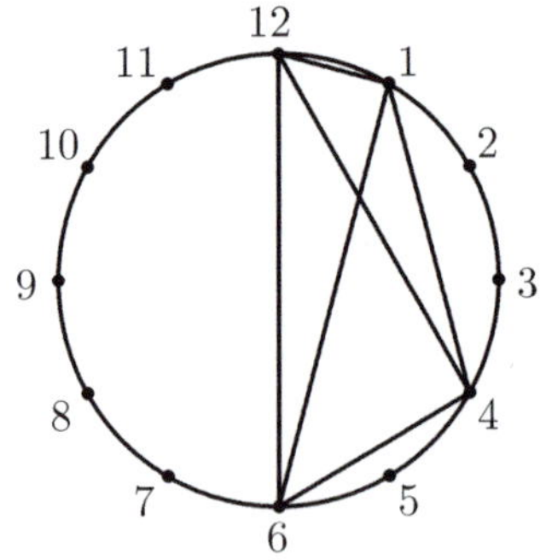

Senior O-Level Paper

1. Let O be the centre of the regular hexagon $ABCDEF$ and let M be the midpoint of BC. Let P be the point such that $PB = PC = 1$ and $PA = 2$. Note that $OA = 2MC$ and $PA = 2PC$. Hence POA and PMC are similar right triangles. It follows that P lies on AC. Hence $AC = 3$ and $AB = \sqrt{3}$.

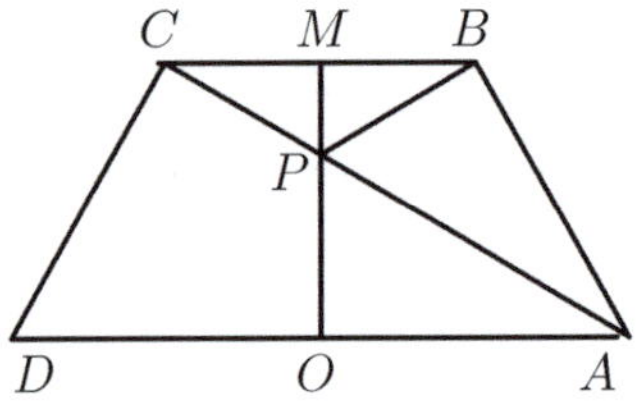

2. Let k be the greatest common divisor of a and b. Let $a = kc$ and $b = kd$, so that c and d are relatively prime. We now have
$$\frac{a^{n+1}+b^{n+1}}{a^n+b^n} = k(c+d) - \frac{kcd(c^{n-1}+d^{n-1})}{c^n+d^n}.$$
Since $c^n + d^n$ is relatively prime to cd, it must divide $k(c^{n-1}+d^{n-1})$. However, this cannot hold for infinitely many n unless $c = d = 1$. It follows that we must have $a = b$.

3. Divide each side of a triangle into 26 equal parts and join the points of division by lines parallel to the sides of the triangle. This divides the triangle into $26^2 = 676$ triangles. If we combine the four at the top, we have 673 triangles. Now divide each into three kites by dropping from its incentre perpendiculars to the sides. A kite always has an incircle, and a kite with two right angles opposite each other has a circumcircle. The number of kites is $673 \times 3 = 2019$.

4. The assistant mentally arrange the boxes in clockwise order, dividing the circle into thirteen unit arcs. The length of a chord is measured by the number of unit arcs on the minor arc it cuts off. Consider the quadrilateral in the diagram below. The lengths of its four sides and two diagonals are 1, 2, 3, 4, 5 and 6 in some order, covering all possible chord lengths. When the audience chooses two of the boxes, the assistant determines the length of the chord joining them. Then he rotates the quadrilateral about the centre of the circle until this chord coincides with a side or a diagonal of the quadrilateral. There is always a unique position for the quadrilateral. The assistant will open the box immediately preceding the two adjacent boxes on the quadrilateral. For example, if the boxes chosen by the audience are numbered 7 and 9, the assistant will rotate the quadrilateral to the position shown in dotted lines in the diagram below, and will open the box numbered 2. When the magician comes in, she look at the number n of the open box. Then she opens the boxes numbered $n+1$, $n+2$, $n+5$ and $n+7$, reduced modulo 13. Continuing the example, she will open the boxes numbered 2+1=3, 2+2=4, 2+5=7 and 2+7=9.

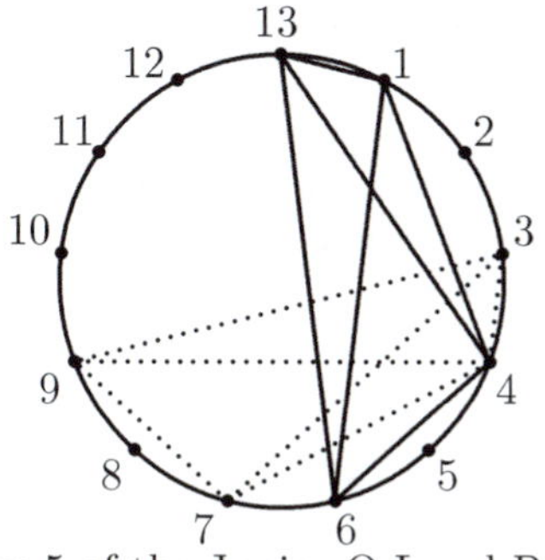

(Compare with Problem 5 of the Junior O-Level Paper.)

5. By a block, we mean 39 copies of 1 followed by a copy of 41. The sum of these 40 numbers is 80. Take 25 blocks followed by 19 copies of 1. The sum is $25 \times 80 + 19 = 2019$ and the length of the row is $25 \times 40 + 19 = 1019$. We claim that this is maximum. We will prove that any sum of any block of 40 adjacent numbers must be at least 80. Let the numbers be a_1, a_2, $\ldots$, a_{40} and consider the 40 sums a_1, $a_1 + a_2$, $\ldots$, $a_1 + a_2 + \cdots + a_{40}$. These sums are increasing and therefore distinct. If all of these sums less than 80, there are at most 78 different values, namely, 1, 2, $\ldots$, 39, 41, 42, $\ldots$, 79. If one of the sums is 39, we cannot also have 79 as a sum, as otherwise there will be a block of adjacent numbers with sum 40. Similarly, we can have at most one sum from each of the pairs (38,78), (37,77), $\ldots$, (1,41). However, we will have at most 39 sums. This is a contradiction. Since each block of 40 adjacent terms has sum at least 80, the first 25 groups will have sum at least 2000. We have $25 \times 40 = 1000$ terms so far, and at most $2019 - 2000 = 19$ more. This justifies our claim.

Junior A-Level Paper

1. If the fourth largest number is 22, then the sum of the largest four numbers is at least 22+23+24+25=94. Hence the sum of the smallest three numbers is at most $100 - 94 = 6$. They must be 1, 2 and 3, and the other four numbers must be 22, 23, 24 and 25.

2. Label the grasshoppers 0 to 2018 from left to right. To construct the new sequence of moves, let grasshopper number k jump over number 0 to become number $-l$, for $1 \leq k \leq 2018$. After that, if number k jumps to the right in the original sequence of moves, then number $-k$ jumps to the right in the new sequence. The desired result follows from symmetry.

3. Let P and Q be the centres of the equal circles. Draw the line through P parallel to AB, and the line through Q parallel to AC, intersecting at D.

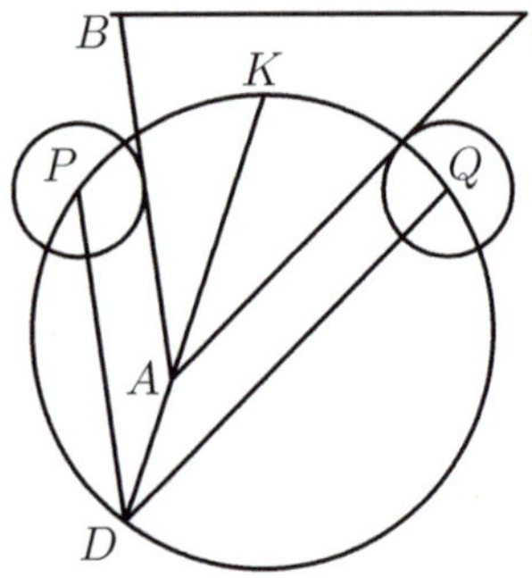

Construct the circumcircle of triangle PDQ. Since P and Q are fixed points while $\angle PDQ = \angle BAC$ is constant, this circumcircle is independent of the position of ABC. Let K be the midpoint of the arc PQ not containing D. Then DK is the bisector of $\angle PDQ$. Points on this line are equidistant from DP and DQ. Hence they are also equidistant from AB and AC since the circles with centres P and Q are of the same size. Hence DK is also the bisector of $\angle BAC$, which means that A lies on DK. It follows that the bisector of $\angle BAC$ always passes through K.

4. For $1 \le k \le 100$, let the kth point in cyclic order be labeled a_k. We define $s = a_1 + a_3 + \cdots + a_{99}$ and $t = a_2 + a_4 + \cdots + a_{100}$. Let S be the sum of the labels of all chords and T be the sum of the labels of all blue chords. Then $2S = (s+t)^2 - 1$ and $2T = s^2 + t^2 - 1$. It follows that the desired difference is given by

$$S - 2T = \frac{1}{2}((s+t)^2 - 1) - (s^2 + t^2 - 1) = \frac{1}{2}(1 - (s-t)^2).$$

Hence the maximum value is $\frac{1}{2}$, attained if and only if $s = t$. For instance, we may take $a_1 = a_{100} = \frac{1}{\sqrt{2}}$ and $a_k = 0$ for $2 \le k \le 99$.

5. Suppose that $n = 2k$ for some positive integer k. Put the numbers 1 to $4k^2$ into the table in the following order. We start with the first row from left to right, then go down the last column, go up the second last column and so on, until we go up the first column and put the number $4k^2$ directly below the number 1. The diagram below on the left illustrates the case $k = 3$.

1	2	3	4	5	6
36	27	26	17	16	7
35	28	25	18	15	8
34	29	24	19	14	9
33	30	23	20	13	10
32	31	22	21	12	11

1	2	3	4	5	0
0	3	2	5	4	1
5	4	1	0	3	2
4	5	0	1	2	3
3	0	5	2	1	4
2	1	4	3	0	5

Clearly, two consecutive integers are adjacent in the same row or the same column. We now reduce the numbers modulo n to $0, 1, \ldots, n-1$. The case $n = 6$ is illustrated in the diagram above on the right. For even-numbered columns, the last number is the same as the first number of the preceding column. For odd-numbered columns after the first, the second number is the same as the first number of the preceding column. Hence no two numbers in the same column are congruent modulo n.

This also implies that no two numbers in the same row are congruent modulo n, since this is the case for the first row.

If n is odd, paint the squares of the table black and white in the standard board pattern, with black squares at the corners. Label the rows 1 to n from the bottom and the columns 1 to n from the left. Then the sum of the row number and the column number is even for any black square and odd for any white square. Since consecutive integers are in adjacent squares, all odd numbers are in black squares and all even numbers are in white squares. Reduce the integers modulo n and consider the n numbers in the same congruence class. By the hypothesis, the sum of their row numbers and columns numbers must be the even number $(1+2+\cdots+n)+(1+2+\cdots+n)$. Hence the number of white squares occupied by these n numbers must be even. Thus the numbers in congruence classes 0 and 1 together occupy an even number of white squares. Since consecutive integers are in adjacent squares, exactly half of these $2n$ numbers must occupy white squares. This is not possible if n is odd.

6. Let L be the point such that BCL is an equilateral triangle, as shown in the diagram below. Perform a counterclockwise $60°$ rotation about L, mapping C into B and K into D. We claim that D coincides with A. Note that $\angle KLC = 60° + \angle KLB = \angle DLB$. Triangles KLC and DLB are congruent since $LC = LB$ and $LK = LD$. Hence $DB = KC = BC$. Moreover, since K and L are symmetric about DC, $\angle KDC = 30°$. This justifies the claim. Hence $\angle AKB = \angle LKB = \frac{1}{2}(360° - 60°) = 150°$.

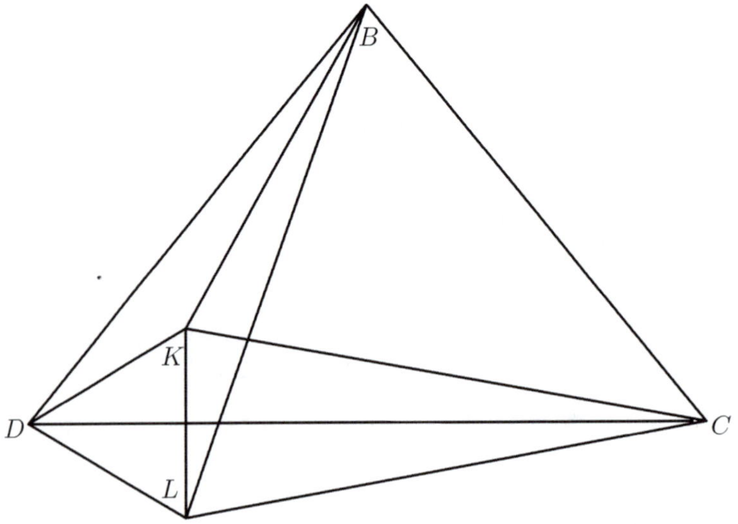

7. Label the counters in each pile 1 to 400 from the bottom. When two counters from different piles are removed, the number of points gained by Pete is equal to their absolute difference. We can assign a + sign to the larger number and a − sign to the smaller one. If they are equal, then it does not matter how the signs are assigned. Hence exactly half of the $100 \times 400 = 20000$ labels are positive. If all the labels between 201 and 400 are positive and all the labels between 1 and 200 are negative, then Pete would gain $100((400-200)+(399-199)+\cdots+(201-1)) = 4000000$ points. However, consider the moment when we remove for the first time a counter with label k between 201 and 400. The other counter removed at the same time must have a label greater than or equal to k. It follows that at least one counter with label k is assigned −. Similarly, at least one counter with label between 1 and 200 is assigned +, the last time such a counter is removed. Hence Pete can gain at most $4000000 - 2(400+399+\cdots+201) + 2(200+199+\cdots+1) = 3920000$ points. This can be achieved as follows. In the first stage, he removes 200 counters from each of the Pile 1 and Pile 2, one at a time. He gains 0 points. In the second stage, he removes 200 counters from each of Pile 1 and Pile 3, and 200 counters from each of Pile 2 and Pile 4. He gains 40000 points. In the subsequent stages, he removes 200 counters from Pile 3 and Pile 5, from Pile 4 and Pile 6, from Pile 5 and Pile 7, from Pile 6 and Pile 8, and so on. He gains 40000 points each stage. In the last stage, he removes the remaining 200 counters from Pile 99 and Pile 100, gaining 0 points. Thus the total gain is $98 \times 40000 = 3920000$ points.

Senior A-Level Paper

1. Let two adjacent digits of a multiple of 7 be chosen. We separate it into two numbers x and y, with x consisting of all the digits from the left up to and including the first of the two chosen digits, and y consisting of the remaining digits. Let n be the number of digits in y. Then the original multiple is $10^n x + y$, and the number obtained by inserting a digit d between the chosen digits is $10^n(10x+d)+y$. The difference between these two numbers is $10^n(9x+d)$. Since every seventh number is a multiple of 7, there exists at least one value for d such that $9x+d$ is a multiple of 7. Then the new number will also be a multiple of 7. We claim that the same digit d can be added k times between the chosen digits for any positive integer k, and we will still have a multiple of 7. We use mathematical induction on k. The basis $k = 1$ has already be established. Suppose the claim holds up to some $k \geq 1$. The difference between the number after k copies of d has been added and the number after $k+1$ copies of d has been added is $10^{n+k}(9x+d)$, which is a multiple of 7. This completes the inductive argument.

2. (See Problem 2 of the Junior A-Level Paper.)

3. Let P be the centre of a circle of radius p and Q be the centre of a circle of radius $q > p$. Let the variable angle be $\angle BAC$ with AB tangent to the first circle and AC to the second. Draw the line through P parallel to AB, and the line through Q parallel to AC, intersecting at D. Construct the circumcircle of triangle PDQ. Since P and Q are fixed points while $\angle PDQ = \angle BAC$ is constant, this circumcircle is independent of the position of ABC. Let K be the point of the arc PQ not containing D such that $PK : QK = p : q$. Let R be the circumradius of PDQ. Then

$$\begin{aligned}\sin\angle PDK : \sin\angle QDK &= 2R\sin\angle PDK : 2R\sin\angle QDK \\ &= PD : QD \\ &= p : q.\end{aligned}$$

Hence the distances from the points on this line to PD and QD are in the ratio $p : q$. It follows that the distances from them to AB and AC are also in the ratio $p : q$, since that is the ratio of the radii of the circles with centres P and Q. Hence A lies on DK. It follows that the ray from A such that the distances of points on it to AB and AC is in the ratio $p : q$ always passes through K.

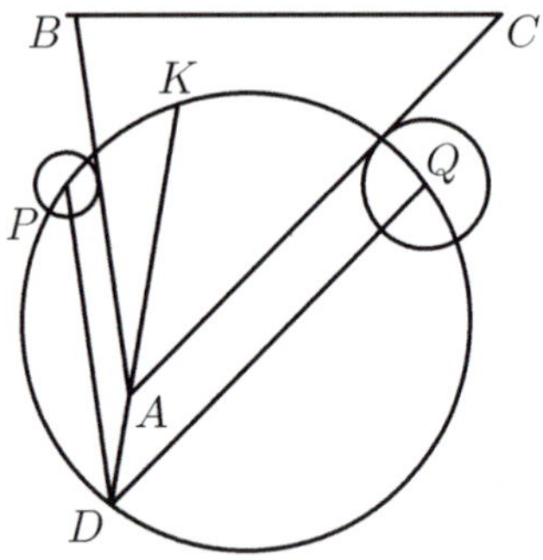

(Compare with Problem 3 of the Junior A-Level Paper.)

4. (See Problem 5 of the Junior A-Level Paper.)

5. Let O be the origin and $OABD$ be the square in the first quadrant of the horizontal xy-plane, with side length a. Let C be a point directly above D with z-coordinate c. We take O as the vertex of our tetrahedron and ABC as its base. The equation of the plane of ABC is $cy + az = ac$ and the line through O perpendicular to this plane is the intersection of the planes $x = 0$ and $ay = cz$. It follows that the point of intersection of this line and the plane of ABC is $P(0, \frac{ac^2}{a^2+c^2}, \frac{a^2c}{a^2+c^2})$. Hence AP and BC are both parallel to the yz-plane.

On the other hand, AB is parallel to the x-axis but CP is not. Hence $ABCP$ has exactly one pair of parallel sides. Note that

$$\begin{aligned} AB &= a, \\ BC &= \sqrt{a^2+c^2}, \\ AP &= \sqrt{\left(\frac{ac^2}{a^2+c^2}-a\right)^2+\left(\frac{a^2c}{a^2+c^2}\right)^2} \\ &= \frac{a^2}{\sqrt{a^2+c^2}}. \end{aligned}$$

Hence the area of $ABCP$ is $\frac{a}{2}(\sqrt{a^2+c^2}+\frac{a^2}{\sqrt{a^2+c^2}})$. On the other hand, the projection of the tetrahedron onto the plane of the lateral face OAB is the square $OABD$ of area a^2.

(a) We want the area of $OABD$ to be 1, so that we choose $a=1$. We now choose c so that the area of $ABCP$ is also 1. We must have $c=0$ since $\frac{1}{2}(\sqrt{1+c^2}+\frac{1}{\sqrt{1+c^2}})=1$. Hence this case is impossible since the tetrahedron $OABC$ degenerates to the square $OABD$.

(b) We choose $a=\frac{1}{\sqrt{2019}}$ because we want the area of $OABD$ to be $\frac{1}{2019}$. In order for area of $ABCP$ to be 1, we take c to be the root of the equation $\frac{1}{2\sqrt{2019}}\left(\sqrt{\frac{1}{2019}+c^2}+\frac{1}{2019\sqrt{\frac{1}{2019}+c^2}}\right)=1$.

6. Boris always chooses $x_1=x_2=x_3=0$ and $x_8=x_9=x_{10}=100$. He will choose $1\le x_4\le x_5\le x_6\le x_7\le 12$ according to Anna's action. Then the only cards which matter are those containing all of x_8, x_9 and x_{10} as well as two of x_4, x_5, x_6 and x_7. Thus the game simplifies to one with six cards containing the pairs (x_4,x_5), (x_4,x_6), (x_4,x_7), (x_5,x_6), (x_5,x_7) and (x_6,x_7). In the first round, Anna will take (x_6,x_7). Boris takes (x_5,x_7). Since both (x_4,x_7) and (x_5,x_6) are better than (x_4,x_6), each will get one of them in the second round. In the final round, Anna takes (x_4,x_6) over (x_4,x_5). If Anna takes (x_4,x_7) in the second round, Boris chooses $x_4=1,$, $x_5=3$, $x_6=4$ and $x_7=5$. Anna's sum is $20+5+4=29$ while that of Boris is $15+12+3=30$. If Anna takes (x_5,x_6) in the second round, Boris chooses $x_4=3$, $x_5=4$, $x_6=5$ and $x_7=12$. Anna's sum is $60+20+15=95$ while that of Boris is $48+36+12=96$. Hence Boris always wins.

7. A polygonal line may be represented by a word in which every letter is either U for up or R for right. We use the notation $f(w)$ to denote the number of ways into which the associated board of the word w may be dissected into dominoes. Thus the word $RRURUURU$ represents the polygonal line in the diagram below on the left.

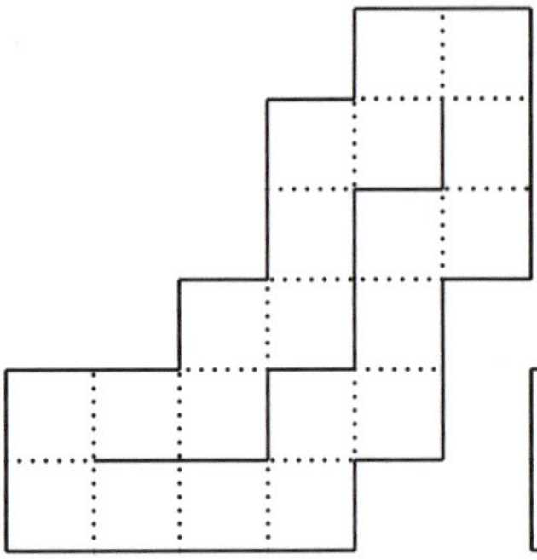

$f(RRURUURU) = 38$

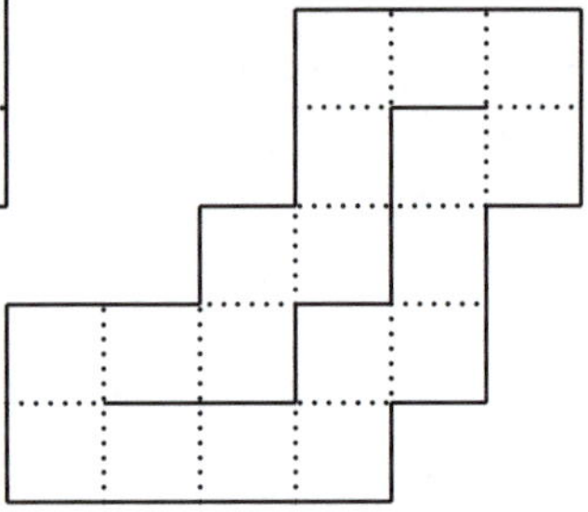

$f(RRURUUR) = 29$

The subsequent diagrams shown the polygonal lines represented by words obtained from $RRURUURU$ by contracting one letter at a time from the end, down to the empty word. In each case, the associated board is also shown, along with the value of f.

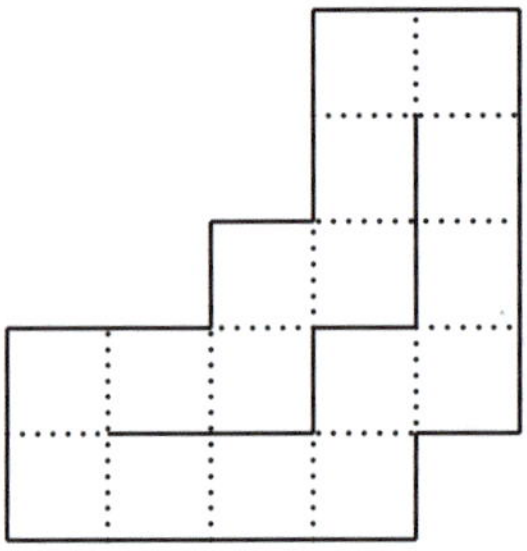

$f(RRURUU) = 20$

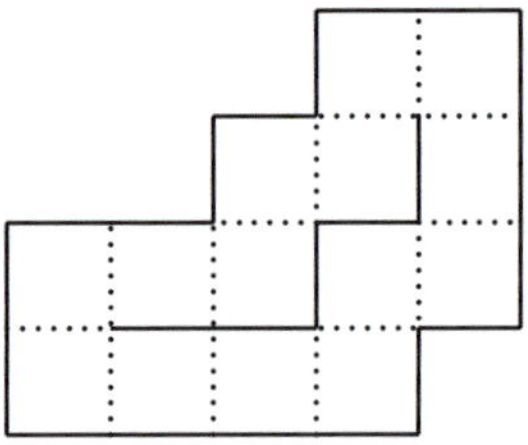

$f(RRURU) = 11$

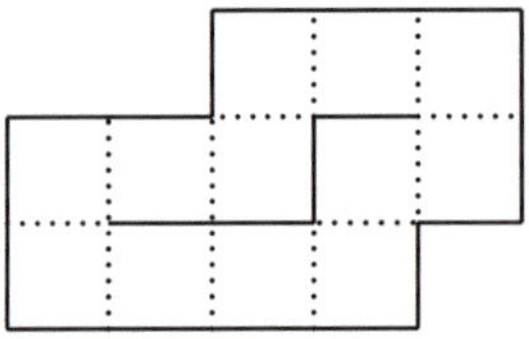

$f(RRUR) = 9$

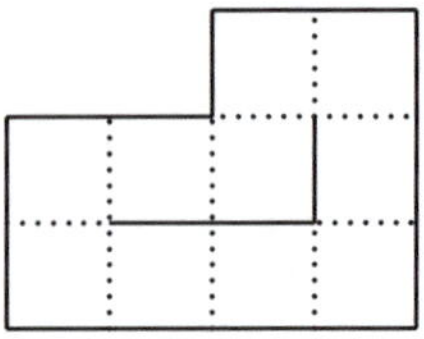

$f(RRU) = 7$

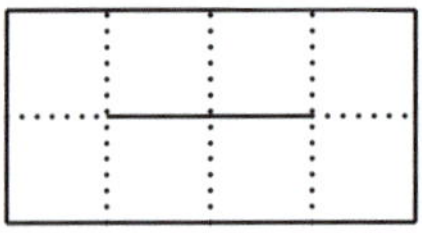

$f(RR) = 5$

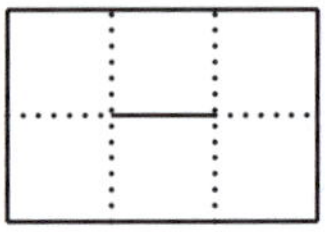

$f(R) = 3$

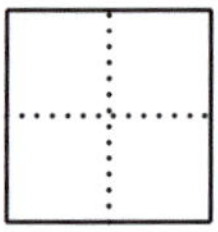

$f(\emptyset) = 2$

From $(\emptyset) = 2$ and $f(R) = 3$, we generate recursively the values 5, 7, 9, 11, 20, 29 and 38 as follows.

$$\begin{aligned}
f(RR) &= f(R) + f(\emptyset);\\
f(RRU) &= f(\emptyset) + f(RR);\\
f(RRUR) &= f(RRU) + f(\emptyset);\\
f(RRURU) &= f(\emptyset) + f(RRUR);\\
f(RRURUU) &= f(RRUR) + f(RRURU);\\
f(RRURUUR) &= f(RRURUU) + f(RRUR);\\
f(RRURUURU) &= f(RRUR) + f(RRURUUR).
\end{aligned}$$

In each case, $f(w)$ is the sum of two terms. The first term is obtained when the top right square of the board associated with w is covered by a vertical domino. The second term is obtained when the top right square of the board associated with w is covered by a horizontal domino. The placement of this domino may force the placement of other dominoes.

In the first and the fifth equations, where the last two letters of w are the same, $f(w)$ is the sum of the preceding two terms. In the other five equations, where the last two letters of w are different, $f(w)$ is the sum of the preceding term and the last term for which the next two letters are the same, if such a term exists. Otherwise, $f(w)$ is just one more than the preceding term.

For the second, the third and the fourth equations, such a term exists, namely $f(\emptyset) = f(RR) - f(R)$. The second equation may be rewritten as $f(RRU) = (f(RR)-f(R))+f(RR) = 2f(RR)-f(R)$. For the third and the fourth equations, we have $f(RRUR) = 2f(RRU) - f(RR)$ and $f(RRURU) = 2f(RRUR) - f(RRU)$.

For the last two equations, such a term is

$$f(RRUR) = f(RRURUU) - f(RRURU).$$

For the sixth equation, we have

$$f(RRURUUR) = 2f(RRURUU) - f(RRURU).$$

Similarly,

$$f(RRURUURU) = 2f(RRURUUR) - f(RRURUU)$$

for the last equation. In other words, $f(w)$ is the difference between twice the preceding term and the term before that.

In every sequence such as $\{2,3,5,7,9,11,20,29,38\}$, the first two terms are always 2 and 3. If a and b are two consecutive terms, then the next term is either $a+b$ or $2b-a$. It follows that a and b are relatively prime, and $a < b < 2a$.

Let the last two terms be m and n. Then m and n are relatively prime, and $m < n < 2m$. We can reconstruct the entire sequence backwards. If $3m > 2n$, as in the case $3 \times 29 > 2$, the preceding term must be $2m - n$ or 20 in our example. If $3m < 2n$, as in $3 \times 11 < 2 \times 20$, the preceding term must be $n - m$, or 9 in our example. Continuing this way, we can trace the sequence back to 3 and 2.

Such a numerical sequence matches exactly two words, one starting with U and the other starting with R. There are exactly $\frac{1}{2}\phi(n)$ values of m that satisfy $m < n < 2m$ and are relatively prime to n. Thus there are exactly $\phi(n)$ words whose associated boards can be dissected into dominoes in exactly n ways.

Tournament 41

Fall 2019

Junior O-Level Paper

1. A magician lays the 52 cards of a standard deck in a row, and announces in advance that the Three of Clubs will be the only card remaining after 51 steps. In each step, the audience points to any card. The magician can either remove that card or remove the card in the corresponding position counting from the opposite end. What are the possible positions for the Three of Clubs at the start in order to guarantee the success of this trick?

2. A and C are fixed points on a circle with centre O and P is a variable point on the same circle. X and Y are the respective midpoints of PA and PC, and H is the orthocentre of triangle OXY. Prove that H is a fixed point.

3. Counters numbered 1 to 100 are arranged in order in a row. It costs 1 dollar to interchange two adjacent counters, but nothing to interchange two counters with exactly 3 other counters between them. What is the minimum cost for rearranging the 100 counters in reverse order?

4. Each of 1000 points on a circle is labeled with the square of an integer. The sum of any 41 adjacent labels is a multiple of 41^2. Is it necessarily true that each of the 1000 integers is a multiple of 41?

5. Basil has sufficiently many copies of the I-tricube, which is a $1 \times 1 \times 3$ block, and of the V-tricube, which is a $1 \times 2 \times 2$ block with a unit cube missing at a corner. Basil builds with these pieces a solid rectangular box each dimension of which is at least 2. For each case, what is the minimum number of copies of the I-tricube which Basil must use?

Note: The problems are worth 4, 4, 4, 5 and 5 points respectively.

Senior O-Level Paper

1. A magician lays the 52 cards of a standard deck in a row, and announces in advance that the Three of Clubs will be the only card remaining after 51 steps. In each step, the audience points to any card. The magician can either remove that card or remove the card in the corresponding position counting from the opposite end. What are the possible positions for the Three of Clubs at the start in order to guarantee the success of this trick?

2. Let $ABCDE$ be a convex pentagon such that $AB = BC$, and AE is parallel to CD. Let K be the point of intersection of the bisectors of $\angle A$ and $\angle C$. Prove that BK is parallel to AE.

3. In each step, we may multiply a positive integer by 3 and then add 1 to the product. If the positive integer is even, we may divide it by 2. If the positive integer is odd, we may subtract 1 from it and then divide the difference by 2. Prove that starting with 1, we can obtain any positive integers in a finite number of steps.

4. In a polygon, not necessarily convex, any two adjacent sides are perpendicular to each other. Prove that for each vertex A, the number of other vertices B such that the bisector of $\angle A$ is perpendicular to the bisector of $\angle B$ is even.

5. Counters numbered 1 to 100 are arranged in order in a row. It costs 1 dollar to interchange two adjacent counters, but nothing to interchange two counters with exactly 4 other counters between them. What is the minimum cost for rearranging the 100 counters in reverse order?

Note: The problems are worth 3, 4, 4, 5 and 5 points respectively.

Junior A-Level Paper

1. The *complexity* of a positive integer is the number of primes, not necessarily distinct, in its prime factorization. Find all integers $n \geq 2$ such that all integers strictly between n and $2n$ have complexity

   (a) not greater than the complexity of n;

   (b) less than the complexity of n.

2. ABC is an acute triangle with area S. K is a point inside ABC while L and M are points on BC such that KLM is also an acute triangle, with area S'. Prove that $\frac{S}{AB+AC} > \frac{S'}{KL+KM}$.

3. The weight of each of 100 coins is 1 gram, 2 grams or 3 grams, and there is at least one of each kind. Is it possible to determine the weight of each coin using at most 101 weighings on a balance?

4. O is the circumcentre of triangle ABC. D and F are respective midpoints of BC and AB. OP and OQ are the perpendiculars from O to the bisectors of the interior and exterior angles at B. Prove that the midpoint of DF lies on PQ.

5. Prove that for each positive integer m, there exists at least one integer $n > m$ such that both mn and $(m+1)(n+1)$ are squares of integers.

6. Peter starts with several 100-ruble bills and no coins. He buys books, each of which costs a positive integral number of rubles. If he buys a book which costs 100 rubles or more, he pays only in 100-ruble bills, and gets change in 1-ruble coins, at most 99 of them. If he buys a book which costs less than 100 rubles, he pays with 1-ruble coins if he has enough; otherwise he pays with a 100-ruble bill and gets change. When he has only coins left, their total value is equal to the total cost of the books he has bought. Can this common total be as high as 5000 rubles?

7. Peter has an $n\times n$ stamp, $n > 10$, such that 102 of the squares are coated with black ink. He presses this stamp 100 times on a 101×101 board, each time leaving a black imprint on 102 unit squares of the board. Is it possible that the board is black except for one square at a corner?

Note: The problems are worth 2+2, 7, 7, 7, 8, 8 and 10 points respectively.

Senior A-Level Paper

1. $P(x, y)$ is a polynomial such that for any integer $n \geq 0$, each of $P(n, y)$ and $P(x, n)$ is either the zero polynomial or a polynomial with degree at most n. Is it possible that the degree of $P(x, x)$ is odd?

2. D, E and F are points on the sides BC, CA and AB, respectively, of an acute triangle ABC. AF, BE and CF are concurrent at a point H. In the circle with diameter AD, draw the chord through H perpendicular to AD. In the circle with diameter BE, draw the chord through H perpendicular to BE. In the circle with diameter CF, draw the chord through H perpendicular to CF. If these three chords have the same length, prove that H is the orthocentre of ABC.

3. The weight of each of 100 coins is unknown, but is one of 1 gram, 2 grams or 3 grams, and there is at least one of each kind. Is it possible to determine the weight of each coin using at most 101 weighings on a balance.?

4. The increasing sequence of positive numbers

$$\cdots < a_{-2} < a_{-1} < a_0 < a_1 < a_2 < \cdots$$

is infinite in both directions. For a positive integer k, let b_k be the smallest integer such that the ratio of the sum of any k consecutive terms of the sequence to the largest of those k terms is not greater than b_k. Prove that eventually, the sequence b_1, b_2, b_3, ... either coincides with the sequence of positive integers, or is constant.

5. M is a point inside a convex quadrilateral $ABCD$ which is equidistant from the lines AB and CD, as well as from the lines BC and AD. The area of $ABCD$ is equal to $MA \cdot MC + MB \cdot MD$. Prove that the quadrilateral $ABCD$ has

   (a) an incircle;

   (b) a circumcircle.

6. A cube consisting of $(2n)^3$ unit cubes is pierced by several needles parallel to the edges of the cube, each piercing exactly $2n$ unit cubes. Each unit cube is pierced by at least one needle. A subset of these needles is *regular* if there are no two needles in the subset that pierce the same unit cube.

   (a) Prove that there exists a regular subset consisting of $2n^2$ needles such that all of them have either the same direction or two different directions.

   (b) What is the maximum size of a regular subset that is guaranteed to exist?

7. Some of the integers $1, 2, 3, \ldots, n$ have been painted red so that for each triple (a, b, c) of red integers, $b = c$ whenever $a(b - c)$ is a multiple of n. Prove that the number of red integers is not greater than the number of positive integers up to n that are relatively prime to n.

Note: The problems are worth 5, 5, 6, 10, 6+6, 6+6 and 12 points respectively.

Spring 2020

Junior O-Level Paper

1. In a 6×6 county, 27 of the squares are cities and the other 9 are villages. Each village is claimed by a city if and only if it shares at least one vertex with the city. Is it possible that the number of cities claiming a village is different for each village?

2. What is the maximum number of distinct integers in a row such that the sum of any 11 adjacent integers is either 100 or 101?

3. Let $ABCD$ be a rhombus. Let $APQC$ be a parallelogram containing the point B, with $AP = AB$. Prove that B is the orthocentre of triangle PDQ.

4. For some integer n, the equation $x^2 + y^2 + z^2 - xy - yz - zx = n$ has an integer solution. Prove that the equation $x^2 + y^2 - xy = n$ also has an integer solution.

5. On the standard 8×8 board, there is a rook in each of the squares $(a, 1)$ and $(c, 3)$. Alice moves first, followed by Bob, and turns alternate thereafter. In each turn, the player moves one of the rooks any number of squares upwards or to the right. The rook may not move through or stop at the square of the other rook. The player who moves either rook into the square $(h, 8)$ wins. Which of Alice and Bob has a winning strategy?

Note: The problems are worth 4, 4, 4, 5 and 5 points respectively.

Senior O-Level Paper

1. Is it possible to fill a 40×41 table with integers so that for each n, there are n copies of n among the squares which share common sides with its square?

2. What is the minimum number of points on the surface of a sphere such that for every hemisphere except possibly one, at least one of the points lies in the interior of the hemisphere?

3. There are 41 letters on a circle; each letter is A or B. We may replace ABA by B and vice versa, as well as replace BAB by A and vice versa. Is it always possible, using these replacements, to obtain a circle containing a single letter?

4. We say that a nonconstant polynomial $p(x)$ with real coefficients is split into two squares if it is represented as $a(x)+b(x)$ where $a(x)$ and $b(x)$ are squares of polynomials with real coefficients. Is there such a polynomial $p(x)$ that it may be split into two squares

   (a) in exactly one way;

   (b) in exactly two ways?

   Two splittings that differ only in the order of the summands are considered to be the same.

5. Two circles intersect at P and Q. An arbitrary line ℓ through Q intersects the circles again at A and B. The tangents to the circles at these points intersect at C. The bisector of $\angle CPQ$ intersects ℓ at D. Prove that D lies on a fixed circle.

Note: The problems are worth 4, 4, 5, 2+3 and 5 points respectively.

Junior A-Level Paper

1. Does there exist a positive multiple of 2020 which contains each of the ten digits the same number of times?

2. A dragon has 41! heads and the knight has three swords. The gold sword cuts off half of the current number of heads of the dragon plus one more. The silver sword cuts off one third of the current number of heads plus two more. The bronze sword cuts off one fourth of the current number of heads plus three more. The knight can can use any of the three swords in any order. However, if the current number of heads of the dragon is not a multiple of 2 or 3, the swords do not work and the dragon eats the knight. Will the knight be able to kill the dragon by cutting off all its heads?

3. Does there exist a cyclic n-gon in which the lengths of any two sides are different and all angles are integral numbers of degrees, where

   (a) $n = 19$;

   (b) $n = 20$?

4. For which integers $n \geq 2$ is it possible to write real numbers into the squares of an $n \times n$ table, so that every integer from 1 to $2n(n-1)$ appears exactly once as the sum of the numbers in two adjacent squares?

5. In $ABCD$, AB is parallel to CD and $AB = 3CD$. The tangents to the circumcircle at A and C intersect at K. Prove that $\angle KDA = 90°$.

6. Alice has a deck of 36 cards, 4 suits of 9 cards each. She picks any 18 cards and gives the rest to Bob. In each of 18 turns, Alice plays one of her cards first and then Bob plays one of his cards. If the two cards are of the same suit or of the same value, Bob gains a point. What is the maximum number of points he can guarantee regardless of Alice's actions?

7. Glen constructs a sequence $\{a_k\}$ of positive integers by choosing a positive integer n and a positive integer $a_0 < n$. For $k \geq 1$, a_k is the remainder when n is divided by a_{k-1}. Eventually, $a_{m+1} = 0$ for some positive integer m. Is it possible that $a_0 + a_1 + \cdots + a_m > 100n$?

Note: The problems are worth 4, 5, 4+3, 8, 9, 9 and 12 points respectively.

Senior A-Level Paper

1. Does the region bounded by the parabolas $y = x^2$ and $y = x^2 - 1$ contain a line segment of length greater than 10^6?

2. Alice finds positive integers a, b and c and tells Bob that there exist unique positive integers x, y and z such that a, b and c are the respective least common multiples of x and y, of x and z, and of y and z. Prove that Bob can determine c if Alice also tells him the values of a and b.

3. Is it possible for a tetrahedron to have two cross-sections, a square of side length at most 1 and a square of side length at least 100?

4. On each of $2n$ children is placed a black hat or a white hat. There are n hats of each colour. The children form one or several dancing circles, with at least two children in each, and the colours of the hats alternate within each circle. Prove that this can be done in exactly $(2n)!$ different ways.

5. $ABCD$ is a cyclic quadrilateral. The circles with diameters AB and CD intersect at P and Q, the circles with diameters BC and AD intersect at R and S, and the circles with diameters AC and BD intersect at T and U. Prove that PQ, RS and TU are concurrent.

6. There are $2n$ consecutive integers on a board. In each move, they are divided into pairs. Each pair is replace by their sum and their difference, which may be taken to be positive or negative. Prove that no $2n$ consecutive integers can appear on the board again.

7. For which k is it possible to place a finite number of queens on the squares of an infinite board, so that the number of queens in each row, each column and each diagonal is either 0 or exactly k?

Note: The problems are worth 4, 5, 8, 9, 9, 10 and 12 points respectively.

Solutions

Fall 2019

Junior O-Level Paper

1. The Three of Club should be the first or the last card. If the audience points to any other card, the magician removes it. If the audience points to the Three of Clubs, the magician removes the card from the opposite end. These are the only possible positions for the Three of Clubs to guarantee the success of the trick. Suppose it is in neither of these positions. The audience points to the Three of Clubs. The magician must remove a card which is neither the first nor the last card. These two cards will remain in place until they are the only cards left. This means that the Three of Clubs has already been removed, and the trick has failed.

2. Let M be the midpoint of AC. Then OM is perpendicular to AC. Since X and Y are the respective midpoints of PA and PC, XY is parallel to AC, and therefore perpendicular to OM. Similarly, OX is perpendicular to YM and OY is perpendicular to XM. It follows that M is the orthocentre of triangle OXY. Hence H coincides with M, which is a fixed point.

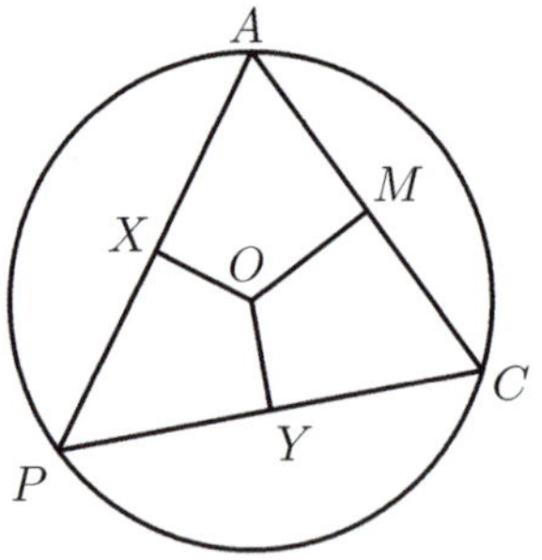

3. Paint the 100 positions red, yellow, blue, green, red, yellow, blue, green, and so on. Then two counters on positions of the same colour can be interchanged for free, while two counters in positions of adjacent colours can be interchanged for 1 dollar. If two counters on positions of adjacent colours are not adjacent themselves, we can bring them next to each other using only free moves. All counters in red positions must go to green positions, and vice versa. All counters in yellow positions must go to blue positions, and vice versa. Since red and green are adjacent colours, we can spend 25 dollars to interchange the 50 counters on them. Similarly, the 50 counters on yellow and blue positions can be interchanged for another 25 dollars, for a total cost of 50 dollars.

We now use free moves to put every counter in its correct position. Since 100 counters must change the colours of their positions, and 1 dollar can only pay for 2 such changes, 50 dollars is the minimum cost.

4. Let the integers be $a_1, a_2, \ldots, a_{1000}$. Then $a_i \equiv a_j \pmod{41}$ if $i \equiv j \pmod{41}$. Note that 1000 and 41 are relatively prime. Hence for every i, $a_i \equiv k \pmod{41}$ for some fixed k. The sum of any 41 adjacent labels will be congruent modulo 41^2 to $41k^2$. This sum will not be divisible by 41^2 unless $k \equiv 0 \pmod{41}$.

5. Basil can easily build a $1\times2\times3$ block with two copies of the V-tricube. He can build a $3\times3\times3$ block with nine copies of the V-tricube by stacking three of them on top of one another, covering the central column, and filling the remaining space with three copies of the $1\times2\times3$ block. Since Basil can build the box with copies of the I-tricube and of the V-tricube, at least one dimension of the box, say the altitude, is divisible by 3. We can cut the box into $m\times n$ horizontal slabs with altitude 3, where $m \geq 2$ and $n \geq 2$. Basil then builds one slab at a time, by filling the $m\times n$ base with copies of the 1×2 rectangle and of the 3×3 square. We consider two cases.
Case 1. Either m or n is even.
Each row has even length and can be filled with copies of the 1×2 rectangle from copies of the $1\times2\times3$ block. Hence the whole base can be filled with copies of the 1×2 rectangle.
Case 2. Both m and n are odd.
Then $m \geq 3$ and $n \geq 3$. Basil can start the bottom 3 rows with a copy of the 3×3 square from a copy of the $3\times3\times3$ block. The remaining part of each of these 3 rows has even length and can be filled with copies of the 1×2 rectangle. Since the number of remaining rows is even, Basil can fill them with copies of the 1×2 rectangle.

Senior O-Level Paper

1. (See Problem 1 of the Junior O-Level Paper.)

2. Let the line through B parallel to AE intersect the bisector of $\angle C$ at the point H.

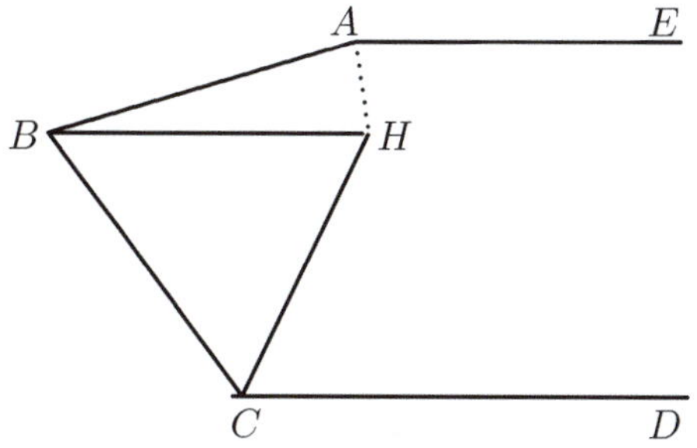

Since BH is parallel to CD, $\angle HCD = \angle CHB$. Hence $\angle BCH = \angle BHC$ so that $BC = BH$. Since $BA = BC$, we have $BA = BH$ so that $\angle BAH = \angle BHA$. Since BH is parallel to AE, $\angle BHA = \angle HAE$. It follows that BH is the bisector of $\angle A$, so that H and K coincide. Hence BK is parallel to AE.

3. In the following table, the first row consists of the positive integers in order. The second row consists of the positive integers congruent modulo 3 to 1, in increasing order starting from 4. The third row consists of the positive integers alternatingly congruent modulo 3 to 2 and 0, in increasing order starting from 2 and 3 respectively. The table extends indefinitely to the right.

1	2	3	4	5	6	7	8	9	...
4	7	10	13	16	19	22	25	28	...
2	3	5	6	8	9	11	12	14	...

Note that each positive integer in the second row fills a gap in the third row which occurs to its right. In each column, the number in the second row is generated by the number in the first row, and the number in the third row is generated by the number in the second row. Since we start with 1, we can obtain all three numbers in the first column, in particular, the 2 in the third row. This allows us to obtain all three numbers in the second column, in particular, the 3 in the third row. This allows us to obtain all three numbers in the third column. The number in the third row is 5, skipping over the gap 4. However, as we have pointed out, the number 4 has already appeared. In this manner, we can obtain all three numbers in every column. Since the first row consists of all the positive integers, we have the desired result.

4. Note that the sides of the polygon are alternatingly horizontal and vertical, and the angles have measures either 90° or 270°. A vertex is painted red if its angle bisector runs northwest to southeast, and green if it runs southwest to northeast. Red vertices of the polygon are marked by black dots in the diagrams below. The desired result is equivalent to the statement that there is an even number of vertices of each colour. If such a polygon is convex, it must be a rectangle, and it has two vertices of each colour.

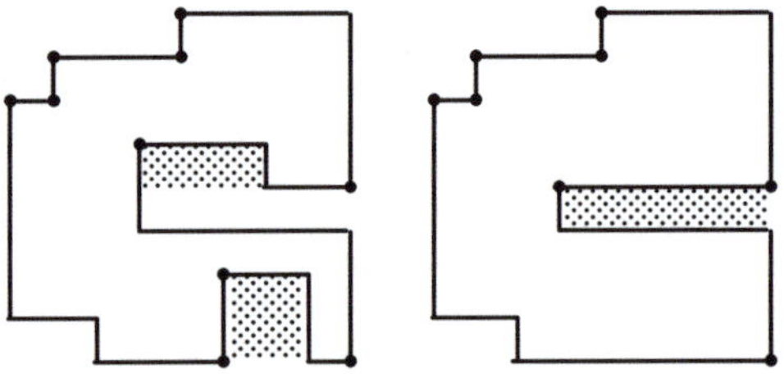

The diagram above on the left shows a typical case where the polygon is non-convex. We shade rectangles outside the polygon which are bounded on three sides by the polygon while the fourth side is completely free. We absorb these rectangles into the polygon, transforming the polygon to the one in the diagram above on the right.

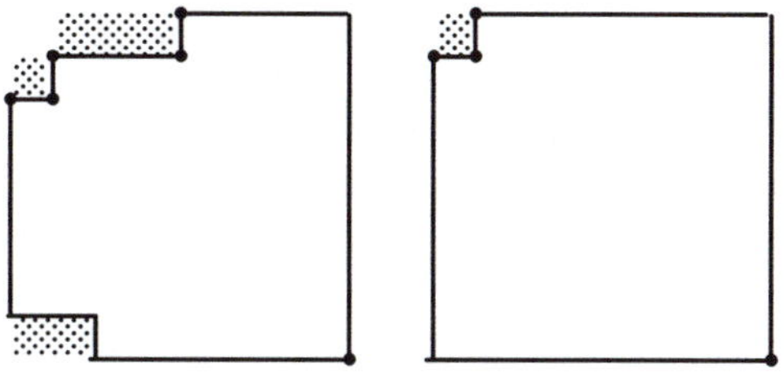

Since there is now another such rectangle, it is shaded, and its absorption yields the polygon in the third diagram. We shade rectangles outside the polygon which are bounded on two sides by the polygon while the other two sides are completely free. We absorb these rectangles into the polygon, transforming the polygon to the one in the fourth diagram. Since there is now another such rectangle, it is shaded, and its absorption completes the transformation of the original polygon into a rectangle. The parity condition for vertices of each colour holds for a rectangle. We claim that it is unchanged by any of our transformations. It will then follow that the parity condition for vertices of each colour also holds for the original polygon.

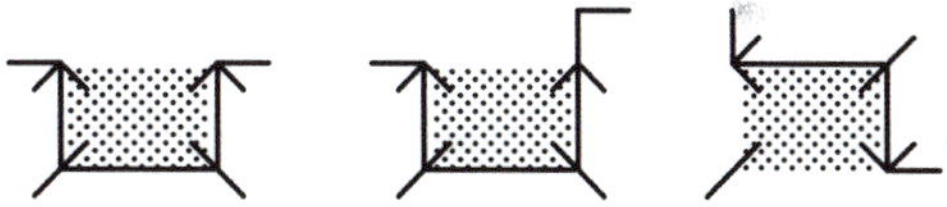

We now justify the claim. A vertex which belongs both to the original polygon as well as to the absorbed rectangle is said to be stable if it has the same colour in both cases, and unstable otherwise. In each of the three diagrams above, the rectangle has two stable vertices and two unstable ones. Since the rectangle has two vertices of each colour, each transformation changes the number of vertices of each colour by an even amount.

5. Paint the 100 positions red, yellow, brown, blue, green, red, yellow, brown, blue, green, and so on. Then two counters in positions of the same colour can be interchanged for free, while two counters in positions of adjacent colours can be interchanged for 1 dollar. If two counters on positions of adjacent colours are not adjacent themselves, we can bring them next to each other using only free moves. All counters in red positions must go to green positions, and vice versa. All counters in yellow positions must go to blue positions, and vice versa. All counters on brown positions must stay in brown positions.

Since red and green are adjacent colours, we can spend 20 dollars to interchange the 40 counters on them. Yellow and brown are adjacent colours, as are blue and brown. We can spend 40 dollars to interchange the following pairs of counters: (2,3), (4,2), (7,4), (9,7), (12,9), (14,12), (17,14), (19,17), ..., (92,89), (94,92), (97,94) and (99,97). Now every counter is in a position of the correct colour, except that counter 3 is in a yellow position while counter 99 is in a brown position. Since yellow and brown are are adjacent colours, these two counters can be interchanged for 1 dollar, bringing the total cost to 61 dollars. We now use free moves to put every counter in its correct position. The 40 counters in red and green positions must change the colours of their positions. Since 1 dollar can only pay for 2 such changes, 20 dollars are required. The 40 counters in yellow and blue positions must also change the colour of their positions. Since yellow and blue are not adjacent colours, 40 dollars are required. However, these changes cannot be made without involving at least one counter in a brown position. It follows that 61 dollars is the minimum cost.

Junior A-Level Paper

1. For any positive integer k, the smallest number with complexity k is 2^k.

   (a) If $n = 2^k$, then $2n = 2^{k+1}$. Any integer strictly between n and $2n$ is less than 2^{k+1}, with complexity at most k. If $2^k < n < 2^{k+1}$, then $2^{k+1} < 2n$. Now 2^{k+1} has complexity $k+1$ but n has complexity at most k. It follows that n must be a power of 2.

   (b) If there is such a number n, we must have $n = 2^k$ for some positive integer k. Then $2n = 2^{k+1}$ and $n < 2^{k-1}3 < 2n$. Now both n and $2^{k-1}3$ have complexity k. It follows that there are no such numbers.

2. Reflect A and K to D and N respectively across the line BC. Since ABC is acute, the kite $ABDC$ has an incircle, with radius r. Similarly, the kite $KLNM$ also has an incircle, with radius $r' < r$ since it is inside $ABDC$. The area of $ABDC$ is $2S = r(AB+AC)$ and the area of $KLNM$ is $2S' = r'(KL+KM)$. Hence $\frac{S}{AB+AC} = \frac{r}{2} > \frac{r'}{2} = \frac{S'}{KL+KM}$.

3. One coin of weight 2 gm or two coins each of weight 1 gm will be called an identifier. The weight of an untested coin can be determined in a weighing against the identifier. Label one of the coins A. Weigh A against other coins until we find coins B and C whose weights differ from the weight of A. Let k be the number of weighings so far. Then there are $99 - k$ untested coins. If one of B and C is heavier than A and the other lighter, then we know their weights, and A is an identifier. The total number of weighings is 99.

Suppose both B and C are heavier than A or both are lighter. By symmetry, we may assume that both are lighter. Then weigh A against B and C. If A is heavier, then we know their weights and B+C is an identifier. If A is lighter, then both B and C are identifiers. The total number of weighings is 100. Finally, if we have equilibrium, weigh B against C. If we have equilibrium again, then B+C is an identifier. Otherwise, the heavier of B and C is an identifier. The total number of weighings is 101.

4. Since $\angle ODB = \angle OFB = \angle OPB = \angle OQB = 90°$, D, F, P and Q all lie on a circle with OB as diameter.

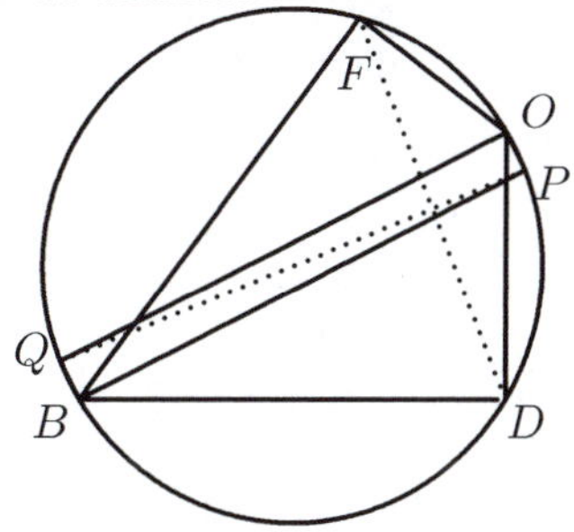

Since $OPBQ$ is a rectangle, $PQ = OB$ so that PQ is also a diameter. Since $\angle DBP = \angle FBP$, we have $DP = FP$. Hence D and F is symmetric about PQ, so that the midpoint of DF lies on PQ.

5. Note that n must be the product of m and the square of an integer. We choose this integer to be a linear function of m, namely $pm+q$ for some positive integers p and q. Then

$$n+1 = m(pm+q)^2 + 1 = p^2m^3 + 2pqm^2 + q^2m + 1.$$

This must be divisible by $m+1$, and the quotient must be the square of an integer. We perforn the following long division.

$$\begin{array}{rllll}
 & & p^2m^2 & +p(2q-p)m & +(q^2-2pq+p^2) \\ \hline
m+1 &)p^2m^3 & +2pqm^2 & +q^2m & +1 \\
 & p^2m^3 & +p^2m^2 & & \\ \hline
 & & p(2q-p)m^2 & +q^2m & \\
 & & p(2q-p)m^2 & +p(2q-p)m & \\ \hline
 & & & (q^2-2pq+p^2)m & +1 \\
 & & & (q^2-2pq+q^2)m & +(q^2-2pq+p^2) \\ \hline
\end{array}$$

It follows that $q^2 - 2pq + p^2 = 1$, so that $|p-q| = 1$. Now the quotient $p^2m^2 + p(2q-p)m + 1$ must be the square of $pm+1$. Hence $2q-p = 2$. Combined with $q-p = 1$, we have $q = 1$ and $p = 0$, which is not acceptable. Combined with $p-q = 1$, we have $q = 3$ and $p = 4$. For any positive integer m, we can choose $n = m(4m+3)^2$. Then $mn = (m(4m+3))^2$ and $n+1 = 16m^3 + 24m^2 + 9m + 1 = (m+1)(4m+1)^2$. Hence $(m+1)(n+1) = ((m+1)(4m+1))^2$.

6. Peter starts with 0 coins and books with total cost 0 rubles. These two totals should be equal when Peter no longer has any bills. When he buys a book which costs 100 rubles or more, his best option is to buy one which costs 101 rubles. This way, the first total increases by 99 and the second by 101, for a net loss of 2 rubles. The loss is greater if the book costs more. Peter can gain by buying books which costs less than 100 rubles. However, the combined gain cannot exceed 98 rubles, and this can be accomplished by buying one book at a cost of 1 ruble. This means that he can buy at most 49 books which cost 101 rubles each. Thus the total number of coins he has at the end is $99 \times 50 = 4950$ and the total cost of the books he has bought is $1 + 101 \times 49 = 4950$ rubles. This is less than 5000 rubles.

7. Remove the square at the intersection of the first row and the first column. Shade the rest of the first column but leave the rest of the first row unshaded. Divide the remaining part of the grid into four 50×50 quadrants and shade the second and the fourth ones. Then the shaded regions can be mapped into the unshaded regions by a 90° rotation. Peter's stamp is 51×101. The inked squares consists of the same row of squares in the two shaded quadrants, along with the shaded squares in the same rows in the first column. By shifting the stamp up and down, Peter can make all shaded squares black. Then he can make the unshaded squares black by rotating the stamp 90°. The diagram below illustrates with a 9×9 grid.

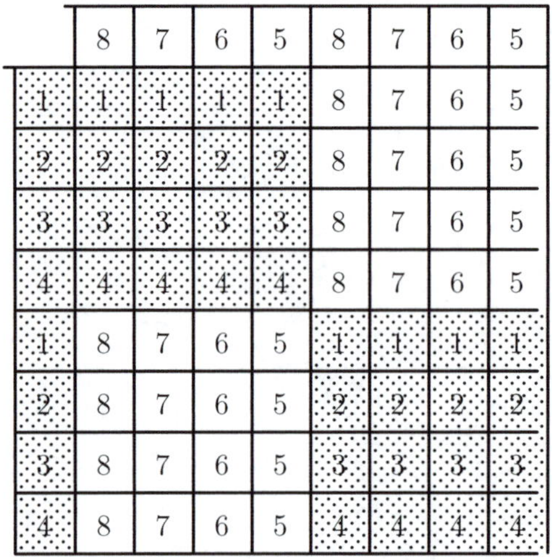

Senior A-Level Paper

1. Let the highest power of x in $P(x,y)$ be a and the highest power of y in $P(x,y)$ be b. By symmetry, we may assume that $a \leq b$, We can write $P(x,y)$ in the form $Q_b(x)y^b + Q_{b-1}(x)y^{b-1} + \cdots + Q_0(x)$, where $Q_b(x)$, $Q_{b-1}(x)$, $\ldots$, $Q_0(x)$ are independent of y. For $0 \leq n \leq b-1$, $P(n,y)$ has degree at most $n \leq b-1$. Hence the term $Q_b(n)y^b$ must vanish, so that $Q_b(x)$ has at least b roots. It follows that the degree of $Q_b(x)$ is at least b, so that the highest power of x in $P(x,y)$ is also at least b. Since this highest power is a, we must have $a \geq b$. Combined with the assumption that $a \leq b$, we have $a = b$. Now $Q_b(x)$ contains a term x^b. It follows that $P(x,y)$ contains a term $x^b y^b$, so that $P(x,x)$ contains a term x^{2b}. Since this is the term with the highest power, the degree of $P(x,x)$ must be even.

2. Let 2ℓ be the common length of the three chords. By the Power of a Point Theorem, each of $AH \cdot HD$, $BH \cdot HE$ and $CH \cdot HF$ is equal to ℓ^2, and hence to one another. It follows that $BCEF$, $CAFD$ and $ABDE$ are cyclic quadrilaterals. Hence $\angle BEC = \angle BFC$, $\angle CFA = \angle CDA$ and $\angle ADB = \angle AEB$. Denote these common values by α, β and γ respectively. Then $\beta + \gamma = \angle CDA + \angle ADB = 180°$. Similarly, we have $\gamma + \alpha = 180°$ and $\alpha + \beta = 180°$. Adding these three equations and dividing by 2, we have $\alpha + \beta + \gamma = 270°$. It follows that each of α, β and γ is $90°$. This means that AD, BE and CF are all altitudes, so that H is the orthocentre.

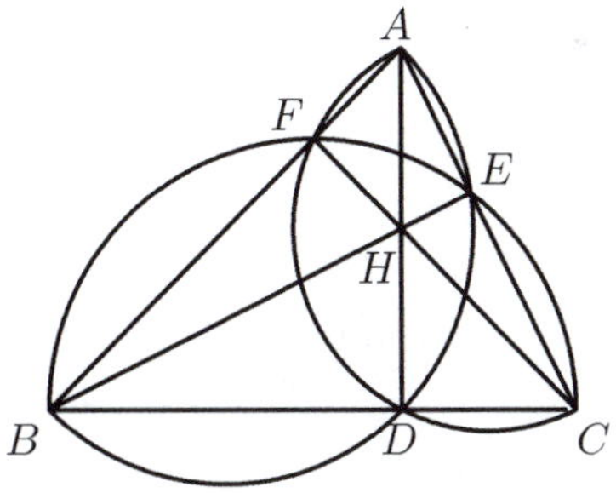

3. (See Problem 3 of the Junior A-Level Paper.)

4. If $b_k = k$ for all k, there is nothing further to prove. Hence we assume that $b_k \leq k-1$ for some k. The for every integer i,

$$\frac{a_{i+1} + a_{i+2} + \cdots + a_{i+k}}{a_{i+k}} \leq k-1,$$

so that $a_{i+1} + a_{i+2} + \cdots + a_{i+k-1} \leq (k-2)a_{i+k}$.

For integer $n > k$,

$$\begin{array}{ccccccccc} a_{i+1} & + & a_{i+2} & +\cdots+ & a_{i+k-1} & \leq & (k-2)a_{i+k}, \\ a_{i+2} & + & a_{i+3} & +\cdots+ & a_{i+k} & \leq & (k-2)a_{i+k+1}, \\ \cdots & + & \cdots & +\cdots+ & \cdots & \leq & \cdots, \\ a_{i+n-k+1} & + & a_{i+n-k+2} & +\cdots+ & a_{i+n-1} & \leq & (k-2)a_{i+n}. \end{array}$$

Then

$$a_{i+1} + \cdots + (k-1)(a_{i+k-1} + \cdots + a_{i+n-k+1}) + (k-2)a_{i+n-k+2}$$
$$+ \cdots + a_{i+n-1} \leq (k-2)(a_{i+k} + a_{i+k+1} + \cdots + a_{i+n}).$$

Let $S = a_{i+1} + a_{i+2} + \cdots + a_{i+n-1}$. Note that

$$(k-1)S - (k-2)a_{i+1} - (k-3)A_{i+2} - \cdots - a_{i+k-2}$$
$$-a_{i+n-k+2} - 2a_{i+n-k+3} - \cdots - (k-2)a_{i+n-1}$$

is less than or equal to

$$(k-2)S - (k-2)(a_{i+1} + a_{i+2} + \cdots + a_{i+k-1}) + (k-2)a_{i+n}.$$

This implies that we have

$$\begin{aligned} S \quad \leq \quad & -a_{i+2} - 2a_{i+3} - \cdots - (k-2)a_{i+k-1} + a_{i+n-k+2} \\ & +2a_{i+n-k+3} + \cdots + (k-2)a_{i+n-1} + (k-2)a_{i+n}. \end{aligned}$$

It follows that

$$\begin{aligned} S & \leq (1 + 2 + 3 + (k-2) + (k-2))a_{i+n} \\ & = \frac{(k-2)(k+1)}{2} a_{i+n}. \end{aligned}$$

Hence $b_n \leq \frac{(k-2)(k+1)}{2} + 1$ for all n. Since $\{b_n\}$ is non-decreasing, it is eventually constant.

5. Let [ ] denote area.
   (a) Drop perpendiculars MP and MQ from M to AB and CD respectively. Then $MP = MQ$. We first prove an auxiliary result.

   Lemma.

   $[MAP] + [MCQ] \leq \frac{1}{2} MA \cdot MC$. Equality holds if and only if $MP = MQ = MA \cdot MC\sqrt{MA^2 + MC^2}$.

   Proof:

   Let $MP = MQ = x$, $MA = r$, $MC = s$. Then the inequality is equivalent to

$$\begin{aligned} & \tfrac{1}{2}x(\sqrt{r^2 - x^2} + \sqrt{s^2 - x^2}) \leq \tfrac{1}{2}rs \\ \Leftrightarrow \quad & x^2(r^2 + s^2 - 2x^2 + 2\sqrt{(r^2 - x^2)(s^2 - x^2)}) \leq r^2 s^2 \\ \Leftrightarrow \quad & 0 \leq r^2 s^2 - (r^2 + s^2)x^2 + 2x^4 - 2x^2\sqrt{(r^2 - x^2)(s^2 - x^2)} \\ \Leftrightarrow \quad & 0 \leq (\sqrt{(r^2 - x^2)(s^2 - x^2)} - x^2)^2. \end{aligned}$$

Equality holds if and only if $\sqrt{(r^2 - x^2)(s^2 - x^2)} = x^2$, which yields $x^2 = r^2s^2/(r^2 + s^2)$.

Drop perpendiculars MU and MV from M to AD and BC respectiovely. Then $MU = MV$. Applying the lemma to MAU and MCV, we have $[MAU] + [MCV] \leq \frac{1}{2}MA \cdot MC$. Equality holds if and only if $MU = MV = MA \cdot MC\sqrt{MA^2 + MC^2} = MP = MQ$. Applying the lemma to the other two pairs of opposite triangles in $ABCD$. Hence $ABCD$ has an incircle.

(b) We have

$$\sin MAP = \sin MAU = \frac{MC}{\sqrt{MA^2 + MC^2}}$$

and

$$\sin MCQ = \sin MCV = \frac{MA}{\sqrt{MA^2 + MC^2}}.$$

It follows that

$$\sin^2 \frac{BAD}{2} + \sin^2 \frac{BCD}{2} = 1,$$

so that $\angle BAD + \angle BCD = 180°$. Hence $ABCD$ has a circumcircle.

6. (a) Call the needles x-needles, y-needles and z-needles according to their directions. Take the larger of the numbers of x-needles and y-needles in each $2n \times 2n$ xy-layer, the larger of the numbers of y-needles and z-needles in each $2n \times 2n$ yz-layer, and the larger of the numbers of z-needles and x-needles in each $2n \times 2n$ zx-layer. Let k be the minimum of all these $6n$ maxima. Consider the layer where maximum is k. We may assume that it is an xy-layer. It contains $2n - k$ rows and $2n - k$ columns free of x-needles and y-needles. The $(2n - k)^2$ unit cubes at their intersection must pierced with z-needles. Paint these z-needles red. There are exactly k yz-layers which do not contain red needles. In each such layer, we can choose at least k y-needles, and the total from these k layers is at least k^2 y-needles. Add them to the red needles, and we have a regular subset of size $k^2 + (2n-k)^2 = k^2 + 4n^2 - 4nk + 2k^2 = 2n^2 + 2(n-k)^2 \geq 2n^2$.

(b) By (a), a regular subset of $2n^2$ needles is guaranteed to exist. We now construct an example in which the maximum regular subset consists of exactly $2n^2$ needles, so that this is the desired value. Divide the $2n \times 2n \times 2n$ cube into eight $n \times n \times n$ cubes. Pierce with needles in each of the three directions all n^3 unit cubes in the southwest $n \times n \times n$ cube in the bottom layer, as well as all n^3 unit cubes in the northeast $n \times n \times n$ cube in the top layer. Then every unit cube in the $2n \times 2n \times 2n$ cube is pierced by at least one of the $6n^2$ needles.

To obtain a regular subset, we must remove at least 2 needles from each of the $2n^3$ cubes that are being pierced in all three directions. This means the removal of at least $4n^2$ needles, leaving behiind the desired regular subset of size $2n^2$.

7. We can have as many as $\phi(n)$ red integers by painting all those less than n and relatively prime to n. We claim that this is the largest possible number of red integers. Otherwise, some prime q must divide both n and a red integer. We may assume that q is the largest such prime. Take a to be a red integer which is a multiple of q. If there exist two red numbers $b \neq c$ such that $b \equiv c \pmod{\frac{n}{q}}$, then $a(b-c)$ will be divisible by n, which is a contradiction. We consider two cases.
Case 1. The largest prime divisor of n is q.
Then $\phi(n) = n\left(\prod_{p|n} \frac{p-1}{p}\right) \geq n\left(\prod_{p=2}^{q} \frac{p-1}{p}\right) = \frac{n}{q}$. Since we have more than $\phi(n)$ red integers and only $\frac{n}{q}$ residue classes modulo $\frac{n}{q}$, b and c exist.
Case 2. The largest prime divisor of n is not q.
Let $n = mk$ where k is the product of all primes divisors of n greater than q. Then the number of residue classes modulo $\frac{n}{q} = k\frac{m}{q}$ is $\phi(k)\frac{m}{q}$. As in Case 1, $\phi(m) \geq \frac{m}{q}$. Hence $\phi(n) = \phi(k)\phi(m) \geq \phi(k)\frac{m}{q}$, so that the same contradiction as in Case 1 is obtained.

Spring 2020

Junior O-Level Paper

1. In the diagram below, the cities are shaded and the numbers of cities claiming the villages are shown.

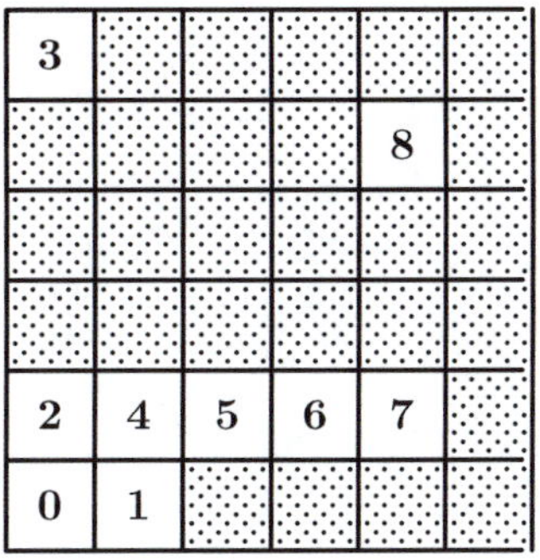

2. Let the numbers be $a_1, a_2, \ldots, a_n$. Define $s_k = a_k + a_{k+1} + \cdots + a_{k+10}$. Then $a_{12} = a_1 + s_2 - s_1$. Since $a_{12} \neq a_1$, $s_2 \neq s_1$. Hence one of them is 100 and the other one is 101. A similar argument shows that $s_1 = s_3 = \cdots = s_{2k-1}$ while $s_2 = s_4 = \cdots = s_{2k}$. Suppose $n \geq 23$. We have $a_{23} = a_{12} + s_{13} - s_{12} = a_1 + s_2 - s_1 + s_{13} - s_{12} = a_1$, which is a contradiction. It follows that $n \leq 22$. The following row of numbers has all the desired properties: 10, 20, 30, 40, 50, 60, 70, 80, 90, 100, −450, 11, 19, 31, 39, 51, 59, 71, 79, 91, 99, −449.

3. The diagonal BD of the rhombus $ABCD$ is perpendicular to AC. Since AC is parallel to PQ, BD is an altitude of triangle PDQ. Complete the rhombus $BCQR$. Then $ABRP$ is also a rhombus and $CDPR$ a parallelogram. The diagonal BQ of $BCQR$ is perpendicular to CR. Since CR is parallel to PD, BQ is also an altitude of triangle PDQ. It follows that B is indeed the orthocentre.

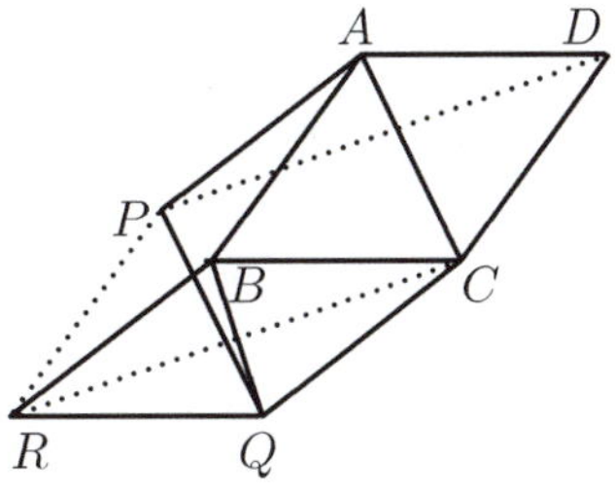

4. Suppose $x^2+y^2+z^2-xy-yz-zx=n$ has an integer solution (a,b,c) with $a \geq b \geq c$. Then

$$\begin{aligned} n &= \frac{1}{2}((a-b)^2+(b-c)^2+(a-c)^2)) \\ &= \frac{1}{2}(2(a-b)^2+2(b-c)^2-2(a-b)(b-c)) \\ &= (a-b)^2+(b-c)^2-(a-b)(b-c). \end{aligned}$$

It follows that $x^2+y^2-xy=n$ has an integer solution $(a-b,b-c)$.

5. The number of available moves is finite, so that the game must end in a win for one of the players. If either rook reaches the top row or the rightmost column, the player whose turn is next wins. The only positions from which an entry into the forbidden zone cannot be delayed are when one of the rooks is in the square $(g,7)$ while the other rook is in the square $(g,6)$ or $(f,7)$. In either case, the rooks occupy adjacent squares. In her first turn, Alice cannot make the rooks occupy adjacent squares. If Bob can do so, he can maintain this advantage and wins the game. If not, he moves the rook which has just been moved back into the main diagonal between the squares $(a,1)$ and $(g,7)$. Eventually, his opportunity will come.

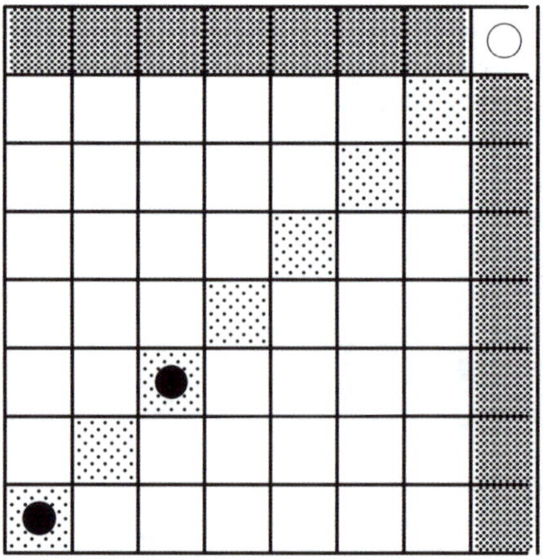

Senior O-Level Paper

1. The first three rows of the table are shown below.
1|22|1|22|1|22|1|22|1|22|1|22|1|22|1|22|1|22|1|22|1|22|1|22|1|22|1
1|22|1|22|1|22|1|22|1|22|1|22|1|22|1|22|1|22|1|22|1|22|1|22|1|22|1
0|11|0|11|0|11|0|11|0|11|0|11|0|11|0|11|0|11|0|11|0|11|0|11|0|11|0
We repeat this cycle 14 times and omit the last row to obtain 41 rows.

2. Three points are not sufficient. They determine a circle which divides the sphere into two parts. If both parts are hemispheres, then neither contains any of these three points in its interior. Otherwise, one part contains a hemisphere whose equator is on a plane parallel to the plane containing the three points. It does not contain any of these three points in its interior. We now show that four points are sufficient. Let N be the North Pole and let ABC be an equilateral triangle inscribed in the Equator. Then the Southern Hemisphere does not contain any of A, B, C and N in its interior. For any other hemisphere, its equator must intersect the Equator in two antipodal points. Hence it must contain in its interior one half of the Equator, and hence at least one of the points A, B and C.

3. The task is always possible since 41 is odd. Since the replacements do not change the parities of the numbers of As amd Bs, we may assume that the number of A's is odd, so that the single letter at the end is an A. We perform $(X, A, A, Y) \to (X, BAB, A, Y) \to (X, B, ABA, Y) \to (X, B, B, Y)$, replacing two adjcent As by two adjacent Bs. The circle now consists of an odd number of isolated As separated by blocks of Bs of various lengths. We can eliminate the shortest block of Bs by performing $(X, B, A, B, Y) \to (X, BAB, Y) \to (X, A, Y)$ a few times. The task can then be accomplished by alternating these two processes.

4. More generally, we prove that number of splittings is either 0 or infinite, and cannot be 1 or 2. Suppose to the contrary that the number is finite. Then

$$\begin{aligned} p(x) &= a(x)^2 + b(x)^2 \\ &= (\cos\theta a(x) + \sin\theta b(x))^2 + (-\sin\theta a(x) + \cos\theta b(x))^2 \end{aligned}$$

for any θ. So if there is only a finite number of splittings, one of them, with $\theta = \alpha$, must coincide with infinitely many splittings with $\theta = \beta$. Note that the parenthesis are permuted if we replace β by $\beta \pm \frac{\pi}{2}$ and the signs in the parenthesis are changed if we replace β by $\beta \pm \pi$. Thus for infinitely many β, we have $\cos\alpha a(x) + \sin\alpha b(x) = \cos\beta a(x) + \sin\beta b(x)$. It follows that

$$\begin{aligned} b(x) &= \frac{\cos\alpha - \cos\beta}{\sin\alpha - \sin\beta} a(x) \\ &= \frac{2\sin\frac{a-b}{2}\sin\frac{\alpha+\beta}{2}}{2\sin\frac{a-b}{2}\cos\frac{\alpha+\beta}{2}} a(x) \\ &= \tan\frac{\alpha+\beta}{2} a(x). \end{aligned}$$

This is impossible unless $a(x) = b(x) = 0$. This contradicts the assumption that $p(x)$ is non-constant.

5. Since CA is a tangent, $\angle CAQ = \angle APQ$. Similarly, $\angle CBQ = \angle BPQ$. Hence $ACBP$ is a cyclic quadrilateral. Now $\angle CPA = \angle CBA = \angle BPQ$, so that PD bisects $\angle APB$ as well as $\angle CPQ$. We claim that $\angle PDQ$ is independent of the choice of ℓ. Note that $\angle PAQ$ and $\angle PBQ$ are constant. It follows that so is $\angle APB$. When D lies on AQ as in the diagram below, $\angle PDQ = \angle PAQ + \frac{1}{2}\angle APB$, which is constant. Hence D lies on a circular arc passing through P and Q. When D lies on BQ, $\angle PDQ = \angle PBQ + \frac{1}{2}\angle APB = 180° - (\angle PAQ + \frac{1}{2}\angle APB)$. It follows that D lies on a circular arc whose union with the early circular arc is a full circle.

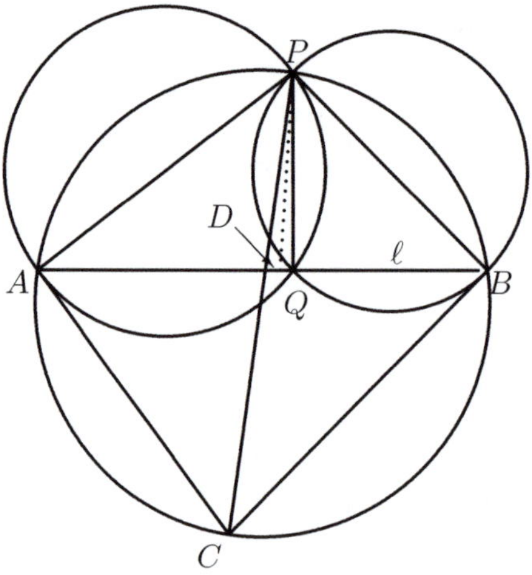

Junior A-Level Paper

1. Note that $2020 = 2^2 \times 5 \times 101$. If the last two digits of a number are 20, it will be a multiple of $2^2 \times 5$. The number 19193838474756562020 is a multiple of 2020, and contains each of the digits exactly twice. An example which contains each of the digits exactly once is 1237548960.

2. If the number of heads is even, the knight can reduce the number of heads again to an even number. Indeed, if the number of heads is $4n - 2$, then after a gold sword strike it will become $2n - 2$. If the number of heads is $4n$, then after bronze sword strike it will become $3n - 3$, and after the next silver sword strike it becomes $2n - 4$. The knight can act this way until there are four or two heads left. The task is completed with one final strike from the bronze or the gold sword, respectively.

3. (a) Suppose such a 19-gon $P_1P_2 \dots P_{19}$ exists. Let θ_i be the angle subtended by the arc P_iP_{i+1} for $1 \le i \le 19$, where $P_{20} = P_1$ and $P_0 = P_{19}$. All θ_i are distinct, and $\theta_1 + \theta_2 + \cdots + \theta_{19} = 180°$. Since $\angle P_{i-1}P_iP_{i+1}$ is integral, $\theta_{i-1} + \theta_i$ is integral.

Case 1. All θ_i are integral.
Then $\theta_1 + \theta_2 + \cdots + \theta_{19} \geq 1° + 2° + \cdots + 19° > 180°$. We have a contradiction.
Case 2. Not all θ_i are integral.
We may assume that in the degree measures, θ_1 has a fractional part ϵ. Then θ_2 has a fractional part $\delta = 1 - \epsilon$, θ_3 has a fractional part ϵ, and so on. Since 19 is odd, we must have $\epsilon = \delta = \frac{1}{2}$. It follows that $\theta_1 + \theta_2 + \cdots + \theta_{19} \geq \frac{1}{2}° + 1\frac{1}{2}° + \cdots + 18\frac{1}{2}° > 180°$. We have a contradiction.

(b) Such a 20-gon $P_1P_2 \ldots P_{20}$ exists. Let θ_i be the angle subtended by the arc P_iP_{i+1} for $1 \leq i \leq 20$, where $P_{21} = P_1$ and $P_0 = P_{20}$. We take $\theta_1 = 4\frac{1}{3}°$, $\theta_2 = 4\frac{2}{3}°$, $\theta_3 = 5\frac{1}{3}°$, $\theta_4 = 5\frac{2}{3}°$, ..., $\theta_{19} = 13\frac{1}{3}°$ and $\theta_{20} = 13\frac{2}{3}°$. Then $\theta_1 + \theta_2 + \cdots + \theta_{20} = 180°$. Moreover, $\theta_{i-1} + \theta_i$ is integral for all i.

4. This is possible for all n. Consider first the case $n = 2k$. The numbers in the first column start with 0, and increase by $2k$ and $2k-1$ alternatingly, resulting in 0, $2k$, $4k-1$, $6k-1$, $8k-2$, ..., $(4k-4)k-(k-1)$ and $(4k-2)k-(k-1)$. In odd-numbered rows, the numbers increase by 1 and 0 alternatingly. In even-numbered rows, they increase by 0 and 1 alternatingly. Thus the numbers in the last column are k, $3k-1$, $5k-1$, $7k-2$, $9k-2$, ..., $(4k-3)k-(k-1)$ and $(4k-1)k-k$. The following table illustrates the construction for $k = 3$.

0	1	1	2	2	3
6	6	7	7	8	8
11	12	12	13	13	14
17	17	18	18	19	19
22	23	23	24	24	25
28	28	29	29	30	30

The pairwise sums along the first row are 1, 2, ..., $2k-1$. The pairwise sums between the first row and the second row are $2k$, $2k+1$, ..., $4k-1$. The pairwise sums along the second row are $4k$, $4k+1$, ..., $6k-2$. The pairwise sums between the second row and the third row are $6k-1$, $6k$, ..., $8k-2$, and so on. The pairwise sums along the last row are $2(4k-2)k-(k-1)) = 8k^2-5k+1$, $8k^2-5k+2$, ..., $8k^2-4k = 2n(n-1)$.

The construction for $n = 2k+1$ is similar. The numbers in the first column start with 0, and increase by $2k+1$ and $2k$ alternatingly, resulting in 0, $2k+1$, $4k+1$, $6k+2$, $8k+2$, ..., $(4k-2)k+k$ and $(4k)k+k$. In odd-numbered rows, the numbers increase by 1 and 0 alternatingly. In even-numbered rows, they increase by 0 and 1 alternatingly.

Thus the numbers in the last column are k, $3k+1$, $5k+1$, $7k+2$, $9k+2$, $\ldots$, $(4k-1)k+k$ and $(4k+1)k+k$. The largest pairwise sum is indeed $2((4k+1)k+k) = 8k^2+4k = 2n(n-1)$. The following table illustrates the construction for $k=3$.

0	1	1	2	2	3	3
7	7	8	8	9	9	10
13	14	14	15	15	16	16
20	20	21	21	22	22	23
26	27	27	28	28	29	29
33	33	34	34	35	35	36
39	40	40	41	41	42	42

5. Let H be the point on AB such that $AH = CD$. Then $AHCD$ is a parallelogram. Its diagonals DH and AC bisect each other at the centre M. Since $KA = KC$, KM is perpendicular to AC. Since $AB = 3CD$, DH is perpendicular to AB. Now

$$\begin{aligned} \angle KMD &= 90^\circ - \angle DMC \\ &= 90^\circ - \angle HMA \\ &= \angle CAH \\ &= \angle DBA \\ &= \angle KAD. \end{aligned}$$

Hence $AKDM$ is cyclic. It follows that $\angle KDA = \angle KMA = 90^\circ$.

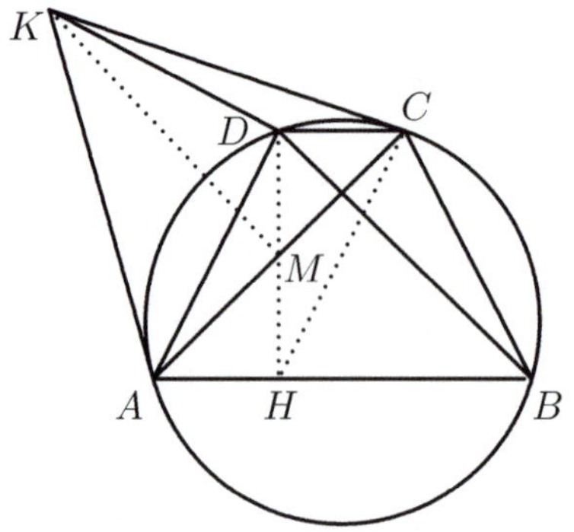

6. If Alice retains all four cards of each of three ranks, and the remaining six cards of one of the suits, the three cards of this suit of the chosen ranks cannot be matched by Bob. Thus Bob can get at most 15 points. We now prove that Bob can guarantee getting 15 points by the following strategy.

Step 1.
Here Bob will not lose any points.
Case 1. Bob holds 2 of the 4 cards of a rank.
They will be used to match Alice's cards of the same rank.
Case 2. Bob holds all 4 cards of a rank and 0 cards of another rank.
The four cards will be used to match Alice's cards of the other rank by suits.
Case 3. Bob holds 3 cards of a rank and 1 card of another rank.
For instance, they may be the Aces of Spades, Hearts and Diamonds. If the four cards is a Club, then Bob can match Alice's cards of these two ranks by suits. If the four cards is not a Club, say the Two of Spades, Bob will use the Ace and Two of Spades to match Alice's Ace and Two of Clubs, and the Aces of Hearts and Diamonds to match Alice's Twos of Hearts and Diamonds.
Step 2.
Suppose there are still some ranks remaining.
Case 1. Bob holds 1 card of a remaining rank.
Then he holds 1 or 4 cards of each remaining rank. Let there be a of the first type and b of the second type. Then $2a + 2b = a + 4b$ so that $a = 2b$. Hence the ranks can be divided into groups of three, with Bob holding all 4 cards of one rank and 1 card of each of the other two ranks. For instance, Bob may be holding all the Aces along with a Two and a Three. If the Two and the Three are of the same suit, say Spades, Bob will use them to match Alice's Two of Hearts and the Three of Diamonds. He then uses the Ace of Hearts to match Alice's Three of Hearts, and the Aces of Diamonds and Clubs to match Alice's Twos of Diamonds and Clubs.
Case 2. Bob does not hold exactly 1 card of any remaining rank.
Then he holds 0 or 3 cards of each remaining rank. The argument is analogous to that in Case 1.
Here Bob will lose 1 point for each group of 3 ranks. Since there are 9 ranks altogether, Bob cannot lose more than $9 \div 3 = 3$ points, and can guarantee getting at least 15 points.

7. Such a sequence exists. Choose m to be a positive integer such that $\frac{1}{2} + \frac{1}{3} + \cdots + \frac{1}{m+2} > 100$. We claim that we can choose the sequence so that $n = a_0 + a_1 = 2a_1 + a_2 = 3a_2 + a_3 = \cdots = ma_{m-1} + a_m = (m+2)a_m$. Then $a_0 > \frac{n}{2}$, $a_1 > \frac{n}{3}$, $a_2 > \frac{n}{4}$, $\ldots$, $a_{m-1} > \frac{n}{m+1}$ and $a_m = \frac{n}{m+2}$. It follows that $a_0 + a_1 + a_2 + \cdots + a_{m-1} + a_m > 100n$.

   We now justify our claim. Choose $a_m = m!$ and $n = (m+2)m!$. From $ma_{m-1} + m! = (m+2)m!$, a_{m-1} is a positive multiple of $(m-1)!$. Similarly, a_{m-2} is a positive multiple of $(m-2)!$, a_{m-3} is a positive multiple of $(m-3)!$, and so on. This leads to a choice of a_0 for which all terms in the sequence are positive integers.

Senior A-Level Paper

1. Note that the equation of the tangent to $y = x^2$ at the point (a, a^2) is $y = 2a(x - a) + a^2$. Solving $x^2 - 1 = 2a(x - a) + a^2$, we have $x = a \pm 1$. Hence this tangent intersects $y = x^2 - 1$ at the points $P(a - 1, a^2 - 2a)$ and $Q(a + 1, a^2 + 2a)$. We have

$$\begin{aligned} PQ^2 &= ((a+1) - (a-1))^2 + ((a^2 + 2a) - (a^2 - 2a))^2 \\ &= 4a^2 + 4. \end{aligned}$$

   If we choose $a > \frac{1}{2}10^6$, then $PQ > 10^6$.

2. For positive integers m and n, $m \bigtriangleup n$ denotes their greatest common divisor and $m \bigtriangledown n$ denotes their least common multiple. Let an arbitrary prime p occur in x, y and z to the powers $k \geq \ell \geq m$. If $\ell > 0$, we can change m within the range from 0 to ℓ without changing a, b or c. Hence we may assume that x, y and z are pairwise relatively prime. Then $a = xy$, $b = xz$ and $c = yz = \frac{yz}{x^2} = \frac{a \bigtriangledown b}{a \bigtriangleup n}$.

3. Let the vertices of the small square cross-section be $PQRS$ with respective coordinates $(0, \frac{1}{3}, \frac{1}{3})$, $(0, \frac{1}{3}, -\frac{1}{3})$, $(0, -\frac{1}{3}, -\frac{1}{3})$ and $(0, -\frac{1}{3}, \frac{1}{3})$. Let two of the vertices of the tetrahedron be $A(\ell, 0, h)$ and $B(\ell, 0, -h)$. The other two vertices are the point C of intersection of the extensions of AP and BQ, along with the point D of intersection of the extensions of AS and BR. Note that $AC = AD = BC = BD > 200$ with a suitable choice of ℓ, and AC is perpendicular to BD with a suitable choice of h. Let K, L, M and N be the respective midpoints of AB, AD, CD and BC. Then $KL = MN = \frac{1}{2}BC > 100$. Moreover, KL and MN are parallel to BC. Hence $KLMN$ is a parallelogram. Since $AC = BD$ and they are perpendicular to each other, $KLMN$ is the desired large square cross-section.

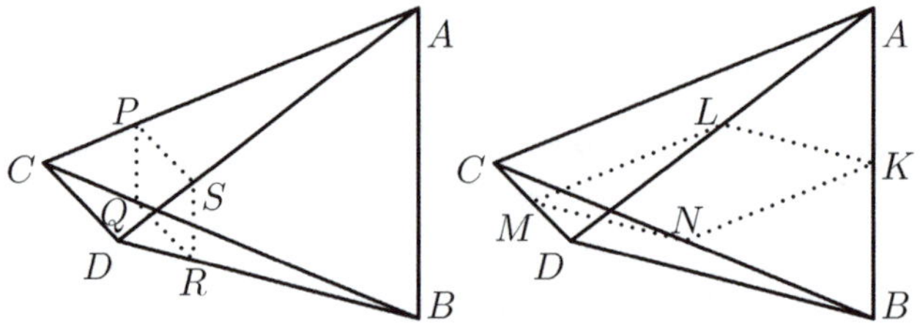

4. There are $\binom{2n}{n}$ ways of placing the hats on the children. We claim that the number of ways of distributing them into dancing circles of even length is $(n!)^2$. We use induction on n. For $n = 1$, the number of ways is $1 = (1!)^2$. Suppose the result holds up to some $n \geq 1$. Consider the next case with $2(n + 1)$ children.

Focus on a particular child wearing a white hat. This child is in a dancing circle of length $2k$ where $1 \le k \le n+1$. We can choose the remaining $k-1$ children in the dancing circle wearing white hats in $\binom{n}{k-1}$ ways, choose the k children in the dancing circle wearing black hats in $\binom{n+1}{k}$ ways and form the dancing circle in $(k-1)!k!$ ways. By the induction hypothesis, the other dancing circles may be formed in $((n-k+1)!)^2$ ways. The number of ways is

$$\binom{n}{k-1}\binom{n+1}{k}(k-1)!k!((n-k+1)!)^2 = n!(n+1)!,$$

which is independent of k. Summing from $k = 1$ to $k = n+1$, we have $((n+1)!)^2$, completing the inductive argument. The desired result follows immediately.

5. Let I, J, K, L, M and N be the respective midpoints of AB, CD, BC, DA, AC and BD. Then $IKJL$, $IMJN$ and $KMLN$ are all parallelograms. Hence their diagonals IJ, KL and MN bisect one another at their common midpoint X. Let Y be the point symmetric about X to the circumcentre O of $ABCD$. (This is known as the Monge point.) Then $IOJY$ is a parallelogram since the diagonals IJ and OY bisect each other at X. Note that

$$\begin{aligned} IY^2 - IA^2 &= OJ^2 - (OA^2 - OI^2) \\ &= OI^2 - (OC^2 - OJ^2) \\ &= JY^2 - JC^2. \end{aligned}$$

The left side is the power of the point Y with respect to the circle with diameter AB while the right side is the power of the point Y with respect to the circle with diameter CD. Hence Y lies on the radical axis PQ of these two circles. Similarly, it lies on RS as well as TU.

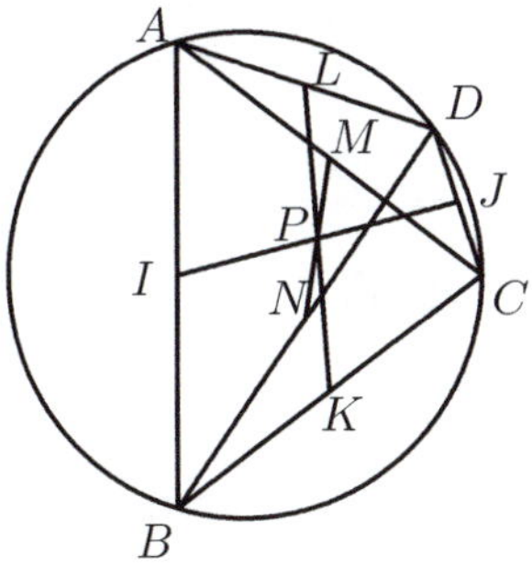

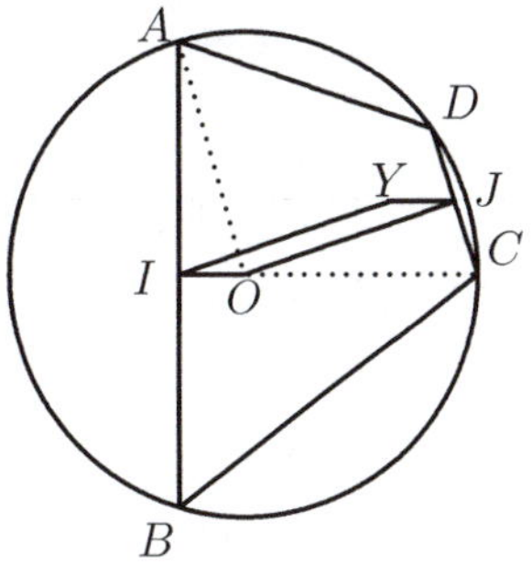

6. Let $n = 2^{k-1}m$ for some positive integer k and some odd positive integer m. The sum of the squares of $2n$ consecutive positive integers starting from $t+1$ is

$$\begin{aligned}&(1^2+2^2+\cdots+(t+2^km)^2)-(1^2+2^2+\cdots+t^2)\\ =\ &\frac{(t+2^km)(t+1+2^km)(2t+1+2^{k+1}m)}{6}-\frac{t(t+1)(2t+1)}{6}\\ =\ &\frac{((2t+1)2^km+2^{2k}m^2)(2t+1+2^{k+1}m)}{6}.\end{aligned}$$

Since $2t+1$ is odd, the second factor in the numerator is odd. The first term in the first factor is divisible by 2^k but not by 2^{k+1}, whereas the second term is divisible by 2^{k+1}. Cancelling the 2 from the denominator, this sum is divisible by 2^{k-1} but not by 2^k. When we replace a and b by $a+b$ and $\pm(a-b)$, we replace a^2+b^2 by $(a+b)^2+(a-b)^2 = 2(a^2+b^2)$. Hence the sum of the squares of the numbers doubles every move. If the $2n$ numbers become consecutive again after $\ell \geq 1$ moves, the sum of their squares will be divisible by $2^{k-1+\ell}$. Since $k-1+\ell \geq k$, it is divisible by 2^k, which is a contradiction.

7. This is possible for every positive integer k. The configuration will consists of k^2 groups each with k^2 queens, for a total of k^4 queens. The groups are well aligned and properly spaced. For $k=1$, a lone queen satisfies the condition. The diagram below shows the construction for $k=2$.

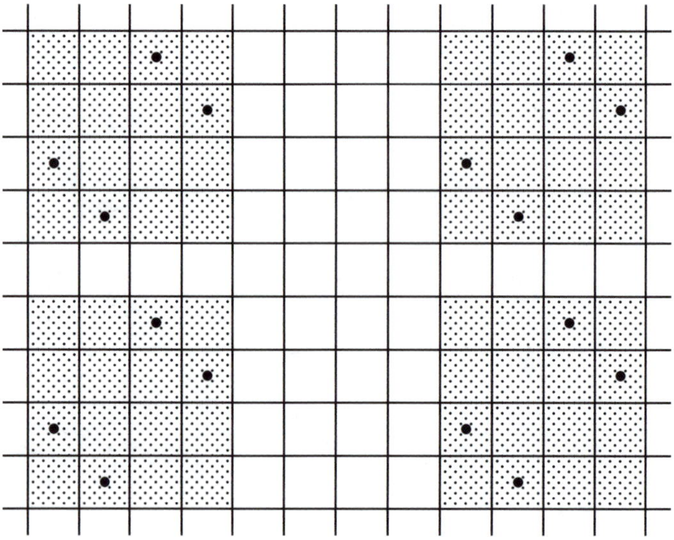

The diagram below shows the construction for $k = 3$. The constructions for higher values of k are analogous.

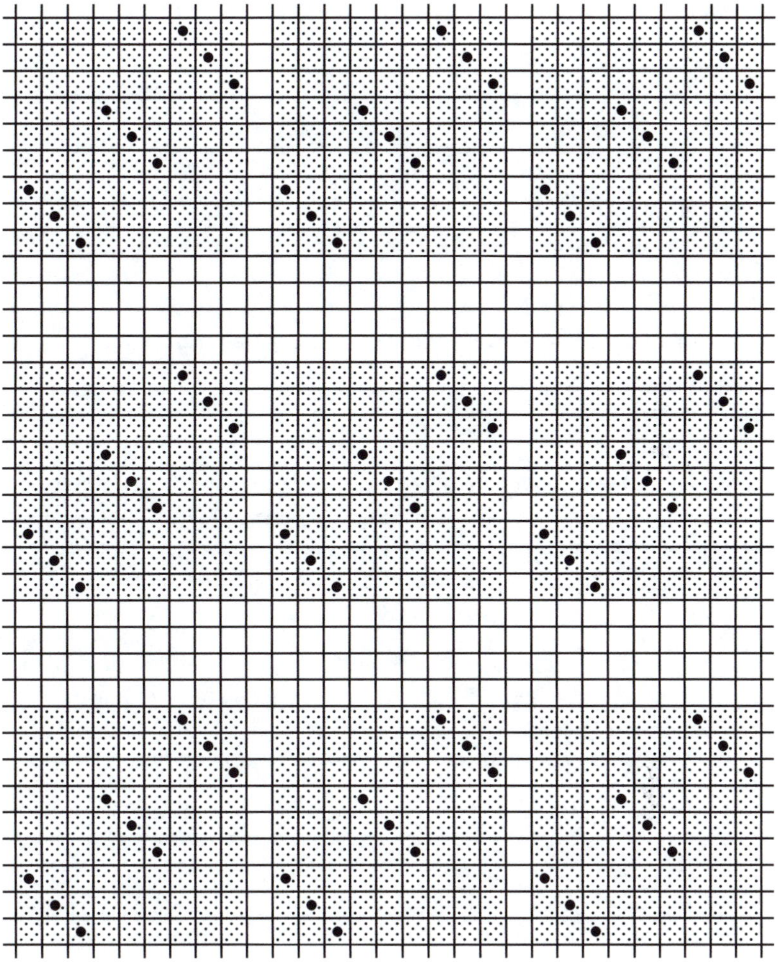

Tournament 42

Fall 2020

Junior O-Level Paper

1. Is it possible to paint 100 points on a circle red so that there are exactly 1000 right triangles with their vertices at the red points?

2. Eight players participate in several tournaments. In each, they pair themselves in four quarterfinals. The winners move onto to two semifinals, and eventually, two of them reach the final. Each player has played every other players exactly once during these tournaments.

   (a) Must each player reach the semifinals more than once?

   (b) Must each player reach the finals at least once?

3. Anna and Boris play a game with n counters. Anna goes first, and turns alternate thereafter. In each move, a player takes either 1 counter or a number of counters equal to a prime divisor of the current number of counters. The player who takes the last counter wins. For which n does Anna have a winning strategy?

4. P is a point inside an equilateral triangle of side length d such that its respective distances from the three vertices are a, b and c. Prove that there exists a point Q and an equilateral triangle of side length a such that its respective distances from the three vertices are b, c and d.

5. Eight elephants are lined up in increasing order of weight. Starting from the third, the weight of each elephant is equal to the total weight of the two elephants immediately in front. One of them tests positive for Covid, and may have lost some weight. The others test negative and their weights remain the same. Is it possible, in two weighings using a balance, to test if there has been a weight loss, and if so, to determine the elephant which has lost weight?

Note: The problems are worth 3, 2+3, 5, 5 and 5 points respectively.

Senior O-Level Paper

1. $P(x)$, $Q(x)$ and $P(x)+Q(x)$ are quadratic polynomials with real coefficients. If each has a repeated root, do these roots have to coincide?

2. X_1, X_2, …, X_{10} are collinear points in that order. For $1 \le i \le 9$, Y_i is a point such that $Y_iX_i = Y_iX_{i+1}$ and $\angle X_iY_iX_{i+1}$ has a fixed value θ. If Y_i lies on a semicircle with diameter X_1X_{10}, determine θ.

3. A positive multiple of 2020 has distinct digits and if any two of them switch positions, the resulting number is not a multiple of 2020. How many digits can such a number have?

4. AD, BE and CF are the bisectors of the angles of triangle ABC. Is it always possible to form two triangles with the six segments BD, DC, CE, EA, AF and FB?

5. There are 101 numbers in a circle. Each is either 10 or 11. Prove that one can choose one of them such that the sum of the k numbers clockwise from it is equal to the sum of the k numbers counterclockwise from it, where

   (a) $k = 50$;

   (b) $k = 49$.

Note: The problems are worth 3, 4, 5, 5 and 3+3 points respectively.

Junior A-Level Paper

1. Is it true that for any convex quadrilateral, there exists a circle which intersects each of its sides at two distinct points?

2. A pair of distinct positive integers is said to be *nice* if their arithmetic mean and their geometric mean are both integer. Is it true that for each nice pair, there exists another nice pair with the same arithmetic mean but a different geometric mean?

3. Anna and Boris play a game in which Anna goes first and turns alternate thereafter. In her turn, Anna suggests an integer. In his turn, Boris writes down on a whiteboard either that number or the sum of that number with all previously written numbers. Is it always possible for Anna to ensure that at some moment among the written numbers there are one hundred copies of the number

   (a) 5;

   (b) 10?

4. An X-pentomino consists of five squares where four of them share a common side with the fifth one. Is it possible to cut nine copies of an X-pentominoes from an 8×8 piece of paper?

5. Do there exist 100 distinct positive integers such that the cube of one of them is equal to the sum of the cubes of the other 99 numbers?

6. Anna and Boris play a game involving two round tables at each of which n children are seated. Each child is a sworn friend of both neighbours but no others. Anna can make two children agree to be sworn friends, whether they sit at the same table or at different tables. She can do so with $2n$ pairs. Boris can then make n of those pairs change their minds. Anna wins if the $2n$ children can be seated at one large round table so that each is a sworn friend of both neighbours. For which n does Anna have a winning strategy?

7. No three sides of a convex quadrilateral can form a triangle. Prove that one of its angles is

   (a) at most $60°$;

   (b) at least $120°$.

Note: The problems are worth 4, 7, 3+4, 7, 8, 10 and 6+6 points respectively.

Senior A-Level Paper

1. There are n positive integers. The arithmetic mean for each pair of these integers is written on a blackboard and their geometric mean on a whiteboard. Given that for each pair, at least one of those means is an integer, prove that all numbers on at least one of the boards are integer.

2. Baron Munchausen claims that if the polynomial $x^n - ax^{n-1} + bx^{n-2} + \cdots$ has n positive integer roots, then there exist a lines in the plane such that they have exactly b points of intersection among them. Is the Baron right?

3. Two circles ω and λ, with centres A and B respectively, intersect at C and D. The segment AB intersects ω at K and λ at L. The ray DK intersects λ again at N, and the ray DL intersects ω again at M. Prove that KM intersects LN at the incentre of triangle ABC.

4. Anna and Boris play a game involving two round tables at each of which n children are seated. Each child is a sworn friend of both neighbours but no others. Anna can make two children agree to be sworn friends, whether they sit at the same table or at different table. She can do so with $2n$ pairs. Boris can then make n of those pairs change their minds. Anna wins if the $2n$ children can be seated at one large round table so that each is a sworn friend of both neighbours. For which n does Anna has a winning strategy?

5. Does there exist a rectangle which can be cut into a hundred rectangles such that all of them are similar to the original one but no two are congruent?

6. Anna and Boris play a game involving Egyptian fractions, which are the reciprocals of the positive integers. Anna goes first, and turns alternate thereafter. In each turn, Anna writes down an Egyptian fraction on the whiteboard. Boris writes down one Egyptian fraction in his first turn, two Egyptian fractions in his second turn, three in his third, and so on. Boris wins if and when the sum of all the Egyptian fractions on the whiteboard is an integer. Can Anna prevent this?

7. A white knight is at the southwest corner square of a $1000 \times n$ board, where $n > 2020$ is an odd integer. There is a black bishop in each of the northwest corner square and the southeast corner square. The white knight wishes to reach the northeast corner square without landing on a square occupied or attacked by a bishop, and visit every square exactly once. The white knight succeeds by cheating sometimes, sneaking into an adjacent square in same row or the same column. Prove that the number of ways for the white knight to succeed does not depend on the value of n.

Note: The problems are worth 4, 5, 6, 7, 7, 10 and 12 points respectively.

Spring 2021

Junior O-Level Paper

1. Do there exist nine consecutive positive integers whose product is equal to the sum of nine other consecutive positive integers?

2. In triangle ABC, BE and CF are angle bisectors while BH and CK are altitudes. If $\angle HBE = \angle KCF$, must ABC be isosceles?

3. There are four coins of weights 1001, 1002, 1004 and 1005 grams respectively. Is it possible to determine the weight of each coin using a balance at most four times?

4. Is it possible to dissect a polygon into four isosceles triangles no two of which are congruent, if the polygon is

   (a) a square;

   (b) an equilateral triangle?

5. In a $100 \times n$ board, an ant starts from the square at the bottom left corner and wishes to reach the square at the top right corner. It may only crawl to an adjacent square immediately above or immediately to the right. Obstacles in the form of dominoes are placed in its way and the ant may not visit either of the two squares occupied by a domino. No two dominoes may have a point in common, and none can cover up the ant's starting square or its destination square. Is the task always possible if

   (a) $n = 101$;

   (b) $n = 100$?

Note: The problems are worth 4, 4, 4, 3+3 and 2+4 points respectively.

Senior O-Level Paper

1. A pentagon is dissected into three triangles by two diagonals. Is it possible for the centroids of these triangles to be collinear if the pentagon is

   (a) convex;

   (b) non-convex?

2. There are four coins of weights 1000, 1002, 1004 and 1005 grams, respectively. Is it possible to determine the weight of each coin using a balance at most four times, if

(a) the balance is normal;

(b) the balance is faulty, in that its left pan is 1 gram lighter than its right pan?

3. For which positive integers n do there exist n consecutive positive integers whose product is equal to the sum of n other consecutive positive integers?

4. Let p and q be real numbers with $p \neq 0$. Is it possible for the equation $\lfloor x^2 \rfloor + px + q = 0$ to have more than 100 roots?

5. The diameter through A of the circumcircle of an acute triangle ABC intersects the altitude from B at H and the altitude from C at K. The circumcircles of triangles BAH and CAK intersect again at P. Prove that the line AP bisects BC.

Note: The problems are worth 2+2, 2+2, 5, 5 and 6 points respectively.

Junior A-Level Paper

1. The number $2021 = 43 \times 47$ is composite. Prove that if we insert any number of copies of the digit 8 between 20 and 21, the resulting number is also composite.

2. In a room are 1000 candies on a table and a line of children. Each child in turn divides the current number of candies on the table by the current number of children in the line. If the quotient is an integer, the child takes that number of candies. Otherwise, a boy rounds the quotient up and a girl rounds it down to the next integer and then takes that number of candies. After taking the candies, the child leaves the room. The process continues until all the children have left the room. Prove that the total number of candies taken by the boys is independent of their positions in the line.

3. E anf F are points on the sides CA and AB, respectively, of an equilateral triangle ABC, and K is a point on the extension of AB such that $AF = CE = BK$. Let P be the midpoint of EF. Prove that PK is perpendicular to PC.

4. Fifty natives stand in a circle. Each announces the age of his left neighbour. Then each announces the age of his right neighbour. Each native is either a knight who tells both numbers correctly, or a knucklehead who increases one of the numbers by 1 and decreases the other by 1. Each knucklehead chooses which number to increase and which to decrease independently. Is it always possible to determine which of the natives are knights and which are knuckleheads?

5. The floor of an $n \times 21$ room is covered with unit square tiles. At the centre of each tile is an unlit candle with negligible radius. In each turn, we choose a straight line on the floor not passing through any candles and with at least one candle on each side. Moreover, all candles on one side of it must still be unlit, and we then light them. The task is to light all the candles taking as many turns as possible. What is the maximum number of turns if

   (a) $n = 21$;

   (b) $n = 20$?

6. A hotel has n unoccupied rooms upstairs, k of which are under renovation. All doors are closed, and it is impossible to tell if a room is occupied or under renovation without opening the door. There are 100 tourists in the lobby downstairs. Each in turn goes upstairs to open the door of some room. If it is under renovation, she closes its door and opens the door of another room, continuing until she reaches a room not under renovation. She moves into that room and then closes the door. Each tourist chooses the doors she opens. For each k, determine the smallest n for which the tourists can agree on a strategy while in the lobby, so that no two of them move into the same room.

7. Let p and q be two relatively prime positive integers. A frog hops along a straight path so that on every hop it moves either p units in one direction or q units in the opposite direction. Eventually, the frog returns to its starting point. Prove that for every positive integer d with $d < p + q$, there are two spots visited by the frog which are at a distance d apart.

Note: The problems are worth 4, 5, 6, 7, 4+4, 10 and 12 respectively.

Senior A-Level Paper

1. In a room there are 1000 candies on a table and a line of children. Each child in turn divides the current number of candies on the table by the number of children currently in the room. If the quotient is an integer, the child takes that number of candies.

Otherwise, a boy rounds the quotient up and a girl rounds it down to the next integer and then takes that number of candies. After taking the candies, the child leaves the room. The process continues until all the children have left the room. Prove that the total number of candies taken by the boys is independent of their positions in the line.

2. Does there exist a positive integer n such that for any real numbers x and y, there exist n real numbers such that x is equal to their sum and y is equal to the sum of their reciprocals?

3. Let D be the midpoint of the side BC of triangle ABC. The circle ω passing through A and tangent to BC at D intersects the side AB at the point F and the side AC at the point E. Let X and Y be the midpoints of BE and CF respectively. Prove that the circumcircle of triangle DXY is tangent to ω.

4. There is a row of $100n$ tuna sandwiches. A boy and his cat take alternate turns, with the cat going first. In her turn, the cat eats the tuna from one sandwich anywhere in the row if there is any tuna left. In his turn, the boy eats the first sandwich from either end, and continues until he has eaten 100 of them, switching ends at any time. Can the boy guarantee that, for every positive integer n, the last sandwich he eats contains tuna?

5. A hotel has n unoccupied rooms upstairs, k of which are under renovation. All doors are closed, and it is impossible to tell if a room is occupied or under renovation without opening the door. There are 100 tourists in the lobby downstairs. Each in turn goes upstairs to open the door of some room. If it is under renovation, she closes its door and opens the door of another room, continuing until she reaches a room not under renovation. She moves into that room and then closes the door. Each tourist chooses the doors she opens. For each k, determine the smallest n for which the tourists can agree on a strategy while in the lobby, so that no two of them move into the same room.

6. Find a real number r such that for any positive integer n the difference between $\lceil r^n \rceil$ and the nearest square of an integer is equal to 2.

7. Peter wants to draw $n > 2$ arcs of length α of great circles on a unit sphere so that they do not intersect each other. Prove that

   (a) this is possible if $\alpha < \pi + \frac{2\pi}{n}$;

   (b) this is impossible if $\alpha > \pi + \frac{2\pi}{n}$.

Note: The problems are worth 4, 5, 5, 8, 8, 10 and 6+7 points respectively.

Solutions

Fall 2020

Junior O-Level Paper

1. In each right triangle, two of its vertices must be diametrically opposite. Two diametrically opposite red points form a right triangle with each of the other red points. It follows that the total number of right triangles is always a multiple of 98. Since 1000 is not a multiple of 98, the desired situation is impossible.

2. Since each player plays every other player exactly once, each has played 7 games, and the total number of games played is $\frac{8\times7}{2} = 28$. Since there are 7 games in each tournament, the number of tournaments is $28 \div 7 = 4$. In each tournament, 4 players play 1 game each, 2 players play 2 game each and 2 players play 3 game each.

   (a) If a player reaches the semifinals at most once, she can have played at most 1+1+1+3=6 games even if she moves from the semifinal onto the final. This is a contradiction.

   (b) There are 8 players and 8 spots in the finals. If one player does not reach any final, some other player must do so at least twice. That player will have played at least 1+1+3+3=8 games even if she did not make it into the semifinals in the other two tournaments. This is a contradiction.

3. Boris wins whenever $n \equiv 0 \pmod 4$ and Anna wins otherwise. We first justify the statement about Boris by mathematical induction. If $n = 0$, Boris has already won. Suppose $n > 0$. If Anna takes 2 counters, Boris does likewise, maintaining $n \equiv 0 \pmod 4$. If Anna takes a number of counters congruent to 3 modulo 4, Boris can maintain $n \equiv 0 \pmod 4$ by taking 1 counter. Suppose Anna takes a number of counters congruent to 1 modulo 4, leaving behind $4k + 3$ counters for some integer k. Not all prime divisors of $4k+3$ can be congruent to 1 modulo 4. Hence Boris can choose a prime divisor congruent to 3 modulo 4, take that many counters and maintaining $n \equiv 0 \pmod 4$. The induction hypothesis can be applied in each case. Suppose $n \equiv 1 \pmod 4$. Anna takes 1 counter and steals the winning strategy of Boris. Suppose $n \equiv 2 \pmod 4$. Anna can still do so by taking 2 counters. Finally, suppose $n \equiv 3 \pmod 4$. Then the number of remaining counters has a prime divisor congruent to 3 modulo 4. Anna takes that many counters and steals the winning strategy of Boris.

4. Let ABC be the equilateral triangle of side length d with $PA = a$, $PB = b$ and $PC = c$. Let R be the points on the opposite side of CA to P such that PAR is an equilateral triangle of side length a. Now $AB = d = AC$, $AP = a = AR$ and $\angle PAB = 60° - \angle PAC = \angle RAC$. Hence triangles PAB and RAC are congruent, so that $RC = PB = b$. Let Q coincide with C. Then $QR = b$, $QP = c$ and $QA = d$ as desired.

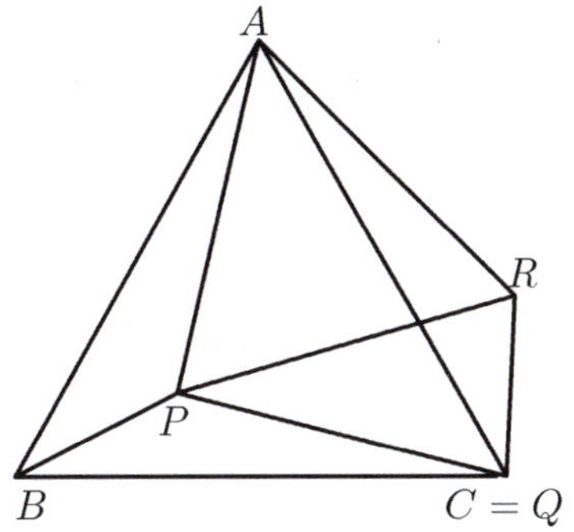

5. Label the elephants from A to H in increasing order of weight. In the first weighing, we put B, C and H against D, F and G. In the second weighing, we put A, B and G against C, E and F. The conclusion is summarized in the chart below, where X means that no elephant has lost any weight.

Conclusion	ABG > CEF	ABG = CEF	ABG < CEF
BCH > DFG	F	D	G
BCH = DFG	E	X	A
BCH < DFG	C	H	B

Senior O-Level Paper

1. Let the repeated root of $P(x)$ be a, so that $P(x) = p(x-a)^2$ for some real number p. Similarly, $Q(x) = q(x-b)^2$ for some real number q, with b as the repeated root. Then

$$P(x) + Q(x) = (p+q)x^2 - 2(pa+qb)x + (pa^2 + qb^2).$$

One-quarter of the discriminant of this quadratic polynomial is

$$(pa + qb)^2 - (p + q)(pa^2 + qb^2) = -pq(a - b)^2.$$

Hence we must have $b = a$. Then $P(x) + Q(x) = (p + q)(x - a)^2$ so that its repeated root is also a.

2. Let O be the centre of the whole circle. Let Z_2, Z_4, Z_6 and Z_8 be the respective reflectional image across X_1X_{10} of Y_2, Y_4, Y_6 and Y_8. Then $\angle X_1OY_2 = \angle X_1OZ_2 = 2\angle X_1Y_1Z_2 = 2\theta$. Similarly, each of $\angle Y_2OY_4$, $\angle Y_4OY_6$, $\angle Y_6OY_8$ and $\angle Y_8OX_{10}$ is equal to 2θ. Since the sum of these five angles is $180°$, we have $\theta = 18°$.

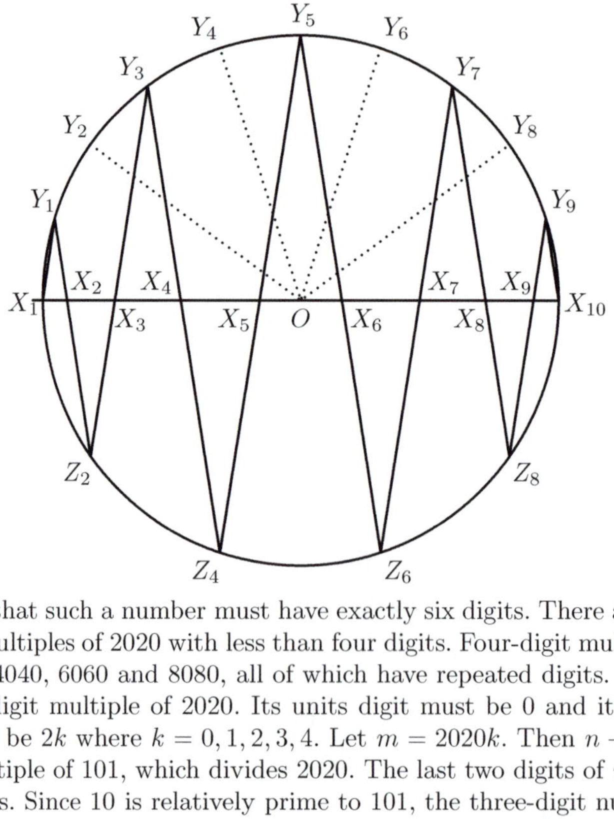

3. We claim that such a number must have exactly six digits. There are no positive multiples of 2020 with less than four digits. Four-digit multiples are 2020, 4040, 6060 and 8080, all of which have repeated digits. Let n be a five-digit multiple of 2020. Its units digit must be 0 and its tens digit must be $2k$ where $k = 0, 1, 2, 3, 4$. Let $m = 2020k$. Then $n - m$ is also a multiple of 101, which divides 2020. The last two digits of $n - m$ are both 0s. Since 10 is relatively prime to 101, the three-digit number obtained after the two 0s have been deleted is also a multiple of 101. Hence it is one of 101, 202, ..., 909. It follows that n has repeated digits. Observe that the shortest string of 9s which is a multiple of 101 is 9999. If a switch of two digits of a multiple of 2020 results in another multiple of 2020, the digits switched must be at least four places apart.

Let n be a multiple of 2020 with at least seven distinct digits. Let a be the millions digit of n and b be the hundreds digit of n. Let m be the number obtained from n by switching the positions of a and b. Then $n-m = 10^6(a-b)+10^2(b-a) = 100(a-b)(10000-1) = 999900(a-b)$. Since it is a multiple of 2020, so is m, and we have a contradiction. Finally, we offer an example of a multiple of 2020 with six distinct digits, $174 \times 2020 = 351480$. Recall that the units digit must be 0 and the tens digit must be even. Hence there are no switches of digits that would result in another multiple of 2020.

4. Let $BC = a$, $CA = b$ and $AB = c$. Then $BD = \frac{ca}{b+c}$, $DC = \frac{ab}{b+c}$, $CE = \frac{ab}{c+a}$, $EA = \frac{bc}{c+a}$, $AF = \frac{bc}{a+b}$ and $FB = \frac{ca}{a+b}$. We may assume that $a \leq b \leq c$. By the Triangle Inequality, $a+b > c$. Now $\frac{ab}{b+c} \leq \frac{ca}{b+c} \leq \frac{ca}{a+b}$. Since $\frac{ab}{b+c} + \frac{ca}{b+c} = a > \frac{ca}{a+b}$, DC, BD and FB form a triangle. Similarly, we have $\frac{ab}{c+a} \leq \frac{bc}{c+a} \leq \frac{bc}{a+b}$. Since $\frac{ab}{c+a} + \frac{bc}{c+a} = b > \frac{bc}{a+b}$, CE, EA and AF also form a triangle.

5. Let the coins be C_0, C_1, $\ldots$, C_{100} in clockwise order, and let their respective weights be a_0, a_1, $\ldots$, a_{100}. Represent the coins by the vertices of a regular 101-gon $C_0C_1\ldots C_{100}$. Draw half of the circumcircle clockwise from the midpoint of the arc $C_{50}C_{51}$ to C_0, and direct the diameter so that its tail is at C_0. For $0 \leq i \leq 100$, define $\Delta(C_i)$ as the total weight of the k adjacent coins of C_i on the semicircle minus the total weight of the k adjacent coins of C_i off the semicircle, when the tail of the directed diameter is at C_i. Define $\delta(C_i)$ similarly but when the head of the directed diameter is at C_i. Clearly, $\delta(C_i) = -\Delta(C_i)$. We shall rotate the semicircle clockwise about its centre, until the head of the directed diameter is at C_0.

   (a) For $k = 50$, we have

$$\begin{aligned}
\Delta(C_0) &= (a_{100}+a_{99}+\cdots+a_{51}) - (a_1+a_2+\cdots+a_{50}),\\
\delta(C_{51}) &= (a_{52}+a_{53}+\cdots+a_{100}+a_0) - (a_{50}+a_{49}+\cdots+a_1),\\
\Delta(C_1) &= (a_0+a_{100}+a_{99}+\cdots+a_{52}) - (a_2+a_3+\cdots+a_{51}),\\
\delta(C_{52}) &= (a_{53}+a_{54}+\cdots+a_{100}+a_0+a_1)\\
&\quad -(a_{51}+a_{50}+\cdots+a_2),\\
\cdots &= \cdots,\\
\delta(C_0) &= (a_1+a_2+\cdots+a_{50}) - (a_{100}+a_{99}+\cdots+a_{51}).
\end{aligned}$$

   In each equations, one bracket of the difference is identical to that of the preceding equation while the other bracket changes by only one term. It follows that the value of successive equations changes by at most 1. If $\Delta(C_0) = 0 = \delta(C_0)$, there is nothing further to prove. Otherwise, $\Delta(C_0)$ and $\delta(C_0)$ have opposite signs, so that some equation in between about C_i satisfies $\Delta(C_i) = 0 = \delta(C_i)$. This C_i is the coin we seek.

(b) For $k = 49$, we have

$$\begin{aligned}
\Delta(C_0) &= (a_{100} + a_{99} + \cdots + a_{52}) - (a_1 + a_2 + \cdots + a_{49}),\\
\delta(C_{51}) &= (a_{52} + a_{53} + \cdots + a_{100}) - (a_{50} + a_{49} + \cdots + a_2),\\
\Delta(C_1) &= (a_0 + a_{100} + a_{99} + \cdots + a_{53})\\
&\quad -(a_2 + a_3 + \cdots + a_{50}),\\
\delta(C_{52}) &= (a_{53} + a_{54} + \cdots + a_{100} + a_0)\\
&\quad -(a_{51} + a_{50} + \cdots + a_3),\\
\cdots &= \cdots,\\
\delta(C_0) &= (a_1 + a_2 + \cdots + a_{49}) - (a_{100} + a_{99} + \cdots + a_{52}).
\end{aligned}$$

In each equations, one bracket of the difference is identical to that of the preceding equation while the other bracket changes by only one term. It follows that the value of successive equations changes by at most 1. If $\Delta(C_0) = 0 = \delta(C_0)$, there is nothing further to prove. Otherwise, $\Delta(C_0)$ and $\delta(C_0)$ have opposite signs, so that some equation in between about C_i satisfies $\Delta(C_i) = 0 = \delta(C_i)$. This C_i is the coin we seek.

Junior A-Level Paper

1. It is not true. The diagram below shows a counterexample, a parallelogram $ABCD$. The centre of a circle intersecting AB at two distinct points must lie in the infinite strip bounded by the lines through A and B perpendicular to AB. Similarly, the centre of a circle intersecting CD at two distinct points must lie in the infinite strip bounded by the lines through C and D perpendicular to CD. Since these two infinite strips are disjoint, there is no circle which intersects each of AB and CD at two distinct points.

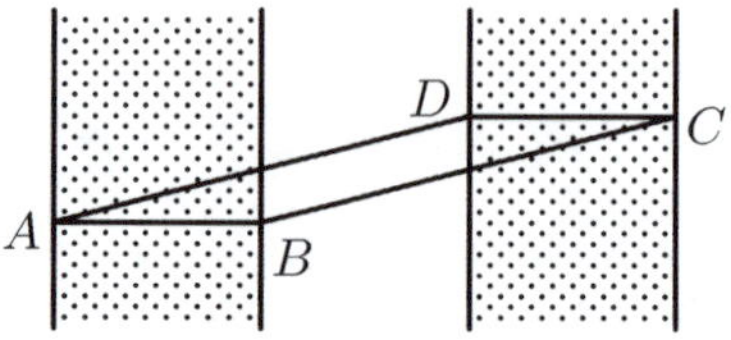

2. It is true. Let the common arithmetic mean be a and the geometric mean of the original pair be g. Then the original pair of numbers are $a-d$ and $a+d$ for some d, and we have $g^2 = (a-d)(a+d) = a^2 - d^2$. The new pair of numbers are $a-g$ and $a+g$, with $a^2 - g^2 = d^2$. If $d = g$, then $a^2 = 2g^2$ and $\sqrt{2} = \frac{a}{g}$. Since $\sqrt{2}$ is irrational, $d \neq g$.

3. (a) It cannot be done. Whenever Anna suggests an even number, Boris writes it down. Whenever Anna suggests an odd number, Boris writes down the sum of all suggested numbers. The current sum is initially even, and remains even as long as Anna suggests even numbers. The first time Anna suggests an odd number, Boris writes down an odd number, making the current sum odd. Thereafter, the current sum remains odd, and Boris will never write down a second odd number.

   (b) It can be done. Anna suggests 10. Since the current sum of the numbers on the whiteboard is 0, Boris is forced to write down the first 10. Next, Anna suggests -15. If Boris writes down -15, then the current sum becomes -5. In her third move, Anna suggests 10 again. Boris may write down $-5+10=5$ on his third move, taking the current sum back to 0. Anna then forces the second 10 in the next turn. Alternatively, Boris can choose to write down the second 10, making the current sum 5. In her next move, Anna suggests -10. If Boris writes down -10, the current sum becomes -5 and Anna can continue as before. If Boris writes down $5-10=-5$, the current sum becomes 0 once again, and Anna can continue as before. Suppose that in the second move, Boris writes down $10-15=-5$, we are back in situations which have already been analyzed. Thus Anna can eventually make one hundred copies of 10 appear on the whiteboard.

4. The diagram below shows nine copies of the X-pentomino enclosed in a square. The total area of the X-pentominoes is 45. Along the edges are eight $(1,2,\sqrt{5})$ triangles with total area 8. Each of the four corner pieces has area less than 2. Hence the area of the square is less than 61 and its side length is less than 8.

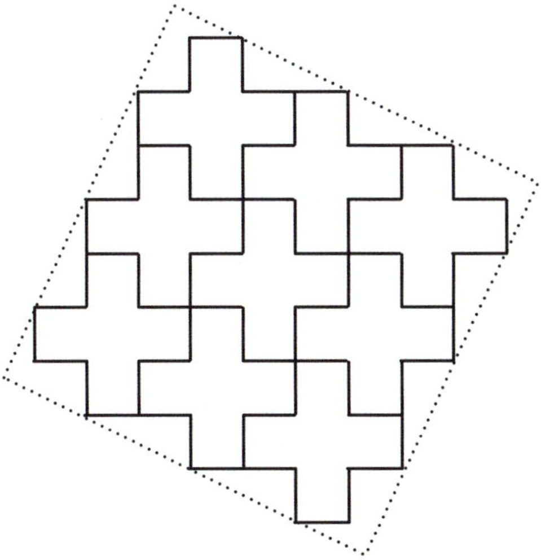

5. We have a cube equal to the sum of three cubes, namely, $3^3+4^3+5^3=6^3$. It follows that $9^3+12^3+15^3=18^3$ and $18^3+24^3+30^3=36^3$. Combining the two results yields a cube equal to the sum of five cubes, namely, $9^3+12^3+15^3+24^3+30^3=18^3+24^3+30^3=36^3$. Combining this with $36^3+48^3+60^3=72^3$, we can obtain a cube equal to the sum of seven cubes. Continuing in this manner, we obtain eventually a cube equal to the sum of 99 cubes, and those 99 integers are distinct.

6. Boris has a winning strategy if n is even. He writes down the $4n$ names in the $2n$ pairs of friendship induced by Anna. Note that the same child's name may appear several times. Boris divides his list into four, according to whether the child sits at the first or the second table, and whether the child is in an odd-numbered seat or an even-numbered seat. By symmetry, we may assume that the list of children in odd-numbered seats at the first table is the shortest, so that there are at most n names on it. Boris will get these children to break up their induced friendship, and he can do so since this involves breaking up at most n pairs. Each of these children retains only the original sworn friends, so that the first table cannot be broken up for adding children from the second table. Anna has a winning strategy if n is odd. The child in the sth seat at the first table and the child at the tth seat at the second table are induced to be friends for $(s,t)=(1,1)$, (1,2), (2,2), (2,3), $\ldots$, $(n-1,n-1)$, $(n-1,n)$, (n,n) and $(n,1)$. She writes down these pairs and divides her list into two. The first consists of (1,1), (2,2), $\ldots$, (n,n) while the second consists of (1,2), (2,3), $\ldots$, $(n,1)$. Since n is odd, Boris cannot break up an equal number of induced friendships from the two lists. By symmetry, we may assume that more than half of the pairs in the second list remain intact. In particular, both $(k-1,k)$ and $(k,k+1)$ survive for some k. Anna can now seat the children from the first table in the order 1 to $k-1$, switch to the children from the second tale in the order k to 1, continue from n to $k+1$, and switch back to the first table with the children in the order k to n.

7. Let BC he the longest side of the convex quadrilateral $ABCD$. In triangle ACD, we have $AD+CD>AC$. By assumption, $BC\geq AD+CD$. Hence $BC>AC$, so that $\angle ABC$ is not the largest angle in triangle ABC. Thus $\angle ABC<90°$. Similarly, $\angle BCD<90°$, so that the projection EF of AD on BC lies within BC. Note that $AD\geq EF$. We may also assume by symmetry that $CD\geq AB$.

   (a) Suppose to the contrary that both $\angle ABC>60°$ and $\angle BCD>60°$. Then we have a contradiction since

$$\begin{aligned} BC &= BE+EF+FC \\ &< \frac{AB}{2}+AD+\frac{CD}{2} \\ &\leq AD+CD. \end{aligned}$$

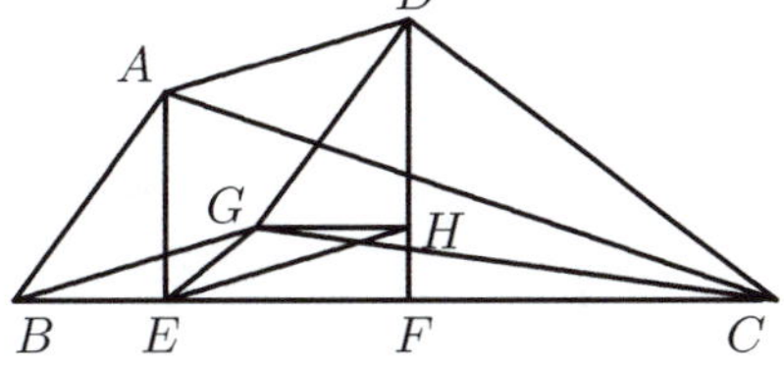

(b) Complete the parallelograms $DABG$ and $DAEH$. Then triangles ABE and DGH are congruent. We have

$$\begin{aligned} CG &> CF + GH \\ &= CF + BE \\ &= BC - EF \\ &\geq BC - AD \\ &\geq CD \\ &\geq AB \\ &= DG. \end{aligned}$$

Hence $\angle CDG$ is the largest angle of triangle CDG, so that

$$60^\circ \leq \angle CDG = \angle CDF + \angle ABE = 180^\circ - (\angle BCD + \angle ABC).$$

Now $\angle CDA + \angle DAB = 360^\circ - (\angle BCD + \angle ABC) \geq 240^\circ$. It follows that we have either $\angle CDA \geq 120^\circ$ or $\angle DAB \geq 120^\circ$.

Senior A-Level Paper

1. If all the integers are odd or all are even, then every arithmetic mean is an integer. Hence we may assume that there are both odd and even integers. Since the arithmetic mean of an odd-even pair is not an integer, their geometric mean must be an integer. Let one of the odd integers be expressed in the form hk^2, where h has no repeated prime factors. Then every even integer must be of the form hm^2 for the same h. It then follows that every odd integer must be of the form hn^2 for the same h also. Hence all geometric means are integers.

2. The Baron is right. Let the positive integer roots be $r_1, r_2, \ldots, r_n$. Then $a = r_1 + r_2 + \cdots + r_n$ and

$$b = r_1r_2 + r_1r_3 + \cdots + r_1r_n + r_2r_3 + r_2r_4 + \cdots + r_{n-1}r_n.$$

Each root r_k corresponds to a group of r_k mutually parallel lines. Lines corresponding to different roots are not parallel. Then the total number of lines is indeed a and the total number of points of intersection is indeed b.

3. Let I be the point of intersection of KM and the bisector of $\angle CAB$. Then triangles CIA and KIA are congruent so that $\angle ACI = \angle AKI$. Let $\angle CAB = \alpha$ and $\angle ABC = \beta$. Then $\angle BCA = 180° - (\alpha + \beta)$. Note that $\angle MAC = 2\angle MDC = 2\angle LDC = \beta$. Since $AM = AK$,

$$\angle AKI = \frac{1}{2}(180° - \angle MAK) = 90° - \frac{\alpha+\beta}{2}.$$

It follows that $\angle ACI = \frac{1}{2}\angle BCA$, so that I is the incentre of triangle ABC. By symmetry, the point of intersection of LN and the bisector of $\angle ABC$ is also I, so that I is indeed the point of intersection of KM and LN.

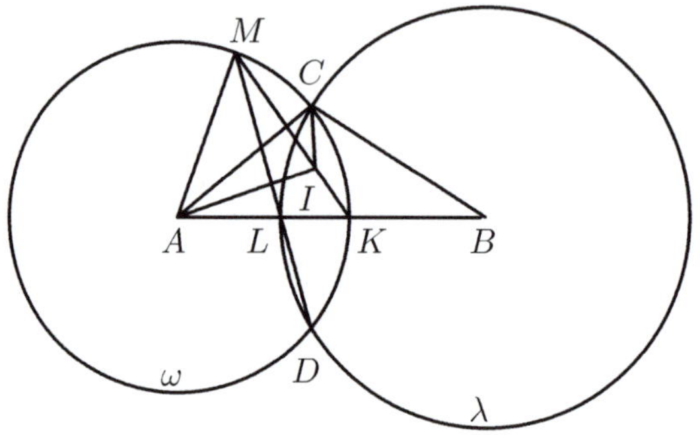

4. (See Problem 6 of the Junior A-Level Paper.)

5. We first find a rectangle R_0 which can be dissected into 4 rectangles R_1, R_2, R_3 and R_4 such that all are similar to R_0 but no two are congruent. R_1 is the right part of R_0, R_2 is the bottom part of $R_0 - R_1$, R_3 is the right part of $R_0 - R_1 - R_2$, and $R_4 = R_0 - R_1 - R_2 - R_3$. Let the horizontal and vertical dimensions of R_i be x_i and y_i respectively, as shown in the diagram below.

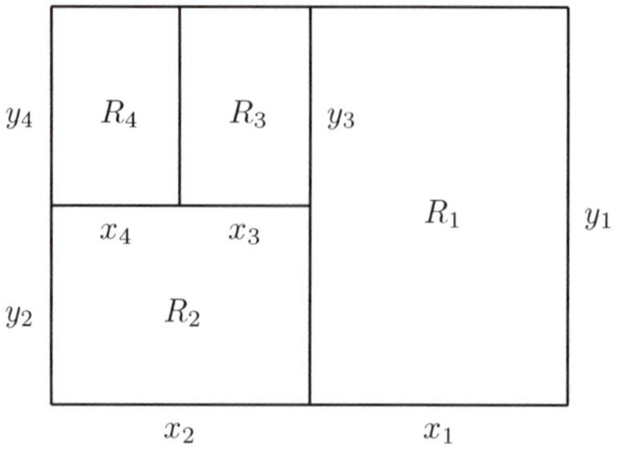

To make all of them similar, we set $y_1 = x_1 t$, $y_2 = x_2 t$, $y_4 = x_4 t$ but $x_3 = y_3 t$. Set $x_4 = 1$. Then $y_3 = y_4 = t$, $x_3 = t^2$, $x_2 = t^2+1$, $y_2 = t^3+t$, $y_1 = t^3 + 2t$ and $x_1 = t^2 + 2$ for some real number $t > 1$, as shown in the diagram below.

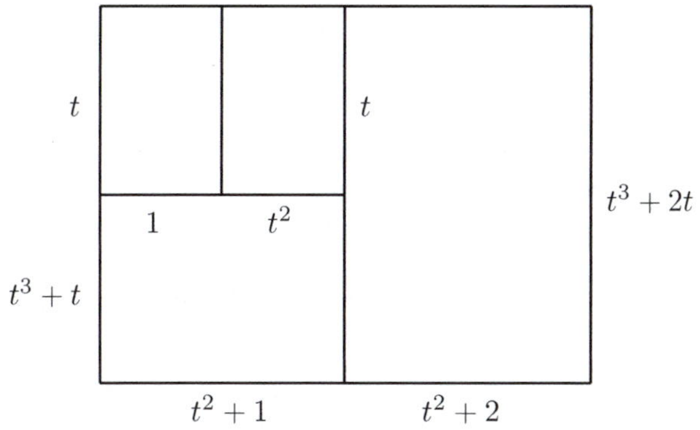

To make R_0 similar to these four rectangles, we need $2t^2+3 = x_1+x_2 = y_1 t = t^4 + 2t^2$. This simplifies to $t^4 = 3$, yielding $t = \sqrt[4]{3}$. The diagram below shows the dissection drawn in very close approximation.

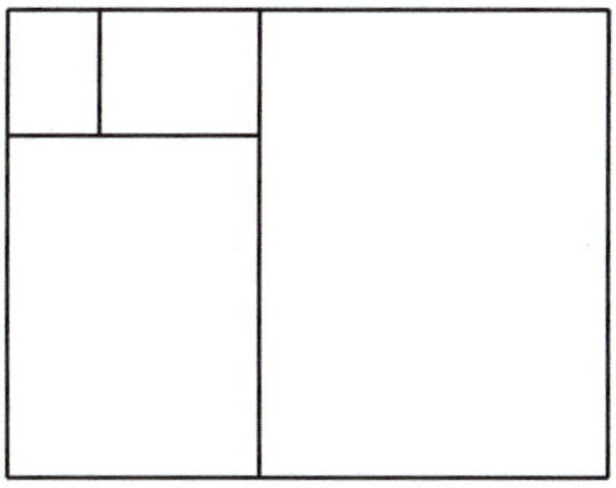

Applying this process with R_4 playing the role of R_0, we obtain a dissection of R_0 into 7 rectangles with the desired properties. Continuing in this manner, we eventually obtain a dissection of R_0 into 100 rectangles with the desired properties.

6. Anna can always prevent Boris from winning. We first prove an auxiliary result.

Lemma.

For any positive integer n and any positive number k, the equation $\frac{1}{x_0} + \frac{1}{x_1} + \cdots + \frac{1}{x_n} = k$ has finitely many solutions in positive integers $(x_0, x_1, \ldots, x_n)$.

Proof:

By symmetry, we may assume that $x_0 \le x_1 \le \cdots \le x_n$. Then $\frac{1}{x_0} \ge \frac{k}{n+1}$ so that $x_0 \le \frac{n+1}{k}$. Hence x_0 has only a finite number of possible values. Similarly, for each fixed value of x_0, x_1 has only a finite number of possible values, and so on.

Returning to the main problem, we use mathematical induction on n to prove that Boris cannot win on his nth turn. For $n = 1$, Boris can only win with $\frac{1}{2} + \frac{1}{2} = 1$. By writing down $\frac{1}{3}$, Anna can stop Boris from winning.

Suppose that Boris wins on his nth move for some $n \geq 2$. Then he has not won on his $(n-1)$st move, and the sum of all numbers on the whiteboard is $m + a$ where m is an integer and $0 < a < 1$. In their nth turn, $n + 1$ additional Egyptian fractions will be written down, with sum $k = \frac{1}{x_0} + \frac{1}{x_1} + \cdots + \frac{1}{x_n}$. Boris can only win if $a+k$ is an integer. Since $k \leq n+1$, $k+a$ has only a finite number of possible values, and so does k. By the lemma, Boris has only finitely many winning combinations $(x_0, x_1, \ldots, x_n)$. All Anna has to do is write down an Egyptian fraction which does not appear in any of these combinations. We have a contradiction.

7. We first illustrate with an 6×17 board. The diagram below shows one successful path for the knight.

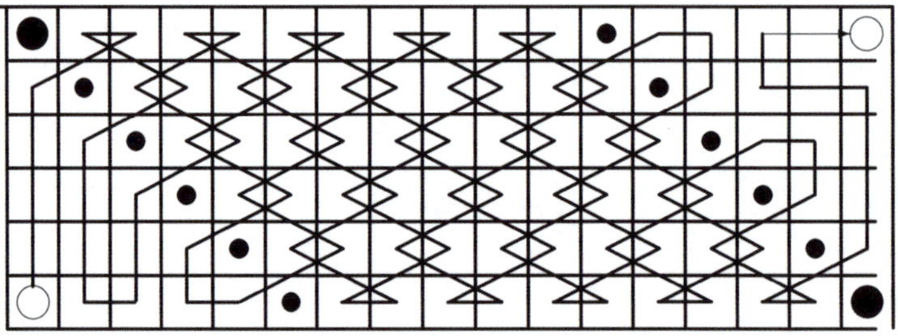

The board is divided by the two diagonals occupied or attacked by the bishops into two end regions and a central region. We claim that for any successful path of the knight, the part in the central region is unique.

Consider any successful path. Draw an arrow from each square of it to the next square. In the first six columns, 6 white squares are either occupied or attacked by a bishop. Hence the number of black squares visited by the knight exceeds the number of white squares visited by the knight by exactly 6. It follows that at least 6 arrows must cross from black squares to the left of the line separating the first six columns from the rest of the board to white square to the right of this line.

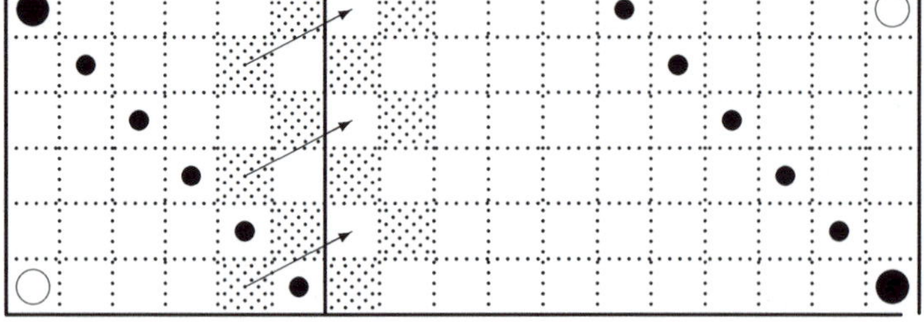

These 6 black squares must be in column five or column six, and there are exactly 6 black squares in these two columns. The arrows from three of them are uniquely determined, as shown in the diagram above.

This forces unique determination of the arrows from the other three black squares, as shown in the diagram below.

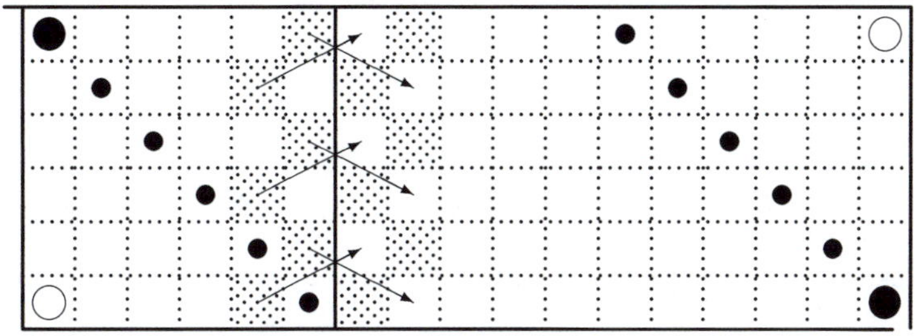

We now separate the first eight columns from the rest of the board. The diagram below shows that exactly the same situation arises. This will also be the case if we separate the first ten columns from the rest of the board. Hence we may widen or narrow the board by two columns at a time without affecting uniqueness. Thus the claim is justified.

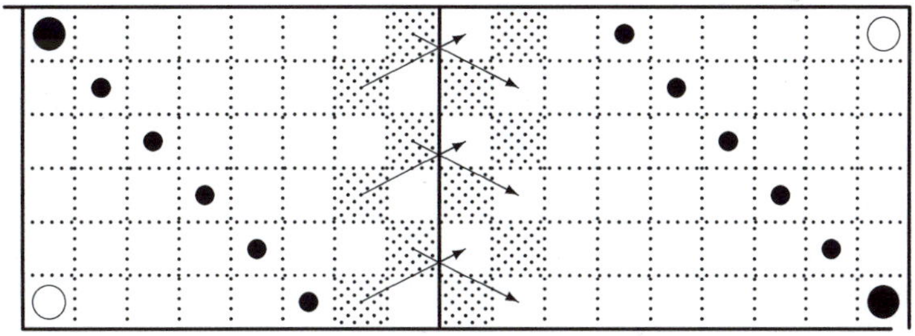

Returning to the original problem, the argument is precisely the same. The part of any successful path in the central region is uniquely determined. Hence the larger dimensions do not affect the desired conlcusion.

Spring 2021

Junior O-Level Paper

1. We have $1 \times 2 \times \cdots \times 9 = 9 \times 40320 = 40316 + 40317 + \cdots + 40324$.

2. ABC may not be isosceles. We may have $\angle B = 30°$ and $\angle C = 90°$. Then $\angle ACF = 45°$ and $\angle ACK = 30°$, so that $\angle KCF = 15°$. Now H coincides with C, and we have $\angle HBE = 15°$ also.

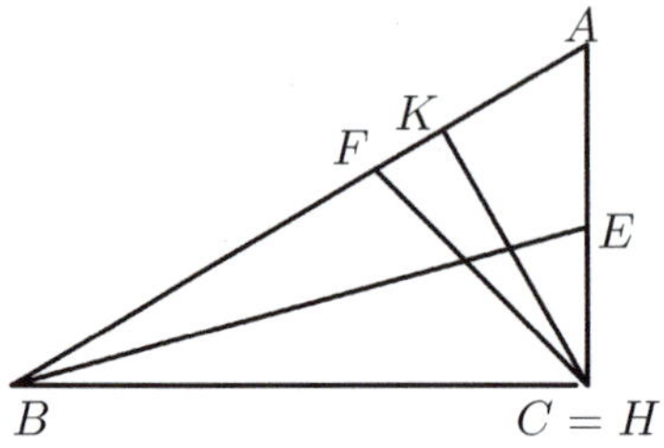

3. Let the four coins be A, B, C and D. We first weigh A and B against C and D. There are two cases.
Case 1. We have equilibrium.
We weigh A against B and C against D. By symmetry, we may assume that A is heavier than B and C is heavier than D. Finally, we weigh A against C. By symmetry, we may assume that A is heavier than C. Then the weights of A, B, C and D are 1005, 1001, 1004 and 1002 grams respectively.
Case 2. We do not have equilibrium.
By symmetry, we may assume that A and B are heavier than C and D. We weigh A against B and C against D. By symmetry, we may assume that A is heavier than B and C is heavier than D. We know that the weight of A is 1005 grams and the weight of D is 1001 grams. Finally, we weigh B against C. The heavier of them has weight 1004 grams and the lighter one has weight 1002 grams.

4. (a) It is possible. In the diagram below on the left, two triangles of different sizes have equal angles of $45°$, a third triangle has equal angles of $22\frac{1}{2}°$ and the last triangle has equal angles of $67\frac{1}{2}°$.

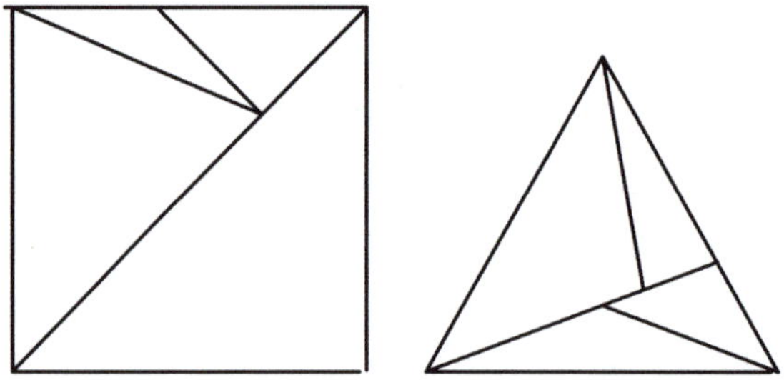

(b) It is possible. In the diagram above on the right, two triangles of different sizes have equal angles of 40°, a third triangle has equal angles of 20° and the last triangle has equal angles of 80°.

5. (a) It is not always possible. The ant's path can be blocked by 50 dominoes. For $1 \leq i \leq 50$, the ith domino occupies the $(2i-1)$st and $2i$th squares from the left on the $2i$th row from the bottom. The diagram below illustrates the case for a 6×7 board.

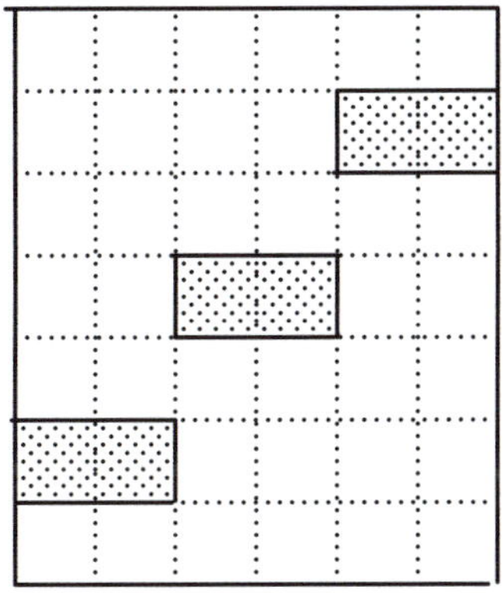

(b) Number the rows 1 to 100 from bottom to top and the columns 1 to 100 from left to right. Let (r,c) denote the square in the rth row and the cth column. We use induction on n to prove that for $2 \leq n \leq 99$, the ant can reach (n,n) unless it is occupied by a domino. For $n=2$, (1,2) and (2,1) cannot both be occupied. If (2,2) is not occupied, the ant can get there. Suppose the ant is at (n,n) for some $n \geq 2$. As before, $(n,n+1)$ and $(n+1,n)$ cannot both be occupied. If $(n+1,n+1)$ is not occupied, the ant can get there. Suppose it is. Then $(n+2,n+2)$ cannot be occupied. It can be reached from (n,n) either via $(n,n+1)$, $(n,n+2)$ and $(n+1,n+2)$ or via $(n+1,n)$, $(n+2,n)$ and $(n+2,n+1)$. The domino which occupies $(n+1,n+1)$ occupies only one of these six squares, but prevents the other five from being occupied. Hence the two possible paths for the ant to reach $(n+2,n+2)$ cannot both be blocked. Since (100,100) is not blocked, the ant can eventually get there.

Senior O-Level Paper

1. (a) Let the convex pentagon $ABCDE$ be partitioned by the diagonals AC and AD into triangles ABC, ACD and ADE. Let P, Q and R be their respective centroids. Let H, K and L be the respective midpoints of BC, CD and DE. Since $AP = 2PH$ and $AQ = 2QK$, PQ is parallel to HK. Similarly, QR is parallel to KL.

If P, Q and R are collinear, then so are H, K and L. This is impossible since $ABCDE$ is convex.

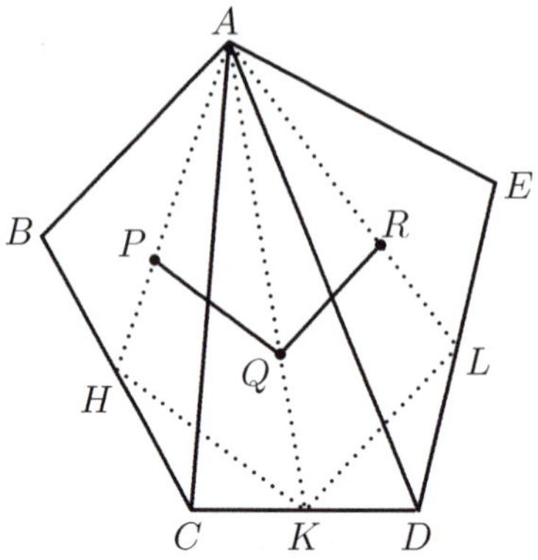

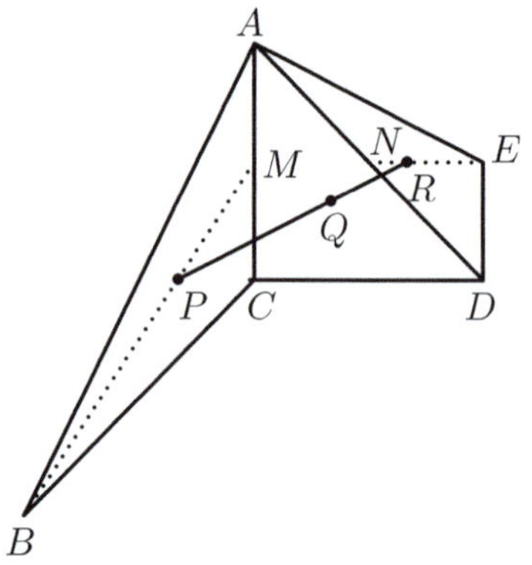

(b) This is possible. Let ACD be any triangle with centroid Q. Draw a line through Q which does not intersect the side CD. Take a point P on the other side of AC as Q, and a point R on the other side of AD as Q. Let M be the midpoint of AC and N be the midpoint of AD. Extend MP to B so that $BP = 2PM$. Extend NR to E so that $ER = 2RN$. Then P is the centroid of triangle ABC and R is the centroid of triangle ADE, and $ABCDE$ is the desired non-convex pentagon.

2. (a) Note that no weighing can result in equilibrium. Hence a weighing can eliminate at most half of the possible rankings. Since $2^4 < 4!$, four weighings will not be sufficient.

(b) Let the four coins be A, B, C and D. In the first weighing, put A and B on the lighter left pan and C and D on the heavier right pan. There are three cases.
Case 1. There is equilibrium.
The weights of A and B are 1005 grams and 1000 grams while the weights of C and D are 1004 grams and 1002 grams. Comparing A with B and C with D in the next two weighings settles the issue.
Case 2. The left pan sinks.
The weight of one of A and B is 1005 grams and the weight of one of C and D is 1000 grams. These can be determined in the next two weighings. The final weighing determines which coin weighs 1004 grams and which coin weighs 1002 grams.
Case 3. The right pan sinks.
The weight of one of A and B is 1000 grams. In the second weighing, compare A with B. By symmetry, we may assume that A weighs 1000 grams. In the third weighing, put C in the left pan and D in the right pan. If there is equilibrium, the issue is completely settled. Otherwise, whichever pan sinks contains the heavier coin. By symmetry, we may assume that D is heavier than C.

In the final weighing, put D in the left pan and B in the right pan. If there is equilibrium, D weighs 1005 grams and B weighs 1004 grams. If the left pan sinks, D weighs 1005 grams and B weighs 1002 grams. If the right pan sinks, B weighs 1005 grams and D weighs 1004 grams.

(Compare with Problem 3 of the Junior O-Level Paper.)

3. Consider the product of any n consecutive positive integers. Since one of them is divisible by n, so is their product. Then the sum of the other n consecutive positive integers is also divisible by n. Hence their average is an integer. This is not possible if n is even. If n is odd, take n copies of the average. Keep the middle copy. Increase those on one side by 1, by 2 and so on, and decrease those on the other side by 1, by 2 and so on. This will result in n consecutive integers with the correct sum. Since the minimum value of the average is $(n-1)!$, all of these integers are positive.

4. It is possible. Let $p=-202$ and $q=10201$. For $1\le k\le 101$, we take $x=100+\frac{k}{202}$. Then we have $-(px+q)=20200+k-10201=9999+k$ while $\lfloor x^2\rfloor=\lfloor 10000+\frac{100k}{101}+\frac{k^2}{40804}\rfloor=9999+k$. Hence the equation has more than 100 roots.

5. Let the diameter AQ intersect the altitudes BE and CF at H and K respectively. Then $\angle AQB=\angle ACB$. Hence QAB and CBE are similar right triangles, so that $\angle QAB=\angle CBE$. It follows that BC is tangent to the circumcircle of triangle BAH. We can prove in an analogous manner that BC is also tangent to the circumcircle of triangle CAK. Let P be the second point of intersection of these two circles, and let the line AP intersect BC at M. Then $MB^2=MP\cdot MA=MC^2$, so that M is indeed the midpoint of BC.

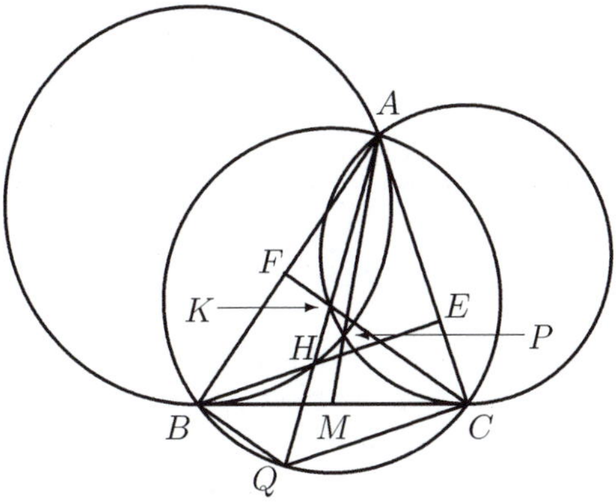

Junior A-Level Paper

1. Note that we have $20821 - 2021 = 1880$, $208821 - 20821 = 18800$, $2088821 - 208821 = 188000$, and so on. Note that $188 = 4 \times 47$ and $2021 = 42 \times 47$. All these numbers are divisible by 47 and are therefore composite.

2. Note that the last child leaving the room takes all remaining candies. Let k be the number of children initially. If k is a divisor of 1000, then every child gets the same number of candies and there is nothing to prove. Otherwise, by the Division Algorithm, there exist integers q and r such that $1000 = kq + r$ with $0 < r < k$. Let there be kq chocolates and r caramels. Every child is guaranteed k chocolates. If the current number of caramels is less than the current number of children, a boy will get a caramel while a girl will not. If the current number of caramels is equal to the current number of children, this situation will remain until the end, and every child from this point on will get a caramel. It follows that every boy gets a caramel, wherever he may be in the line up. The desired conclusion follows.

3. Extend AC to L so that $EC = CL$. Then AKL is also equilateral. Since $EP = PF$, $FL = 2PC$ and is parallel to PC. Since $KL = AK$, $FK = FK - BK + AF = AB = CA$ and $\angle FKL = 60° = \angle CAK$, FKL and CAK are congruent triangles. Hence $CK = FL = 2PC$. Moreover, triangles FAL and CLK are also congruent, so that $\angle FLA = \angle CKL$. Now $\angle KCP = 60° - \angle PCA + \angle BCK = 60° - \angle FLA + \angle CKL = 60°$. Hence KPC is half an equilateral triangle, so that PK is perpendicular to PC.

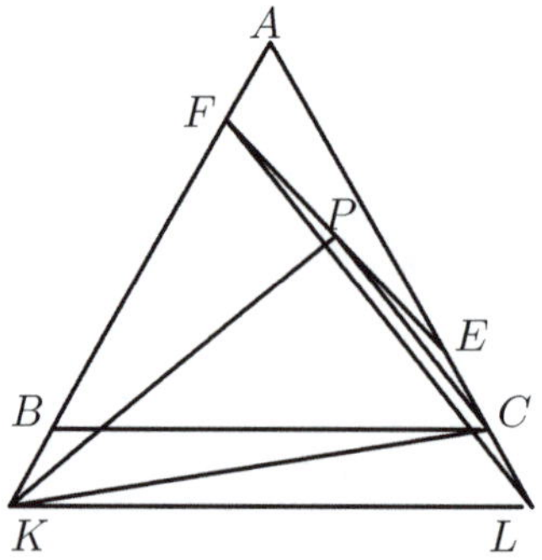

4. The sum of the two announcements of each native, knight or knucklehead, is equal to the sum of the ages of the two neighbours. Let the natives be #1 to #50 in cyclic order. If we add up the sums from #2, #6, #10, …, #46 and #50, we have the total age of the 25 odd-numbered natives plus the age of #1.

If we add up the sums from #4, #8, ..., #48, we have the total age of the 24 odd-numbered natives other than #1. Subtracting the second total from the first and dividing the difference by 2, we have the age of #1. The age of every native can be determined in a similar manner. Knowing the age of each native, we can tell whether any particular native is a knight or a knucklehead.

5. It is always possible to take three turns. Join the top candle of the leftmost column to the bottom candle of the column second from the left. Displace this line slightly so that on one side of it are all the candles of the leftmost column except for the top candle. The second line isolates this candle, and the third line separates the leftmost column from the rest. If there is at least one candle on each side of a line, then each side must contain at least one corner candle. Hence at most four turns can be taken.

   (a) The side of any line containing the central candle must contain at least two corner candles. Hence at most three turns can be taken.

   (b) Join the top left candle to the bottom right candle. Since 20 and 21 are relatively prime, this line does not pass through any other candle. Displace this line slightly in two ways so that the top left candle and the bottom right candle are on the same side of it. The third and the fourth lines isolate these two candles respectively. Thus four turns can be taken.

6. We consider two separate cases.
   Case 1. $k = 2t$ for some $t \geq 1$.
   We assign $t + 1$ rooms to each tourist. When a tourist goes upstairs, she opens the rooms assigned to her in order. If each finds a room not under renovation, all is well. Note that since $2(t+1) > 2t$, the number of tourists whose assigned rooms are all under renovation is at most one. There are only $2t - (t+1) = t - 1$ other rooms under renovation, and she can move into the last room assigned to any of the other tourists. Thus $100(t + 1) = 50k + 100$ rooms are sufficient. We now show that these many rooms are necessary. If $k = 0$, clearly 100 rooms are necessary. Let $k > 0$. Any strategy requires each tourist to open $k + 1$ rooms on a list. We can organize these lists as a $2(t + 1) \times 100$ table. Suppose there are at most $100(t + 1) - 1$ rooms. The top $t + 1$ rows of the table contains $100(t + 1)$ entries. Since the number of rooms is 1 less, there exists two equal entries in different columns. For definiteness, say that the ith entry from the top in the first column and the jth entry from the top in the second column both indicate Room m. Since $i \leq t + 1$ and $j \leq t + 1$, the total number of entries in the two columns above m is at most $t + t = k$. Let all the rooms listed above Room m in these two columns be renovated. Then the first tourist and the second tourist will both move into Room m.

Case 2. $k = 2t + 1$ for some $t \geq 0$.
Again, we assign $t + 1$ rooms to each tourist, but there is one other unassigned room. When a tourist goes upstairs, she opens the rooms assigned to her in order. If each finds a room not under renovation, all is well. Since $2(t+1) > 2t+1$, the number of tourists whose assigned rooms are all under renovation is at most one. This tourist now opens the door of the unassigned room. If it is not under renovation, she moves in. This will be the case if $t = 0$. If $t > 0$ and this room is under renovation, then there are only $(2t + 1) - (t + 1) - 1 = t - 1$ other rooms under renovation, and she can move into the last room assigned to any of the other tourists. Thus $100(t + 1) + 1 = 50k + 51$ rooms are sufficient.

We now show that these many rooms are necessary. Any strategy requires each tourist to open $k + 1$ rooms on a list. We can organize these lists as a $2(t + 1) \times 100$ table. The top $t + 1$ rows of the table contains $100(t + 1)$ entries, equal to the number of rooms. If there exists two equal entries in different columns, we can argue as in Case 1 that two tourists will move into the same room. Otherwise, one of the entries must be identical to the entry in the $(t + 2)$nd row of another column. Since $t + (t + 1) = 2t + 1 = k$, the same argument can also be applied.

7. Since p and q are relatively prime, the frog can only return to its starting point by hopping in the first direction kq times and in the second direction kp times for some positive integer k. Construct a lattice path P starting from the origin. Each time the frog hops to the right, P extends to the right by a unit. Each time the frog hops to the left, P extends above by a unit. The diagram below shows P for $p = 7$, $q = 11$ and $k = 1$ via the sequence of hops $7 - 11 + 7 - 11 + 7 + 7 - 11 + 7 - 11 + 7 + 7 - 11 + 7 - 11 + 7 + 7 - 11 + 7 = 0$.

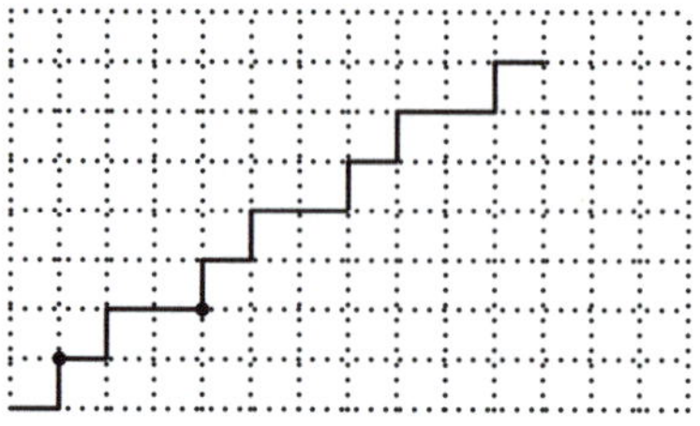

For any positive integer d, there exist positive integers a and b such that $d = pa - qb$ by the Euclidean Algorithm. We shift P a units to the right and b units up, resulting in a lattice path Q. The diagram below shows Q for $d = 10$, $a = 3$ and $b = 1$ obtained from the equation $10 = 7(3) - 11(1)$. Suppose P and Q have a common point (x, y). Then the point $(x - a, y - b)$ also lies on P. The points visited by the frog on these two occasions are $p(x - a) - q(y - b)$ and $px - qy$ respectively, and we have $(px - qy) - (p(x - a) - q(y - b)) = pa - qb = d$ as desired.

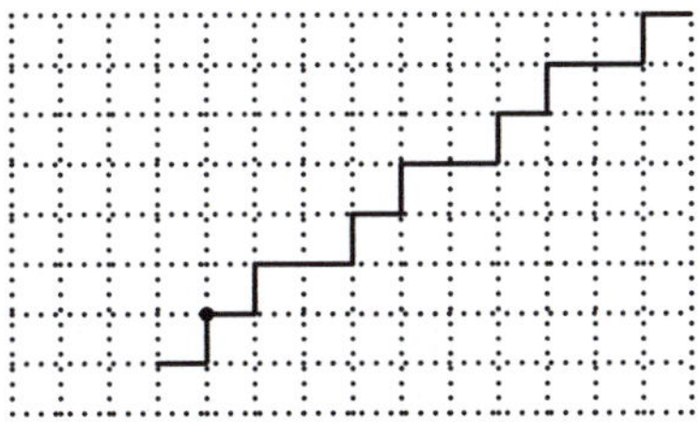

In the diagram below, a second copy of P is added from its upper right endpoint. A lattice path R is obtained from P for $d = 20$, $a = 6$ and $b = 2$ via the equation $20 = 7(6) - 11(2)$. It does not intersect the expanded P. Consider the shaded area above R and below the expanded P, enclosed in a $kq \times (kp + \ell)$ rectangle for some positive integer ℓ. It is 0 when R coincides with P. Shifting R a units to the right increases this area by kpa and shifting R b units up decreases it by kqb. Hence the shaded area is equal to $k(pa - qb) = kd$.

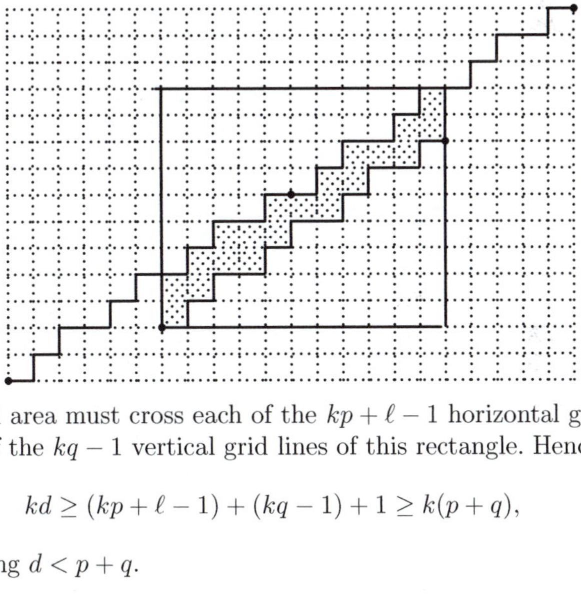

The shaded area must cross each of the $kp + \ell - 1$ horizontal grid lines and each of the $kq - 1$ vertical grid lines of this rectangle. Hence

$$kd \geq (kp + \ell - 1) + (kq - 1) + 1 \geq k(p + q),$$

contradicting $d < p + q$.

Senior A-Level Paper

1. (See Problem 2 of the Junior A-Level Paper.)

2. Such a value is $n = 6$. If $x = 0 = y$, take three copies of 1 and three copies of -1. If $x = 0 \neq y$, take four copies of $\frac{3}{y}$ and two copies of $-\frac{6}{y}$. We have $\frac{4\times3}{y} - \frac{2\times6}{y} = 0$ while $\frac{4y}{3} - \frac{2y}{6} = y$. Similarly, if $x \neq 0 = y$, take four copies of $\frac{3}{x}$ and two copies of $-\frac{6}{x}$.

Finally, suppose that $x \neq 0 \neq y$. Take two copies of each of $\frac{2x}{3}$ and $\frac{3}{2y}$ along with $-\frac{x}{3}$ and $-\frac{3}{y}$. We have

$$\frac{4x}{3} - \frac{x}{3} + \frac{2 \times 3}{2y} - \frac{3}{y} = x \text{ and } \frac{2 \times 3}{2x} - \frac{3}{x} + \frac{4y}{3} - \frac{y}{3} = y.$$

3. We have $BF \cdot BA = BD^2 = CD^2 = CE \cdot CA$. Now $CE = 2DX$ and CE is parallel to DX. Also, $BF = 2DY$ and BF is parallel to DY. Hence $\frac{DX}{DY} = \frac{CE}{BF} = \frac{BA}{CA}$. Moreover, we have

$$\angle CAB = 180^\circ - \angle ABC - \angle BCA = 180^\circ - \angle YDC - \angle XDB = \angle XDY.$$

It follows that triangles ABC and DXY are similar. Since $\angle YDC = \angle ABC = \angle DXY$, BC is tangent to the circumcircle of triangle DXY at D. Hence the two circles are also tangent at D.

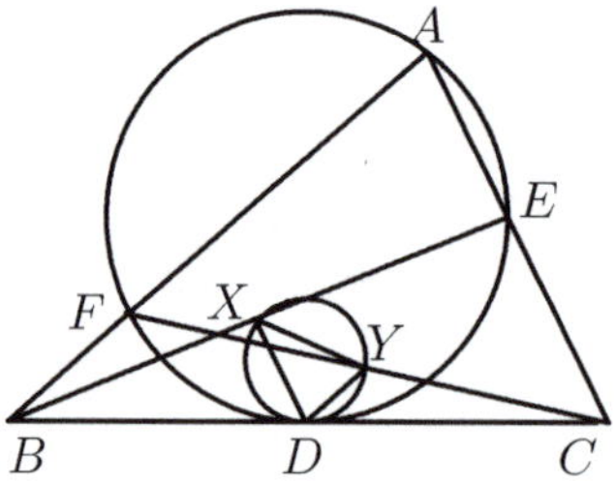

4. The cat can win if $n = 3^{100}$, which is the total number of moves. Number the tuna sandwiches 1 to 100×3^{100} from left to right. Note that

$$1 + 2(1 + 3 + 3^2 + \cdots + 3^{99}) = 1 + 2\left(\frac{3^{100} - 1}{3 - 1}\right) = 3^{100}.$$

The cat's strategy is divided into 100 stages. For $1 \leq k \leq 100$, the kth stage consists of $2 \times 3^{100-k}$ moves, reducing the number of sandwiches from $100 \times 3^{101-k}$ to $100 \times 3^{100-k}$. There is a final move after stage 100. In stage 1, the cat divides the sandwiches into a left part, a middle part and a right part, each consisting of 100×3^{99} sandwiches. In the first 3^{99} moves, the cat eats all the tuna from the sandwiches in the middle part whose numbers are congruent to 1 modulo 100. Meanwhile, in each move, the boy eats 100 sandwiches from one end or the other, or both. In any case, the numbers of these sandwiches are not congruent to one another modulo 100, and none of them are from the middle part. At the midway point of stage 1, the boy has eaten 100×3^{99} sandwiches, leaving behind 200×3^{99} sandwiches. The tuna in all sandwiches in the middle part whose numbers are congruent to 1 modulo 100 has been eaten. In the next 3^{99} moves, the cat eats the tuna from the remaining sandwiches whose numbers are congruent to 1 modulo 100. At the end of stage 1, there are 100×3^{99} sandwiches. The tuna from all those whose numbers are congruent to 1 modulo 100 has been eaten.

The remaining stages are conducted in exactly the same manner. At the end of stage k, $1 \le k \le 100$, there are $100 \times 3^{100-k}$ sandwiches, and the tuna from all those whose numbers are congruent to 1, 2, $\ldots$, k modulo 100 has been eaten. At the end of stage 100, there are 100 sandwiches, and the tuna from all of them has been eaten. The cat wins on the next move.

5. (See Problem 6 of the Junior A-Level Paper.)

6. Define $a_0 = a_1 = 2$ and $a_{n+2} = 2a_{n+1} + a_n$ for $n \ge 0$. The characteristic equation of this recurrence relation is $x^2 - 2x - 1 = 0$, and the characteristic roots are $x = 1 \pm \sqrt{2}$. Hence $a_n = \alpha(1+\sqrt{2})^n + \beta(1-\sqrt{2})^n$ for some real numbers α and β. From the initial conditions, we have $2 = \alpha + \beta = \alpha(1+\sqrt{2}) + \beta(1-\sqrt{2}) = (\alpha+\beta) + \sqrt{2}(\alpha - \beta)$. Hence $\alpha - \beta = 0$ so that $\alpha = \beta = 1$. Now

$$\begin{aligned} a_{2n} &= (1+\sqrt{2})^{2n} + (1-\sqrt{2})^{2n} \\ &= ((1+\sqrt{2})^n + (1-\sqrt{2})^n)^2 - 2(1+\sqrt{2})^n(1-\sqrt{2})^n \\ &= a_n^2 - 2(-1)^n. \end{aligned}$$

Note that $(1 \pm \sqrt{2})^2 = 3 \pm 2\sqrt{2}$ so that $a_{2n} = (3+2\sqrt{2})^n + (3-2\sqrt{2})^n$. Since $0 < 3 - 2\sqrt{2} < 1$, we have $0 < (3-2\sqrt{2})^n < 1$ for any n, so that $\lceil (3+2\sqrt{2})^2 \rceil = a_{2n}$. Hence $3 + 2\sqrt{2}$ has the desired property since a_{2n} is alternatingly 2 more or 2 less than a_n^2.

7. (a) Let $P_1P_2 \ldots P_n$ be a regular n-gon inscribed in the Tropic of Cancer. Let $Q_1Q_2 \ldots Q_n$ be the regular n-gon inscribed in the Tropic of Capricorn such that Q_k is diametrically opposite to P_k for $1 \le k \le n$. P_kQ_k is the diameter of a great circle ω_k which lies entirely within the Tropics. Every two of these great circles intersect in two points. By symmetry, the points P_k, Q_k and the $2n-2$ points of intersection with the other $n-1$ circles are evenly distributed. Now each point of intersection is between the left half of a circle with a right half of another circle. If we take all the left halves, no two of them intersect. Moreover, we can extend each at either end short into the right half, short of the first point of intersection on it. This adds an arc of length just less than $\frac{\pi}{n}$ at each end, bringing the total to the desired value.

   (b) Let $A_1, A_2, \ldots, A_n$ be pairwise non-intersecting closed arcs of great circles with lengths $\pi + \alpha_1$, $\pi + \alpha_2$, $\ldots$, $\pi + \alpha_n$, where $\alpha_k > 0$ for $1 \le k \le n$. We claim that $\displaystyle\sum_{i=1}^{n} \alpha_i \le 2\pi$, which implies the desired result.

For $1 \leq k \leq n$, let B_i be the closed arc such that its union with A_K is the whole great circle. Its length is $\pi - \alpha_i$. Consider all open hemispheres containing B_k. Let X_k and Y_k be dividing points between A_k and B_k. For any hemisphere, its pole is the point on it farthest away from its base. Any hemisphere contains B_k if and only if it contains X_k and Y_k, that is, its pole belongs to open hemispheres with poles at X_k and Y_k. Therefore the set S_k of such poles is an intersection of two hemispheres, that is the spherical slice with an angle α_k. The area of the slice is $2\alpha_i$. Let us prove that the S_k are pairwise non-intersecting. Since the area of the sphere is 4π we would get $\sum_{i=1}^{n} \alpha_i \leq 2\pi$.

Assume that some point Z belongs to $S_i \cup S_j$, $1 \leq i < j \leq n$. Then the hemisphere with a pole Z contains both B_i and B_j. Hence its complementary closed hemisphere H intersects A_i and A_j along whole semicircles rather than parts of them. But two any such semicircles on H intersect (since their ends are antipodal points on the base of H. This contradiction justifies the claim.